Frontiers in Probability and the Statistical Sciences

The “Frontiers” is a series of books (edited volumes and monographs) in probability and statistics designed to capture exciting trends in current research as they develop. Some emphasis will be given on novel methodologies that may have interdisciplinary applications in scientific fields such as biology, ecology, economics, environmental sciences, finance, genetics, material sciences, medicine, omics studies and public health.

Mamikon S. Ginovyan

Random Toeplitz Functionals and Their Applications

Springer

Mamikon S. Ginovyan
Department of Mathematics and Statistics
Boston University
Boston, MA, USA

ISSN 2624-9987 ISSN 2624-9995 (electronic)
Frontiers in Probability and the Statistical Sciences
ISBN 978-3-032-00960-9 ISBN 978-3-032-00961-6 (eBook)
https://doi.org/10.1007/978-3-032-00961-6

Mathematics Subject Classification: 47B10, 47B35, 60F05, 60F17, 60G22, 62F12, 62G20, 15B05, 60F10, 60G20, 60G51, 62F10, 62G10, 60G10, 60G15, 62G05

This Springer imprint is published by the registered company Springer Nature Switzerland AG
The registered company address is: Gewerbestrasse 11, 6330 Cham, Switzerland

*In loving memory of my parents,
Smbat and Lisa Ginovyan.*

Preface

A significant part of large-sample statistical inference relies on the limit theorems of probability theory. In this book, we focus on the limit theorems for random Toeplitz and tapered Toeplitz quadratic functionals (forms) of stationary observations, as well as their applications in statistical inference problems.

Depending on the memory (or dependence) structure of the underlying processes, these functionals, once normalized, can exhibit different limits. The primary focus of this book is to describe these limiting distributions in cases where the underlying processes are Gaussian, OIP-driven, or Lévy-driven linear processes with memory, defined in either discrete or continuous time. The results are applied in both parametric and nonparametric statistical estimation, as well as in hypothesis testing related to the spectrum of the observed processes.

Toeplitz matrices and operators, which have significant independent interest and a wide range of applications across various fields of science, arise naturally in the context of stationary processes. This is because the covariance matrix of a discrete time stationary process is a truncated Toeplitz matrix generated by the spectral density of that process. Conversely, any non-negative summable function generates a Toeplitz matrix, which can be considered the spectral density of some discrete time stationary process. Consequently, the corresponding truncated Toeplitz matrix will serve as the covariance matrix of that process. In the continuous time case, the Toeplitz matrix is replaced by a Toeplitz operator.

The trace approximation problem and its applications in the statistical analysis and prediction of discrete time stationary processes trace back to the classical monograph by Grenander and Szegő *'Toeplitz Forms and Their Applications'*, University of California Press, Berkeley and Los Angeles (1958).

In recent years, there has been renewed interest in this subject, particularly in connection with models exhibiting long-range dependence and statistical inferences based on tapered data. Several books on the topic have been published over the past 20 years. For instance, in Chap. 6 of the monograph by Giraitis et al. [147], central limit theorems for random Toeplitz quadratic forms in discrete time linear models were obtained. Results concerning the Whittle estimation procedure for discrete time short memory processes are presented in Chapters II and III of the monograph by Dzhaparidze [93]. The asymptotic theory of parametric estimation under the local asymptotic normality (LAN) condition for discrete time short memory and long memory models has been developed in Chaps. 3 and 6 of the monograph by Taniguchi and Kakizawa [295].

Many features set this book apart. First, several topics are not discussed elsewhere, including most of Chap. 4 (a unified treatment of the trace approximation problem), Chap. 5 (limit theorems in the continuous time case), Chap. 6 (limit theorems for Toeplitz processes), and Chap. 7 (central limit theorems for tapered functionals). Second, some topics are up to date. For example, the estimation of nonlinear spectral functionals (Chap. 8), the results concerning Whittle estimators (Chap. 9), and the goodness-of-fit tests based on tapered data (Chap. 10) have been supplemented by recent work in the area. Third, some specialized topics, such as the continuous time and tapered cases, reflect our perspective.

Although the book contains a relatively large number of early and up-to-date references, we had to be quite selective and could not include every single relevant work. We assume that the reader has a background in basic time series analysis (e.g., at the level of Brockwell and Davis [57] and stochastic processes (e.g., at the level of Cramer and Leadbetter [69]).

Almost all results stated in the text are proved in detail. A precise reference to the literature is provided in cases where the proof is omitted.

Broadly speaking, the material in this book is broken down into the following four parts: (i) Preliminaries ant model specification (Chaps. 2 and 3) (ii) The trace approximation problem for Toeplitz and tapered Toeplitz matrices and operators (Chap. 4) (iii) Central and non-central limit theorems for random Toeplitz and tapered Toeplitz quadratic functionals (Chaps. 5–7) (iv) Statistical inferences (estimation and hypothesis testing) about the spectrum of the model (Chaps. 8–10).

In more detail, the contents of this book are as follows. The book is divided into chapters, sections, and subsections. We use triple numeration for assertions and formulas. Each chapter begins with a brief overview and concludes with comments, complements, and bibliographical notes.

In Chap. 1, we provide an overview of the basic problems that constitute the body of the book. The somewhat protracted Chap. 2 provides definitions of important auxiliary concepts and presents basic results necessary for the rest of the book. This arrangement avoids interruptions in the main text. For the same reason, bibliographic remarks and other comments have been placed at the end of each chapter. In Chap. 3, we introduce the model of interest—a second-order stationary process—and review some key concepts and results from the theory of stationary processes. In Chap. 4, we provide a unified treatment of the trace approximation problem for both Toeplitz and tapered Toeplitz matrices and operators. In Chap. 5, we present results on central limit theorems for random Toeplitz type quadratic functionals of stationary processes. The underlying processes are either Gaussian or linear processes with memory and are defined in discrete or continuous time. In Chap. 6, we establish both central and non-central functional limit theorems for processes generated by Toeplitz functionals. In Chap. 7, we analyze the asymptotic distributions of tapered Toeplitz type quadratic functionals. Chapter 8 is devoted to the nonparametric statistical estimation of both linear and some nonlinear smooth spectral functionals for linear models with tapered data. In Chap. 9, we consider the Whittle parametric estimation procedure based on the tapered data. Chapter 10 focuses on constructing goodness-of-fit tests using tapered data. The book contains two appendices. Appendix A focuses on the motivations and benefits of the tapering technique, while Appendix B briefly discusses several important problems closely related to the topics covered in the book.

The proposed book serve as a valuable resource for researchers in time series analysis, econometrics, finance, and related fields, as well as for practitioners and applied researchers who need to analyze data (or tapered data) coming from models where the dependence (memory) structure is particularly relevant. Researchers working on the theory of Toeplitz matrices and operators will also find a variety of interesting results in this book.

Furthermore, it can be utilized as a foundational textbook for a graduate-level course in time series analysis or the statistics of stochastic processes. It may also serve as a supplemental resource for those studying advanced statistics, probability, econometrics, finance, or physics.

Some of the material included in the book was presented in the "Seminar Course" on time series analysis, as well as supplementary content for the graduate-level class "Estimation Theory" at Boston University.

I have many people to thank. First and foremost, I would like to express my deep gratitude to my teachers, Professors I.A. Ibragimov and V.N. Solev, from the St. Petersburg Department of the Steklov Mathematical Institute of the Russian Academy of Sciences, for introducing me to this beautiful area of mathematical statistics. I also wish to extend special thanks to my co-authors: Murad Taqqu from Boston University, Artur Sahakyan from Yerevan State University, and Shuyang (Ray) Bai from the University of Georgia. In preparing the final text, I had fruitful discussions with many of my colleagues: Youri Kutoyants (Le Mans University), Luis Carvalho, Ashis Gangopadhyay, Tasso Kaper, Mark Kon, and Konstantinos Spiliopoulos (Boston University), Mikhail Lifshits (St. Petersburg State University), Nikolai Leonenko (Cardiff University), and Liudas Giraitis (Queen Mary University of London). I sincerely thank all of them.

I would like to extend my gratitude to the Department of Mathematics and Statistics at Boston University for providing excellent support and a conducive working environment.

The author acknowledges support from the National Science Foundation under grants DMS-0706786 and DMS-1309009.

I would like to thank the editors at Springer-Verlag for their patience, support, and assistance in bringing this book to its final form.

Finally, I am deeply grateful to my family for their unwavering support throughout my academic pursuits.

I am responsible for any remaining errors and typos, and I would appreciate it if readers could send a quick email to the author if they find any.

Boston, MA, USA Mamikon S. Ginovyan
April, 2025

Declarations

Competing Interests The authors have no competing interests to declare that are relevant to the content of this manuscript.

Contents

Glossary

Acronyms, Notation, and Conventions

The following acronyms, notation, and conventions are used throughout the book.

The letters C, c, M, and m, with or without indices, are used to denote positive constants. The values of these constants may vary from one instance to another.

a.e.	almost everywhere
a.s.	almost surely
c.t.	continuous time
d.t.	discrete time
s.d.	spectral density
c.f.	covariance function
r.k.	reproducing kernel
r.v.	random variable
r.h.s.	right-hand side
l.h.s.	left-hand side
i.i.d.	independent identically distributed
n.n.d.	non-negative definite
resp.	respectively
ARMA	autoregressive moving average
ARFIMA	autoregressive fractionally integrated moving average
BLUE	best linear unbiased estimator
CARMA	c.t. autoregressive moving average
CARFIMA	c.t. autoregressive fractionally integrated moving average
CLT	central limit theorem
NCLT	non-central limit theorem
FCLT	functional central limit theorem
FNCLT	functional non-central limit theorem
fBm	fractional Brownian motion

fGn	fractional Gaussian noise
fRBm	fractional Riesz-Bessel motion
LAN	local asymptotic normality
LAM	locally asymptotically minimax
LLN	law of large numbers
LDP	large deviation principle
MDP	moderate deviation principle
MLE	maximum likelihood estimator
OIP	orthogonal-increments process
SM, IM, LM	short memory, intermediate memory, long memory, respectively
SVF	slowly varying function
RKHS	reproducing kernel Hilbert space
:=	by definition
$[x]$	integer part (floor) of x
$x \wedge y, x \vee y$	$\min\{x, y\}$, $\max\{x, y\}$, respectively
$\mathbb{N}$	set of naturals (positive integers); $\mathbb{N}_0 := \mathbb{N} \cup \{0\}$
$\mathbb{Z}$	set of integers
$\mathbf{C}$	set of complex numbers
$\mathbb{T}$	unit circle
$\mathbb{R}^m$	m-dimensional Euclidean space: $\mathbb{R}^1 = \mathbb{R} := (-\infty, \infty)$; $\mathbb{R}^+ := [0, \infty)$
Λ	frequency domain; $\Lambda := \mathbb{R}$ in the c.t. case, and $\Lambda := [-\pi.\pi)$ in the d.t. case
$\mathbb{U}$	time domain; $\mathbb{U} = \mathbb{R}$ in the c.t. case, and $\mathbb{U} = \mathbb{Z}$ in the d.t. case
$L^p(\mu) := L^p(\Lambda, \mu)$ $(p \geq 1)$	weighted Lebesgue space with respect to the measure μ; $L^p(\mu) := L^p$ if $d\mu(\lambda) = d\lambda$
$\|\| \cdot \|\|_{p,\mu}$	norm in $L^p(\mu)$; $\|\| \cdot \|\|_{p,\mu} := \|\| \cdot \|\|_p$ if $d\mu(\lambda) = d\lambda$
L^0	space of measurable functions
L^∞	space of essentially bounded functions
L^1_{loc}	space of locally integrable functions
$\mathbb{C}_{loc}(\mathbb{R}^m)$	space of continuous functions on $\mathbb{R}^m$ with bounded support
l^p $(p \geq 1)$	space of p-summable sequences
$\mathbb{C}[0, 1]$	space of continuous functions on [0, 1]
$\mathbb{D}[0, 1]$	Skorohod space = space of functions on [0, 1] that are right-continuous and have left-hand limits ((càdlàg) functions)
$\mathrm{Lip}(\Lambda, \alpha)$	Lipschitz class
$\mathrm{Lip}(\Lambda, p, \alpha)$	L^p-Lipschitz class
$H_p(\beta) := H_p(\Lambda, \beta)$	L^p-Hölder class
$H^{2\pm}(\mathbb{T})$	Hardy spaces
$H^{1/2}$	Sobolev space
$P_\pm(\mathbb{T})$	orthogonal projections of $L^2(\mathbb{T})$ onto $H^{2\pm}(\mathbb{T})$

$SV_0(\mathbb{R})$	class of slowly varying at zero functions
$\mathcal{S}_p$	Schatten class
$(\cdot,\cdot)_H$, $\\|\cdot\\|_H$	inner product and norm in a Hilbert space H, respectively
$\omega(\cdot,\delta)$	modulus of continuity
$\omega_p(\cdot,\delta)$	L^p–modulus of continuity
$\mathbf{1}_A(\cdot)$	indicator function of a set $A \subset \Lambda$
$\overline{z}$, $(\overline{\varphi}(z))$	complex conjugate of a number $z \in \mathbf{C}$ and a complex-valued function $\varphi(z)$, respectively
$\psi^{(k)}$	k-th derivative of a function ψ; $\psi^{(1)} = \psi'$
$\psi\|_A$	restriction of a function ψ on a set $A \subset \Lambda$
$\widehat{\phi}$	Fourier transform of a function ϕ
$\phi * \psi$	convolution of functions $\phi, \psi \in L^1(\Lambda)$
$\phi \bar{*} \psi$	reversed convolution
$E_{A,p}(f)$	best approximation in the metric of L^p
$D_T(u)$	Dirichlet kernel
$F_T(u)$	Fejér kernel
$\mathrm{WN}(0,\sigma^2)$	white-noise with variance σ^2
$\mathrm{IID}(0,\sigma^2)$	i.i.d.-noise with variance σ^2
$N(\mu,\sigma^2)$	normal distribution with mean μ and variance σ^2
$N_k(\mu,\Sigma)$	multivariate normal distribution with k-dimensional mean vector μ and $k \times k$ covariance matrix Σ
$B(\cdot)$ $(W(\cdot))$	Brownian motion
$B_H(\cdot)$	fractional Brownian motion with Hurst index H
$\mathbb{E}[\cdot]$	expectation operator
$\mathbb{P}$, $\mathbb{P}(A)$	probability measure, probability of an event A, respectively
$\mathrm{Var}[\xi]$, $\mathrm{Cov}[\xi]$	variance, covariance of a r.v. ξ, respectively
$\\|A\\|_p$	p-Schatten norm of an operator (or matrix) A
$\mathrm{tr}[A]$	trace of an operator (or matrix) A
$\det[A]$	determinant of an operator (or matrix) A
$\\|A\\|_2$	Euclidean norm of an operator (or matrix) A
$\\|A\\|_\infty$	spectral norm of an operator (or matrix) A
Q_T^h	tapered Toeplitz quadratic functional
$Q_T := Q_T^h$	for $h(t) = \mathbf{1}_{[0,1]}(t)$
$\chi_k(Q_T)$	k-th order cumulant of Q_T
$\mathrm{ARMA}(p,q)$	d.t. Autoregressive Moving Average process
$\mathrm{CARMA}(p,q)$	c.t. Autoregressive Moving Average process
$\mathrm{ARFIMA}(p,d,q))$	d.t. Fractionally Integrated $\mathrm{ARMA}(p,q)$
$\mathrm{CARFIMA}(p,d,q))$	c.t. Fractionally Integrated $\mathrm{CARMA}(p,q)$
$B_T^h(\psi)$	$(T \times T)$-truncated tapered Toeplitz matrix generated by ψ
$B_T(\psi) := B_T^h(\psi)$	for $h(t) = \mathbf{1}_{[0,1]}(t)$

$W_T^h(\psi)$	T-truncated tapered Toeplitz operator generated by ψ
$W_T(\psi) := W_T^h(\psi)$	for $h(t) = \mathbf{1}_{[0,1]}(t)$
$I_T^h(\lambda)$	tapered periodogram; $I_T(\lambda) := I_T^h(\lambda)$ for $h(t) = \mathbf{1}_{[0,1]}(t)$
$\overset{d}{=}$, $\sim$	equality in distribution, as in $Z \sim N(0, 1)$
$\overset{d}{\to}$	convergence in distribution
$\overset{P}{\to}$	convergence in probability
$\overset{f.d.d.}{\longrightarrow}$	convergence of finite-dimensional distributions
$\Longrightarrow$	weak convergence
$\overset{C}{\Longrightarrow}$	weak convergence in the space $\mathbb{C}[0, 1]$
$\overset{D}{\Longrightarrow}$	weak convergence in the space $\mathbb{D}[0, 1]$
$p - \lim$	limit in probability
$L^2 - \lim$	limit in the space L^2
$a_n \sim b_n$ $(n \to \infty)$	will mean that $a_n/b_n \to 1$ $(n \to \infty)$
$u(x) \sim v(x)$ $(x \to x_0)$	will mean that $u(x)/v(x) \to 1$ $(x \to x_0)$
$o(\cdot)$, $O(\cdot)$	small and big O, respectively
$o_P(\cdot)$, $O_P(\cdot)$	small and big O in probability, respectively
$\oplus$	orthogonal sum
$\operatorname{int} \Theta$	interior of a set Θ
δ_{ij}	Kronecker's delta ($\delta_{ij} = 1$ if $i = j$ and $\delta_{ij} = 0$ if $i \neq j$).

Basic Assumptions

The following assumptions apply throughout the book unless stated otherwise.

1. The stochastic processes are assumed to be either discrete or continuous time, centered (with a mean equal to zero), non-degenerate (with a strictly positive variance), and second-order stationary. In the case of continuous time, these processes are assumed to be measurable and mean square continuous.
2. In the discrete time case, the functions considered are assumed to be 2π-periodic and are periodically extended to the real line $\mathbb{R}$.
3. The taper function $h(\cdot)$ satisfies Assumption 4.8.1.
4. All Hilbert spaces are assumed to be separable.
5. An operator refers to a bounded linear operator.

Chapter 1
Introduction and Overview

A significant part of large-sample statistical inference relies on the limit theorems of probability theory, which involve random quadratic functionals of stationary observations. Depending on the memory (or dependence) structure of the underlying processes, these functionals, once normalized, can converge to different limits. The limiting random variable can be either Gaussian, meaning that the functional, with appropriate normalization, satisfies the central limit theorem, or non-Gaussian.

The focus of this text is to describe these limiting distributions. We prove central and non-central limit theorems for Toeplitz and tapered Toeplitz-type random quadratic functionals of second-order stationary processes. The results are applied in both parametric and nonparametric statistical estimation, as well as in hypothesis testing related to the spectrum of the observed processes. The underlying processes are Gaussian, OIP-driven, or Lévy-driven linear processes with memory, and they are defined in either discrete or continuous time.

We also discuss a question concerning the approximations of the products of Toeplitz and tapered Toeplitz matrices and operators. These approximations are a key tool for obtaining central limit theorems for the corresponding random quadratic functionals, but they are also of interest in themselves.

The term "central limit theorem" (CLT) refers to the statement that a suitably standardized quadratic functional converges in distribution to a Gaussian random variable. Limit theorems in which a suitably standardized quadratic functional converges in distribution to a non-Gaussian random variable are termed "non-central limit theorems" (NCLT).

The present book addresses the following three fundamental problems that are key to its content:

(a) Central and non-central limit theorems for Toeplitz and tapered Toeplitz-type random quadratic functionals of stationary processes for both discrete and continuous time cases

M. S. Ginovyan, *Random Toeplitz Functionals and Their Applications*,
Frontiers in Probability and the Statistical Sciences,
https://doi.org/10.1007/978-3-032-00961-6_1

(b) Statistical inferences regarding the spectrum of the underlying processes, focusing on nonparametric and parametric statistical estimation, as well as hypothesis testing
(c) The trace approximation problem for Toeplitz and tapered Toeplitz-type matrices and operators

We will now provide a concise overview of these topics.

1.1 Limit Theorems for Toeplitz and Tapered Toeplitz Random Quadratic Functionals

1.1.1 Toeplitz Random Quadratic Functionals

Let $\{X(u),\ u \in \mathbb{U}\}$ be a centered real-valued weakly stationary process with spectral density (s.d.) $f(\lambda)$, $\lambda \in \Lambda$, and covariance function (c.f.) $r(t)$, $t \in \mathbb{U}$. We consider both the continuous time (c.t.) case, where $\mathbb{U} = \mathbb{R} := (-\infty, \infty)$, and the discrete time (d.t.) case, where $\mathbb{U} = \mathbb{Z} := \{0, \pm 1, \pm 2, \ldots\}$. The domain Λ of the frequency variable λ is $\Lambda = \mathbb{R}$ in the c.t. case, and $\Lambda := [-\pi.\pi)$ in the d.t. case.

We first discuss a question concerning the asymptotic distribution (as $T \to \infty$) of the following Toeplitz-type random quadratic functionals of the process $X(u)$:

$$Q_T := \begin{cases} \sum_{t=1}^{T} \sum_{s=1}^{T} \widehat{g}(t-s)X(t)X(s) & \text{in the d.t. case,} \\ \\ \int_0^T \int_0^T \widehat{g}(t-s)X(t)X(s)\,dt\,ds & \text{in the c.t. case,} \end{cases} \tag{1.1.1}$$

where

$$\widehat{g}(t) := \int_{\Lambda} e^{i\lambda t}\, g(\lambda)\, d\lambda, \quad t \in \mathbb{U} \tag{1.1.2}$$

is the Fourier transform of some real, even, integrable function $g(\lambda)$, $\lambda \in \Lambda$. We will refer to $g(\lambda)$ and its Fourier transform $\widehat{g}(t)$ as the *generating function* and *generating kernel* for the functional Q_T, respectively. In the discrete time case, the functions $f(\lambda)$ and $g(\lambda)$ are assumed to be 2π-periodic and periodically extended to $\mathbb{R}$. In the continuous time case, the process $X(u)$ is assumed to be measurable and mean square continuous.

The limiting distributions of the functionals in (1.1.1) are completely determined by the spectral density $f(\lambda)$ (or covariance function $r(t)$) and the generating function $g(\lambda)$ (or generating kernel $\widehat{g}(t)$). Depending on the properties of these functions, the limits can be either Gaussian (meaning that Q_T, with the appropriate normalization, follows the central limit theorem) or non-Gaussian.

The following two questions arise naturally:

(a) Under what conditions on $f(\lambda)$ (resp. $r(t)$) and $g(\lambda)$ (resp. $\widehat{g}(t)$) will the limits be Gaussian? This is referred to as the CLT problem.
(b) Describe the limiting distributions, if they are non-Gaussian. This is known as the NCLT problem.

Conditions expressed in terms of the spectral density $f(\lambda)$ and the generating function $g(\lambda)$ will be referred to as *frequency domain conditions*. In contrast, conditions expressed in terms of the covariance function $r(t)$ and the generating kernel $\widehat{g}(t)$ will be termed *time domain conditions*.

We discuss these and related questions for both discrete and continuous time stationary processes in Chap. 5.

The next problem addressed in the text pertains to functional central and non-central limit theorems for the following stochastic processes generated by the quadratic functionals in (1.1.1):

$$Q_T(\tau) := \begin{cases} \sum_{t=1}^{[T\tau]} \sum_{s=1}^{[T\tau]} \widehat{g}(t-s)X(t)X(s) & \text{in the d.t. case,} \\ \int_0^{T\tau} \int_0^{T\tau} \widehat{g}(t-s)X(t)X(s)\,dt\,ds & \text{in the c.t. case,} \end{cases} \tag{1.1.3}$$

where $\tau \in [0, 1]$ and $[\,\cdot\,]$ stands for the greatest integer.

We say that a *functional central limit theorem (FCLT)* for $Q_T(\tau)$ holds if the process $Q_T(\tau)$, with an appropriate normalization, converges weakly in the space $\mathbb{C}[0, 1]$ in the continuous time case, and in the space $\mathbb{D}[0, 1]$ in the discrete time case, to Brownian motion. In contrast, we say that a *functional non-central limit theorem (FNCLT)* for the process $Q_T(\tau)$ holds if the limit is non-Gaussian, or if it is Gaussian, it is not Brownian motion.

Both the functional central and non-central limit theorems for the processes $Q_T(\tau)$ are established for various models in Chap. 6.

1.1.2 Tapered Toeplitz Random Quadratic Functionals

1.1.2.1 Tapering: Statistical Motivation

Much of the statistical inference—both parametric and nonparametric estimation, as well as hypothesis testing—about the spectrum or covariance of a stationary process $\{X(u), u \in \mathbb{U}\}$ relies on the standard (non-tapered) sample. This sample consists of a segment $\mathbf{X}_T$ of length T taken from one realization of the process $X(u)$, specifically

$$\mathbf{X}_T := \begin{cases} \{X(t),\ t = 1, \dots, T\} & \text{in the d.t. case,} \\ \{X(t),\ 0 \le t \le T\} & \text{in the c.t. case,} \end{cases} \tag{1.1.4}$$

where T can take different values. In other words, statistical inference is based on a family of experiments consisting of the observations of the values of $\mathbf{X}_T$, where T varies. These experiments can be identified with a family of probability spaces:

$$\mathcal{E}_T := (\mathcal{X}_T, \mathcal{A}_T, \mathbb{P}_T),$$

where $\mathcal{X}_T$ is the set of all observable events $\mathbf{X}_T$, $\mathcal{A}_T$ is a basic collection of measurable subsets of $\mathcal{X}_T$, and $\mathbb{P}_T$ is a probability measure defined for all sets in $\mathcal{A}_T$ (see, e.g., Ibragimov and Has'minskii [185], Section 1.1, and Dzhaparidze and Yaglom [94]).

In the statistical analysis of stationary processes, however, data are often tapered before calculating the statistics of interest. Consequently, the statistical inference procedure relies on the *tapered data* rather than the original data $\mathbf{X}_T$ given by (1.1.4). The tapered data, denoted as $\mathbf{X}_T^h$, is of the following form:

$$\mathbf{X}_T^h := \begin{cases} \{h_T(t)X(t),\ t = 1, \dots, T\} & \text{in the d.t. case,} \\ \{h_T(t)X(t),\ 0 \le t \le T\} & \text{in the c.t. case,} \end{cases} \tag{1.1.5}$$

where $h_T(t) := h(t/T)$ with a taper function $h(t)$, $t \in [0, 1]$.

The benefits of tapering data have been widely reported in the literature (see, e.g., Bloomfield [44], Brillinger [55], Dahlhaus [70–72], Dahlhaus and Künsch [78], Ginovyan and Sahakyan [140, 143], Guyon [165], and references therein). For example, data tapers are introduced to reduce the so-called leakage effects, that is, to obtain better estimation of the spectrum of a model in the case where it contains high peaks. Another application of data tapers occurs when some data values are missing. Additionally, the use of tapers leads to bias reduction, which is especially important when dealing with spatial data. In this context, tapers can be employed to mitigate the so-called edge effects. For more details about motivations and benefits of tapering, see Sect. 4.8.1 and Appendix A.

1.1.2.2 Tapered Toeplitz Random Quadratic Functionals

In the tapered case, to establish the asymptotic properties of the corresponding statistical estimators, we need to study the asymptotic distributions (as $T \to \infty$) of the following tapered Toeplitz-type random quadratic functionals of the process $X(u)$:

$$Q_T^h := \begin{cases} \sum_{t=1}^T \sum_{s=1}^T \widehat{g}(t-s)h_T(t)h_T(s)X(t)X(s) & \text{in the d.t. case,} \\ \int_0^T \int_0^T \widehat{g}(t-s)h_T(t)h_T(s)X(t)X(s)\,dt\,ds & \text{in the c.t. case,} \end{cases} \tag{1.1.6}$$

where $\widehat{g}(t)$, $t \in \mathbb{U}$, is the Fourier transform of some integrable even function $g(\lambda)$, $\lambda \in \Lambda$, and $h_T(t) := h(t/T)$ with a taper function $h(t)$, $t \in [0, 1]$. Central limit theorems and a functional central limit theorem for the tapered functionals Q_T^h are established in Chap. 7.

1.2 Statistical Inferences

Quadratic functionals of the forms mentioned in Eqs. (1.1.1) and (1.1.6) naturally arise in the context of statistical inference problems. These functionals are relevant for both nonparametric and parametric statistical estimation, as well as for hypothesis testing related to the spectra of stationary processes, utilizing both non-tapered and tapered data. Understanding their limiting distributions is essential for establishing the asymptotic properties of the corresponding statistical estimators and for determining the sampling distributions of test statistics when conducting hypothesis tests.

1.2.1 The Nonparametric Estimation Problem

We assume that the spectral density $f(\lambda)$ of the observed model $X(u)$ is *unknown* and belongs to a given (infinite-dimensional) class $\mathcal{F} \subset L^p := L^p(\Lambda)$ $(p \geq 1)$ of spectral densities possessing some specified smoothness properties. The objective is to estimate the value $J(f)$ of a given known functional $J(\cdot)$ at an *unknown* "point" $f \in \mathcal{F}$ based on observations (1.1.4) (in the non-tapered case) and (1.1.5) (in the tapered case), and investigate the asymptotic properties of the proposed estimators as $T \to \infty$, taking into account the dependence structure of the model $X(u)$ and the smoothness structure of the "parametric" set $\mathcal{F} \subset L^p(\Lambda)$ $(p \geq 1)$.

It is important to note that the solution to the problem depends significantly on the type of *estimand* functional, $J(f)$, whether it is linear or nonlinear.

In the case where the estimand functional $J(f)$ is a linear integral functional in $L^p(\Lambda)$, $p \geq 1$, then a natural statistical estimator for $J(f)$ is the linear integral functional of the empirical spectral density (periodogram) of the process $X(t)$. This is represented as a quadratic functional of the form (1.1.1) in the non-tapered case, and of the form (1.1.6) in the tapered case. When the functional $J(f)$ is nonlinear, we consider the statistic $J(\widehat{f_T})$ as an estimator for $J(f)$, where $\widehat{f_T}$ is a suitable kernel estimator for the unknown f.

Results on nonparametric estimation of both linear and nonlinear spectral functionals—specifically, the asymptotic properties of the estimators, the construction of efficient estimators under the local asymptotic normality condition, and bounding the minimax mean square risks of estimators—are established in Chap. 8.

It is important to note that linear and nonlinear functionals of the periodogram play a key role in the parametric estimation of the spectral parameters of stationary processes when using the minimum contrast estimation method with various contrast functionals (see Chap. 9).

1.2.2 The Parametric Estimation Problem

We assume here that the spectral density f of the model belongs to a given parametric family of spectral densities $\mathcal{F} := \{f(\lambda, \theta) : \theta \in \Theta\}$, where $\theta := (\theta_1, \ldots, \theta_p)$ is an *unknown* vector parameter and Θ is a subset of the Euclidean space $\mathbb{R}^p$. The problem of interest is to estimate the unknown parameter θ based on the non-tapered data (1.1.4) (or the tapered data (1.1.5)) and to investigate the asymptotic properties (as $T \to \infty$) of the proposed estimators. These properties depend on the memory structure of the model $X(t)$ and the smoothness of its spectral density f.

There are various statistical methods for estimating the unknown spectral parameter θ, including the method of moments, maximum likelihood estimation, the Whittle method, and the minimum contrast method, among others. Here, we focus on the Whittle method, which is based on smoothed periodogram analysis in the frequency domain, and the minimization of quadratic functionals of the form (1.1.1) and (1.1.6). Results on parametric estimation of spectral parameters for both tapered and non-tapered cases are presented in Chap. 9. Specifically, we outline the Whittle estimation procedure and discuss the asymptotic properties of both the Whittle and simplified Whittle estimators. Additionally, we discuss robust estimation procedures for linear models, encompassing both parametric and nonparametric scenarios.

1.2.3 Hypothesis Testing: Goodness-of-Fit Tests

The problem of interest here is the construction of goodness-of-fit tests for testing the hypothesis H_0 that the hypothetical spectral density of a stationary Gaussian process $X(t)$ conforms to a specified form, utilizing the tapered data given in (1.1.5). It turns out that in the case where the hypothetical spectral density $f(\lambda)$ of $X(t)$ does not depend on unknown parameters (i.e., when the hypothesis H_0 is simple), the proposed test statistic has a limiting chi-square distribution. In the case where the null hypothesis H_0 is composite, meaning that the hypothetical spectral density of $X(t)$ depends on an unknown parameter, we select an appropriate estimator for this unknown parameter and describe the limiting distribution of the proposed test statistic. This distribution is similar to those obtained by Chernoff and Lehmann [62] for independent observations, as well as by Dzhaparidze [93] for discrete time short memory stationary observations. The testing procedure applies to both short and long memory models and is discussed in Chap. 10.

1.3 The Trace Approximation Problem for Toeplitz Matrices and Operators

Toeplitz matrices and operators, which have significant independent interest and a wide range of applications in various fields of science, arise naturally in the context of stationary processes. This is because the covariance matrix of a discrete time stationary process is a truncated Toeplitz matrix generated by the spectral density of that process. Conversely, any non-negative summable function generates a Toeplitz matrix, which can be considered the spectral density of some discrete time stationary process. Consequently, the corresponding truncated Toeplitz matrix will serve as the covariance matrix of that process. In the continuous time case, the Toeplitz matrix is replaced by a Toeplitz operator.

Truncated Toeplitz matrices and operators are of particular importance and serve as tools for studying many topics in the spectral and statistical analysis of discrete and continuous time stationary processes. These topics include central and non-central limit theorems, and large deviations of Toeplitz-type random quadratic forms and functionals, estimation of spectral parameters and functionals, asymptotic expansions of estimators, hypothesis testing about the spectrum, prediction of future values based on the observed past of the process, etc.

The trace approximation problem and its applications in the statistical analysis and prediction of discrete time stationary processes can be traced back to the classical monograph by Grenander and Szegő [162].

To state the trace approximation problem addressed in this text, we first need to recall two important facts.

First, let f and g be two integrable real symmetric functions defined on Λ, and let $A(f)$ and $A(g)$ be the infinite Toeplitz matrices (or operators) generated by the functions f and g, respectively. As a consequence of the Parseval-Plancherel theorem, we have (see Grenander and Szegő [162], Section 8.1)

$$A(f)A(g) = A(fg). \tag{1.3.1}$$

Now, let $A_T(f)$ and $A_T(g)$ be the finite (truncated) Toeplitz matrices (or operators) generated by the functions f and g, respectively (see Sect. 4.3 for definitions). The relation in (1.3.1) does not hold for $A_T(f)$ and $A_T(g)$, making it necessary to seek an analogous form of (1.3.1).

Second, for the truncated Toeplitz matrix (or operator) $A_T(f)$, we have

$$\frac{1}{T}\operatorname{tr}[A_T(f)] = \int_\Lambda f(\lambda)d\lambda, \tag{1.3.2}$$

where $\operatorname{tr}[A_T(f)]$ denotes the trace of $A_T(f)$. What happens to the relation (1.3.2) when $A_T(f)$ is replaced by a product of truncated Toeplitz matrices (or operators)? It is important to note that the product of Toeplitz matrices (or operators) is not necessarily a Toeplitz matrix (or operator).

The goal is to approximate the trace of the product of Toeplitz matrices (or operators) by using the trace of a *single* Toeplitz matrix (or operator) generated by the product of their respective generating functions.

More specifically, let $\mathcal{H} = \{h_1, h_2, \ldots, h_m\}$ be a collection of integrable real symmetric functions defined on Λ. Define

$$S_A(T) := \frac{1}{T}\text{tr}\left[\prod_{i=1}^{m} A_T(h_i)\right] \quad \text{and} \quad M_A := \frac{1}{T}\text{tr}\left[A_T\left(\prod_{i=1}^{m} h_i(\lambda)\right)\right],$$

and let

$$\Delta(T) := S_A(T) - M_A.$$

How accurately is $S_A(T)$ approximated by M_A? What is the rate of convergence of the approximation error $\Delta(T)$ to zero as $T \to \infty$? These questions are known as the two trace approximation problems, referred to as Problem (A) and Problem (B). Similar issues arise in the context of tapered Toeplitz matrices and operators.

Following Szegő (see Grenander and Szegő [162], Section 8.1), the class of truncated Toeplitz matrices (or operators) satisfying $\Delta(T) \to 0$ as $T \to \infty$ is called the *trace complete class*.

Problems (A) and (B) concerning both Toeplitz and tapered Toeplitz matrices and operators are discussed in Chap. 4.

1.4 A Link Between Stationary Processes and the Trace Approximation Problem

To understand the relevance of the trace approximation problem to stationary processes, consider a question regarding the asymptotic distribution (as $T \to \infty$) of the quadratic functional in (1.1.1) in the case where $X(u)$ is a Gaussian process.

Let $A_T(f)$ be the covariance matrix (or operator) of the process $\{X(u), u \in \mathbb{U}\}$. Specifically, $A_T(f)$ denotes either the $T \times T$ Toeplitz matrix or the T-truncated Toeplitz operator generated by the spectral density f. Similarly, let $A_T(g)$ denote either the $T \times T$ Toeplitz matrix or the T-truncated Toeplitz operator generated by an even integrable function g. The following result establishes a link between the distribution of the quadratic functional Q_T in (1.1.1) and the trace problem for Toeplitz matrices and operators (for more details, see Sect. 5.4):

1. The quadratic functional Q_T in (1.1.1) has the same distribution as the sum $\sum_{k=1}^{\infty} \lambda_k \xi_k^2$ ($\sum_{k=1}^{T} \lambda_k \xi_k^2$ in the discrete time case), where $\{\xi_k, k \ge 1\}$ are independent $N(0, 1)$ Gaussian random variables and $\{\lambda_k, k \ge 1\}$ are the eigenvalues of the operator $A_T(f)A_T(g)$.

2. The k-th-order cumulant $\chi_k(\cdot)$ of Q_T is given by

$$\chi_k(Q_T) = 2^{k-1}(k-1)! \sum_{j=1}^{\infty} \lambda_j^k = 2^{k-1}(k-1)! \operatorname{tr}[A_T(f)A_T(g)]^k. \quad (1.4.1)$$

Thus, to describe the asymptotic distributions of the quadratic functionals in (1.1.1), we need to analyze the corresponding traces of the products of Toeplitz matrices (or operators). Similarly, to determine the asymptotic distributions of the tapered quadratic functionals in (1.1.6), we must examine the traces of the products of tapered Toeplitz matrices (or operators).

1.5 The Structure of the Book

The book consists of ten chapters and two appendices. Each chapter starts with a brief overview and ends with bibliographical notes and comments.

Chapter 1 provides an overview of the main problems that constitute the body of the book. These problems include: (a) Central and non-central limit theorems for Toeplitz and tapered Toeplitz random quadratic functionals of stationary processes, in both discrete and continuous time cases (b) Statistical inference problems related to the spectrum of observed processes, including nonparametric and parametric statistical estimation, and hypotheses testing (c) The trace approximation problem for Toeplitz and tapered Toeplitz matrices and operators

In Chap. 2, we introduce several well-known concepts and results from real analysis, functional analysis, and matrix analysis. These include function spaces, inequalities, embedding and interpolation theorems, Fourier transforms and convolutions, Dirichlet and Fejér kernels and singular integrals, Szegő's and power counting theorems, Schatten classes and norms for compact operators, trace class and Hilbert-Schmidt operators, and reproducing kernel Hilbert spaces. Additionally, we present some technical results involving the properties of slowly and regularly varying functions, which are utilized to state and prove the results in this text.

In Chap. 3, we introduce the model of interest, which is a second-order (weakly) stationary process. We recall some key concepts and results from the theory of stationary processes, including spectral representations, Hilbert spaces associated with stationary processes, and Kolmogorov's isometric isomorphism theorem. Additionally, we cover linear processes and their spectral characterizations, the dependence (memory) structures of the model, single and multiple stochastic integrals with respect to random measures, and various representations of L^2-processes, such as the Karhunen-Loéve orthogonal series representation and the RKHS representation.

Chapter 4 presents a comprehensive approach to the trace approximation problem, covering both Toeplitz and tapered Toeplitz matrices and operators. We discuss their applications to discrete and continuous time stationary processes. Additionally, the chapter provides estimates for the Schatten norms of these matrices and operators.

In Chap. 5, we present results on central limit theorems for random Toeplitz-type quadratic functionals of stationary processes. The underlying processes are either Gaussian or linear processes with memory, defined in discrete or continuous time. We provide results that include sufficient conditions for the CLT in both the frequency and time domains, with a primary focus on the frequency domain results. As applications, we provide Berry-Esséen bounds, as well as large and moderate deviation results for these functionals.

In Chap. 6, we establish weak convergence in the space $\mathbb{C}[0, 1]$ for the continuous time case and in the space $\mathbb{D}[0, 1]$ for the discrete time case of normalized stochastic processes. These processes are generated by random Toeplitz quadratic functionals of Gaussian and Lévy-driven linear processes that exhibit long-range dependence. Both central and non-central functional limit theorems have been established.

In Chap. 7, we examine the asymptotic distributions of tapered random Toeplitz-type quadratic functionals and processes. We prove central limit theorems for tapered Toeplitz-type quadratic functionals in both discrete and continuous time cases, as well as a functional central limit theorem for a suitably normalized stochastic process generated by a tapered Toeplitz-type quadratic functional of a Lévy-driven linear process. Additionally, we establish the moderate deviation principle (MDP) for tapered quadratic functionals.

Chapter 8 focuses on nonparametric statistical estimation of spectral functionals for stationary linear models based on tapered data. We consider averaged tapered periodograms as estimators for unknown functionals. We analyze the bias of these estimators and establish their consistency and asymptotic normality using the central limit theorems presented in Chaps. 5–7. Additionally, we obtain results on constructing asymptotically efficient estimators under the local asymptotic normality (LAN) condition and on bounding the minimax mean square risks of these estimators. The results presented in this chapter are utilized in Chap. 9 to establish the consistency and asymptotic normality of the minimum contrast estimators based on the Whittle contrast functionals for stationary linear models with tapered data.

In Chap. 9, we discuss the statistical parametric estimation of spectral parameters based on tapered data for stationary linear models in both discrete and continuous time. Our focus is on the Whittle method, which utilizes smoothed periodogram analysis in the frequency domain. We prove that, under broad regularity conditions on the model's spectral density, the Whittle estimator based on tapered data is consistent and asymptotically normal. Additionally, we examine simplified Whittle estimators and robust estimation procedures for linear models, including both parametric and nonparametric scenarios.

Chapter 10 focuses on constructing goodness-of-fit tests for testing the hypothesis that the hypothetical spectral density of a stationary Gaussian process has a specified form based on tapered data. We show that when the hypothetical spectral density of the model does not depend on unknown parameters, then the proposed test statistic follows a limiting chi-square distribution. In contrast, when the null hypothesis is composite, indicating that the hypothetical spectral density of the observed process depends on an unknown parameter, we select an appropriate estimator for this parameter and describe the limiting distribution of the test statistic.

Appendix A discusses the motivations and benefits of data tapering. It addresses several issues that can arise in statistical estimation for stationary models, including leakage, trough, variance, bias, and edge effects. Additionally, we offer a brief overview of Dahlhaus's approach to these effects, as outlined in his papers Dahlhaus [72, 73].

In Appendix B, we briefly discuss several important problems that are closely related to the topics covered in the book. These include the Szegő and Avram-Parter-type theorems, limit theorems for quadratic forms in bivariate Appell polynomials, estimation of spectral density and spectral functions, the asymptotic behavior of the prediction error for stationary sequences, and large-sample estimation for the mean of a stationary sequence.

Chapter 2
Preliminaries

In this chapter, we introduce several well-known concepts and results from real, functional, and matrix analysis. These include function spaces, inequalities, embedding and interpolation theorems, Fourier transforms and convolutions, Dirichlet and Fejér kernels and singular integrals, Schatten classes and norms of compact operators, and congruent and reproducing kernel Hilbert spaces. Additionally, we present technical results concerning the properties of slowly and regularly varying functions, which are essential for stating and proving the results in this text.

In Sect. 2.1, we introduce the basic function spaces utilized in this text and present some well-known results that are frequently cited throughout the text. In Sect. 2.2, we discuss the properties of the Dirichlet and Fejér kernels, as well as singular integrals. Section 2.3 is devoted to Fourier transforms and convolutions, along with some important well-known theorems such as the Parseval-Plancherel theorem, the Riesz-Thorin interpolation theorem, and Young's convolution inequality. In Sect. 2.4, we present Szegő's weak and strong theorems, which are used in Chap. 9 for approximating the likelihood function. Section 2.5 focuses on the properties of slowly and regularly varying functions. In Sect. 2.6, we outline power counting theorems that are used in the proofs of non-central limit theorems. Section 2.7 briefly summarizes results from functional and matrix analysis, including the Schatten classes and norms of compact operators, trace class and Hilbert-Schmidt operators, and congruent and reproducing kernel Hilbert spaces.

2.1 Function Spaces

The following standard notation will be used throughout the book.

By $\mathbb{N}$, $\mathbb{Z}$, $\mathbb{R}$, and $\mathbf{C}$ we denote the sets of natural, integer, real, and complex numbers, respectively. Also, $\mathbb{N}_0 := \mathbb{N} \cup \{0\}$ and $\mathbb{R}^+ := [0, \infty)$.

M. S. Ginovyan, *Random Toeplitz Functionals and Their Applications*,
Frontiers in Probability and the Statistical Sciences,
https://doi.org/10.1007/978-3-032-00961-6_2

In this text, Λ always denotes one of the sets $\mathbb{R} = (-\infty, \infty)$ or $[-\pi, \pi)$. Functions defined on the interval $[-\pi, \pi)$ are considered to be 2π-periodic and are extended periodically to the entire real line $\mathbb{R}$. We identify 2π-periodic functions both with those defined on $[-\pi, \pi)$ and with functions defined on the torus $\mathbb{T} := \mathbb{R}/2\pi\mathbb{Z}$ (see Remark 3.1.1).

By $\mathbb{C}(\Lambda)$, we denote the set of all functions ψ that are uniformly continuous and bounded on Λ, endowed with the following norm: $\|\psi\|_{\mathbb{C}} := \sup_{\lambda\in\Lambda} |\psi(\lambda)|$. We also denote by $\mathbb{C}_{\text{loc}}(\Lambda)$ the space of continuous functions on Λ that have bounded support.

Let $(\Lambda, \mathcal{F}, \mu)$ be a measure space, and $1 \le p < \infty$. By $L^p(\mu) := L^p(\Lambda, \mu)$, we denote the Lebesgue space of functions defined on Λ with respect to the measure μ:

$$L^p(\Lambda, \mu) := \{\psi : \ ||\psi||_{p,\mu} < \infty\},$$

where

$$||\psi||_{p,\mu} := \left(\int_\Lambda |\psi(\lambda)|^p d\mu(\lambda)\right)^{1/p}$$

is the norm in $L^p(\Lambda, \mu)$. For $p = \infty$, the space $L^\infty(\Lambda, \mu)$ is defined as the set of essentially bounded (μ-a.e. bounded) functions:

$$L^\infty(\Lambda, \mu) := \{\psi : \ ||\psi||_\infty := \text{ess sup}_{\lambda\in\Lambda}|\psi(\lambda)| < \infty\}.$$

By $L^0(\Lambda, \mu)$, we denote the class of all μ-a.e. finite μ-measurable functions. By $L^1_{\text{loc}}(\Lambda, \mu)$, we denote the set of locally integrable functions, which are those functions for which the Lebesgue integral is finite on all compact subsets of Λ.

In the case where μ is the Lebesgue measure ($d\mu(\lambda) = d\lambda$), we will use the following notation: $L^p(\Lambda)$, $L^\infty(\Lambda)$, $L^0(\Lambda)$, $L^1_{\text{loc}}(\Lambda)$, and $||\psi||_p$.

Observe that for $\Lambda = \mathbb{T}$, the following embedding holds:

$$L^q(\mathbb{T}, \mu) \subset L^p(\mathbb{T}, \mu) \quad \text{for} \quad 1 \le p < q.$$

Since $\mathbb{R}$ is non-compact, the condition $f \in L^p(\mathbb{R}, \mu)$ does not necessarily imply $f \in L^q(\mathbb{R}, \mu)$ for any other value of q, where $1 \le q \le \infty$ (for counterexamples, see, e.g., Butzer and Nessel [60], Problem 5.2.1). However, the following interpolation theorem holds (see, e.g., Butzer and Nessel [60], p. 2):

Proposition 2.1.1 (**L^p-Interpolation Theorem**) *If a function f is in $L^p(\mathbb{R}, \mu) \cap L^q(\mathbb{R}, \mu)$ with $1 \le p < q \le \infty$, then $f \in L^r(\mathbb{R}, \mu)$ for every r such that $p \le r \le q$.*

By $l^p := l^p(\mathbb{Z})$, we denote the set of those sequences $x := \{x_n = x(n),\ n \in \mathbb{Z}\}$ for which

$$\|x\|_p := \left(\sum_{n=-\infty}^{\infty} |x_n|^p\right)^{1/p} < \infty \quad \text{for} \quad 1 \le p < \infty,$$

and $\|x\|_\infty := \sup_{n\in\mathbb{Z}} |x_n| < \infty$ for $p = \infty$. A similar definition may be given when dealing with sequences on $\mathbb{N}$.

Proposition 2.1.2 (Dominated Convergence Theorem) *Let* $\{f_n, n \in \mathbb{N}\} \subset L^1(\Lambda)$*, and let* $\lim_{n\to\infty} f(\lambda) = f(\lambda)$ *a.e. If there exists a function* $g \in L^1(\Lambda)$ *such that* $|f(\lambda)| \le g(\lambda)$ *a.e. for all* n*, then* $f \in L^1(\Lambda)$*, and*

$$\lim_{n\to\infty} \int_\Lambda f_n(\lambda)d\lambda = \int_\Lambda f(\lambda)d\lambda.$$

Recall Lusin's theorem, which provides an approximation of a measurable function using continuous functions (see Natanson [226], Section 4.4, Theorem 4).

Proposition 2.1.3 (Lusin's Theorem) *Let* $f(x)$ *be a measurable function defined on* $[a, b]$*, which is a.e. finite. Then for every* $\varepsilon > 0$*, there exists a continuous function* $g(x)$ *such that* $\text{mes}\{x : f(x) \ne g(x)\} < \varepsilon$*. Moreover, if* $|f(x)| \le C$*, then* $|g(x)| \le C$ *as well.*

For our purposes, the following corollary of Lusin's theorem is important because it provides an approximation for integrable functions (see, e.g., Hong et al. [176], Theorem 3.3.7).

Proposition 2.1.4 *If* $f \in L^1(\Lambda)$*, then for every* $\varepsilon > 0$*, there exists a continuous function* g *on* $\mathbb{R}$ *such that* $\|f - g\|_{L^1(\Lambda)} \le \varepsilon$*.*

Proposition 2.1.4 states that if $f \in L^1(\Lambda)$, then for any $\varepsilon > 0$, the function f can be decomposed as $f = f_1 + f_2$, where f_1 is continuous on Λ and $\int_\Lambda |f_2(\lambda)|d\lambda < \infty$.

If p is such that $1 \le p \le \infty$, then the Hölder conjugate number q is defined through $1/p + 1/q = 1$ in case $1 < p < \infty$, $q = \infty$ if $p = 1$, and $q = 1$ if $p = \infty$.

In the following proposition, we present several versions of the Hölder inequality that are frequently used in this text:

Proposition 2.1.5 (Hölder's Inequalities)

(a) Let $1 \le p, q \le \infty$ *with* $p^{-1} + q^{-1} = 1$*. If* $f \in L^p(\Lambda, \mu)$ *and* $g \in L^q(\Lambda, \mu)$*, then* $fg \in L^1(\Lambda, \mu)$ *and*

$$\|fg\|_{1,\mu} \le \|f\|_{p,\mu}\|g\|_{q,\mu}.$$

For $p = q = 2$*, the inequality is also known as the Cauchy-Schwarz inequality.*

(b) *Let* $1 \le p, q, r \le \infty$ *with* $p^{-1} + q^{-1} = r^{-1}$. *If* $f \in L^p(\Lambda, \mu)$ *and* $g \in L^q(\Lambda, \mu)$, *then* $fg \in L^r(\Lambda, \mu)$ *and*

$$\|fg\|_{r,\mu} \le \|f\|_{p,\mu} \|g\|_{q,\mu}.$$

(c) *Let* $f_i \in L^{p_i}(\Lambda, \mu)$, $1 < p_i < \infty$, $i = 1, 2, \ldots, m$, *with* $\sum_{i=1}^m p_i^{-1} = 1$. *Then* $\prod_{i=1}^m f_i \in L^1(\Lambda, \mu)$ *and*

$$\Big\| \prod_{i=1}^m f_i \Big\|_{1,\mu} \le \prod_{i=1}^m \|f_i\|_{p_i,\mu}.$$

(d) *Let* $x := \{x_n, n \in \mathbb{Z}\} \in l^p$, $1 \le p \le \infty$, *and* $y := \{y_n, n \in \mathbb{Z}\} \in l^q$. *Then* $xy \in l^1$ *and*

$$\sum_{n=-\infty}^{\infty} |x_n y_n| \le \|x\|_p \|y\|_q.$$

For a function $\psi \in \mathbb{C}(\Lambda)$, we denote the modulus of continuity of ψ by $\omega(\psi, \delta)$, which is defined as follows:

$$\omega(\psi, \delta) := \sup_{0<h\le\delta} \|\psi(\cdot + h) - \psi(\cdot)\|_{\mathbb{C}}, \quad \delta > 0.$$

Given a number $0 < \alpha \le 1$, by $\mathrm{Lip}(\Lambda, \alpha)$, we denote the Lipschitz class of functions defined on Λ:

$$\mathrm{Lip}(\Lambda, \alpha) = \{\psi \in \mathbb{C}(\Lambda) : \omega(\psi; \delta) = O(\delta^\alpha) \text{ as } \delta \to 0\}.$$

For a function $\psi \in L^p(\Lambda)$, $1 \le p \le \infty$, let $\omega_p(\psi, \delta)$ denote the L^p-modulus of continuity of ψ:

$$\omega_p(\psi, \delta) := \sup_{0<h\le\delta} \|\psi(\cdot + h) - \psi(\cdot)\|_p, \quad \delta > 0.$$

Given numbers $0 < \alpha \le 1$ and $1 \le p \le \infty$, we denote by $\mathrm{Lip}(\Lambda, p, \alpha)$ the L^p-Lipschitz class of functions defined on Λ:

$$\mathrm{Lip}(\Lambda, p, \alpha) = \{\psi \in L^p(\Lambda) : \omega_p(\psi; \delta) = O(\delta^\alpha) \text{ as } \delta \to 0\}.$$

Observe that if $\psi \in \mathrm{Lip}(\Lambda, p, \alpha)$, then there exists a constant C such that

$$\omega_p(\psi; \delta) \le C\,\delta^\alpha \quad \text{for all} \quad \delta > 0.$$

Given numbers $p \geq 1$, $0 < \alpha \leq 1$, $r \in \mathbb{N}_0 := \mathbb{N} \cup \{0\}$, where $\mathbb{N}$ is the set of natural numbers, we set $\beta = \alpha + r$ and denote by $H_p(\beta) := H_p(\Lambda, \beta)$ the L^p-Hölder class. This class consists of those functions $\psi(\lambda) \in L^p(\Lambda)$ that have r-th derivatives in $L^p(\Lambda)$ and satisfy Hölder's condition with some positive constant C:

$$||\psi^{(r)}(\cdot + h) - \psi^{(r)}(\cdot)||_p \leq C|h|^\alpha.$$

For the next result, see, e.g., Nikol'skii [227], Section 5.2.

Proposition 2.1.6 *Let $f \in H_p(\beta)$, $p > 1, \beta > 0$, and let $E_{A,p}(f)$ denote the best approximation (in the metric of L^p) of a function f by entire analytic functions of exponential type A in the case $\Lambda = \mathbb{R}$ and by polynomials of degree A in the case $\Lambda = \mathbb{T}$. Then, with a positive constant C, we have*

$$E_{A,p}(f) \leq CA^{-\beta}.$$

In the next proposition, we present Bernstein's inequality (see, e. g., Nikol'skii [227], Section 3.2).

Proposition 2.1.7 *Let ψ_A be an entire function of exponential type A that belongs to the class $L^p(\mathbb{R})$ for real λ, where $1 \leq p \leq \infty$. Then, for any $k \in \mathbb{N}$, the following inequality holds for the kth-order derivative $\psi_A^{(k)}$ of ψ_A:*

$$\|\psi_A^{(k)}\|_p \leq A^k \|\psi_A\|_p. \tag{2.1.1}$$

The next proposition is the well-known Hardy-Littlewood-type embedding theorem for the classes $H_p(\beta)$ (see, e. g., Nikol'skii [227], Section 6.3).

Proposition 2.1.8 *Let $\psi \in H_p(\beta)$ with $\beta > 0$ and $p \geq 1$. The following assertions hold:*

(a) If $\beta \leq 1/p$ and $p < p_1 < p/(1 - \beta p)$, then $\psi \in H_{p_1}(\beta - 1/p + 1/p_1)$.
(b) If $\beta > 1/p$, then ψ is continuous and $||\psi||_\infty < \infty$.

The condition ($\mathcal{H}$), defined below, plays a crucial role in this text.

Definition 2.1.1 (Condition ($\mathcal{H}$)) We say that a pair of integrable functions (f, g) satisfies condition ($\mathcal{H}$), and write $(f, g) \in (\mathcal{H})$, if $f \in H_p(\beta_1)$ for $\beta_1 > 0$, $p > 1$ and $g \in H_q(\beta_2)$ for $\beta_2 > 0$, $q > 1$ with $1/p + 1/q = 1$, and one of the conditions (a)–(d) is satisfied:

(a) $\beta_1 > 1/p$, $\beta_2 > 1/q$.
(b) $\beta_1 \leq 1/p$, $\beta_2 \leq 1/q$, and $\beta_1 + \beta_2 > 1/2$.
(c) $\beta_1 > 1/p$, $1/q - 1/2 < \beta_2 \leq 1/q$.
(d) $\beta_2 > 1/q$, $1/p - 1/2 < \beta_1 \leq 1/p$.

Proposition 2.1.9 *If $(f, g) \in (\mathcal{H})$, then there exist numbers p_1 $(p_1 > p)$ and q_1 $(q_1 > q)$, such that $H_p(\beta_1) \subset L_{p_1}$, $H_q(\beta_2) \subset L_{q_1}$, and $1/p_1 + 1/q_1 \leq 1/2$.*

Proof The case $\beta_1 > 1/p$, $\beta_2 > 1/q$ is obvious, since in view of Proposition 2.1.8 (b), we have $H_p(\beta_1) \subset L_\infty$ and $H_q(\beta_2) \subset L_\infty$.

Let $\beta_1 \le 1/p$, $\beta_2 \le 1/q$, and $\beta_1 + \beta_2 > 1/2$. For an arbitrary number $\varepsilon > 0$ satisfying $\beta_1 > \varepsilon$ and $\beta_2 > \varepsilon$, we set

$$\frac{1}{p_1} = \frac{1}{p} - \beta_1 + \varepsilon \quad \text{and} \quad \frac{1}{q_1} = \frac{1}{q} - \beta_2 + \varepsilon.$$

It is easy to see that $p < p_1 < p/(1 - \beta_1 p)$ and $q < q_1 < q/(1 - \beta_2 q)$. Hence, by Proposition 2.1.8 (a), we obtain $H_p(\beta_1) \subset L_{p_1}$ and $H_q(\beta_2) \subset L_{q_1}$. On the other hand, we have

$$\frac{1}{p_1} + \frac{1}{q_1} = \frac{1}{p} + \frac{1}{q} - (\beta_1 + \beta_2) + 2\varepsilon = 1 - (\beta_1 + \beta_2) + 2\varepsilon.$$

Since $\beta_1 + \beta_2 > 1/2$, choosing ε sufficiently small, we obtain $1/p_1 + 1/q_1 \le 1/2$.

Now let $\beta_1 > 1/p$ and $1/q - 1/2 < \beta_2 \le 1/q$. By Proposition 2.1.8 (b), we have $H_p(\beta_1) \subset L_\infty$. For an arbitrary number $\varepsilon > 0$ satisfying $\beta_2 > \varepsilon$, we set

$$\frac{1}{q_1} = \frac{1}{q} - \beta_2 + \varepsilon.$$

Obviously, $q < q_1 < q/(1 - \beta_2 q)$, and hence, $H_q(\beta_2) \subset L_{q_1}$ by Proposition 2.1.8 (a). Further, we have

$$\frac{1}{p_1} + \frac{1}{q_1} = \frac{1}{q} - \beta_2 + \varepsilon.$$

Since $1/q - \beta_2 < 1/2$, choosing ε sufficiently small, we obtain $1/p_1 + 1/q_1 \le 1/2$.

The case $\beta_2 > 1/q$ and $1/p - 1/2 < \beta_1 \le 1/p$ can be treated similarly. □

2.2 The Dirichlet and Fejér Kernels and Singular Integrals

2.2.1 *Properties of the Dirichlet Kernel*

We denote by $D_T(u)$ the Dirichlet kernel both for the real line ($u \in \mathbb{R}$) and for the periodic case ($u \in \mathbb{T}$), defined by

$$D_T(u) := \begin{cases} \frac{\sin(Tu/2)}{\sin(u/2)} & \text{for } u \in \mathbb{T},\ T \in \mathbb{N}, \\ \\ \frac{\sin(Tu/2)}{u/2} & \text{for } u \in \mathbb{R},\ T \in \mathbb{R}^+. \end{cases} \tag{2.2.1}$$

In the next proposition, we list some well-known properties of the kernel $D_T(u)$ (see, e.g., Giraitis et al. [147], p. 8).

Proposition 2.2.1 *The Dirichlet kernel $D_T(u)$ possesses the following properties:*

(a) $|D_T(u)| \le \pi \min\{T, |u|^{-1}\}, \quad u \in \Lambda.$
(b) $|D_T(u)| \le 2\pi T \psi_T(u)$, $u \in \Lambda$, *where* $\psi_T(u) = (1 + T|u|)^{-1}$.
(c) $|D_T(u)| \le \pi\, T^{\delta} |u|^{\delta-1}$ *for any* $\delta \in [0, 1]$ *and* $u \in \Lambda$.

The next proposition presents an important property of the function $\psi_T(u)$ as stated in Proposition 2.2.1 (b) (cf. Giraitis et al. [147], p. 162).

Proposition 2.2.2 *Let $\psi_T(u)$ be as in Proposition 2.2.1 (b). Then, for any $\delta \in (0, 1)$, there exists a constant $C_\delta > 0$ such that*

$$T \int_{\Lambda} \psi_T(x - y)\psi_T(x - z)dx \le C_\delta \psi_T^{1-\delta}(y - z)$$

for all $y, z \in \Lambda$.

2.2.2 Properties of the Fejér Kernel

We denote by $F_T(u)$ the Fejér kernel both for the real line ($u \in \mathbb{R}$) and for the periodic case ($u \in \mathbb{T}$), defined by

$$F_T(u) := \begin{cases} \frac{1}{2\pi T} \cdot \left[\frac{\sin(Tu/2)}{\sin(u/2)}\right]^2 & \text{for } u \in \mathbb{T},\ T \in \mathbb{N}, \\ \\ \frac{1}{2\pi T} \cdot \left[\frac{\sin(Tu/2)}{u/2}\right]^2 & \text{for } u \in \mathbb{R},\ T \in \mathbb{R}^+. \end{cases} \tag{2.2.2}$$

In the next proposition, we list some properties of the kernel $F_T(u)$, demonstrating that $F_T(u)$ is an approximate identity (see, e.g., Butzer and Nessel [60] and Zygmund [321]).

Proposition 2.2.3 *The Fejér kernel $F_T(u)$ possesses the following properties:*

$$\int_{\Lambda} F_T(u)\, du = 1; \tag{2.2.3}$$

$$\sup_T \int_{\Lambda} F_T(u)\, du = C < \infty; \tag{2.2.4}$$

$$\lim_{T\to\infty} \int_{\Lambda\setminus[-\delta,\delta]} F_T(u)du = 0 \quad \textit{for any} \quad \delta > 0; \tag{2.2.5}$$

$$\int_{u\geq 1} F_T(u)du \leq C\,T^{-1}; \tag{2.2.6}$$

$$\int_0^1 F_T(u)u^\alpha du \leq \begin{cases} CT^{-\alpha} & \textit{if } 0 < \alpha < 1 \\ CT^{-1}\ln T & \textit{if } \alpha = 1 \\ CT^{-1} & \textit{if } \alpha > 1. \end{cases} \tag{2.2.7}$$

2.2.3 The Dirichlet and Fejér Singular Integrals

The Dirichlet singular integral corresponding to a function $\psi(\lambda) \in L^1(\Lambda)$ is defined by the following equation:

$$\psi_A(\lambda) := \begin{cases} \frac{1}{2\pi}\int_{-\pi}^{\pi} \frac{\sin A(\lambda-x)}{\sin(\lambda-x)}\psi(x)dx & \text{for } \lambda \in \mathbb{T}, \\ \\ \frac{1}{\pi}\int_{-\infty}^{+\infty} \frac{\sin A(\lambda-x)}{\lambda-x}\psi(x)dx & \text{for } \lambda \in \mathbb{R}. \end{cases} \tag{2.2.8}$$

Note that $\psi_A(\lambda)$ is an entire analytic function of exponential type A in the case $\Lambda = \mathbb{R}$ and is a trigonometric polynomial of degree A in the case $\Lambda = \mathbb{T}$.

In the next proposition, we collect several properties of the Dirichlet singular integral $\psi_A(\lambda)$.

Proposition 2.2.4 *The following assertions hold:*

(a) Let $\psi(\lambda) \in H_p(\beta)$, $p \geq 1$, $\beta > 0$. Then $||\psi_A||_p \leq C(p)||\psi||_p$ and

$$||\psi - \psi_A||_p \leq C(p,\beta)\,A^{-\beta}.$$

(b) Let $\psi(\lambda) \in L^p$, $p \geq 1$. Then for $p < q \leq \infty$

$$||\psi_A||_q \leq 2\,A^{1/p-1/q}||\psi_A||_p.$$

(c) Let $\psi(\lambda) \in H_p(\beta)$, where $\beta = r + \alpha$, $r \in \mathbb{N}_0$, $0 < \alpha \leq 1$ and $p \geq 1$. Then

$$||\psi^{(j)}||_p \leq C < \infty, \quad j = 1, \ldots, r.$$

(d) Let $\psi(\lambda) \in H_p(\beta)$, $p \geq 1$, $\beta > 0$, $q > p$, and $\beta \neq 1/p - 1/q$. Then

$$||\psi_A||_q \leq C \cdot \max\{1; A^{1/p-1/q-\beta}\}.$$

Proofs of assertions (a)–(c) can be found in Nikol'skii [227], Section 8.10 (see also Butzer and Nessel [60], Chapters 1 and 3). For the proof of assertion (d), we refer to the paper by Ibragimov and Has'minskii [186].

2.3 Fourier Transforms and Convolutions

The convolution of functions $\phi, \psi \in L^1(\Lambda)$, denoted by $\phi * \psi$, is defined by

$$(\phi * \psi)(u) = \int_\Lambda \phi(u - x)\psi(x)dx.$$

The symbol $\bar{*}$ will be used to denote the reversed convolution:

$$(\phi^{\bar{*}2})(u) = (\phi \bar{*} \phi)(u) = \int_\mathbb{R} \phi(u + x)\phi(x)dx.$$

By $\mathcal{F}$ and $\mathcal{F}^{-1}$, we denote the Fourier and the inverse Fourier transforms:

$$(\mathcal{F}\phi)(u) = \widehat{\phi}(u) = \int_\mathbb{R} e^{ixu}\phi(x)dx, \quad (\mathcal{F}^{-1}\phi)(u) = \frac{1}{2\pi}\int_\mathbb{R} e^{-ixu}\phi(x)dx.$$

We will use the following well-known identities:

$$\mathcal{F}(\phi * \psi) = \mathcal{F}(\phi) \cdot \mathcal{F}(\psi) \quad \text{and} \quad \mathcal{F}(\phi \bar{*} \psi) = \mathcal{F}(\phi) \cdot \overline{\mathcal{F}(\psi)}. \tag{2.3.1}$$

In the next proposition, we present the Parseval-Plancherel theorem, also referred to as an equality or identity (see, e.g., Bogachev [45], Section 3.10).

Proposition 2.3.1 (Parseval-Plancherel Theorem) *The mapping $\phi \to \widehat{\phi}$ is a bijection of $L^2(\Lambda)$, and for all $\phi, \psi \in L^2(\Lambda)$,*

$$\int_\Lambda \phi(x)\overline{\psi(x)}dx = \frac{1}{2\pi}\int_\Lambda \widehat{\phi}(x)\overline{\widehat{\psi}(x)}dx. \tag{2.3.2}$$

The next proposition presents the Riesz-Thorin interpolation theorem for L^p-spaces (see, e.g., Bergh and Löfström [36], p. 2, and Simon [279], Theorem 6.6.1).

Proposition 2.3.2 (Riesz-Thorin Interpolation Theorem) *Let $p_0, p_1, q_0, q_1 \in [1, \infty]$ and $t \in (0, 1)$. Define*

$$\frac{1}{p_t} = \frac{1-t}{p_0} + \frac{t}{p_1}, \quad \text{and} \quad \frac{1}{q_t} = \frac{1-t}{q_0} + \frac{t}{q_1}.$$

Let T be a linear transformation from $L^{p_0}(\Lambda) \cap L^{p_1}(\Lambda)$ to $L^{q_0}(\Lambda) \cap L^{q_1}(\Lambda)$ such that $\|Tf\|_{q_j} \le C_j\|f\|_{p_j}$ for $f \in L^{p_0} \cap L^{p_1}$ and constants $C_j > 0$, $j = 0, 1$. Then

$$\|Tf\|_{q_t} \le C_0^{1-t}C_1^t\|f\|_{p_t} \quad \text{for all} \quad f \in L^{p_0} \cap L^{p_1}.$$

The next proposition presents Young's convolution inequality (see, e.g., Bogachev [45], Theorem 3.9.4, and Simon [279], Theorem 6.6.3).

Proposition 2.3.3 (Young's Convolution Inequality) *For any numbers p, q, r satisfying $1 \le p, q \le r \le \infty$ and $\frac{1}{r} = \frac{1}{p} + \frac{1}{q} - 1$, and for any functions $f \in L^p(\mathbb{R})$ and $g \in L^q(\mathbb{R})$, the function $f * g$ is defined almost everywhere on $\mathbb{R}$, $f * g \in L^r(\mathbb{R})$, and*

$$\|f * g\|_r \le \|f\|_p \|g\|_q. \tag{2.3.3}$$

In the next proposition, we present the Hausdorff-Young inequality (see, e.g., Simon [279], Theorem 6.6.2).

Proposition 2.3.4 (Hausdorff-Young Inequality) *Let $f \in L^p(\mathbb{R})$, $1 \le p \le 2$. Then $\widehat{f} \in L^q(\mathbb{R})$, where $q = p/(p-1)$ and*

$$\|\widehat{f}\|_q \le (2\pi)^{1/2-1/p} \|f\|_p. \tag{2.3.4}$$

The next proposition presents the Riesz representation theorem for functionals in the spaces L^p (see, e.g., Rudin [264], Theorem 6.16).

Proposition 2.3.5 (Riesz Representation Theorem) *Let $(\Lambda, \mathcal{F}, \mu)$ be a measure space with a positive σ-finite measure μ, and let Φ be a bounded linear functional on $L^p(\Lambda)$ $(1 \le p < \infty)$. Then there exists a unique function $g \in L^q(\Lambda)$ with $1/p + 1/q = 1$, such that*

$$\Phi(f) = \int_\Lambda f g d\mu, \quad f \in L^p(\Lambda).$$

Moreover, we have $\|\Phi\| = \|g\|_q$.

2.4 Szegő's Theorems

The next result, known as *Szegő's weak theorem*, proved by Szegő in 1915 (see, e.g., Grenander and Szegő [162], Section 5.2), involves approximation of the determinant of a Toeplitz matrix by the integral of the logarithm of its generating function.

Proposition 2.4.1 (Szegő's Weak Theorem) *Let $f(\lambda)$ be a non-negative integrable function on $\mathbb{T}$. The following limiting relation holds:*

$$\log\det B_T(f) = \frac{T}{2\pi}\int_{\mathbb{T}} \log f(\lambda)\, d\lambda + o(T) \quad \text{as} \quad T \to \infty. \tag{2.4.1}$$

where $B_T(f)$ is the $T \times T$ Toeplitz matrix generated by the function f (see Sect. 4.3 for the definition of $B_T(f)$).

Note that since $\log(x) \le x-1$ and $f \in L^1(\mathbb{T})$, we have $\int_{\mathbb{T}} \max(0, \log f(\lambda))d\lambda < \infty$. Thus, the integral on the right-hand side of (2.4.1) is either convergent or diverges to $-\infty$, in which case (2.4.1) says $(\det B_T(f))^{1/T} \to 0$ as $T \to \infty$.

Assuming that the derivative $f'(\lambda)$ exists and satisfies a Lipschitz condition with arbitrary exponent α $(0 < \alpha < 1)$, in 1952 Szegő obtained the leading term of the expression $o(T)$ in (2.4.1). This result is known as *Szegő's strong theorem* (see, e.g., Grenander and Szegő [162], Section 5.5).

The next result, known as *Ibragimov's version or a sharp version of Szegő's strong theorem*, was proved by Ibragimov in 1968 (see, e.g., Ibragimov [184] and Simon [278], p. 24).

Proposition 2.4.2 (Szegő's Strong Theorem) *Let $f(\lambda)$ be a non-negative integrable function on $\mathbb{T}$ such that $\ln f(\lambda) \in L^1(\mathbb{T})$. The following limiting relation holds:*

$$\log \det B_T(f) = \frac{T}{2\pi} \int_{\mathbb{T}} \log f(\lambda)\, d\lambda + \sum_{k=1}^{\infty} k|d_k|^2 + o(1) \quad \text{as} \quad T \to \infty, \tag{2.4.2}$$

where d_k are the Fourier coefficients of the function $\ln f(\lambda)$.

Let $H^{1/2}$ denote the Sobolev space of functions defined by

$$H^{1/2} := \left\{ h \in L^2(\mathbb{T}) : \|h\|_{1/2} := \sum_{t=1}^{\infty} t|\widehat{h}(t)|^2 < \infty \right\}, \tag{2.4.3}$$

where $\widehat{h}(t)$ denotes the Fourier transform of a function $h(\lambda) \in L^1(\mathbb{T})$.

Then, the asymptotic relation (2.4.2) holds whenever the series on the right-hand side is defined, specifically when $\log f$ belongs to the Sobolev space $H^{1/2}$.

2.5 Slowly and Regularly Varying Functions

In this section, we introduce the concepts of slowly and regularly varying functions and briefly discuss some properties of these functions that are necessary for our purposes. For a comprehensive treatment of slowly and regularly varying functions, see the monographs by Bingham et al. [43] and Seneta [272].

First, we recall the definition of slowly varying functions (SVF). There are two slightly different standard definitions, one by Karamata and the other by Zygmund.

A measurable function $u : (c, \infty) \to \mathbb{R}$ $(c \ge 0)$ is called *slowly varying at infinity (in Karamata's sense)* if it is positive for x large enough, and for any $t > 0$

$$\lim_{x\to\infty} \frac{u(xt)}{u(x)} = 1.$$

A slowly varying at infinity function $u(x)$ on $[0, \infty)$ is called *quasi-monotone* if it is of bounded variation on any compact interval of $[0, \infty)$ and if, for some $\delta > 0$,

$$\int_0^y x^{\delta}|du(x)| = O\left(y^{\delta}u(y)\right) \quad \text{as} \quad y \to \infty.$$

An example of a slowly varying at infinity function that is quasi-monotone is $u(x) = (1+a/x)^p$, where $a > 0$ and $p \in \mathbb{R}$ (see Pipiras and Taqqu [246], Example 2.2.13). This example will be used in the proof of Lemma 4.7.1.

A measurable function $u(x)$ is called *slowly varying at infinity in Zygmund's sense* if for x large enough, it is positive, and for any $\delta > 0$, there exists a finite number $x_0(\delta) > 0$ such that $x^{\delta}u(x)$ monotonically increases, and $x^{-\delta}u(x)$ monotonically decreases for $x \geq x_0(\delta)$.

It is known that the class of functions that are slowly varying in the Zygmund sense is a subclass of those that are slowly varying in the Karamata sense. Additionally, the Zygmund class is a subset of the class of quasi-monotone slowly varying functions (see, e.g., Theorem 1.5.5 and Corollary 2.7.4 in Bingham et al. [43]).

A function $u(x)$ is called *slowly varying at zero* if the function $\tilde{u}(x) := u(1/x)$ is slowly varying at infinity.

A measurable function $V : \mathbb{R}_+ \to \mathbb{R}$ is called *regularly varying at infinity with exponent* $\alpha \in \mathbb{R}$ if $V(x) = x^{\alpha}u(x)$, where $u(x)$ is slowly varying at infinity. It is *regularly varying at zero* if $\tilde{V}(x) := V(1/x)$ is regularly varying at infinity.

Slowly varying functions are regularly varying functions with exponent $\alpha = 0$.

The following property of slowly varying functions is well-known (see, e.g., Seneta [272], pp. 68 and 94).

Proposition 2.5.1 *Let $u(x)$ and $v(x)$, $x \in \mathbb{R}$, be slowly varying at zero functions. Then the following assertions hold:*

(a) For any $p < 1$

$$\int_0^y x^{-p}u(x)dx = O\left(y^{1-p}u(y)\right) \quad \text{as} \quad y \to 0.$$

(b) The function $x^p u(x)$ increases on some interval $(0, \delta)$ if $p > 0$, and decreases, if $p < 0$.

(c) $u(x)v(x)$ and $\frac{u(x)}{v(x)}$ are slowly varying at zero functions.

The next proposition provides a bound for slowly varying functions, called *Potter's bound* (see, e.g., Bingham et al. [43], Theorem 1.5.6 (i), and Giraitis et al. [147], Section 2.2). This property of slowly varying functions will be used in the proof of the functional non-central limit theorems, given in Sect. 6.5.

Proposition 2.5.2 *Let $u(x) : (0, \infty) \to \mathbb{R}$ be a function slowly varying at $x = 0$ and bounded on intervals $[c, \infty)$ for any $c > 0$. Then for any $\epsilon > 0$, there exists a constant $C > 0$, so that if T is large enough, then for any $x \in (0, \infty)$*

$$\frac{u(x/T)}{u(1/T)} \le C(|x|^{\epsilon} + |x|^{-\epsilon}). \tag{2.5.1}$$

An important consequence of the inequality (2.5.1) is that for any arbitrarily small $\epsilon > 0$, there are constants C_1, C_2, and $x_1 > 0$ such that

$$C_1 x^{-\epsilon} \le u(x) \le C_2 x^{\epsilon} \tag{2.5.2}$$

for all $x \ge x_1$ (see Pipiras and Taqqu [246], Section 2.2).

Now we introduce a special class of slowly varying at zero functions, denoted by $SV_0(\mathbb{R})$, which plays a key role in this book.

Definition 2.5.1 Define $SV_0(\mathbb{R})$ to be the class of slowly varying at zero functions $u(x)$, $x \in \mathbb{R}$, satisfying the following conditions: for some $a > 0$, $u(x)$ is bounded on $[-a, a]$, $\lim_{x\to 0} u(x) = 0$, $u(x) = u(-x)$, and $0 < u(x) < u(y)$ for $0 < x < y < a$.

An example of a function belonging to the class $SV_0(\mathbb{R})$ is $u(x) = |\ln|x||^{-\gamma}$ with $\gamma > 0$ and $a = 1$.

Proposition 2.5.3 *For any functions $u, v \in SV_0(\mathbb{R})$ and numbers $p, q < 1$, $p+q > 1$, there exists a constant $M > 0$ such that for $y \in \mathbb{T}$*

$$\int_{\mathbb{T}} |x|^{-p}|x-y|^{-q} u(x) v^{-1}(x-y)dx \le M|y|^{1-p-q} u(y) v^{-1}(y). \tag{2.5.3}$$

Proof Denote

$$R(x, y) := |x|^{-p}|x-y|^{-q} u(x) v^{-1}(x-y).$$

It is easy to verify that for any $\delta > 0$

$$\sup_{|y|>\delta} \int_{\mathbb{T}} R(x, y)dx < \infty \text{ and } \min_{|y|>\delta} y^{1-p-q} u(y) v^{-1}(y) > 0.$$

Therefore, it is enough to prove (2.5.3) for $y \in (-\delta, \delta)$ with a sufficiently small $\delta > 0$. Applying Proposition 2.5.1 (a), we obtain the following inequalities:

$$\int_{0<|x|<|y|/2} R(x, y)dx \le \left(\frac{|y|}{2}\right)^{-q} v^{-1}\left(\frac{y}{2}\right) \int_{0<|x|<|y|/2} |x|^{-p} u(x)dx$$
$$\le C y^{1-p-q} u(y) v(y), \tag{2.5.4}$$

$$\int_{|y|/2<|x|<2|y|} R(x,y)dx \le \left(\frac{|y|}{2}\right)^{-p} u(2|y|) \int_{|y|/2<|x|<2|y|} |x-y|^{-q} v^{-1}(x-y)dx$$

$$\le C|y|^{-p} u(|y|) \int_{0<|x|<4|y|} |x|^{-q} v^{-1}(x)dx \le C y^{1-p-q} u(y)v(y), \tag{2.5.5}$$

and

$$\int_{2|y|<|x|<\pi} R(x,y)dx \le |y|^{-p} v^{-1}(y) \int_{2|y|<|x|<\pi} |x|^{-q} u(x)dx$$

$$\le C y^{1-p-q} u(y)v(y). \tag{2.5.6}$$

From (2.5.4) to (2.5.6), we obtain (2.5.3). □

The next proposition can be proved similarly.

Proposition 2.5.4 *Given functions $u, w \in SV_0(\mathbb{R})$ satisfying*

$$\int_{\mathbb{T}} x^{-1} u(x) w^{-3}(x) dx < \infty.$$

For any $q \in (0, 1)$, there exists a constant $M > 0$ such that for $y \in \mathbb{T}$

$$\int_{\mathbb{T}} |x|^{-1} |x-y|^{-q} u(x) w^{-2}(x) w^{-1}(x-y) dx \le M |y|^{-q} w^{-3}(y).$$

The following proposition provides an estimate of the Dirichlet kernel in terms of functions from the class $SV_0(\mathbb{R})$.

Proposition 2.5.5 *Let $D_T(x)$, $x \in \mathbb{T}$, be the Dirichlet kernel defined by* (2.2.1). *For any function $w \in SV_0(\mathbb{R})$ and a number $t \in (0, 1)$, there exists a constant $M > 0$ such that*

$$|D_T(x)| \le M w(1/T)\, T^t |x|^{t-1} w^{-1}(x), \quad x \in \mathbb{T}. \tag{2.5.7}$$

Proof According to Proposition 2.5.1 (b), the functions $x^{t-1} w^{-1}(x)$ and $x^{-t} w(x)$ are decreasing in some interval $(0, \delta)$. We can assume that $T^{-1} < \delta$ and $|x| < \delta$. Now, if $|x| \le T^{-1}$, then by the inequality

$$T^{1-t} w^{-1}(1/T) \le x^{t-1} w^{-1}(x)$$

and Proposition 2.2.1 (a), we obtain

$$|D_T(x)| \le T = w(1/T)\, T^t T^{1-t} w^{-1}(1/T) \le w(1/T)\, T^t |x|^{t-1} w^{-1}(x),$$

yielding (2.5.7). The proof in the case $|x| > T^{-1}$ is similar. □

2.6 Power Counting Theorems

Power counting theorems provide convergence conditions for certain classes of integrals on $\mathbb{R}^n$ whose integrands are products of functions that are bounded near zero by powers of linearly dependent affine functionals and are bounded near infinity by different powers of those functionals. These theorems are useful for studying the asymptotic distributions of various statistics in time series models with long-range dependence. The results presented below were used by Terrin and Taqqu [301] to establish non-central limit theorems for quadratic forms Q_T of discrete time stationary processes with long-range dependence (see Theorem 6.2.5).

First, we introduce some notation. Let $\mathfrak{L} := \{L_1, \ldots, L_m\}$ be a collection of linear functionals $L_i(\mathbf{x})$ on $\mathbb{R}^n$, $i = 1, \ldots, m$. For numbers $0 < a_i \leq b_i$, $c_i > 0$, and real constants α_i and β_i, we define

$$P_1(\mathbf{x}) := f_1(L_1(\mathbf{x}))f_2(L_2(\mathbf{x}))\cdots f_m(L_m(\mathbf{x})), \tag{2.6.1}$$

where the functions f_i, $(i = 1, \ldots, m)$ satisfy the following condition:

$$|f_i(y)| \leq \begin{cases} c_i|y_i|^{\alpha_i} \text{ if } & |y| < a_i \\ c_i|y_i|^{\beta_i} \text{ if } & |y| > b_i, \end{cases} \tag{2.6.2}$$

and $|f_i(y)|$ is bounded above in the interval (a_i, b_i), $i = 1, \ldots, m$. The constants α_i and β_i are called the *exponents* of $|y|$ around 0 and ∞, respectively.

For $W \subset \mathfrak{L}$, we set $s(W) := \text{span}(W) \cap \mathfrak{L}$, where $\text{span}(W)$ denotes the linear span of W. Define

$$d_0(W) := r(W) + \sum_{s(W)} \alpha_i,$$

$$d_\infty(W) := r(\mathfrak{L}) - r(W) + \sum_{\mathfrak{L}\backslash s(W)} \beta_i.$$

A summation over a set E refers to summing over the set $\{i : L_i \in E\}$, $|E|$ denotes the cardinality of E, while $r(E)$ represents the rank of E, defined as the number of linearly independent functionals in E. We call a set W *padded* if, for every L in W, L is also in $s(W \setminus \{L\})$, that is, L can be expressed as a linear combination of other elements in W. Note that $W = \emptyset$ is padded. Additionally, if W is linearly independent, then

$$d_0(W) := |W| + \sum_{s(W)} \alpha_i.$$

The proofs of the following results can be found in Terrin and Taqqu [301] (see also Fox and Taqqu [101] and Terrin and Taqqu [303]).

Proposition 2.6.1 *Let $r(\mathfrak{L}) = n$ and assume (a) $d_0(W) > 0$ for every nonempty subset W of $\mathfrak{L}$ with $s(W) = W$, while (b) $d_\infty(W) < 0$ for every proper subset W of $\mathfrak{L}$ with $s(W) = W$, including the empty set. Then*

$$\int_{\mathbb{R}^n} |P_1(\mathbf{x})| d\mathbf{x} < \infty, \tag{2.6.3}$$

where $P_1(\mathbf{x})$ is as defined in (2.6.1).

Proposition 2.6.1 can be extended to the class of functionals of the form $L_i(\mathbf{x}) + \theta_i$, where θ_i is a constant. For $i = 1, \ldots, m$, let α_i and θ_i be real constants, and let $M_i(\mathbf{x})$ be a linear functional on $\mathbb{R}^n$. Put $L_i(\mathbf{x}) := M_i(\mathbf{x}) + \theta_i$ and set

$$P_2(\mathbf{x}) := |L_1(\mathbf{x})|^{\alpha_1} \cdots |L_m(\mathbf{x})|^{\alpha_m}. \tag{2.6.4}$$

Proposition 2.6.2 *Let $s(W)$ and $d_0(W)$ be defined as above. Suppose that $d_0(W) > 0$ for every nonempty subset W of $\mathfrak{L}$ such that $s(W) = W$. Then*

$$\int_{[-t,t]^n} |P_2(\mathbf{x})| d\mathbf{x} < \infty \quad \textit{for all} \quad t > 0, \tag{2.6.5}$$

where $P_2(\mathbf{x})$ is as defined in (2.6.4).

Remark 2.6.1 If $\alpha_i > -1$ and $\beta_i \geq -1$ in Proposition 2.6.1, then it suffices to verify the conditions (a) and (b) for sets W that are also padded. If $\alpha_i > -1$ in Proposition 2.6.2, then it suffices to consider subsets W that are also padded (for more details, see Terrin and Taqqu [301, 303]).

2.7 Some Matrix Analysis and Operator Theory

In this section, we provide a brief summary of key concepts and results from matrix analysis and operator theory that are essential for this text. We will use the terms "operators" and "matrices" interchangeably.

2.7.1 The Basic

Let H be a separable Hilbert space with inner product $(\cdot, \cdot)$ and norm $\|\cdot\|$, respectively. By $\mathcal{L}(H)$, we denote the space of all bounded linear operators from H to itself. The uniform norm of an operator $A \in \mathcal{L}(H)$ is the number

$$\|A\| = \|A\|_\infty := \sup_{\varphi \in H,\, \|\varphi\|=1} (\varphi, A\varphi).$$

For an operator $A \in \mathcal{L}(H)$ by A' and A^*, we denote the transpose and the adjoint of A, respectively. An operator A is called self-adjoint or Hermitian if $A = A^*$, normal if $AA^* = A^*A$, and unitary if $AA^* = A^*A = I$, where I is the identity operator.

A Hermitian operator A is said to be non-negative definite, written $A \geq 0$, if $(\varphi, A\varphi) \geq 0$ for all $\varphi \in H$. If $(\varphi, A\varphi) > 0$ for all $\varphi \in H$ $(\varphi \neq 0)$, we will say that A is positive definite and will write $A > 0$.

If $A \in \mathcal{L}(H)$ with $A \geq 0$, then there exists a unique operator $B \geq 0$ so that $B^2 = A$, and we define the operators $A^{1/2} := B$ and $|A| := (A^*A)^{1/2}$.

If $\{\varphi_k, k \in \mathbb{N}\}$ is an orthonormal basis of H, the matrix of any operator A is given by $a_{ij} = (\varphi_i, A\varphi_j)$, and will write $A = ||a_{ij}||$, $i, j \in \mathbb{N}$. Notice that an operator A is self-adjoint if and only if $a_{ij} = \overline{a}_{ji}$. In the case where $a_{ij} = a_{i-j}$, the matrix $A = ||a_{i-j}||$, $i, j \in \mathbb{N}$ is called a *Toeplitz matrix*.

An operator $A \in \mathcal{L}(H)$ is called *finite rank* if and only if $dim(\mathcal{R}(A)) < \infty$, where $\mathcal{R}(A)$ denotes the range of A. The dimension is called its *rank*, written $r(A)$. The next proposition contains the spectral theorem for finite dimensions.

Proposition 2.7.1 *Let A be a self-adjoint operator on a finite-dimensional space H. Then A has an orthonormal basis of eigenvectors, specifically there exists an orthonormal basis $\{\varphi_k, k = 1, \dots, n\}$ and corresponding numbers $\{\lambda_k, k = 1, \dots, n\}$ in $\mathbb{R}$, called eigenvalues of A, such that*

$$A\varphi_k = \lambda_k \varphi_k, \quad k = 1, \dots, n.$$

For an operator A with eigenvalues $\{\lambda_k, k = 1, \dots, n\}$, counted with their algebraic multiplicities, the trace of A, denoted by $\mathrm{tr}(A)$, is defined by

$$\mathrm{tr}(A) := \sum_{k=1}^{n} \lambda_k.$$

By $\|A\|_2$, we denote the Euclidean (or Frobenius) norm defined by

$$\|A\|_2 := \left[\mathrm{tr}(AA^*)\right]^{1/2}.$$

In the next proposition, we list some well-known properties of the above defined norms (see, e.g., Dahlhaus [73], Davies [80], Graybill [158], Section 5.6).

Proposition 2.7.2 *If A and B are $n \times n$ matrices, then the following assertions hold.*

(a) $\mathrm{tr}(AB) \leq \|A\|_2 \|B\|_2$.
(b) $\|AB\|_2 \leq \|A\|_\infty \|B\|_2$.
(c) $\|AB\|_2 \leq \|A\|_2 \|B\|_\infty$.
(d) $\|A\|_\infty \leq \|A\|_2 < \sqrt{n}\|A\|_\infty$.
(e) $\|AB\|_\infty \leq \|A\|_\infty \|B\|_\infty$.

(f) $\|A^*\|_\infty = \|A\|_\infty$.
(g) $\operatorname{tr}(A) \le \sqrt{n}\|A\|_2$.

For the proof of the following result, see, e.g., Rao [252], Section 14.

Proposition 2.7.3 *If A and B are $k \times s$ matrices, then the nonzero eigenvalues of the matrices AB' and $B'A$ coincide.*

Proposition 2.7.4 *Let A and B be two linear bounded operators; moreover, A is non-negative definite. Then the sets of nonzero eigenvalues of the operators AB, BA, and $A^{1/2}BA^{1/2}$ coincide, where $A^{1/2}$ is the positive definite square root of A.*

Proof Let $AB\varphi = \lambda\varphi$, $\varphi \neq 0$, and $\lambda \neq 0$. If $\psi = A^{1/2}B\varphi$, then we have

$$A^{1/2}BA^{1/2}\psi = A^{1/2}BAB\varphi = \lambda A^{1/2}B\varphi = \lambda\psi,$$

where $\psi \neq 0$, because $A^{1/2}\psi = \lambda\varphi$.

Conversely, if $A^{1/2}BA^{1/2}\varphi = \lambda\varphi$, then denoting $\psi = A^{1/2}\varphi$, we have

$$AB\psi = A^{1/2}A^{1/2}B\psi = A^{1/2}A^{1/2}BA^{1/2}\varphi = \lambda A^{1/2}\varphi = \lambda\psi,$$

and $A^{1/2}Ba\psi = \lambda\varphi$. Thus, the sets of nonzero eigenvalues of the operators AB and $A^{1/2}BA^{1/2}$ coincide. Similarly, we can show that the sets of nonzero eigenvalues of the operators BA and $A^{1/2}BA^{1/2}$ also coincide. □

2.7.2 The Schatten Classes of Compact Operators

By $\mathcal{S}(H)$, we denote the class of completely continuous (compact) operators, which consists of those operators $A \in \mathcal{L}(H)$ that transform every infinite and bounded set in H into a compact set (see, e.g., Riesz and Sz.-Nagy [254], Section 85, p. 206).

Let $B \in \mathcal{S}(H)$; then $|B| = (B^*B)^{1/2} \in \mathcal{S}(H)$. The eigenvalues of the operator $|B| = (B^*B)^{1/2}$ are called the *singular values* (or *s-numbers*) of the operator B. We enumerate the nonzero s-numbers in decreasing order, taking into account their multiplicities. Thus, $s_1(B) \ge s_2(B) \cdots$, and

$$s_j(B) = \lambda_j(|B|), \quad j = 1, \ldots r(|B|),$$

where $r(|B|) = \dim(\mathcal{R}(|B|))$.

Note that $s_1(B) = ||B||_\infty$, $s_j(B) = s_j(B^*)$, $j \ge 1$, and if B is self-adjoint or normal (i.e., if B and B^* commute), then

$$s_j(B) = |\lambda_j(B)|, \quad j \ge 1,$$

where $\lambda_j(B)$, $j \geq 1$, are the eigenvalues of the operator B (see Gohberg and Krein [155], Section 2.2, p. 27).

For a number p, $1 \leq p \leq \infty$, the *p-Schatten norm* of a compact operator $B \in \mathcal{S}(H)$ is defined by

$$||B||_p = \begin{cases} (\sum_{j=1}^{\infty} s_j^p(B))^{1/p}, & \text{for } 1 \leq p < \infty, \\ \sup_j s_j(B), & \text{for } p = \infty. \end{cases} \tag{2.7.1}$$

For $1 \leq p \leq \infty$, by $\mathcal{S}_p := \mathcal{S}_p(H)$, we denote the *Schatten class* of all compact operators $B \in \mathcal{S}(H)$ with $||B||_p < \infty$.

Observe that $\mathcal{S}_\infty(H) = \mathcal{S}(H)$. The class $\mathcal{S}_1$ is referred to as the trace class (or nuclear) operators, while the class $\mathcal{S}_2$ is known as the Hilbert-Schmidt class of operators.

We say that an operator $A \in \mathcal{L}(H)$ has a *finite matrix trace* if for any orthonormal basis $\{\varphi_k, k \in \mathbb{N}\}$ of the space H the series

$$\sum_{k=1}^{\infty}(A\varphi_k, \varphi_k) \tag{2.7.2}$$

converges. If $A \in \mathcal{S}_1$, then the sum in (2.7.2) does not depend upon the orthonormal basis $\{\varphi_k, k \in \mathbb{N}\}$ of the space H, and this sum, denoted by $\mathrm{tr}(A)$, is called the *(matrix) trace* of the operator A. Thus,

$$\mathrm{tr}(A) := \sum_{k=1}^{\infty}(A\varphi_k, \varphi_k). \tag{2.7.3}$$

The following proposition outlines the properties of the trace (see Gohberg and Krein [155], Section 3.7, pp. 96, 101, and Simon [280], Sections 3.6 and 3.7).

Proposition 2.7.5 *The following assertions hold:*

(a) *A necessary and sufficient condition for an operator $A \in \mathcal{L}(H)$ to have a finite matrix trace is that $A \in \mathcal{S}_1$.*
(b) *(Lidskii's theorem). If $A \in \mathcal{S}_1$, then the matrix trace of the operator A coincides with its spectral trace:*

$$\mathrm{tr}(A) = \sum_{k=1}^{r(A)} \lambda_k(A), \tag{2.7.4}$$

where $r(A)$ is the rank of A and $\lambda_k(A)$ are the eigenvalues of A, counted with their algebraic multiplicities.

(c) *If $AB \in \mathcal{S}_1$ and $BA \in \mathcal{S}_1$, then*

$$tr(AB) = \mathrm{tr}(BA). \tag{2.7.5}$$

(d) $A \in \mathcal{S}_p$ *if and only if* $|A|^p \in \mathcal{S}_1$, *and*

$$||A||_p^p = \mathrm{tr}(|A|^p). \tag{2.7.6}$$

(e) *Let* $1 < p < \infty$ *and* $q = p/(p-1)$. *Then for any* $A \in \mathcal{S}_p$ *and* $B \in \mathcal{S}_q$, *we have* $AB \in \mathcal{S}_1$ *and*

$$|\mathrm{tr}(AB)| \leq ||AB||_1 \leq ||A||_p ||B||_q. \tag{2.7.7}$$

(f) *If* $A \in \mathcal{S}_2$, *then* $A^*A \in \mathcal{S}_1$ *and*

$$\mathrm{tr}(A^*A) = ||A||_2^2. \tag{2.7.8}$$

In the next proposition, we collect several properties of the Schatten norms (see Gohberg and Krein [155], Sections 3.7 and 3.8, pp. 92 and 104).

Proposition 2.7.6 *The following assertions hold:*

(a) $\mathrm{tr}(B) \leq ||B||_1$ *and the equality sign is attained for a non-negative* B.
(b) *If* $0 < p_1 < p_2 \leq \infty$ *and* $B \in \mathcal{S}_{p_1}$, *then* $B \in \mathcal{S}_{p_2}$ *and*

$$||B||_{p_2} \leq ||B||_{p_1}.$$

(c) *If* $B_j \in \mathcal{S}_{p_j}$, $p_j \geq 1$, $j = 1, \ldots, n$, *and* $p^{-1} := \sum_{k=1}^{n} (p_k)^{-1} \leq 1$, *then* $B := B_1 \times \cdots \times B_n \in \mathcal{S}_p$, *and*

$$||B||_p \leq ||B_1||_{p_1} \cdot \cdots \cdot ||B_n||_{p_n}.$$

In particular, if $B_1 \in \mathcal{S}_p$ $(p \geq 1)$ *and* $B_2 \in \mathcal{S}_q$, *where* $1/p + 1/q = 1$, *then* $B_1 B_2 \in \mathcal{S}_1$, $B_2 B_1 \in \mathcal{S}_1$, *and*

$$||B_1 B_2||_1 \leq ||B_1||_p ||B_2||_q, \quad ||B_2 B_1||_1 \leq ||B_1||_p ||B_2||_q$$

(d) $||B||_\infty = ||B||$, *where* $||B||$ *is the operator norm of* B, *defined as*

$$||B|| = \sup_{||u||_2 = 1} |(Bu, u)|.$$

Remark 2.7.1 The Schatten classes $\mathcal{S}_p$ serve as analogs of L^p spaces for compact operators. The Riesz-Thorin interpolation theorem (Proposition 2.3.2) is applicable to the spaces $\mathcal{S}_p$ (see Gohberg and Krein [155], Theorem 13.1, or Zhu [320], Theorem 2.6).

2.7.3 *Integral Operators: Decomposition of Hilbert-Schmidt Kernels*

Let $T := [a, b]$ be a closed finite interval, and let $K(t, s)$ be a Hermitian kernel defined on $T \times T$:

$$K(t, s) = \overline{K(t, s)}.$$

Consider a linear integral operator A_K defined on the weighted space $L^2(f, T)$ generated by the kernel $K(t, s)$: for $\phi(t) \in L^2(f, T)$

$$[A_K\phi](t) = \int_a^b K(t, s)\phi(s)f(s)ds, \quad t \in [a, b]. \tag{2.7.9}$$

The operator A_K is called a *Hilbert-Schmidt operator* if

$$\|A_K\|_{HS}^2 := \int_a^b \int_a^b |K(t, s)|^2 f(t)f(s)\text{dtds} < \infty. \tag{2.7.10}$$

The kernel $K(t, s)$ satisfying (2.7.10) is called *Hilbert-Schmidt kernel*, and the norm $\|A_K\|_{HS}$ is referred to as the Hilbert-Schmidt norm of A_K.

Thus, the operator A_K defined by (2.7.9) with the kernel $K(t, s)$ satisfying (2.7.10) belongs to the Hilbert-Schmidt class $\mathcal{S}_2$. The converse also is true, if $A \in \mathcal{S}_2$, then there exists a unique function $K(t, s) \in L^2(f \times f, T \times T)$ such that $A = A_K$, where A_K is defined as in (2.7.9). Moreover, we have

$$\|A_K\|_{HS}^2 = \|A_K\|_2^2 = \|K\|_{L^2(f\times f, T\times T)}^2.$$

Thus, the relation (2.7.9) establishes an isometric isomorphism between the spaces $\mathcal{S}_2(L^2(f, T))$ and $L^2(f \times f, T \times T)$.

In the following proposition, we list some well-known properties of integral operators defined in (2.7.9) (see, e.g., Gohberg and Krein [155], Sections 3.9 and 3.10; Riesz and Sz.-Nagy [254], Section 98; and Simon [280], Chapter 3).

Proposition 2.7.7 *The following assertions hold:*

(a) If the kernel $K(t, s)$ is non-negative definite, that is, for all $n \in \mathbb{N}$ and $t_k \in T$, $z_k \in \mathbf{C}$, $k = 1, \dots n$,

$$\sum_{k,j=1}^{n} K(t_k, t_j)z_k\bar{z}_j \geq 0,$$

then the integral operator A_K defined in (2.7.9) is a trace class (nuclear) operator if and only if

$$\int_a^b K(t,t)f(t)dt < \infty. \tag{2.7.11}$$

Moreover, under (2.7.11) the trace of the operator A_K is given by

$$\mathrm{tr}[A_K] = \int_a^b K(t,t)f(t)dt. \tag{2.7.12}$$

(b) If A_K is a trace class operator, then for $k \in \mathbb{N}$

$$\mathrm{tr}[A_K^k] = \int_{T^k} K(t_1,t_2)\cdots K(t_{k-1},t_k)K(t_k,t_1) \times f(t_1)\cdots f(t_k)dt_1\cdots dt_k. \tag{2.7.13}$$

(c) The kernel $K^*(t,s)$ of the adjoint operator A_K^* is given by

$$K^*(t,s) = \overline{K(s,t)}.$$

(d) If the functions $K_1(t,s)$ and $K_2(t,s)$ are the kernels of the operators A_{K_1} and A_{K_2}, respectively, then the kernel $K(t,s)$ of the operator $A_{K_1}A_{K_2}$ is given by

$$K(t,s) = \int_a^b K_1(t,u)K_2(u,s)f(u)du. \tag{2.7.14}$$

In the next two propositions, we present Schmidt's and Mercer's theorems on the decomposition of Hilbert-Schmidt kernels. These theorems will be used in Chap. 3 to obtain a representation of a general second-order process in terms of orthogonal series, known as the Karhunen-Loéve expansion.

Proposition 2.7.8 (Schmidt's Theorem) *Every Hermitian Hilbert-Schmidt kernel $K(t,s)$ defined on $T \times T$ admits the representation:*

$$K(t,s) = \sum_{i=1}^{\infty} \lambda_i e_i(t)\overline{e_i(s)}, \quad t,s \in [a,b] \tag{2.7.15}$$

where $\{e_i(t), i \in \mathbb{N}\}$ is the orthonormal sequence of eigenfunctions and $\{\lambda_i, i \in \mathbb{N}\}$ is the sequence of the corresponding eigenvalues of the Hilbert-Schmidt operator A_K generated by the kernel $K(t,s)$, specifically

$$\int_a^b K(t,s)e_i(s)f(s)ds = \lambda_i e_i(t) \quad \text{and} \quad \int_a^b e_i(t)\overline{e_j(t)}f(t)dt = \delta_{ij}, \tag{2.7.16}$$

where δ_{ij} is the Kronecker delta. The series in (2.7.15) *converges in the mean, and in this case,*

$$\|A_K\|_{HS}^2 = \|A_K\|_2^2 = \sum_{i=1}^{\infty} \lambda_i^2.$$

Observe that the series in (2.7.15) is not necessarily uniformly convergent, even for continuous on $T \times T$ kernel $K(t, s)$. However, for an important special class of *non-negative definite and continuous kernels*, the convergence is uniform on $T \times T$. This is Mercer's theorem that follows.

Proposition 2.7.9 (Mercer's Theorem) *If the kernel $K(t, s)$ in Proposition 2.7.8 is continuous (on $T \times T$) and non-negative definite, then the series in* (2.7.15) *converges absolutely and uniformly on $T \times T$. The eigenfunctions $e_i(t)$ ($i \in \mathbb{N}$) are continuous functions on T, and the eigenvalues are non-negative, $\lambda_i \geq 0$, $i \in \mathbb{N}$. Moreover, the operator A_K, defined by* (2.7.9)*, is trace class, and tr(A_K) is given by the formula in* (2.7.12).

2.7.4 *Congruent Hilbert Spaces*

In this section, we define the concepts of a Hilbert space spanned by a family of elements and the congruence (isometric isomorphism) between two Hilbert spaces. We also present the basic congruence theorem, which is crucial for the various representations of second-order stochastic processes.

2.7.4.1 Congruent Hilbert Spaces

Recall that an abstract Hilbert space H is a complete inner product space. Consider two abstract Hilbert spaces H_1 and H_2 with inner products $(\cdot, \cdot)_1$ and $(\cdot, \cdot)_2$, respectively.

The spaces H_1 and H_2 are called *isomorphic* if there exists a mapping V from H_1 onto H_2 which is one-to-one, so that the inverse V^{-1} exists as a mapping from H_2 onto H_1, satisfying the following linearity property: for any $u_1, u_2 \in H_1$ and constants $c_1, c_2 \in \mathbb{R}$

$$V(c_1u_1 + c_2u_2) = c_1V(u_1) + c_2V(u_2).$$

The mapping V is called *isomorphism* between H_1 and H_2.

The spaces H_1 and H_2 are called *congruent* or *isometrically isomorphic* if there exists an isomorphism V between H_1 and H_2 that preserves the inner products. Specifically, for any $u_1, u_2 \in H_1$,

$$(u_1, u_2)_1 = (V(u_1), V(u_2))_2. \tag{2.7.17}$$

In this case, the mapping V is referred to as a *congruence* or *isometric isomorphism.*

Note that a congruence not only maps linear combinations of vectors from one space to corresponding linear combinations in another space (as does an isomorphism) but also preserves limits. Specifically, if V is a congruence between H_1 and H_2, then $u = \lim_n u_n$ if and only if $V(u) = \lim_n V(u_n)$. This means that $\|u - u_n\|_1 \to 0$ if and only if $\|V(u) - V(u_n)\|_2 \to 0$ as $n \to \infty$.

2.7.4.2 Hilbert Space Spanned by a Family of Elements

Let H be a Hilbert space with a norm $\|\cdot\|_H$ and an inner product $(\cdot, \cdot)_H$, $\mathbf{T}$ be an index set, and $\{u(t), t \in \mathbf{T}\}$ be a family of members of H. We define the linear manifold *spanned* by the family $\{u(t), t \in \mathbf{T}\}$, denoted by

$$L := \text{sp}\{u(t),\ t \in \mathbf{T}\} = L(u(t),\ t \in \mathbf{T}),$$

as the set consisting of all elements $u(t) \in H$, which can be represented in the form of a linear combination:

$$u = \sum_{k=1}^{n} c_k u(t_k), \quad n \in \mathbb{N} \tag{2.7.18}$$

for some real constants $c_1, \ldots, c_n$ and $t_1, \ldots, t_n \in \mathbf{T}$.

We define the *Hilbert space spanned by the family* $\{u(t), t \in \mathbf{T}\}$ to be the closure of L in H:

$$\overline{L} = \overline{L}(u(t),\ t \in \mathbf{T}) = \overline{sp}\{u(t),\ t \in \mathbf{T}\}_H. \tag{2.7.19}$$

Observe that the space $\overline{L}$ consists of all finite linear combinations of the form given in (2.7.18), as well as their limits in the norm of H.

Generally, $\overline{L}$ is a subspace of H. It is possible for $\overline{L}$ to be equal to H. In this case, we say that the family $\{u(t), t \in \mathbf{T}\}$ *spans* H.

Note that the family $\{u(t),\ t \in \mathbf{T}\}$ spans H if and only if the only vector in H that satisfies the condition $(g, u(t)) = 0$ for every $t \in \mathbf{T}$ is $g = 0$ (see, e.g., Parzen [238]).

A family of vectors $\{u(t),\ t \in \mathbf{T}\}$ in a Hilbert space H is called a *basis* for H, if it spans H, meaning $\overline{L} = H$, and no proper subset of the family spans H. Specifically, for any subset $\mathbf{T}' \subset \mathbf{T}$, we have $\overline{sp}\{u(t),\ t \in \mathbf{T}'\}_H \neq H$.

It is well-known that any two abstract Hilbert spaces H_1 and H_2 with the same dimension are isometrically isomorphic (congruent).

The next proposition plays a key role in the proof of this assertion, as well as in the representations of second-order stochastic processes discussed in Sect. 3.5.

Proposition 2.7.10 (Basic Congruence Theorem) *Let H_1 and H_2 be two abstract Hilbert spaces with inner products $(\cdot, \cdot)_1$ and $(\cdot, \cdot)_2$, respectively. Let $\{\varphi(t),\ t \in \mathbf{T}\}$ and $\{\psi(t),\ t \in \mathbf{T}\}$ be families of vectors that span H_1 and H_2, respectively. Suppose that, for every $s, t \in \mathbf{T}$,*

$$(\varphi(s), \varphi(t))_1 = (\psi(s), \psi(t))_2. \tag{2.7.20}$$

Then the spaces H_1 and H_2 are congruent, and one can define a congruence V from H_1 and H_2 that has the property that

$$V[\varphi(t)] = \psi(t) \text{ for all } t \in \mathbf{T}. \tag{2.7.21}$$

Recall that a family of vectors $\{\varphi(t), t \in \mathbf{T}\}$ in a Hilbert space H is called orthonormal if, for every $s, t \in \mathbf{T}$,

$$(\varphi(s), \varphi(t)) = \delta_{ts} = \begin{cases} 1, & t = s \\ 0, & t \neq s \end{cases}$$

By the well-known Gramm-Schmidt orthogonalization method, it can be proven that any Hilbert space H possesses an orthonormal basis. From this fact and Proposition 2.7.10, we can conclude that any two Hilbert spaces of the same dimension are congruent. More precisely, let H_1 and H_2 be two Hilbert spaces of the same dimension; then there exist an index set $\mathbf{T}$ and families of vectors $\{\varphi(t), t \in \mathbf{T}\}$ and $\{\psi(t), t \in \mathbf{T}\}$ that form orthonormal bases for H_1 and H_2, respectively. Then

$$(\varphi_1, \varphi_2)_1 = (V[\varphi_1], V[\varphi_2])_2$$

holds, from which it follows that H_1 and H_2 are congruent.

2.7.5 Reproducing Kernel Hilbert Spaces (RKHSs)

The concept of *reproducing kernel Hilbert space (RKHS)* was first introduced by S. Zaremba in his 1907 work on boundary value problems for harmonic functions. Since then, the subject has been systematically developed by many authors. RKHSs have wide-ranging applications in various fields, including complex function theory, harmonic analysis, and quantum mechanics. They are particularly important in probability and mathematical statistics. Specifically, RKHSs provide valuable mathematical tools for addressing problems in statistical signal processing, limit theorems, representations of stochastic processes, nonparametric estimation, and prediction theory, among others (see, e.g., Aronszajn [11], Berlient and Thomas-Agnan [37], Hida and Hidsuda [175], Parzen [239], Saitoh [265], and references therein).

2.7.5.1 Definition and Properties of RKHSs

Let H be a class of functions defined on a set $\mathbf{T}$, forming a Hilbert space (real or complex) with a norm $\|\cdot\|$ and an inner product $(\cdot,\cdot)$, and let $K(t,s)$ be a function defined on $\mathbf{T}\times\mathbf{T}$.

Definition 2.7.1 The function $K(t,s)$, $t,s\in\mathbf{T}$, is called a reproducing kernel (r.k.) of the space H if it satisfies the following conditions:

(a) For every $s\in\mathbf{T}$, $K(t,s)$ as a function of t belongs to H; that is, $K(\cdot,s)\in H$.
(b) The reproducing property holds: for every $s\in\mathbf{T}$ and every $f\in H$,

$$(f(\cdot),K(\cdot,t))=f(s). \tag{2.7.22}$$

The Hilbert space H with r.k. $K(t,s)$ is called *reproducing kernel Hilbert space (RKHS)* and often is denoted by $H(K)$.

From formula (2.7.22) with $f(\cdot)=K(\cdot,t)$, we obtain the following important equality:

$$(K(\cdot,t),K(\cdot,s))=K(s,t). \tag{2.7.23}$$

Properties of Reproducing Kernels

1. **Uniqueness.** If a r.k. $K(t,s)$ exists, then it is unique.
2. **Existence.** For existence of a r.k. $K(t,s)$, it is necessary and sufficient that for every $s\in\mathbf{T}$, the function $s(f)\equiv f(s)=$ "the point evaluator functional" be continuous functional of $f\in H$.
3. **Non-negative definiteness (n.n.d.).** The r.k. $K(t,s)$ is a non-negative definite kernel, meaning that for all $n\in\mathbb{N}$, $t_1,\ldots,t_n\in\mathbf{T}$, and $c_1,\ldots,c_n\in\mathbf{C}$,

$$\sum_{kj=1}^{n}K(t_k,t_j)c_k\overline{c}_j\geq 0. \tag{2.7.24}$$

The converse is also true: for every non-negative definite kernel $K(t,s)$, there is a (unique) Hilbert space H for which $K(t,s)$ is the reproducing kernel. Specifically, the following proposition holds, known as the Moore-Aronszajn theorem.

Proposition 2.7.11 (Moore-Aronszajn Theorem) *Let* $\mathbf{T}$ *be a given set, and let* $K(t, s)$, $t, s \in \mathbf{T}$, *be a non-negative definite kernel defined on* $\mathbf{T} \times \mathbf{T}$. *Then there exists a Hilbert space* H *of functions defined on* $\mathbf{T}$ *such that* H *is generated by the family of functions* $\{K(\cdot, t),\ t \in \mathbf{T}\}$:

(a) The system $\{K(\cdot, t),\ t \in \mathbf{T}\}$ *spans the entire space* H*:*

$$H(K) = \overline{sp}\{K(\cdot, t),\ t \in \mathbf{T}\}_H;$$

(b) The inner product in H *satisfies the following condition:*

$$(f(\cdot), K(\cdot, t))_H = f(t) \textit{ for any } f \in H.$$

Thus, $H = H(K)$.

4. **Projection.** If $H(K)$ is a subspace of a larger Hilbert space H, then the formula

$$f(s) = (h(\cdot), K(\cdot, s)), \quad s \in \mathbf{T}$$

gives the projection (orthogonal) f of the element $h \in H$ on $H(K)$.
5. **Orthogonal complement.** Every closed linear subspace H_1 of a RKHS $H(K)$ possesses a r.k. K_1. In particular, if $H_1 = H_1(K_1)$ and $H_2 = H_2(K_2)$ are complementary subspaces of $H(K)$, such that $H(K) = H_1(K_1) \oplus H_2(K_2)$, then $K_1(t, s) + K_2(t, s) = K(t, s)$.

2.7.5.2 Examples of Reproducing Kernel Hilbert Spaces

1. Suppose that the space H is of finite dimension n and that $\{\psi_k(t),\ k = 1, \ldots, n\}$ is an orthonormal system in H. Then the r.k. $K(t, s)$ of H has the form

$$K(t, s) = \sum_{k=1}^{n} \psi_k(t)\overline{\psi_k(s)}.$$

2. Let H_n be the space of trigonometric polynomials of degree at most n considered as a subspace of $L^2[-\pi, \pi)$. Then H_n is a RKHS with r.k. equal to the discrete Dirichlet kernel (see formula (2.2.1)).
3. Let $H :=$ PWE(a) be the Paley-Wiener space of entire functions of exponential type a. Specifically, H consists of complex-valued functions $f(z)$ that satisfy the following condition: for each $\varepsilon > 0$, there exists a constant A_ε such that

$$|f(z)| \leq A_\varepsilon \exp\{(a + \varepsilon)|z|\},$$

and whose restrictions to $\mathbb{R}$ belongs to $L^2(\mathbb{R})$. It is known that

$$H \equiv H_1 = \{f(z) = \frac{1}{2\pi}\int_{-a}^{a} \varphi(t)e^{-itz}dt, \ \varphi \in L^2(-a,a)\}.$$

By Parseval-Plancherel identity, we have

$$f(z) = \frac{1}{2\pi}\int_{-a}^{a} \varphi(t)e^{-itz}dt = \int_{-\infty}^{\infty} f(\lambda)\overline{\frac{\sin a(\lambda - z)}{\pi(\lambda - z)}}d\lambda,$$

showing that PWE(a) is a RKHS with r.k. equal to the continuous Dirichlet kernel (see formula (2.2.1)).

2.8 Bibliographical Notes

Section 2.1 For function spaces and related notions and results, we refer to the books by Butzer and Nessel [60], Nikol'skii [227], and Simon [279].

Condition $(\mathcal{H})$ and Proposition 2.1.9, which play a crucial role in this text, were considered by Ginovyan [120].

Sections 2.2 and 2.3 The properties of Dirichlet and Fejér kernels and singular integrals, and Fourier transforms and convolutions can be found in many books; see, for example, the books by Bogachev [45], Butzer and Nessel [60], Nikol'skii [227], and Zygmund [321]. For the Riesz-Thorin interpolation theorem, which is applied frequently in this text, we refer to the books by Bergh and Löfström [36], Section 1.1, and Simon [279], Section 6.6.

Section 2.4 Both Szegő's weak and strong theorems can be found, for example, in Grenander and Szegő [162] and Simon [278]. It is important to note that both theorems were first obtained by Szegő—the weak theorem in 1915 and the strong theorem in 1952. The stated sharp version of Szegő's strong theorem was first proved by Ibragimov in 1968 (see, e.g., Ibragimov [184] and Simon [278], p. 24).

Section 2.5 For a comprehensive treatment on slowly and regularly varying functions, we refer to the books by Bingham et al. [43] and Seneta [272] (see also Beran et al. [33], Section 1.3; Giraitis et al. [147], Section 2.3; Pipiras and Taqqu [246], Section 2.2; and the paper by Seneta [273]).

The class $SV_0(\mathbb{R})$ of slowly varying at zero functions possessing the specified properties and the proved propositions can be found in the papers by Ginovyan and Sahakyan [135, 136].

Section 2.6 Power counting theorems provide a useful tool for studying the asymptotic distributions of statistics in time series models with long-range dependence. The presented results can be found in the papers by Fox and Taqqu [101] and Terrin and Taqqu [301, 303].

Section 2.7 For matrix analysis results, see, for example, the papers by Dahlhaus [73] and Davies [80], as well as the books by Bhatia [38], Graybill [158], and Rao [252]. For operator theory results, we refer to the books by Gohberg and Krein [155], Riesz and Sz.-Nagy [254], and Simon [280].

Propositions 2.7.5 and 2.7.6, which detail several properties of the trace and Schatten norms, can be found in Sections 3.7 and 3.8 of Gohberg and Krein [155] (see also Simon [280], Chapter 3).

For more information on Hilbert-Schmidt and trace class (nuclear) operators, as well as the proofs of Schmidt's and Mercer's theorems, we refer to Gohberg and Krein [155], Section 3.10; Riesz and Sz.-Nagy [254], Section 98; and Simon [280], Chapter 3.

The Riesz-Thorin interpolation theorem for the Schatten classes $\mathcal{S}_p$ is presented in Gohberg and Krein [155] (Theorem 13.1) and in Zhu [320] (Theorem 2.6).

The proof of the basic congruence theorem (Proposition 2.7.10) can be found in Parzen [239]. For results concerning reproducing kernel Hilbert spaces, we refer to the seminal paper by Aronszajn [11] and the books by Berlient and Thomas-Agnan [37], Hida and Hidsuda [175], Parzen [239], and Saitoh [265].

Chapter 3
The Model: Second-Order (Weakly) Stationary Processes

In this chapter, we introduce the model of interest—a second-order stationary process. We will recall some key concepts and results from the theory of stationary processes, including spectral representations of the covariance function and stationary processes, Hilbert spaces associated with stationary processes, Kolmogorov's isometric isomorphism theorem, linear processes, and dependence (memory) structures of the model. Additionally, we will discuss single and multiple stochastic integrals with respect to random measures, as well as various representations of second-order processes, including the Karhunen-Loève orthogonal series representation and RKHS representation, among others.

A detailed account of the pertinent properties of second-order stationary processes can be found in many books, including those by Ash and Gardner [12], Brockwell and Davis [57], Cramér and Leadbetter [69], Doob [85], Gihman and Skorokhod [110], Ibragimov and Linnik [188], Lindgren [217], Rozanov [263], and Shiryaev [277], among others. Here, we provide an account of the bare essentials needed for our purposes.

In Sect. 3.1, we review some key concepts and results from the theory of stationary processes, including spectral representations of the covariance function and stationary processes, Hilbert spaces associated with stationary processes, and Kolmogorov's isometric isomorphism theorem. Section 3.2 contains linear transformations of stationary processes, spectral characterizations of discrete time linear processes, as well as orthogonal-increments process (OIP)-driven and Lévy-driven continuous time linear processes. In Sect. 3.3, we discuss the memory (dependence) structure of the model for both discrete time and continuous time cases. Depending on the memory structure, we will distinguish short memory (or short-range dependent), long memory (or long-range dependent), and intermediate memory (or antipersistent) models. Section 3.4 is devoted to the definitions and properties of single and multiple stochastic integrals with respect to random measures. We also present a representation of the normalized quadratic functional as a double

M. S. Ginovyan, *Random Toeplitz Functionals and Their Applications*,
Frontiers in Probability and the Statistical Sciences,
https://doi.org/10.1007/978-3-032-00961-6_3

Wiener-Itô stochastic integral. In Sect. 3.5, we present several representations of second-order processes: the orthogonal-series representation (Karhunen-Loéve theorem), the general integral representation (Karhunen theorem), the representation by means of RKHSs (Loéve theorem), and the spectral representation for stationary processes (Cramér-Kolmogorov theorem). In Sect. 3.6, we introduce the concept of generalized stationary processes and briefly discuss some of their properties.

3.1 Key Concepts and Some Basic Results

3.1.1 *Stochastic (Random) Processes and Time Series*

Let $(\Omega, \mathcal{F}, \mathbb{P})$ be a probability space, and let $(E, \mathcal{E})$ be an arbitrary measurable space, where E is an arbitrary set and $\mathcal{E} := \mathcal{E}(E)$ is a σ-field of subsets of E. Let $\mathbb{U} \subset \mathbb{R}$ be an arbitrary set. A stochastic process

$$\{X(u) := X(u, \omega), \ u \in \mathbb{U}, \omega \in \Omega\}$$

defined on $(\Omega, \mathcal{F}, \mathbb{P})$ with state space $(E, \mathcal{E})$ and index set $\mathbb{U}$ is a family of measurable functions (random variables):

$$X(u) : (\Omega, \mathcal{F}) \to (E, \mathcal{E}), \quad u \in \mathbb{U}.$$

In this text, we will consider the cases, where the index set $\mathbb{U}$ consist of the integers: $\mathbb{U} = \mathbb{Z} := \{0, \pm 1, \pm 2, \ldots\}$ (the discrete time (d.t.) case), or is the real line: $\mathbb{U} = \mathbb{R} := (-\infty, \infty)$ (the continuous time (c.t.) case). The state space E will be either $\mathbb{R}$ (the real-valued process) or $\mathbf{C}$ (the complex-valued process).

The law of a stochastic process $\{X(u), u \in \mathbb{U}\}$ is completely characterized by the family of its *finite-dimensional distributions (f.d.d.)*, that is, by the probability distributions

$$\{\mathbb{P}(X(u_1) \le x_1, \ldots, X(u_n) \le x_n), \ n \in \mathbb{N}, x_k \in \mathbb{R}, u_k \in \mathbb{U}\}$$

of the random vectors $(X(u_1), \ldots, X(u_n))'$, $n \in \mathbb{N}$, $u_k \in \mathbb{U}$. Here, the prime indicates the transpose, and all vectors are column vectors throughout.

A stochastic process is often referred to as a *time series*, particularly when it is in discrete time, and the focus is on its mean and covariance functions.

3.1.2 *Gaussian Process*

A stochastic process $\{X(u), u \in \mathbb{U}\}$ is called *Gaussian* if the finite-dimensional distributions of $X(u)$ are multivariate Gaussian, or, equivalently, if for any $n \in \mathbb{N}$,

$c_k \in \mathbb{R}$, and $u_k \in \mathbb{U}$, $k = 1, \ldots, n$, the linear combination $c_1 X(u_1) + \cdots + c_n X(u_n)$ is a Gaussian random variable.

The law of a Gaussian stochastic process $X(u)$ is completely determined by its mean function $m(u) := \mathbb{E}[X(u)]$ and covariance function

$$R(u, v) := \mathbb{E}[X(u) - m(u)][X(v) - m(v)], \quad u, v \in \mathbb{U}.$$

An important example of a Gaussian process is *Brownian motion* (or *Wiener process*), which is defined as a Gaussian process $\{X(t), t \in \mathbb{R}^+\}$ with zero mean ($\mathbb{E}[X(t)] = 0$) and a covariance function:

$$R(t, s) = \sigma^2 \min\{t, s\}, \quad t, s \in \mathbb{R}^+, \sigma > 0,$$

or, equivalently, it is a Gaussian process with independent increments $x(t_k) - x(t_{k-1})$, $k = 1, \ldots, n$, with $t_0 \leq t_1 \cdots \leq t_n$ such that

$$X(t) - X(s) \sim \sigma N(0, t - s), \quad t \geq s > 0.$$

Brownian motion is often denoted by $B(t)$ or $W(t)$. Brownian motion $B(t)$ ($t \in \mathbb{R}$) on the real line is defined as follows: for $t \geq 0$, $B(t) = B_1(t)$, and for $t < 0$, $B(t) = B_2(-t)$, where B_1 and B_2 are two independent Brownian motions defined on the positive half-line. Additionally, a complex-valued Brownian motion $W(t)$ is defined by the equation $W(t) = B_1(t) + iB_2(t)$, where B_1 and B_2 are again two independent real-valued Brownian motions.

3.1.3 Second-Order (Weakly) Stationary Process

A stochastic process $\{X(u), u \in \mathbb{U}\}$ is called a *second-order process* (or L^2-process) if $\mathbb{E}[X^2(u)] < \infty$ for any $u \in \mathbb{U}$. A second-order process $\{X(u), u \in \mathbb{U}\}$ is called a *second-order* (or *weakly*) *stationary* process if the mean function $m(t)$ is constant and the covariance function $R(t, s)$ depends only on the difference of the arguments. Specifically, for any $t, s \in \mathbb{U}$,

$$m(t) = \mathbb{E}[X(t)] = m = \text{constant}, \tag{3.1.1}$$

$$r(u) := R(t - s) = \mathrm{Cov}(X(t), X(s)) = \mathrm{Cov}(X(t - s), X(0)). \tag{3.1.2}$$

The time difference $u = t - s$ above is called the *time lag*. Second-order stationary processes are often called *time series*.

In this text, we will assume that the process $X(u)$ is centered: $\mathbb{E}[X(u)] = 0$. We will also assume that $X(u)$ is a *nondegenerate process*, meaning that $\mathrm{Var}[X(u)] = \mathbb{E}|X(u)|^2 = r(0) > 0$. Without loss of generality, we can take $r(0) = 1$. In the continuous time case, the process $X(u)$ is also assumed to be mean-square (or L^2) continuous, that is,

$$\mathbb{E}[X(t) - X(s)]^2 \to 0 \quad \text{as } t \to s. \tag{3.1.3}$$

This assumption is equivalent to the continuity of the covariance function $r(u)$ at $u = 0$ (see, e.g., Cramér and Leadbetter [69], Section 5.2). The continuity assumption (3.1.3) implies that a standard modification of the process $X(t)$ is separable and measurable (see, e.g., Doob [85], Chapter XI, Section 3).

In this text, we will focus exclusively on second-order stationary processes, which we will simply refer to as "stationary."

A trivial example of a second-order discrete time stationary process is a *white noise* process $\{\xi(u), u \in \mathbb{Z}\}$, denoted as $\xi(u) \sim \text{WN}(0, \sigma^2)$. This process consists of centered, uncorrelated (orthogonal) random variables $\xi(u)$ all with a common variance σ^2. White noise processes serve as "building blocks" in the construction of more complex correlated stationary processes (see Sects. 3.2 and 3.3).

If the random variables $\xi(u)$, $u \in \mathbb{Z}$, are independent and identically distributed (i.i.d.) with mean 0 and variance σ^2, then the process $\{\xi(u), u \in \mathbb{Z}\}$ is referred to as *i.i.d. noise*, and we will use the notation $\xi(u) \sim \text{IID}(0, \sigma^2)$.

3.1.4 *Spectral Representations*

By the Herglotz theorem in the discrete time case and the Bochner-Khintchine theorem in the continuous time case (see, e.g., Ash and Gardner [12] (Chapter 1), Cramér and Leadbetter [69] (Chapter 7), Doob [85] (Chapters 10 and 11), Ibragimov and Linnik [188] (Chapter 16), Shiryaev [277], (Chapter 6)), there exists a finite measure μ on the measure space $(\Lambda, \mathfrak{B}(\Lambda))$, where $\Lambda = [-\pi.\pi)$ in the discrete time case and $\Lambda = \mathbb{R}$ in the continuous time case, and $\mathfrak{B}(\Lambda)$ is the Borel σ-algebra on Λ, such that for any $u \in \mathbb{U}$, the covariance function $r(u)$ of a stationary process $X(u)$ admits the following *spectral representation*:

$$r(u) = \int_{\Lambda} \exp\{i\lambda u\} d\mu(\lambda), \qquad u \in \mathbb{U}. \tag{3.1.4}$$

The measure μ in (3.1.4) is referred to as the *spectral measure* of the process $X(u)$. The function $F(\lambda) := \mu[-\pi, \lambda]$ in the discrete time case and $F(\lambda) := \mu[-\infty, \lambda]$ in the continuous time case, is called the *spectral function* of the process $X(u)$. If $F(\lambda)$ is absolutely continuous with respect to the Lebesgue measure, then the function $f(\lambda) := dF(\lambda)/d\lambda$ is called the *spectral density function* of the process $X(u)$.

Notice that if the spectral density $f(\lambda)$ exists, then $f(\lambda) \geq 0$, $f(\lambda) \in L^1(\Lambda)$, and Eq. (3.1.4) becomes

$$r(u) = \int_{\Lambda} \exp\{i\lambda u\} f(\lambda) d\lambda, \quad u \in \mathbb{U}, \tag{3.1.5}$$

which means that the covariance function $r(u)$ is the Fourier transform of the spectral density $f(\lambda)$ in the continuous time case, and the Fourier coefficients of $f(\lambda)$ in the discrete time case.

Thus, the covariance function $r(u)$ and the spectral function $F(\lambda)$ (or the spectral density function $f(\lambda)$) are equivalent specifications of the second-order properties for a stationary process $\{X(u),\ u \in \mathbb{U}\}$.

To avoid trivial cases, we assume that the spectral measure μ is *nontrivial*, meaning that μ has infinite support. We will employ the standard Lebesgue decomposition of the measure μ:

$$d\mu(\lambda) = d\mu_a(\lambda) + d\mu_s(\lambda) = f(\lambda)d\lambda + d\mu_s(\lambda), \tag{3.1.6}$$

where μ_a is the absolutely continuous part of μ (with respect to the Lebesgue measure), and μ_s is the singular part of μ, representing the sum of the discrete and continuous singular components of μ.

Remark 3.1.1 The parametrization of the unit circle $\mathbb{T}$ by the formula $z = e^{i\lambda}$ establishes a bijection between $\mathbb{T}$ and the interval $\Lambda = [-\pi, \pi)$. Through this bijection, the measure μ on Λ generates the corresponding measure on the unit circle $\mathbb{T}$, which we also denote by μ. Thus, depending on the context, the measure μ will be supported either on Λ or on $\mathbb{T}$, and the corresponding integrals will be over $\Lambda = [-\pi, \pi)$ or over $\mathbb{T}$.

It turns out that the process $X(t)$ itself admits a spectral representation. To present the result, known as Cramér-Kolmogorov theorem, we first define the concept of an orthogonal-increments process.

A complex-valued process $\{Z(\lambda),\ \lambda \in \Lambda\}$ is called an *orthogonal-increments process (OIP)* if it satisfies the following conditions: $\mathbb{E}|Z(\lambda)|^2 < \infty$, $\mathbb{E}[Z(\lambda)] = 0$, $\mathbb{E}|Z(\lambda) - Z(\lambda_0)|^2 \to 0$ as $\lambda \downarrow \lambda_0$, and for any $\lambda_1 < \lambda_2 < \lambda_3 < \lambda_4$,

$$\mathbb{E}[(Z(\lambda_4) - Z(\lambda_3))\overline{([Z(\lambda_2) - Z(\lambda_1))}] = 0. \tag{3.1.7}$$

For an OIP $\{Z(\lambda),\ \lambda \in \Lambda\}$, there exists a unique distribution function $F(\lambda)$ such that for $\lambda_1 < \lambda_2$

$$\mathbb{E}[|Z(\lambda_2) - Z(\lambda_1)|^2] = F(\lambda_2) - F(\lambda_1). \tag{3.1.8}$$

It is common practice in time series analysis to use the shorthand notation:

$$\mathbb{E}[dZ(\lambda)\overline{dZ(\mu)}] = \delta_{\lambda,\mu} dF(\lambda), \tag{3.1.9}$$

where $\delta_{\lambda,\mu}$ is the Kronecker delta function:

$$\delta_{\lambda,\mu} := \begin{cases} 1, & \text{if } \lambda = \mu, \\ 0, & \text{if } \lambda \neq \mu. \end{cases} \tag{3.1.10}$$

The well-known Cramér-Kolmogorov theorem (see, e.g., Cramér and Leadbetter [69], Section 7.5, Ibragimov and Linnik [188], Section 16.5, Shiryaev [277], Section 6.3) states that for any centered stationary process $\{X(u),\ u \in \mathbb{U}\}$ with spectral function F, there exists a unique centered complex-valued orthogonal-increments process $\{Z(\lambda),\ \lambda \in \Lambda\}$ such that for every $u \in \mathbb{U}$, the process $X(u)$ admits the following *spectral representation*:

$$X(u) = \int_\Lambda \exp\{i\lambda u\} dZ(\lambda), \quad u \in \mathbb{U} \tag{3.1.11}$$

and

$$\mathbb{E}[dZ(\lambda)\overline{dZ(\mu)}] = \delta_{\lambda,\mu} dF(\lambda), \quad \lambda, \mu \in \Lambda.$$

The OIP $\{Z(\lambda),\ \lambda \in \Lambda\}$ is called the *spectral process* of $X(u)$. Recall that in the continuous time case, the process $X(u)$ is assumed to be L^2-continuous (see 3.1.3).

It is important to note that even if the process $X(u)$ is real-valued, the corresponding spectral process $Z(\lambda)$ will be complex-valued.

If the spectral function F is absolutely continuous with spectral density $f = F'$, then the spectral representation (3.1.11) can be replaced by (see, e.g., Doob [85], Chapter XI, Section 8):

$$X(u) = \int_\Lambda \exp\{i\lambda u\}\sqrt{f(\lambda)} d\tilde{Z}(\lambda), \quad u \in \mathbb{U} \tag{3.1.12}$$

and

$$\mathbb{E}[d\tilde{Z}(\lambda)\overline{d\tilde{Z}(\mu)}] = \delta_{\lambda,\mu}, \quad \lambda, \mu \in \Lambda.$$

For the definition and properties of stochastic integrals of the type (3.1.11), see Sect. 3.4. Additional references include the books by Ash and Gardner [12] (Chapter 2), Ibragimov and Linnik [188] (Chapter 16), Major [222], Peccati and Taqqu [240] (Chapter 5), and Shiryaev [277] (Chapter 6).

3.1.5 Time and Frequency Domains: Kolmogorov's Congruence Theorem

Given a probability space $(\Omega, \mathcal{F}, \mathbb{P})$, define the L^2-space of random variables $\xi = \xi(\omega)$ with $\mathbb{E}[\xi] = 0$:

$$L^2(\mathbb{P}) = L^2(\mathbb{P}, \Omega) := \left\{\xi : ||\xi||^2 := \int_\Omega |\xi(\omega)|^2 d\mathbb{P}(\omega) < \infty\right\}. \tag{3.1.13}$$

Then $L^2(\mathbb{P})$ becomes a Hilbert space with the following inner product: for $\xi, \eta \in L^2(\mathbb{P})$

$$(\xi, \eta) = \mathbb{E}[\xi\eta] = \int_\Omega \xi(\omega)\eta(\omega)d\mathbb{P}(\omega). \tag{3.1.14}$$

For $a, b \in \mathbb{U}$, $-\infty \le a \le b \le \infty$, we define the space $\mathbb{H}_a^b(X)$ to be the closed linear subspace of the space $L^2(\mathbb{P})$ spanned by the random variables $X(u, \omega)$, $u \in [a, b]$:

$$\mathbb{H}_a^b(X) := \overline{sp}\{X(u),\ a \le u \le b\}_{L^2(\Omega)}. \tag{3.1.15}$$

Note that the subspace $\mathbb{H}_a^b(X)$ consists of all finite linear combinations $\sum_{k=1}^n c_k X(u_k)$ $(a \le u_k \le b)$, as well as their $L^2(\mathbb{P})$-limits (see Sect. 2.7.4).

The space $L^2(X) := \mathbb{H}_{-\infty}^{\infty}(X)$ is called the *Hilbert space generated by the process* $X(u)$, or the *time domain* of the process $X(u)$.

Let μ be the spectral measure of the process $\{X(u),\ u \in \mathbb{U}\}$. Consider the weighted L^2-space $L^2(\mu) := L^2(\Lambda, \mu)$ of complex-valued functions $\varphi(\lambda)$, $\lambda \in \Lambda$, defined by

$$L^2(\mu) := \left\{\varphi(\lambda) : ||\varphi||_\mu^2 := \int_\Lambda |\varphi(\lambda)|^2 d\mu(\lambda) < \infty\right\}. \tag{3.1.16}$$

Then $L^2(\mu)$ becomes a Hilbert space with the following inner product: for $\varphi, \psi \in L^2(\mu)$

$$(\varphi, \psi)_\mu = \int_\Lambda \varphi(\lambda)\overline{\psi}(\lambda)d\mu(\lambda). \tag{3.1.17}$$

The Hilbert space $L^2(\mu)$ is called the *frequency (or spectral) domain* of the process $X(u)$.

The next proposition, known as Kolmogorov's congruence theorem, establishes an isometric isomorphism between the time domain $L^2(X)$ and the frequency $L^2(\mu)$ (see also Sect. 3.5.5, and Grenander and Szegő [162], Section 10.5, Ibragimov and Linnik [188], Section 16.6).

Proposition 3.1.1 (Kolmogorov's Congruence Theorem) *For any stationary process $X(u)$, $u \in \mathbb{U}$, with spectral measure μ, there exists a unique isometric isomorphism V between the time domain $L^2(X)$ and the frequency domain $L^2(\mu)$, such that for any $u \in \mathbb{U}$,*

$$V[X(u)] = e^{iu\lambda}. \tag{3.1.18}$$

The congruence V between the spaces $L^2(X)$ and $L^2(\mu)$, defined by (3.1.18), is referred to as *Kolmogorov's congruence.*

Kolmogorov's congruence illustrates that any linear problem in the time domain $L^2(X)$ can be translated into a corresponding problem in the frequency domain $L^2(\mu)$ and vice versa. This fact enables us to study stationary processes using analytic methods.

3.2 Linear Transformations and Linear Processes

3.2.1 *Linear Transformations of Stationary Processes*

Let $\{X(t), t \in \mathbb{U}\}$ be a stationary process with spectral measure μ, spectral function F, and spectral decomposition:

$$X(t) = \int_\Lambda e^{i\lambda t} dZ_X(\lambda), \quad t \in \mathbb{U}. \tag{3.2.1}$$

We say that a process $\{Y(t), t \in \mathbb{U}\}$ is obtained from the process $X(t)$ through a *linear transformation* with a *transfer* (or *spectral characteristic*) function $\varphi(\lambda)$ if $Y(t)$ admits the spectral representation:

$$Y(t) = \int_\Lambda e^{i\lambda t} \varphi(\lambda) d\, Z_X(\lambda), \quad \varphi(\lambda) \in L^2(\mu). \tag{3.2.2}$$

It is clear that the process $Y(t)$ is also stationary and therefore admits a spectral representation:

$$Y(t) = \int_\Lambda e^{i\lambda t} dZ_Y(\lambda), \quad t \in \mathbb{U}. \tag{3.2.3}$$

It follows from (3.2.2) and (3.2.3) that

$$Z_Y(\lambda) = \int^\lambda \varphi(s) dZ_X(s), \tag{3.2.4}$$

where the symbol $\int^\lambda$ stands for $\int_{-\pi}^\lambda$ in the discrete time case and for $\int_{-\infty}^\lambda$ in the continuous time case.

Therefore, the spectral function $F_Y(\lambda)$ of the process $Y(t)$ is given by

$$F_Y(\lambda) = \int^{\lambda} |\varphi(s)|^2 dF(s). \tag{3.2.5}$$

In particular, if $X(t)$ has a spectral density $f(\lambda)$, then $Y(t)$ also has a spectral density $f_Y(\lambda)$, which can be expressed as

$$f_Y(\lambda) = |\varphi(\lambda)|^2 f(\lambda). \tag{3.2.6}$$

3.2.2 Linear Processes: Existence of Spectral Density Functions

In this text, we consider stationary processes that possess spectral density functions. The theorems that follow, due to A.N. Kolmogorov [200, 201], provide necessary and sufficient conditions for the existence of spectral density functions in both discrete and continuous time cases. These theorems can also be viewed as time domain characterizations of linear processes (see, e.g., Cramér and Leadbetter [69] (Chapter 7) and Ibragimov and Linnik [188] (Chapter 16)).

Theorem 3.2.1 *The following assertions hold:*

(a) *The spectral function $F(\lambda)$ of a discrete time stationary process $\{X(u),\ u \in \mathbb{Z}\}$ is absolutely continuous (with respect to the Lebesgue measure), expressed as $dF(\lambda) = f(\lambda)d\lambda$, if and only if it can be represented as an infinite moving average:*

$$X(u) = \sum_{k=-\infty}^{\infty} a(u-k)\xi(k), \qquad \sum_{k=-\infty}^{\infty} |a(k)|^2 < \infty, \tag{3.2.7}$$

where $\{\xi(k), k \in \mathbb{Z}\} \sim WN(0, \sigma^2)$ is a white noise process with variance σ^2.

(b) *The covariance function $r(u)$ and the spectral density $f(\lambda)$ of $X(u)$ are defined by the following formulas:*

$$r(u) = \mathbb{E}[X(u)X(0)] = \sum_{k=-\infty}^{\infty} a(u+k)a(k), \tag{3.2.8}$$

and

$$f(\lambda) = \frac{\sigma^2}{2\pi}\left|\sum_{k=-\infty}^{\infty} a(k)e^{-ik\lambda}\right|^2 = \frac{\sigma^2}{2\pi}|\widehat{a}(\lambda)|^2, \quad \lambda \in [-\pi, \pi). \tag{3.2.9}$$

(c) *In the case where* $\{\xi(k), k \in \mathbb{Z}\}$ *is a sequence of Gaussian random variables, the process* $\{X(u),\ u \in \mathbb{Z}\}$ *is also Gaussian.*

Similar results apply to continuous time processes that use an orthogonal-increments process instead of white noise.

Theorem 3.2.2 *The following assertions hold:*

(a) *The spectral function* $F(\lambda)$ *of a continuous time stationary process* $\{X(u),\ u \in \mathbb{R}\}$ *is absolutely continuous (with respect to Lebesgue measure), expressed as* $dF(\lambda) = f(\lambda)d\lambda$*, if and only if it can be represented as an infinite continuous moving average:*

$$X(u) = \int_{\mathbb{R}} a(u-t)d\xi(t), \quad \int_{\mathbb{R}} |a(t)|^2 dt < \infty, \tag{3.2.10}$$

where $\{\xi(t), t \in \mathbb{R}\}$ *is a real-valued OIP satisfying* $\mathbb{E}|d\,\xi(t)|^2 = \sigma^2 dt$.

(b) *The covariance function* $r(u)$ *and the spectral density* $f(\lambda)$ *of* $X(u)$ *are given by the following formulas:*

$$r(u) = \mathbb{E}[X(u)X(0)] = \int_{\mathbb{R}} a(u+x)a(x)dx, \tag{3.2.11}$$

and

$$f(\lambda) = \frac{\sigma^2}{2\pi}\left|\int_{\mathbb{R}} e^{-i\lambda t}a(t)dt\right|^2 = \frac{\sigma^2}{2\pi}|\widehat{a}(\lambda)|^2, \quad \lambda \in \mathbb{R}. \tag{3.2.12}$$

(c) *In the case where* $\{\xi(t), t \in \mathbb{R}\}$ *is a Gaussian process, the process* $\{X(u),\ u \in \mathbb{R}\}$ *is also Gaussian.*

Theorems 3.2.1 and 3.2.2 on the existence of spectral densities can be viewed as time domain characterizations of linear processes.

Corollary 3.2.1

(a) *A discrete time stationary process* $X(t)$ *with spectral function* $F(\lambda)$ *is a linear process if and only if* $F(\lambda)$ *is absolutely continuous with respect to the Lebesgue measure. In other words, a discrete time stationary process* $X(t)$ *is a linear process if and only if it can be obtained from a white noise* $\varepsilon(t) \sim \mathrm{WN}(0, \sigma^2)$ *through a linear transformation with transfer function* $\varphi(\lambda)$*:*

$$\varphi(\lambda) := \sum_{k=-\infty}^{\infty} a(k)e^{-i\lambda k} \in L^2(\mathbb{T}).$$

(b) *A continuous time stationary process* $X(t)$ *with spectral function* $F(\lambda)$ *is a linear process if and only if* $F(\lambda)$ *is absolutely continuous with respect to* *the Lebesgue measure. Put another way, a continuous time stationary process $X(t)$ is a linear process if and only if it is derived from an OIP $\{\xi(t), t \in \mathbb{R}\}$ that satisfies $\mathbb{E}|d\,\xi(t)|^2 = \sigma^2 dt$, through a linear transformation with transfer function $\varphi(\lambda)$:*

$$\varphi(\lambda) := \int_{\mathbb{R}} e^{-i\lambda t} a(t)dt \in L^2(\mathbb{R}).$$

The function $a(\cdot)$ in representations (3.2.7) and (3.2.10) serves as a *time-invariant filter*. The linear processes defined by (3.2.7) and (3.2.10) can be viewed as the output of the *linear filter* $a(\cdot)$ applied to the process $\{\xi(u),\ u \in \mathbb{U}\}$, which is referred to as the *driving* or *innovation* process of $X(u)$.

3.2.3 Lévy-Driven Linear Process

To define a Lévy-driven linear process, we first need to define a Lévy process.

A *Lévy process*, $\{\xi(t),\ t \in \mathbb{R}\}$, is defined as a stochastic process with independent and stationary increments, which is continuous in probability, and has sample paths that are right-continuous with left limits (*càdlàg*). Additionally, it satisfies the condition $\xi(0) = \xi(0-) = 0$.

The Wiener process $\{B(t),\ t \ge 0\}$ and the centered Poisson process $\{N_c(t) := N(t)-\mathbb{E}[N(t)],\ t \ge 0\}$ are typical examples of centered Lévy processes. It is important to note that the processes $B(t)$ and $N_c(t)$, like all centered Lévy processes with finite second moments, are also processes with stationary orthogonal-increments. For a detailed account of the relevant properties of Lévy processes, see, for example, the book by Protter [251].

A *Lévy-driven linear process* $\{X(t),\ t \in \mathbb{R}\}$ is a real-valued continuous time stationary process defined by (3.2.10), where the driving process $\xi(t)$ is a Lévy process that satisfies the following conditions:

$$\mathbb{E}[\xi(t)] = 0,\ \mathbb{E}[\xi^2(1)] = 1 \text{ and } \mathbb{E}[\xi^4(1)] < \infty.$$

In the case where $\xi(t) = B(t)$, $X(t)$ is a Gaussian process. The distinction between OIP-driven and Lévy-driven stationary processes is analogous to the difference in discrete time between weak moving averages of uncorrelated white noise and strong moving averages of independent white noise.

OIP-driven and Lévy-driven stationary processes of the form (3.2.10) appear in various fields of science, including economics, finance, and physics. These processes encompass a broad class of popular models used in continuous time time series modeling. For example, the continuous time autoregressive moving average (CARMA) models, defined in Sect. 3.3, serve as continuous time analogs to the classical autoregressive moving average (ARMA) models in the discrete time case.

CARMA models are of the form (3.2.10) and play a central role in the representation of continuous time processes (see Brockwell [56] and Brockwell and Davis [57]).

3.3 Dependence (Memory) Structure of the Model

In the frequency domain setting, the statistical and spectral analysis of stationary processes requires two types of conditions on the spectral density $f(\lambda)$ of the observed process. The first type addresses the *singularities* of $f(\lambda)$ and involves the *linear dependence (or memory) structure* of the process. The second type governs the *smoothness* of $f(\lambda)$.

The memory structure of a stationary process is essentially a measure of the dependence among all the variables in the process, accounting for the effects of all correlations simultaneously. Traditionally, the memory structure has been defined in the time domain by the decay rates of the autocorrelations or, in the frequency domain, by the rates of explosion of low-frequency spectra (see, e.g., Beran [31], Beran et al. [33], Giraitis et al. [147], Guégan [163], Pipiras and Taqqu [246], Robinson [256], and references therein).

For our purposes, it will be convenient to characterize the memory structure in terms of the spectral density function.

Depending on the memory structure, we will distinguish the following types of stationary models:

(a) Short memory (or short-range dependent)
(b) Long memory (or long-range dependent)
(c) Intermediate memory (or antipersistent)

3.3.1 Definition of Different Types of Memory

Let $\{X(t),\ t \in \mathbb{Z}\}$ be a discrete time stationary process with covariance function $r(t)$ $(t \in \mathbb{Z})$ and spectral density $f(\lambda)$ $(\lambda \in [-\pi, \pi))$. A heuristic definition of long memory, short memory, and antipersistence is given as follows: the process $X(t)$ has (a) long memory, (b) short memory, or (c) antipersistence if, as $|\lambda| \to 0$, the spectral density $f(\lambda)$ (a) diverges to infinity, (b) converges to a finite constant, or (c) converges to zero, respectively. Since $2\pi f(0) = \sum_{t\in\mathbb{Z}} r(t)$, this is essentially (in a sense specified more precisely below) equivalent to (a) $\sum_{t\in\mathbb{Z}} r(t) = \infty$, (b) $0 < \sum_{t\in\mathbb{Z}} r(t) < \infty$, and (c) $\sum_{t\in\mathbb{Z}} r(t) = 0$, respectively.

To provide formal definitions of different types of memory (dependence), we recall the concept of slowly varying functions (see Sect. 2.5).

Definition 3.3.1 A stationary process $\{X(t),\ t \in \mathbb{Z}\}$ with spectral density $f(\lambda)$ ($\lambda \in [-\pi, \pi)$) is said to exhibit (a) long memory, (b) intermediate memory, (c) short memory, or (d) antipersistence if

$$f(\lambda) = L_f(\lambda)|\lambda|^{-2d}, \tag{3.3.1}$$

where $L_f(\lambda)$ is a non-negative symmetric function that is slowly varying at zero, and (a) $d \in (0, 1/2)$, (b) $d = 0$ and $\lim_{\lambda\to 0} L_f(\lambda) = \infty$, (c) $d = 0$ and $\lim_{\lambda\to 0} L_f(\lambda) = c_f \in (0, \infty)$, and (d) $d \in (-1/2, 0)$, respectively.

The definitions of memory in both the time and frequency domains are generally not equivalent due to the presence of slowly varying functions. However, they can become equivalent depending on the choice of these functions, as demonstrated in the following proposition (see Beran et al. [33], Theorem 1.3, and Samorodnitsky [268], Theorem 6.2.11).

Proposition 3.3.1 *Let $\{X(t),\ t \in \mathbb{Z}\}$ be a stationary process with covariance function $r(t)$ ($t \in \mathbb{Z}$) and spectral density $f(\lambda)$ ($\lambda \in [-\pi, \pi)$). Then the following assertions hold:*

(a) If the covariance function $r(t)$ has the form

$$r(t) = L_r(t)|t|^{2d-1}, \tag{3.3.2}$$

where $L_r(t)$ is slowly varying at infinity in Zygmund's sense, and either $d \in (0, 1/2)$ or $d \in (-1/2, 0)$ and $\sum_{t\in\mathbb{Z}} r(t) = 0$, then the process $X(t)$ has a spectral density f that satisfies

$$f(\lambda) \sim L_f(\lambda)|\lambda|^{-2d} \quad \text{as} \quad \lambda \to 0 \tag{3.3.3}$$

with

$$L_f(\lambda) = L_r(1/\lambda)\pi^{-1}\Gamma(2d)\sin(\pi/2 - \pi d). \tag{3.3.4}$$

(b) If the spectral density $f(\lambda)$ has the form

$$f(\lambda) = L_f(\lambda)|\lambda|^{-2d}, \quad 0 < \lambda < \pi, \tag{3.3.5}$$

where $d \in (-1/2, 0) \cup (0, 1/2)$, and $L_f(\lambda)$ is slowly varying at zero in Zygmund's sense and of bounded variation on (a, π) for any $a > 0$, then

$$r(t) \sim L_r(t)|t|^{2d-1} \quad \text{as} \quad t \to \infty, \tag{3.3.6}$$

where

$$L_r(t) = 2L_f(1/t)\Gamma(1 - 2d)\sin(\pi d). \tag{3.3.7}$$

In light of Definition 3.3.1 and Proposition 3.3.1, we can state the following general statement that encompasses all four types of memory structures concerning the sum of the covariances (see Beran et al. [33], Corollary 1.1).

Proposition 3.3.2 *Let $\{X(t),\ t \in \mathbb{Z}\}$ be a stationary process with covariance function $r(t)$ and spectral density $f(\lambda)$ of the following form:*

$$f(\lambda) = L_f(\lambda)|\lambda|^{-2d}, \quad 0 < \lambda < \pi \tag{3.3.8}$$

with $d \in (-1/2, 1/2)$, and $L_f(\lambda)$ is slowly varying at zero in Zygmund's sense and of bounded variation on (a, π) for any $a > 0$. Then the following assertions hold:

(a) For $d \in (-1/2, 0)$, we have $\sum_{t\in\mathbb{Z}} r(t) = 2\pi f(0) = 0$.
(b) For $d \in (0, 1/2)$, we have $\sum_{t\in\mathbb{Z}} r(t) = 2\pi \lim_{\lambda\to 0} f(\lambda) = \infty$.
(c) For $d = 0$, we have the following:

(c1) $0 < \sum_{t\in\mathbb{Z}} r(t) = 2\pi f(0) = 2\pi c_f < \infty$ if $0 < \lim_{\lambda\to 0} L_f(\lambda) = c_f < \infty$.
(c2) $0 < \sum_{t\in\mathbb{Z}} r(t) = 2\pi \lim_{\lambda\to 0} f(\lambda) = \infty$ if $\lim_{\lambda\to 0} L_f(\lambda) = \infty$.

The above propositions demonstrate that characterizing memory structure using the spectral density f is more elegant than using the covariance function r. This is because Eq. (3.3.8) applies to all four cases: long memory, short memory, intermediate memory, and antipersistence.

3.3.2 Short Memory Models

Much of statistical and spectral analysis is focused on *short memory* stationary models, where the spectral density $f(\lambda)$ of the model is bounded away from both zero and infinity.

A typical example of a discrete time short memory model is the stationary autoregressive moving average (ARMA) (p, q) process, $X(t)$, which is defined as a stationary solution to the difference equation:

$$\psi_p(B)X(t) = \theta_q(B)\varepsilon(t), \quad t \in \mathbb{Z},$$

where ψ_p and θ_q are polynomials respectively of degrees p and q having no zeros on the unit circle $\mathbb{T}$, B is the backshift operator defined by $BX(t) = X(t-1)$, and $\varepsilon(t) \sim WN(0, \sigma^2)$ is a white noise. The covariance function $r(t)$ of (ARMA)(p, q) process is exponentially bounded:

$$|r(t)| \le Ch^{-k}, \quad t = 1, 2, \ldots; \quad 0 < C < \infty;\ 0 < h < 1,$$

and the spectral density $f(\lambda)$ is a rational function (see, e.g., Brockwell and Davis [57], Section 3.1):

$$f(\lambda) = \frac{\sigma^2}{2\pi} \cdot \frac{|\theta_q(e^{-i\lambda})|^2}{|\psi_p(e^{-i\lambda})|^2}. \tag{3.3.9}$$

A typical example of a continuous time short memory model is a stationary continuous time ARMA(p, q) process, denoted as CARMA(p, q). This process is defined as the solution to a pth-order stochastic differential equation, with a suitable initial condition, driven by standard Brownian motion and its derivatives up to and including order q $(0 \le q < p)$. Specifically, for $t \ge 0$,

$$\begin{aligned} X^{(p)}(t) - \alpha_p X^{(p-1)}(t) - \cdots - \alpha_1 X(t) - \alpha_0 \\ = \sigma\{B^{(1)}(t) + \beta_1 B^{(2)}(t) + \cdots + \beta_q B^{(q+1)}(t)\}, \end{aligned} \tag{3.3.10}$$

where $\{B(t), t \in \mathbb{R}^+\}$ is the standard Brownian motion; the superscript (j) denotes j-fold differentiation with respect to t: $dX^{(j-1)}(t) = X^{(j)}(t)dt$, $j = 1, \dots p-1$. It is also assumed that $\sigma > 0$ and $\beta_q \neq 0$.

The spectral density $f(\lambda)$ of the CARMA(p, q) process is given by the following formula (see, e.g., Brockwell [56]):

$$f(\lambda) = \frac{\sigma^2}{2\pi} \cdot \frac{|\beta(i\lambda)|^2}{|\alpha(i\lambda)|^2}, \tag{3.3.11}$$

where $\alpha(z) = z^p - \alpha_p z^{(p-1)} - \cdots - \alpha_1$ and $\beta(z) = 1 + \beta_1 z + \cdots + \beta_q z^q$. We assume that all roots of the polynomials $\alpha(z)$ and $\beta(z)$ have negative real parts. This assumption guarantees the uniqueness of the polynomials $\alpha(z)$ and $\beta(z)$ corresponding to the given spectral density $f(\lambda)$.

An important special case of the CARMA(p, q) model is the Gaussian continuous time AR(1) process $X(t)$, commonly referred to as the *Ornstein-Uhlenbeck (OU)* process. Specifically, $X(t)$ is a continuous time Gaussian stationary process with spectral density:

$$f(\lambda) = \frac{\sigma^2}{2\pi} \cdot \frac{\alpha^2}{\lambda^2 + \alpha^2}, \quad \alpha > 0,\ \lambda \in \mathbb{R}. \tag{3.3.12}$$

Observe that the Ornstein-Uhlenbeck process $X(t)$ is the only Gaussian stationary Markov process. It satisfies the Langevin stochastic differential equation:

$$dX(t) = -\alpha X(t)dt + \sigma dB(t),$$

where $B(t)$ is a Wiener process.

3.3.3 Discrete Time Long Memory and Antipersistent Models

Data in many fields of science (economics, finance, hydrology, etc.) are often well modeled by stationary processes whose spectral densities are either *unbounded* or *vanishing* at certain fixed points (see, e.g., Beran [31], Beran et al. [33], Giraitis et al. [147], Guégan [163], Palma [234], Pipiras and Taqqu [246], Robinson [256], Taqqu [296], and references therein).

In the discrete context, a basic long memory model is the autoregressive fractionally integrated moving average (ARFIMA$(0, d, 0)$) process $X(t)$, which is defined as a stationary solution of the following difference equation (see, e.g., Brockwell and Davis [57], Section 13.2):

$$(1 - B)^d X(t) = \varepsilon(t), \quad 0 < d < 1/2,$$

where B is the backshift operator and $\varepsilon(t)$ is the standard white noise WN(0,1). The spectral density $f(\lambda)$ of $X(t)$ is given by

$$f(\lambda) = |1 - e^{-i\lambda}|^{-2d} = (2\sin(\lambda/2))^{-2d}, \quad 0 < \lambda \leq \pi, \quad 0 < d < 1/2. \tag{3.3.13}$$

Notice that

$$f(\lambda) \sim |\lambda|^{-2d} \quad \text{as} \quad \lambda \to 0, \tag{3.3.14}$$

indicating that $f(\lambda)$ blows up at the frequency $\lambda = 0$ like a power function, which is a typical behavior for a long memory model.

A typical example of an *antipersistent* model is the ARFIMA$(0, d, 0)$ process $X(t)$, which is characterized by the spectral density $f(\lambda)$ as defined in Eq. (3.3.13) with $d < 0$. In this case, $f(\lambda)$ vanishes at $\lambda = 0$.

Note that the condition $d < 1/2$ ensures that $\int_{-\pi}^{\pi} f(\lambda)d\lambda < \infty$, implying that the process $X(t)$ is well-defined because $\mathbb{E}[|X(t)|^2] = \int_{-\pi}^{\pi} f(\lambda)d\lambda$.

Data can also be represented as a realization of a "mixed" short-long-intermediate-memory stationary process $X(t)$. A well-known example of such a process, commonly encountered in various applied problems, is an ARFIMA(p, d, q) process $X(t)$, defined as a stationary solution to the difference equation:

$$\psi_p(B)(1 - B)^d X(t) = \theta_q(B)\varepsilon(t), \quad d < 1/2,$$

where B is the backshift operator, $\varepsilon(t)$ is a white noise, and ψ_p and θ_q are polynomials of degrees p and q, respectively. The spectral density $f_X(\lambda)$ of $X(t)$ is given by

$$f_X(\lambda) = |1 - e^{-i\lambda}|^{-2d} f(\lambda), \quad d < 1/2, \tag{3.3.15}$$

where $f(\lambda)$ is the spectral density of an ARMA(p, q) process, as defined in (3.3.9).

The parameter d in (3.3.13) and (3.3.15) is referred to as the *memory parameter* of the model process $X(t)$. Note that for $0 < d < 1/2$, the model $X(t)$ specified by the spectral density of the form (3.3.15) displays long memory; for $d < 0$, it exhibits intermediate memory; and for $d = 0$, it shows short memory. For $d \geq 1/2$, the function $f_X(\lambda)$ in (3.3.15) is not integrable, and thus it cannot represent the spectral density of a stationary process. Furthermore, if $d \leq -1$, then the series $X(t)$ is not invertible, indicating that it cannot be used to recover a white noise $\varepsilon(t)$ by passing $X(t)$ through a linear filter (see, e.g., Bondon and Palma [46], and Brockweel and Davis [57], Section 13.2).

The ARFIMA(p, d, q) processes, first introduced by Granger and Joyeux [157] and Hosking [177], have gained popularity due to their effectiveness in characterizing the long-run properties of various economic and financial time series. Additionally, they are useful for modeling multivariate time series, as they can capture a greater number of long-term equilibrium relationships among economic variables compared to traditional multivariate ARIMA models (see, e.g., Guégan [163] and Henry and Zaffaroni [172] for a survey on this topic).

Another important discrete time long memory model is fractional Gaussian noise (fGn). To define fGn, we must first define fractional Brownian motion.

A *fractional Brownian motion* (fBm) with *Hurst index* H, where $0 < H \leq 1$, denoted by $\{B_H(t), t \in \mathbb{R}\}$, is defined as a centered Gaussian H-self-similar process with stationary increments. Specifically, $B_H(t)$ satisfies the following conditions:

(a) $B_H(0) = 0, \mathbb{E}[B_H(t)] = 0, t \in \mathbb{R}$.
(b) $\{B_H(at), t \in \mathbb{R}\} \stackrel{d}{=} \{a^H B_H(t), t \in \mathbb{R}\}$ for any $a > 0$.
(c) $\{B_H(t+u) - B_H(u), t \in \mathbb{R}\} \stackrel{d}{=} \{B_H(t), t \in \mathbb{R}\}$ for each fixed $u \in \mathbb{R}$.
(d) The covariance function is given by

$$\mathrm{Cov}(B_H(s), B_H(t)) = \frac{\sigma_0^2}{2}\left[|t|^{2H} - |s|^{2H} - |t-s|^{2H}\right],$$

where the symbol $\stackrel{d}{=}$ stands for equality of the finite-dimensional distributions, and $\sigma_0^2 = \mathrm{Var}(B_H(1))$. It is called standard if $\sigma_0^2 = 1$.

Then the increment process

$$\{X(k) := B_H(k+1) - B_H(k), k \in \mathbb{Z}\},$$

called *fractional Gaussian noise* (fGn), is a discrete time centered Gaussian stationary process with a covariance function:

$$r(k) = \frac{\sigma_0^2}{2}\left[|k+1|^{2H} - |k|^{2H} - |k-1|^{2H}\right], \quad k \in \mathbb{Z}, \tag{3.3.16}$$

and spectral density function:

$$f(\lambda) = c\,|1 - e^{-i\lambda}|^2 \sum_{k=-\infty}^{\infty} |\lambda + 2\pi k|^{-(2H+1)}, \quad -\pi \le \lambda \le \pi, \tag{3.3.17}$$

where c is a positive constant.

It follows from (3.3.17) that

$$f(\lambda) \sim |\lambda|^{1-2H} \quad \text{as} \quad \lambda \to 0, \tag{3.3.18}$$

which indicates that $f(\lambda)$ blows up if $H > 1/2$ and approaches zero if $H < 1/2$. Furthermore, by comparing (3.3.14) and (3.3.18), we observe that the spectral density of fGn exhibits the same behavior at the origin as ARFIMA$(0, d, 0)$ with the memory parameter $d = H - 1/2$.

Thus, the fGn $\{X(k), k \in \mathbb{Z}\}$ exhibits long memory when $1/2 < H < 1$ and is antipersistent when $0 < H < 1/2$. The variables $X(k)$, $k \in \mathbb{Z}$ are independent when $H = 1/2$. For more details on this topic, we refer to Pipiras and Taqqu [246], Samorodnisky and Taqqu [269], and Taqqu [296, 299].

3.3.4 Continuous Time Long Memory and Antipersistent Models

In the continuous context, a basic process commonly used to model long-range dependence is fractional Brownian motion (fBm) B_H with a Hurst index H. This process can be regarded as a Gaussian process with a "spectral density":

$$f(\lambda) = c|\lambda|^{1-2H}, \quad c > 0, \ \ 0 < H < 1, \ \ \lambda \in \mathbb{R}. \tag{3.3.19}$$

The form (3.3.19) can be understood in a generalized sense (see Flandrin [98], Solo [285], and Yaglom [318], Section 24) because the fBm B_H is a nonstationary process.

A proper stationary model in place of fBm is the *fractional Riesz-Bessel motion* (fRBm), introduced by Anh et al. [5]. It is defined as a continuous time Gaussian process $X(t)$ with a spectral density:

$$f(\lambda) = c\,|\lambda|^{-2\alpha}(1 + \lambda^2)^{-\beta}, \quad \lambda \in \mathbb{R},\ 0 < c < \infty,\ 0 < \alpha < 1,\ \beta \ge 0. \tag{3.3.20}$$

The exponent α determines the long-range dependence, while the exponent β indicates the second-order intermittency of the process (for details, see Anh et al. [6] and Gao et al. [106]).

Notice that the process $X(t)$, specified by the spectral density (3.3.20), is stationary if $0 < \alpha < 1/2$ and is nonstationary with stationary increments if $1/2 \leq \alpha < 1$.

Additionally, the function $f(\lambda)$ in (3.3.20) behaves as $O(|\lambda|^{-2\alpha})$ as $|\lambda| \to 0$ and as $O(|\lambda|^{-2(\alpha+\beta)})$ as $|\lambda| \to \infty$. Thus, under the conditions $0 < \alpha < 1/2$, $\beta > 0$, and $\alpha+\beta > 1/2$, the function $f(\lambda)$ in (3.3.20) is well-defined for both $|\lambda| \to 0$ and $|\lambda| \to \infty$ due to the presence of the factor $(1+\lambda^2)^{-\beta}$, $\beta > 0$, which is the Fourier transform of the Bessel potential (see Anh et al. [6]).

Comparing (3.3.19) and (3.3.20), we observe that the spectral density of fBm is the limiting case as $\beta \to 0$ for that of fRBm with Hurst index $H = \alpha - 1/2$.

Remark 3.3.1 Recall (see Yaglom [318], Section 23) that a centered mean-square continuous process $\{X(t),\ t \in \mathbb{R}\}$ is said to have second-order stationary increments if its structure function

$$D(\tau_1, \tau_2) := E\{[(X(s+\tau_1) - X(s)][(X(s+\tau_2) - X(s)]\}$$

is independent of s for all $s, \tau_1, \tau_2 \in \mathbb{R}$. Then, the function $D(\tau_1, \tau_2)$ has the spectral representation:

$$D(\tau_1, \tau_2) = \int_{-\infty}^{\infty} \left(e^{i\tau_1\lambda} - 1\right)\left(e^{i\tau_2\lambda} - 1\right) dF(\lambda), \tag{3.3.21}$$

where $F(\lambda)$ is a left-continuous real-valued nondecreasing function on $\mathbb{R}$, called the *spectral distribution function* of $X(t)$, such that for any $\varepsilon > 0$,

$$\int_0^{\varepsilon} \lambda^2 dF(\lambda) + \int_{\varepsilon}^{\infty} dF(\lambda) < \infty. \tag{3.3.22}$$

If the spectral distribution function $F(\lambda)$ has a derivative $f(\lambda)$ (with respect to the Lebesgue measure), then f is referred to as the spectral density function of $X(t)$.

In particular, if $f(\lambda)$ has the form given in (3.3.20), then for $\alpha + \beta > 1/2$, the condition (3.3.22) is satisfied, and the model $X(t)$ is a fractional Riesz-Bessel motion.

Another important continuous time long memory model is the continuous time ARFIMA(p, H, q) process, denoted as CARFIMA(p, H, q). This process is defined as the solution of a pth-order stochastic differential equation with an appropriate initial condition, driven by fractional Brownian motion with Hurst parameter H, along with its derivatives up to and including order q $(0 \leq q < p)$. Specifically, for $t \geq 0$,

$$\begin{aligned} X^{(p)}(t) - \alpha_p X^{(p-1)}(t) - \cdots - \alpha_1 X(t) - \alpha_0 \\ = \sigma\{B_H^{(1)}(t) + \beta_1 B_H^{(2)}(t) + \cdots + \beta_q B_H^{(q+1)}(t)\}, \end{aligned}$$

where $\{B_H(t), t \geq 0\}$ is the fractional Brownian motion with Hurst index H such that $1/2 < H < 1$. It is also assumed that $\sigma > 0$ and $\beta_q \neq 0$.

The spectral density $f(\lambda)$ of the CARFIMA(p, H, q) process is given by the following formula (see Tsai and Chan [304]):

$$f(\lambda) = \frac{\sigma^2}{2\pi}\Gamma(2H+1)\sin(\pi H)|\lambda|^{1-2H}\frac{|\beta(i\lambda)|^2}{|\alpha(i\lambda)|^2}, \tag{3.3.23}$$

where $\alpha(z)$ and $\beta(z)$ are polynomials of degrees p and q, respectively.

Note that for $H = 1/2$, the spectral density function given by (3.3.23) becomes that of the short memory CARMA(p, q) process, as specified in (3.3.11).

3.3.5 The Muckenhoupt Model

The following definition can be found in the paper by Hunt et al. [179] (see also the books by Böttcher and Karlovich [51], Garnett [107], and Nikol'skii [228]).

Definition 3.3.2 We say that a non-negative locally integrable function $f(\lambda)$ $(\lambda \in \Lambda)$ satisfies the *Muckenhoupt condition* (A_2) (or has Muckenhoupt's type singularities) if

$$\sup \frac{1}{|J|^2}\int_J f(\lambda)\,d\lambda \int_J \frac{1}{f(\lambda)}\,d\lambda < \infty, \tag{3.3.24}$$

where the supremum is taken over all intervals J, and $|J|$ denotes the length of J.

We denote the class of functions $f(\lambda)$ that satisfy condition (A_2) by $\mathcal{A}_2$.

Definition 3.3.3 We say that a stationary process $X(t)$ is a Muckenhoupt process if its spectral density function $f(\lambda)$ satisfies the Muckenhoupt condition.

It is clear that if a function $f(\lambda)$ is the spectral density of a short memory process, then it belongs to the class $\mathcal{A}_2$. Moreover, the class $\mathcal{A}_2$ encompasses spectral densities that exhibit singularities. In particular, functions of the form

$$f_0(\lambda) := \begin{cases} \prod_{k=1}^n |e^{i\lambda} - e^{i\lambda_k}|^{2\alpha_k}, \ \lambda_k \in \mathbb{T} \text{ (d.t. case)}, \\ \prod_{k=1}^n |\lambda - \lambda_k|^{2\alpha_k}, \quad \lambda_k \in \mathbb{R} \text{ (c.t. case)} \end{cases} \tag{3.3.25}$$

belong to $\mathcal{A}_2$ if and only if $-1/2 < \alpha_k < 1/2$ for all $k = 1, \ldots, n$ (see Böttcher and Karlovich [51], Theorem 2.2).

Notice also that for any slowly varying at zero function $l(\lambda)$, we have $f_0(\lambda)l(\lambda) \in \mathcal{A}_2$, where $f_0(\lambda)$ is defined as in (3.3.25).

Condition (A_2) controls the singularities of the spectral density $f(\lambda)$ and describes the dependence structure of the model $X(t)$. It serves as a regularity condition for the process $X(t)$ and indicates that the maximal coefficient of correlation between the past and future of the process $X(t)$ is less than 1 (see Proposition 3.3.3).

For more information on the regularity conditions for stationary processes, we refer to Chapter 6 of the book by Ibragimov and Rozanov [189].

Definition 3.3.4 The minimal angle between the subspaces H_1 and H_2 of a Hilbert space H, denoted by $u(H_1, H_2)$ $(0 \le u(H_1, H_2) \le \frac{\pi}{2})$, is defined by the equality:

$$\cos u(H_1, H_2) = \sup_{\psi_1 \in H_1, \psi_2 \in H_2} \frac{|(\psi_1, \psi_2)|}{||\psi_1||\,||\psi_2||}, \tag{3.3.26}$$

where $(\cdot, \cdot)$ and $||\cdot||$ denote the inner product and the norm in H, respectively.

The following result presents various characterizations of the Muckenhoupt class $\mathcal{A}_2$ (see, e.g., Garnett [107] and Nikol'skii [228], p. 254).

Proposition 3.3.3 *For any non-negative integrable function $f(\lambda)$ $(\lambda \in \Lambda)$, the following assertions are equivalent:*

(a) $f(\lambda)$ satisfies the Muckenhoupt condition (3.3.24), specifically $f(\lambda) \in \mathcal{A}_2$.

(b) The minimal angle $u_f(\mathbb{H}^0_{-\infty}(X), \mathbb{H}^\infty_1(X))$ in the space $L^2(\mathbb{P})$ between the subspaces $\mathbb{H}^0_{-\infty}(X)$ and $\mathbb{H}^\infty_1(f)$ is positive. This means that

$$\rho(X) := \cos u_f(\mathbb{H}^0_{-\infty}(X), \mathbb{H}^\infty_1(X)) < 1,$$

where (see (3.1.15))

$$\mathbb{H}^b_a(X) := \overline{sp}\{X(u),\ a \le u \le b\}_{L^2(\mathbb{P})} \subset L^2(\mathbb{P}).$$

(c) The minimal angle $u_f(H^0_{-\infty}(f), H^\infty_1(f))$ in the space $L^2(f)$ between the subspaces $H^0_{-\infty}(f)$ and $H^\infty_1(f)$ is positive. This means that

$$\rho(f) := \cos u_f(H^0_{-\infty}(f), H^\infty_1(f)) < 1,$$

where

$$H^b_a(f) := \overline{\mathrm{sp}}\left\{e^{i\lambda t},\ a \le t \le b\right\}_{L_2(f)} \subset L_2(f).$$

(d) $f(\lambda)$ satisfies the Helson-Szegő condition, which means that the function $\ln f(\lambda)$ can be represented as

$$\ln f(\lambda) = u(\lambda) + \widetilde{v}(\lambda),$$

where u and v are real-valued bounded functions, $\widetilde{v}$ is the harmonic conjugate of the function v, and $||v||_\infty < \pi/2$.

3.4 Stochastic Integrals with Respect to Random Measures

In this section, we introduce single and multiple stochastic integrals with respect to random measures that are used throughout the book. Single stochastic integrals with respect to orthogonal-increments measure and Gaussian measure are considered in Sect. 3.4.2. Multiple Itô-Lévy-, Wiener-Itô-, and Stratonovich-type stochastic integrals are discussed in Sect. 3.4.3. The techniques related to these integrals will be employed to prove the central and non-central functional limit theorems for Toeplitz processes presented in Chap. 6.

A comprehensive treatment of the pertinent properties of stochastic integrals with respect to random measures can be found in several books, including Major [222], Nualart [230], Peccati and Taqqu [240], and Pipiras and Taqqu [246]. We will provide only the essential information needed for our specific purposes.

3.4.1 *Random Measures*

Let $(E, \mathcal{E}, \mu)$ be an arbitrary measure space, where E is a set, $\mathcal{E} := \mathcal{B}(E)$ is a σ-field of subsets of E, and μ is a measure on $(E, \mathcal{E})$. We assume that μ is σ-finite and nonatomic. The term "nonatomic" means that $\mu(\{x\}) = 0$ for any $x \in E$.

We shall now define a *random measure* $Z := Z_\mu$, which is associated with the deterministic measure μ, referred to as a *control measure* of Z_μ. A random measure Z has two characteristics: it is a measure and it takes values in a space of random variables. Thus, a random measure Z is a set function of two arguments: $Z = Z(A) = Z(A, \omega)$, where $A \in \mathcal{E}$ is a Borel set and ω is a point in the underlying probability space Ω.

The random measure Z will be used to define both single and multiple stochastic integrals, including off-diagonal and with-diagonal cases. We will now present the formal definition of the random measure Z.

Let $(\Omega, \mathcal{F}, \mathbb{P})$ be a probability space, and let $L^2(\mathbb{P}) = L^2(\mathbb{P}, \Omega)$ be the L^2-space of centered random variables $\xi = \xi(\omega)$, as defined in (3.1.13).

Definition 3.4.1 Let μ be a positive, σ-finite, and nonatomic measure on a measure space $(E, \mathcal{E})$, and let $\mathcal{E}_\mu := \{A \in \mathcal{E} : \mu(A) < \infty\}$. A centered random measure (in $L^2(\mathbb{P})$) on $(E, \mathcal{E})$ with control measure μ, is a function $Z(\cdot, \cdot)$, from $\mathcal{E}_\mu \times \Omega$ to $\mathbb{R}$, such that:

(a) For every fixed $A \in \mathcal{E}_\mu$, the application $\omega \to Z(A, \omega)$ is a random variable
(b) For every fixed $A \in \mathcal{E}_\mu$, $Z(A) \in L^2(\mathbb{P})$ and $\mathbb{E}[Z(A)] = 0$
(c) Z is independently scattered in the sense that if $A_i \in \mathcal{E}_\mu, i = 1, \ldots, n$, are pairwise disjoint, then the random variables $Z(A_i), i = 1, \ldots, n$, are independent.

We will also assume that the random measure Z is σ-additive:

$$Z\left(\bigcup_{n=1}^{\infty} A_n\right) = \sum_{n=1}^{\infty} Z(A_n) \quad \text{a. s.} \tag{3.4.1}$$

for any pairwise disjoint $A_n \in \mathcal{E}_\mu$, $n \in \mathbb{N}$, such that $\cup_{n=1}^{\infty} A_n \in \mathcal{E}_\mu$. Additional assumptions are made based on the types of measures Z being considered.

3.4.2 Single Wiener-Itô Stochastic Integrals

Single stochastic integrals with respect to random measures are expressed as

$$\int_E f(x) Z(dx) =: I(f), \tag{3.4.2}$$

where f is a deterministic (nonrandom), possibly complex-valued measurable function on E, and $Z = Z_\mu$ is a suitable, possibly complex-valued random measure on a measure space $(E, \mathcal{E})$ with control measure μ.

The definition of stochastic integrals of type (3.4.2) involves two steps. First, it is defined for simple functions. Next, using the properties of the integral, it is extended to the general case.

Recall that a function f is called *simple* (or *elementary*) if it can be expressed as

$$f(x) = \sum_{k=1}^{n} c_k \mathbf{1}_{A_k}(x), \tag{3.4.3}$$

where $c_k \in \mathbf{C}$, $A_k \in \mathcal{E}_\mu$, $k = 1, \ldots, n$, are pairwise disjoint, and $\mathbf{1}_A(\cdot)$ is the indicator function of A. Denote the set of simple functions by $\mathcal{S}$.

3.4.2.1 Integrals with Respect to Random Measures with Orthogonal-Increments

A centered random measure $Z(A)$, $A \in \mathcal{E}_\mu$, with control measure μ, is called a *random measure with orthogonal-increments* if Z is σ-additive and for any $A_1, A_2 \in \mathcal{E}_\mu$,

$$\mathbb{E}[Z(A_1)\overline{Z(A_2)}] = \mu(A_1 \cap A_2). \tag{3.4.4}$$

Our aim is to define the single Wiener-Itô integral $I(f)$ with respect to the random measure Z for all functions $f \in L^2(E, \mu)$.

We first define $I(f)$ for an elementary function $f \in \mathcal{S}$ of the form (3.4.3) as follows:

$$I(f) := \sum_{k=1}^{n} c_k Z(A_k). \tag{3.4.5}$$

Observe that this definition does not depend (in the a.s. sense) on the representation (3.4.3) of the simple function f, provided that Z is σ-additive. Moreover, $I(f)$ is linear

$$I(c_1 f_1 + c_2 f_2) = c_1 I(f_1) + c_2 I(f_2) \quad \text{a. s.} \tag{3.4.6}$$

for any $c_1, c_2 \in \mathbf{C}$ and $f_1, f_2 \in \mathcal{S}$.

For $f_1, f_2 \in \mathcal{S}$, using Eq. (3.4.4), we obtain the isometric property:

$$\mathbb{E}[I(f_1)\overline{I(f_2)}] = \int_E f_1(x)\overline{f_2(x)}\mu(dx) = (f_1, f_2)_{L^2(E,\mu)}. \tag{3.4.7}$$

In particular, we have $\mathbb{E}[I(f)] = 0$ and

$$\text{Var}[I(f)] = ||f||^2_{L^2(E,\mu)}. \tag{3.4.8}$$

The definition of the integral $I(f)$ can now be extended to functions $f \in L^2(E, \mu)$ by using the fact that the class of simple functions is dense in the space $L^2(E, \mu)$.

Specifically, for any $f \in L^2(E, \mu)$, there exists a sequence of simple functions $\{f_n, \ n \in \mathbb{N}\}$ such that $||f - f_n||^2_{L^2(E,\mu)} \to 0$ as $n \to \infty$. Consequently, the sequence $\{f_n, n \in \mathbb{N}\}$ is also a Cauchy sequence in $L^2(E, \mu)$. Therefore, by applying Eqs. (3.4.6) and (3.4.7), one can write

$$\mathbb{E}[|I(f_n) - I(f_m)|^2] = \mathbb{E}[|I(f_n - f_m)|^2] = ||f_n - f_m||^2_{L^2(E,\mu)} \to 0$$

as $m, n \to \infty$. This shows that $\{I(f_n), \ n \in \mathbb{N}\}$ is a Cauchy sequence in the space $L^2(\mathbb{P})$. Therefore, it has a limit in $L^2(\mathbb{P})$, which is defined as the integral $I(f)$:

$$I(f) := L^2 - \lim_{n\to\infty} I(f_n). \tag{3.4.9}$$

Observe that the limit in (3.4.9) does not depend on the approximating sequence f_n, and hence the integral $I(f)$ is well-defined. Additionally, the properties in (3.4.6) and (3.4.7) continue to hold for all functions $f_1, f_2 \in L^2(E, \mu)$.

The above defined integral $I(f)$, also written as $\int_E f dZ$ or $I_1^Z(f)$, is called the *single Wiener-Itô stochastic integral* of f with respect to the random measure Z.

3.4.2.2 Integrals with Respect to Gaussian Measures

A Gaussian random measure is a particular case of random measures with orthogonal increments. A random measure Z on $(E, \mathcal{E})$ with orthogonal-increments and control measure μ is called *Gaussian* if any vector $(Z(A_1), \ldots, Z(A_n))$, $A_k \in \mathcal{E}_\mu$, $k = 1, \ldots, n$, is multivariate Gaussian. The stochastic integral $I(f)$ with respect to Gaussian measure Z is defined as in Sect. 3.4.2.1 for all $f \in L^2(E, \mu)$. An additional property of these integrals is that any vector $(I(f_1), \ldots, I(f_n))$, $f_k \in L^2(E, \mu)$, $k = 1, \ldots, n$, is also multivariate Gaussian.

Among complex-valued random measures, the *Hermitian Gaussian random measure* on the measure space $(\mathbb{R}, \mathcal{B}(\mathbb{R}))$ is of particular interest. Let μ be a symmetric measure on $(\mathbb{R}, \mathcal{B}(\mathbb{R}))$ in the sense that $\mu(dx) = \mu(-dx)$. The Hermitian Gaussian random measure Z on $(\mathbb{R}, \mathcal{B}(\mathbb{R}))$ is defined as a complex-valued random measure on $(\mathbb{R}, \mathcal{B}(\mathbb{R}))$ with a symmetric control measure μ such that $\overline{Z(dx)} = Z(-dx)$.

The Hermitian Gaussian random measure Z possesses several important properties. Here, we will mention only two of them (additional properties can be found in Major [222] and in Pipiras and Taqqu [246], Appendix B).

If $g \in L^2(\mathbb{R}, \mu)$ is a Hermitian deterministic function, meaning that $\overline{g(x)} = g(-x)$, $x \in \mathbb{R}$, then the integral $\int_{\mathbb{R}} g(x)Z(dx)$ is always real-valued.

Let B and $\widehat{B}$ be two Gaussian random measures on $(\mathbb{R}, \mathcal{B}(\mathbb{R}))$ with Lebesgue control measures du and dx, respectively. Then, for $f \in L^2(\mathbb{R}, du)$, we have

$$\int_{\mathbb{R}} f(u)B(du) \stackrel{d}{=} \int_{\mathbb{R}} \widehat{f}(x)\widehat{B}(dx), \tag{3.4.10}$$

where $\widehat{f}$ is the Fourier transform of f. The relation follows by observing that both sides of (3.4.10) are centered real-valued Gaussian variables with the same variance.

The integral on the left-hand side of (3.4.10) is referred to as being in the *time domain*, while the integral on the right-hand side of (3.4.10) is in the *frequency (spectral) domain*.

3.4.3 Multiple Itô-Lévy and Wiener-Itô Integrals

We recall the definitions and basic properties of multiple Itô-Lévy and Wiener-Itô integrals. These definitions follow the same two-step process as single Wiener-Itô integrals. First, the integrals are defined for simple functions. Then, using the properties of the integrals and an L^2-approximation argument, they are extended to the general case.

Let f be a real-valued function in $L^2(\mathbb{R}^k)$, and let $\{\xi(t),\ t \in \mathbb{R}\}$ be a Lévy process with $\mathbb{E}[\xi(t)] = 0$ and $\mathrm{Var}[\xi(t)] = \sigma_\xi^2 t$ (see Sect. 3.2.3). The following *off-diagonal* multiple stochastic integral, called *Itô-Lévy integral*, is well-defined:

$$I_k^\xi(f) = \int_{\mathbb{R}^k}' f(x_1, \ldots, x_k)\xi(dx_1)\ldots\xi(dx_k), \tag{3.4.11}$$

where the prime $'$ indicates that we do not integrate on the diagonals $x_i = x_j, i \neq j$.

Indeed, the integral $I_k^\xi(f)$ is first defined on the set $\mathcal{S}^k$ of simple functions. A function $f : \mathbb{R}^k \to \mathbb{R}$ is simple if it has the form

$$f(x_1, \ldots, x_k) = \sum_{i_1,\ldots,i_k=1}^{n} a_{i_1,\ldots,i_k}\mathbf{1}_{A_{i_1}\times\ldots\times A_{i_k}}(x_1, \ldots, x_k), \tag{3.4.12}$$

where $A_1, \ldots, A_k \in \mathcal{B}(\mathbb{R})$ are pairwise disjoint Borel sets, and the real-valued coefficients $a_{i_1,\ldots,i_k}$ are zero if any two of the indices $i_1, \ldots, i_k$ are equal. This property ensures that the function f vanishes on the diagonals $x_i = x_j, i \neq j$.

For $f \in \mathcal{S}^k$, one sets

$$I_k^\xi(f) := \sum_{i_1,\ldots,i_k=1}^{n} a_{i_1,\ldots,i_k}\xi(A_{i_1})\ldots\xi(A_{i_k}). \tag{3.4.13}$$

It can be shown (see, e.g., Nualart [230], p. 9) that the definition in (3.4.13) does not depend on the particular representation of f in the form of (3.4.12). Additionally, the following properties hold: for functions $f, g \in \mathcal{S}^k$ and numbers $a, b \in \mathbb{R}$,

$$I_k^\xi(af + bg) = aI_k^\xi(f) + bI_k^\xi(g) \tag{3.4.14}$$

and

$$I_k^\xi(f) = I_k^\xi(\widetilde{f}), \tag{3.4.15}$$

where $\widetilde{f}$ denotes the symmetrization of f, defined by

$$\widetilde{f}(x_1, \ldots, x_k) := \frac{1}{k!}\sum_\sigma f(x_{\sigma(1)}, \ldots, x_{\sigma(k)}) \tag{3.4.16}$$

with the sum over all permutations σ of the set $\{1, \ldots, k\}$. Note that the symmetrization $\widetilde{f}$ of the function f is symmetric, meaning it remains unchanged when any two of its arguments are interchanged. Moreover, we have

$$\mathbb{E}[I_k^\xi(f)] = 0 \tag{3.4.17}$$

and

$$\mathbb{E}[I_k^\xi(f)I_j^\xi(g)] = \begin{cases} k!\sigma_\xi^{2k}(\widetilde{f}, \widetilde{g})_{L^2(\mathbb{R}^k)}, & \text{if } k = j, \\ 0, & \text{if } k \neq j. \end{cases} \tag{3.4.18}$$

The multiple integral $I_k^\xi(\cdot)$ satisfies the following inequality:

$$\|I_k^\xi(f)\|^2_{L^2(\Omega)} \leq k!\sigma_\xi^{2k}\|f\|^2_{L^2(\mathbb{R}^k)}, \tag{3.4.19}$$

and the inequality in (3.4.19) becomes an equality if f is symmetric:

$$\|I_k^\xi(f)\|^2_{L^2(\Omega)} = k!\sigma_\xi^{2k}\|f\|^2_{L^2(\mathbb{R}^k)}. \tag{3.4.20}$$

The definition (3.4.13) of the integral $I_k^\xi(\cdot)$ can now be extended to functions $f \in L^2(\mathbb{R}^k)$ by using linearity and a standard L^2-approximation argument, as follows.

For $f \in L^2(\mathbb{R}^k)$, there exists a sequence of simple functions $f_n \in \mathcal{S}^k$ such that $||f - f_n||^2_{L^2(\mathbb{R}^k)} \to 0$ as $n \to \infty$. We then show that the sequence $\{I_k^\xi(f_n),\ n \in \mathbb{N}\}$ converges in the space $L^2(\mathbb{P})$. Consequently, it has a limit in $L^2(\mathbb{P})$, which is defined as the integral $I_k^\xi(f)$ for $f \in L^2(\mathbb{R}^k)$:

$$I_k^\xi(f) := L^2 - \lim_{n\to\infty} I_k^\xi(f_n). \tag{3.4.21}$$

Observe that the limit in (3.4.21) does not depend on the approximating sequence f_n, and hence the integral is well-defined. Furthermore, the properties of $I_k^\xi(f)$, initially stated for simple functions, also hold for all functions from $L^2(\mathbb{R}^k)$.

If we take $\xi(\cdot) = B(\cdot)$ in the integral (3.4.11), where $B(\cdot)$ is the real-valued Brownian motion, then the corresponding integral $I_k^B(f)$:

$$I_k^B(f) := \int'_{\mathbb{R}^k} f(x_1, \ldots, x_k)B(dx_1)\ldots B(dx_k) \tag{3.4.22}$$

is called *multiple Wiener-Itô integral.*

The Wiener-Itô multiple integral can also be defined with respect to complex-valued Brownian motions. Specifically, let $W(\cdot)$ be a complex-valued Brownian motion (with real and imaginary parts being independent), and let $g \in L^2(\mathbb{R}^k)$ be a deterministic Hermitian function:

$$g(-u_1, \ldots, -u_k) = \overline{g(u_1, \ldots, u_k)}.$$

Proceeding as with the multiple integrals (3.4.11), we can similarly define a multiple integral of g with respect to W by starting with a Hermitian simple function:

$$g(u_1, \dots, u_k) = \sum_{i_1,\dots,i_k=-n}^{n} a_{i_1,\dots,i_k} \mathbf{1}_{A_{i_1}\times\dots\times A_{i_k}}(u_1, \dots, u_k), \tag{3.4.23}$$

where $A_1, \dots, A_k \in \mathcal{B}(\mathbb{R})$ are pairwise disjoint Borel sets satisfying $A_{i_k} = -A_{-i_k}$. The coefficients $a_{i_1,\dots,i_k}$ are assumed to be zero if any of the indices $i_1, \dots, i_k$ are either equal or the negative of the other. The integral is defined as

$$I_k^W(g) := \int_{\mathbb{R}^k}'' g(u_1, \dots, u_k) W(du_1) \dots W(du_k), \tag{3.4.24}$$

where the double prime $''$ indicates that the hyper-diagonals $\{u_p = \pm u_q,\ p \neq q\}$ are excluded in integration.

The definition (3.4.24) of the integral $I_k^W(g)$ can now be extended to functions $g \in L^2(\mathbb{R}^k)$ by employing linearity and a standard L^2-approximation argument. The multiple integral $I_k^W(g)$ is real-valued and has properties similar to those of the integral $I_k^B(g)$.

Another useful property of Wiener-Itô multiple integrals is the formula for the *change of variables* (see Proposition 4.2 of Dobrushin [83], Major [222], Theorem 4.5, and Pipiras and Taqqu [246], p. 599).

Proposition 3.4.1 *Let W_1 and W_2 be two Hermitian Gaussian measures on $(\mathbb{R}, \mathcal{B}(\mathbb{R}))$ with symmetric control measures μ_1 and μ_2, respectively. Denote the corresponding multiple integrals by $I_{1,k}^W(g_1)$ and $I_{2,k}^W(g_2)$. Suppose μ_1 is absolutely continuous with respect to μ_2. Let $h : \mathbb{R} \to \mathbf{C}$ be a Hermitian function such that*

$$|h(x)|^2 = \frac{d\mu_1}{d\mu_2}(x), \quad x \in \mathbb{R}.$$

Let also $g_{1,j} \in L^2\left(\mathbb{R}^j, \mu_1^j\right)$, $j = 1, \dots, k$, be Hermitian functions. We set

$$g_{2,j}(x_1, \dots, x_j) = g_{1,j}(x_1, \dots, x_j)h(x_1) \cdots h(x_j), \quad j = 1, \dots, k.$$

Then,

$$\left(I_{1,1}^W(g_{1,1}), I_{1,2}^W(g_{1,2}), \dots, I_{1,k}^W(g_{1,k})\right) \stackrel{d}{=} \left(I_{2,1}^W(g_{2,1}), I_{2,2}^W(g_{2,2}), \dots, I_{2,k}^W(g_{2,k})\right).$$

The next result, which can be deduced from Proposition 3.4.1 and Proposition 9.3.1 of Peccati and Taqqu [240], establishes a relationship between the integrals $I_k^B(\cdot)$ and $I_k^W(\cdot)$, as defined by Eqs. (3.4.22) and (3.4.24), respectively.

Proposition 3.4.2 *Let $f_j(\cdot) \in L^2(\mathbb{R}^{k_j})$, $j = 1, \ldots, J$, be real-valued functions, and let*

$$\widehat{f_j}(w_1, \ldots, w_{k_j}) = \int_{\mathbb{R}^{k_j}} f_j(x_1, \ldots, x_{k_j}) e^{i\left(x_1 w_1 + \ldots + x_{k_j} w_{k_j}\right)} dx_1 \ldots dx_{k_j}$$

be the L^2-Fourier transform of $f_j(\cdot)$. Then

$$\begin{aligned}&\left(I_{k_1}^B(f_1), \ldots, I_{k_J}^B(f_J)\right)\\ &\stackrel{d}{=} \left((2\pi)^{-k_1/2} I_{k_1}^W\left(\widehat{f_1} A^{\otimes k_1}\right), \ldots, (2\pi)^{-k_J/2} I_{k_J}^W\left(\widehat{f_J} A^{\otimes k_J}\right)\right),\end{aligned}$$

for any function $A(u) : \mathbb{R} \to \mathbb{C}$ such that $|A(u)| = 1$ and $A(w) = \overline{A(-w)}$ almost everywhere, where $A^{\otimes k}(w_1, \ldots, w_k) := A(w_1) \cdots A(w_k)$.

In the next proposition, we present the *hypercontractivity property* of the multiple Wiener-Itô integrals. This property, which can be found in Major [222], Corollary 5.6, and in Peccati and Taqqu [240], p. 92, will be used in Chap. 6 to prove the tightness of the processes generated by quadratic functionals.

Proposition 3.4.3 (Hypercontractivity Inequality) *If ξ is a Gaussian measure with nonatomic control measure μ, then for every $p > 2$ and every $k \geq 2$, there exists a universal constant $c_{p,k} > 0$, such that the following hypercontractivity property takes place:*

$$\mathbb{E}\left[\left(|I_k^\xi(f)|^p\right)^{1/p}\right] \leq c_{p,k} \mathbb{E}\left[\left(|I_k^\xi(f)|^2\right)^{1/2}\right] \tag{3.4.25}$$

for every $f \in L^2(\mu^k)$.

In the next proposition, we present a stochastic Fubini's lemma that allows for the interchange of deterministic and stochastic integrals (see Peccati and Taqqu [240], Theorem 5.13.1).

Proposition 3.4.4 (Stochastic Fubini's Lemma) *Let (E, μ) be a measure space with $\mu(E) < \infty$, and let $f(e, x_1, \ldots, x_k)$ be a deterministic function on $E \times \mathbb{R}^k$ such that*

$$\int_E \int_{\mathbb{R}^k} f^2(e, x_1, \ldots, x_k) dx_1 \ldots dx_k \mu(de) < \infty.$$

Then we can interchange the order of multiple stochastic integration $I_k^\xi(\cdot)$ and deterministic integration $\int_E f(e, \cdot)\mu(de)$. Specifically, we have

$$\int_E I_k^\xi(f(e, \cdot))\mu(de) = I_k^\xi\left(\int_E f(e, \cdot)\mu(de)\right).$$

This lemma will be used in Chap. 6.

3.4.4 Multiple Stratonovich-Type Integrals

The *with-diagonal* counterpart of the Lévy-Itô integral $I_k^{\xi}(f)$, referred to as a *Stratonovich-type* stochastic integral, is defined by

$$\mathring{I}_k^{\xi}(f) := \int_{\mathbb{R}^k} f(x_1, \ldots, x_k)\xi(dx_1)\ldots\xi(dx_k), \tag{3.4.26}$$

where the integration includes all the diagonals $x_i = x_j, i \neq j$.

We refer to the paper by Farré et al. [97] for a comprehensive treatment of Stratonovich-type integrals $\mathring{I}_k^{\xi}(f)$. Here, we mention some properties of $\mathring{I}_k^{\xi}(f)$ that are necessary for our purposes.

First, observe that for the with-diagonal integral $\mathring{I}_k^{\xi}(f)$ to be well-defined, the integrand f must be square-integrable over all the diagonals of $\mathbb{R}^k$ (see Bai et al. [22], Farré et al. [97], and Peccati and Taqqu [240], Section 5.1). More precisely, it is required that $f \in L^2(\Lambda_n)$, where $\Lambda_n = \sum_{\sigma \in \Pi_n} \lambda_\sigma$, with Π_n denoting all the partitions of the set $\{1, \ldots, n\}$, and λ_σ representing the Lebesgue measure on the diagonals specified by the partition σ, provided that the variables in the same block of σ are identified. For example, if $\sigma = \{\{1, 2\}, \{3\}\}$, then λ_σ is the two-dimensional Lebesgue measure on $\{x_1 = x_2, x_3\}$, and

$$\|f\|_{L^2(\lambda_\sigma)}^2 = \int_{\mathbb{R}^3} f^2(x_1, x_2, x_3)d\lambda_\sigma(x_1, x_2, x_3) = \int_{\mathbb{R}^2} f^2(x_1, x_1, x_3)dx_1dx_3.$$

Next, for the with-diagonal integrals, we have the following simple product formula:

$$\mathring{I}_p^{\xi}(f)\mathring{I}_q^{\xi}(g) = \mathring{I}_{p+q}^{\xi}(f \otimes g).$$

The with-diagonal Stratonovich integral $\mathring{I}_k^{\xi}(f)$ can be expressed in terms of off-diagonal Lévy-Itô integrals of lower orders using the *Hu-Meyer formula*. For a standard discussion of this topic in a Gaussian setting, see the book by Nualart [230]. For general Lévy processes, see the paper by Farré et al. [97]. We will focus only on the special case when $k = 2$, in which case we have

$$\mathring{I}_2^{\xi}(f) = \int_{\mathbb{R}^2}' f(x_1, x_2)\xi(dx_1)\xi(dx_2) + \int_{\mathbb{R}} f(x, x)\xi_c^{(2)}(dx) + \int_{\mathbb{R}} f(x, x)dx, \tag{3.4.27}$$

where

$$\xi_c^{(2)}(t) = \xi^{(2)}(t) - \mathbb{E}\xi^{(2)}(t) = \xi^{(2)}(t) - |t|, \tag{3.4.28}$$

and $\xi^{(2)}(t)$ is the quadratic variation of $\xi(t)$, which is non-deterministic if $\xi(t)$ is non-Gaussian (see Farré et al. [97], and Nualart and Schoutens [231]).

The centered process $\xi_c^{(2)}(t)$ is called a second-order *Teugels martingale*. It is a Lévy process with the same filtration as $\xi(t)$, and its quadratic variation is deterministic:

$$[\xi_c^{(2)}(t), \xi_c^{(2)}(t)] = \kappa_4 t,$$

where κ_4 is the fourth cumulant of $\xi(1)$.

For any $f, g \in L^2(\mathbb{R})$, we have (see Farré et al. [97])

$$\mathbb{E}\left[\int_{\mathbb{R}} f(x)\xi_c^{(2)}(dx) \int_{\mathbb{R}} g(x)\xi_c^{(2)}(dx)\right] = \kappa_4 \int_{\mathbb{R}} f(x)g(x)dx. \tag{3.4.29}$$

The relation (3.4.27) implies that

$$\mathbb{E}\left[\overset{\circ}{I}{}_k^{\xi}(f)\right] = \int_{\mathbb{R}} f(x,x)dx.$$

For $f \in L^2(\mathbb{R}^2)$ and $g \in L^2(\mathbb{R})$, consider the following integrals: the first is an off-diagonal double integral, and the second is a single integral with respect to the Teugels martingale $\xi_c^{(2)}(t)$:

$$\int_{\mathbb{R}^2}' f(x_1, x_2)\xi(dx_1)\xi(dx_2) \quad \text{and} \quad \int_{\mathbb{R}} g(x)\xi_c^{(2)}(dx), \tag{3.4.30}$$

and observe that for any $f \in L^2(\mathbb{R}^2)$ and $g \in L^2(\mathbb{R})$, these integrals are uncorrelated.

This can be easily verified for the indicator functions $f = \mathbf{1}_{A\times B}$ and $g = \mathbf{1}_C$, where A and B are any disjoint Borel sets, and C is any Borel set.

Indeed, by treating $\xi_c^{(2)}(\cdot)$ as a random measure, we have

$$\begin{aligned} &\mathbb{E}[\xi(A)\xi(B)\xi_c^{(2)}(C)] = \qquad (3.4.31)\\ &\mathbb{E}\left[\xi(A)\xi(B)\left(\xi_c^{(2)}(C\cap A^c\cap B) + \xi_c^{(2)}(C\cap A\cap B^c) + \xi_c^{(2)}(C\cap A^c\cap B^c)\right)\right] = 0 \end{aligned}$$

since, for example, $\xi(A)$ is independent of $\xi_c^{(2)}(C\cap A^c\cap B)$ and $\xi(B)$, and $\mathbb{E}[\xi(A)] = 0$. Using linearity and an L^2-approximation argument, it can be easily shown that the integrals in (3.4.30) are uncorrelated for any $f \in L^2(\mathbb{R}^2)$ and $g \in L^2(\mathbb{R})$.

3.4.5 Representation of the Quadratic Functional as a Double Wiener-Itô Stochastic Integral

In this section, we use Itô's formula to show that the normalized quadratic functional $Q_T - \mathbb{E}[Q_T]$, where Q_T is defined in (1.1.1), can be represented as a double Wiener-Itô integral.

Let $\{X(u), u \in \mathbb{U}\}$ be a stationary process with covariance function $r(t)$, spectral measure μ, and random spectral measure Z_μ. For simplicity, we consider the discrete time case. The normalized quadratic functional $Q_T - \mathbb{E}[Q_T]$ can be expressed as

$$Q_T - \mathbb{E}[Q_T] = \sum_{t=1}^{T}\sum_{s=1}^{T} \widehat{g}(t-s)[X(t)X(s) - r(t-s)] \tag{3.4.32}$$

Let $\varphi \in L^2(\Lambda, \mu)$. The following identity is known as Itô's formula (see, e.g., Major [222], Theorem 4.3):

$$\left(\int_\Lambda \varphi(x) Z_\mu(dx)\right)^2 = \int_\Lambda \int_\Lambda \varphi(x)\varphi(y) Z_\mu(dx) Z_\mu(dy) + \int_\Lambda \varphi^2(x)\mu(dx). \tag{3.4.33}$$

Now, using the spectral representation of the process $X(u)$ and Ito's formula, we express $X(t)X(s) - r(t-s)$ as a double Wiener-Itô integral. Indeed, taking into account Eqs. (3.1.4) and (3.1.11), as well as the equality

$$\mathbb{E}\left[\left(\int_\Lambda \varphi(x) Z_\mu(dx)\right)^2\right] = \int_\Lambda \varphi^2(x)\mu(dx) \quad \text{for any } \varphi \in L^2(\Lambda, \mu), \tag{3.4.34}$$

we can apply Ito's formula to the functions $x \to \exp\{ixt\}$ and $x \to \exp\{ixs\}$ to obtain

$$X(t)X(s) - r(t-s) = \int_\Lambda \int_\Lambda e^{ixt+iys} Z_\mu(dx) Z_\mu(dy). \tag{3.4.35}$$

After summation, in view of (3.4.32) and (3.4.35), we obtain

$$Q_T - \mathbb{E}[Q_T] = \int_\Lambda \int_\Lambda \psi_T(x, y) Z_\mu(dx) Z_\mu(dy), \tag{3.4.36}$$

where

$$\psi_T(x, y) = \int_\Lambda G_T(x+u) G_T(y-u) g(u) du \tag{3.4.37}$$

with $G_T(u) = \sum_{t=1}^{T} e^{itu}$.

A similar representation we have in the continuous time case, with $G_T(u) = \int_0^T e^{itu}dt$.

Note that since the function g is assumed to be even, the function $\psi_T(x, y)$ in (3.4.37) is symmetric with respect to x and y. This symmetry is important when taking expectations, as the identity

$$\mathrm{Var}\left[\int_\Lambda \cdots \int_\Lambda \varphi(x_1, \ldots, x_n) Z_\mu(dx_1) \cdots Z_\mu(dx_n)\right]$$
$$= n! \int_\Lambda \cdots \int_\Lambda |\varphi(x_1, \ldots, x_n)|^2 \mu(dx_1) \cdots \mu(dx_n)$$

holds only for symmetric functions (see 3.4.18).

3.5 Representations of L^2-Processes

In this section, we present various representations of L^2-processes, including the orthogonal-series representation (Karhunen-Loéve theorem), the general integral representation (Karhunen theorem), the representation via RKHSs (Loéve theorem), and the spectral representation for stationary processes (Cramér-Kolmogorov theorem).

3.5.1 The Karhunen-Loéve Expansion

We apply Schmidt's and Mercer's theorems on the decomposition of Hilbert-Schmidt kernels (see Propositions 2.7.8 and 2.7.9) to state the following theorem, which provides a representation of a general L^2-process in terms of orthogonal series. This representation is known as the Karhunen-Loéve expansion or the orthogonal decomposition theorem for L^2-processes (see Ash and Gardner [12], Section 1.4.1, and Loéve [219], Section 34.5). The Karhunen-Loéve theorem will be used to prove the central limit theorems for Toeplitz and tapered Toeplitz quadratic functionals, as stated in Chaps. 5 and 7.

Proposition 3.5.1 (Karhunen-Loéve Theorem) *Let $X(t)$ be a real-valued centered L^2-continuous second-order process defined on a closed finite interval $T = [a, b]$ with a covariance function $R(t, s)$. Then the following assertions hold:*

(a) The process $X(t)$ admits the orthogonal series representation:

$$X(t) = \sum_{k=1}^{\infty} \sqrt{\lambda_k} e_k(t) Z_k, \quad t \in [a, b], \tag{3.5.1}$$

where

$$Z_k = \frac{1}{\sqrt{\lambda_k}} \int_a^b X(t) e_k(t) dt \tag{3.5.2}$$

are centered, variance one, and uncorrelated random variables.

(b) The sequence $\{e_k(t), k \in \mathbb{N}\}$ *is the orthonormal sequence of eigenfunctions, and* $\{\lambda_k, k \in \mathbb{N}\}$ $(\lambda_k \geq 0)$ *is the sequence of the corresponding eigenvalues of the integral operator on* $L^2[a, b]$ *generated by the covariance kernel* $R(t, s)$*, that is, the equations in* (2.7.16) *are satisfied with* $R(t, s)$ *instead of* $K(t, s)$*.*

(c) The series in (3.5.1) *converges in* $L^2[a, b]$ *to* $X(t)$ *uniformly in* $t \in T$*, that is,*

$$\sup_{t \in T} \mathbb{E} \left| X(t) - \sum_{k=1}^{n} \sqrt{\lambda_k} e_k(t) Z_k \right|^2 \to 0 \quad as \quad n \to \infty.$$

Notice that if the process $X(t)$ is Gaussian, then the sequence of random variables $\{Z_k, \ k \in \mathbb{N}\}$ in the Karhunen-Loéve theorem forms a Gaussian system, that is, for each $m \in \mathbb{N}$, the random variables $\{Z_k, \ k = 1, \ldots, m\}$ are jointly Gaussian (see Ash and Gardner [12], Section 1.4.3).

One benefit of the Karhunen-Loéve theorem is that it clarifies how the stochastic process $X(t, \omega)$ depends on both arguments: $t \in \mathbf{T}$ and $\omega \in \Omega$.

Example 3.5.1 Let $X(t) = B(t)$ be the standard Brownian motion restricted to $[0, 1]$, that is, $B(t)$ is a centered Gaussian process with covariance function $R(t, s) = \min\{t, s\}$, $t, s \in [0, 1]$ (see Sect. 3.1.2). It can be shown that in this case, the Karhunen-Loéve expansion is given by

$$X(t) = \sqrt{2} \sum_{k=1}^{\infty} Z_k \frac{\sin(n - \frac{1}{2})\pi t}{(n - \frac{1}{2})\pi}, \qquad t \in [0, 1],$$

where Z_k are defined as in (3.5.1) (for details, see Ash and Gardner [12], Section 1.4.4).

3.5.2 *Hilbert Space Representations of* L^2*-Processes*

The purpose of the Hilbert space representation of an L^2-process is to identify specific Hilbert spaces of functions that are congruent (isometrically isomorphic) to the Hilbert space generated by that process.

Recall the Hilbert space $L^2(\mathbb{P})$ of all centered random variables $\xi = \xi(\omega)$ defined on a probability space $(\Omega, \mathcal{F}, \mathbb{P})$ (see Sect. 3.1.5). Given an index set $\mathbf{T}$ and a centered L^2-process $\{X(t), t \in \mathbf{T}\}$ with covariance function $R(t,s),\ t,s \in \mathbf{T}$, let $L^2(X,\mathbf{T})$ denote the Hilbert space spanned by the process $X(t)$:

$$L^2(X,\mathbf{T}) := \overline{sp}\{X(t),\ t \in \mathbf{T}\}_{L^2(\mathbb{P})}. \tag{3.5.3}$$

Note that the space $L^2(X,\mathbf{T})$ consists of all finite linear combinations of the form $\sum_{k=1}^{n} c_k X(t_k)$, $t_k \in \mathbf{T}$, as well as their $L^2(\mathbb{P})$-limits.

The goal is to identify various concrete Hilbert spaces of functions that are congruent to the space $L^2(X,\mathbf{T})$.

Definition 3.5.1 Let H be a Hilbert space with inner product $(\cdot,\cdot)_H$. A family of vectors $\{f(t),\ t \in \mathbf{T}\}$ in the Hilbert space H is said to be a representation of the process $\{X(t),\ t \in \mathbf{T}\}$ if for every $s,t \in \mathbf{T}$, we have

$$(f(s), f(t))_H = R(s,t) = E[X(s)X(t)] = (X(s), X(t))_{L^2(\mathbb{P})}. \tag{3.5.4}$$

In other words, the family of functions $\{f(t), t \in \mathbf{T}\}$ represents the L^2-process $\{X(t), t \in \mathbf{T}\}$ if the Hilbert spaces $L^2(X,\mathbf{T})$ and

$$\overline{L}(f,\mathbf{T}) := \overline{sp}\{f(t),\ t \in \mathbf{T}\}_H \tag{3.5.5}$$

are congruent. The Hilbert space $\overline{L}(f,\mathbf{T})$ is also referred to as a representation of the process $\{X(t),\ t \in \mathbf{T}\}$.

Then, according to the basic congruence theorem (Proposition 2.7.10), there exists a congruence V from $\overline{L}(f,\mathbf{T})$ onto $L^2(X,\mathbf{T})$ satisfying

$$V(f(t)) = X(t). \tag{3.5.6}$$

Moreover, every random variable $Y \in L^2(X,\mathbf{T})$ can be expressed in the following form:

$$Y = V(g) \tag{3.5.7}$$

for some unique function (vector) $g(t) \in \overline{L}(f,\mathbf{T})$.

3.5.3 Integral Representation with Respect to Orthogonal Random Set Functions: Karhunen's Theorem

3.5.3.1 Karhunen's Class of Covariance Functions

Let $\{X(t),\ t \in \mathbf{T}\}$ be a centered L^2-process defined on a probability space $(\Omega,\mathcal{F},\mathbb{P})$ with covariance function $R(t,s),\ t,s \in \mathbf{T}$. Let $(E,\mathcal{E},\mu)$ be an arbitrary measure space, where E is an abstract set (often taken to be the real line $\mathbb{R}$ or the segment

$[-\pi, \pi))$, $\mathcal{E} := \mathcal{B}(E)$ is a σ-field of subsets of E, and μ is a measure on $(E, \mathcal{E})$ that is assumed to be finite and nonatomic.

Let $L^2(\mu) := L^2(E, \mathcal{E}, \mu)$ be the Hilbert space of all $\mathcal{E}$-measurable complex-valued functions defined on E with inner product $(\cdot, \cdot)_\mu$ and norm $\|\cdot\|_\mu$ (see (3.1.16)).

Definition 3.5.2 We say that the covariance function $R(t, s)$, $t, s \in \mathbf{T}$, belongs to the Karhunen class if there exist a measurable space $(E, \mathcal{E}, \mu)$ and a family of functions

$$\{f(t) = f(t, \lambda),\ t \in \mathbf{T},\ \lambda \in E\}$$

from the class $L^2(\mu)$ such that $R(t, s)$ can be represented as

$$R(t, s) = \int_E f(t, \lambda)\overline{f(s, \lambda)}d\mu(\lambda). \tag{3.5.8}$$

It is clear that the covariance function $R(t, s) = r(t - s)$ of an L^2-continuous stationary process $\{X(t), t \in \mathbb{U}\}$ belongs to the Karhunen class with $\mathbf{T} = \mathbb{U}$, $E = \Lambda$, $\mathcal{B}(\Lambda)$ as the σ-field of Borel sets, μ as the Lebesgue-Stieltjes measure on $(\Lambda, \mathcal{B}(\Lambda))$, and $f(t, \lambda) = e^{i\lambda t}$, where $t \in \mathbb{U}$ and $\lambda \in \Lambda$.

The Hilbert-Schmidt kernel $K(t, s)$ (see (2.7.15)):

$$K(t, s) = \sum_{i=1}^{\infty} \lambda_i e_i(t)\overline{e_i(s)}, \quad t, s \in [a, b] \tag{3.5.9}$$

provides another example of a covariance function belonging to the Karhunen class.

Indeed, if we let $\mathbf{T} = [a, b]$, $E = \mathbb{N}$, and $\mathcal{B}(E)$ be the family of all subsets of E, and let μ be the measure on $(E, \mathcal{B}(E)$ satisfying $\mu(\{n\}) = \lambda_n$ for $n \in \mathbb{N}$, and $f(t) = e_n(t)$ for $n \in \mathbb{N}$, then the representation (3.5.9) can be expressed in the form (3.5.8) (see Parzen [239], p. 290).

3.5.3.2 Karhunen's Theorem

Let $(E, \mathcal{B}(E), \mu)$ be a measure space, and let $\{Z(B), B \in \mathcal{B}(E)\}$ be a centered random measure with orthogonal-increments and control (structure) measure μ. Specifically, $\mathbb{E}[Z(B)] = 0$ for $B \in \mathcal{B}(E)\}$, and for any $B_1, B_2 \in \mathcal{B}(E)\}$, we have (see Sect. 3.4.2):

$$\mathbb{E}[Z(B_1)\overline{Z(B_2)}] = \mu(B_1 \cap B_2).$$

Similar to the Hilbert space $L^2(X, \mathbf{T})$, we define the Hilbert space $L^2(Z)$ spanned by $\{Z(B), B \in \mathcal{B}(E)\}$ as the subspace of $L^2(\mathbb{P})$ consisting of all random variables of the form $\sum_{k=1}^{n} c_k Z(B_k)$, $B_k \in \mathcal{B}(E)$, along with their $L^2(\mathbb{P})$-limits.

The definition of the stochastic integral provided in Sect. 3.4.2 establishes a congruence between the Hilbert spaces $L^2(\mu)$ and $L^2(Z)$. Specifically, the mapping $I : L^2(\mu) \to L^2(Z)$, defined by

$$I(\mathbf{1}_B) = Z(B), \tag{3.5.10}$$

where $\mathbf{1}_B$ is the indicator of $B \in \mathcal{B}$, provides a congruence from $L^2(\mu)$ onto $L^2(Z)$. Moreover, for all functions $f_1, f_2 \in L^2(E, \mu)$ and constants $c_1, c_2 \in \mathbf{C}$, we have

$$I(c_1 f_1 + c_2 f_2) = c_1 I(f_1) + c_2 I(f_2) \quad \text{a. s.} \tag{3.5.11}$$

and

$$\mathbb{E}[I(f_1)\overline{I(f_2)}] = \int_E f_1(x)\overline{f_2(x)}\mu(dx) = (f_1, f_2)_\mu. \tag{3.5.12}$$

We will now use the definition of stochastic integrals to obtain a general integral representation of an L^2-process $\{X(t), t \in \mathbf{T}\}$.

We first assume that there exist a measure space $(E, \mathcal{B}(E), \mu)$ with finite measure μ, a stochastic measure $\{Z(B),\ B \in \mathcal{B}(E)\}$ with control measure μ, and a family of functions $\{f(t) = f(t, \lambda),\ t \in \mathbf{T}\}$ satisfying $f(t, \cdot) \in L^2(E, \mu)$ for each $t \in \mathbf{T}$, such that $X(t)$ can be represented in the following form:

$$X(t) = I(f) = \int_E f(t, \lambda) dZ(\lambda). \tag{3.5.13}$$

Then, the covariance function $R(t, s)$ of $X(t)$ can be represented in the following form:

$$\begin{aligned} R(t, s) &= \mathbb{E}[X(t)X(s)] = (X(t), X(s))_{L^2(\mathbb{P})} \\ &= (f(t, \cdot), f(s, \cdot))_{L^2(\mu)} = \int_E f(t, \lambda)\overline{f(s, \lambda)} d\mu(\lambda). \end{aligned}$$

Thus, in view of (3.5.8), the covariance function $R(t, s)$ belongs to Karhunen's class of covariance functions.

Karhunen's theorem states that the converse assertion also holds; that is, (3.5.8) implies (3.5.13). This means that any L^2-process with a covariance function from Karhunen's class can be represented in the form of (3.5.13). Specifically, we have the following result.

Proposition 3.5.2 (Karhunen's Theorem) *Let* $\{X(t), t \in \mathbf{T}\}$ *be a centered* L^2-*process defined on a probability space* $(\Omega, \mathcal{F}, \mathbb{P})$ *with a covariance function* $R(t, s)$, $t, s \in \mathbf{T}$. *Assume that* $R(t, s)$ *belongs to the Karhunen class, meaning there exist a measure space* $(E, \mathcal{B}(E), \mu)$ *with finite measure* μ *and a family of functions*

$\{f(t,\lambda),\ t \in \mathbf{T},\ \lambda \in E\}$ from $L^2(\mu)$ such that (3.5.8) *holds. Then the following assertions hold:*

(a) The family $\{f(t,\cdot),\ t \in \mathbf{T}\}$ is a representation of $\{X(t),\ t \in \mathbf{T}\}$.
(b) If, further, $\{f(t,\cdot),\ t \in \mathbf{T}\}$ spans $L^2(\mu)$, then there exists an orthogonal stochastic measure $\{Z(B),\ B \in \mathcal{B}(E)\}$ with control measure μ such that for each $t \in \mathbf{T}$, the process $X(t)$ admits the integral representation:

$$X(t) = \int_E f(t,\lambda)dZ(\lambda). \tag{3.5.14}$$

(c) Any random variable $Y \in L^2(X,\mathbf{T})$ may be represented as

$$Y = \int_E g(\lambda)dZ(\lambda), \tag{3.5.15}$$

with some unique function $g \in L^2(\mu)$.
(d) The equality (3.5.14) *is satisfied with probability 1 for each $t \in \mathbf{T}$.*

3.5.4 RKHS Representation: The Loéve Theorem

In this section, we demonstrate that an L^2-process $\{X(t), t \in \mathbf{T}\}$ can be represented by the RKHS generated by its covariance function.

Since the covariance function $R(t,s)$ is a non-negative definite kernel, the following result is an immediate consequence of the Moore-Aronszajn theorem (see Proposition 2.7.11).

Proposition 3.5.3 *Let $\{X(t), t \in \mathbf{T}\}$ be a centered L^2-process with a covariance function $R(t,s)$, $t,s \in \mathbf{T}$. The following assertions hold:*

(a) There exists a unique RKHS, denoted by $H(R)$, for which $R(t,s)$ is the reproducing kernel, that is:
   1. *$R(\cdot,s) \in H(R)$ for every $s \in \mathbf{T}$*
   2. *$(f(\cdot), R(\cdot,s)) = f(s)$ for every $s \in \mathbf{T}$ and $f \in H(R)$*

(b) The space $H(R)$ is spanned by the family $\{R(\cdot,s),\ s \in \mathbf{T}\}$.

Now, we establish an isometric isomorphism between the RKHS $H(R)$, generated by the covariance $R(t,s)$, and $L^2(X,\mathbf{T})$, a Hilbert space generated by the underlying L^2-process $\{X(t), t \in \mathbf{T}\}$. This result, proved by Loéve [219] in 1948, provides an RKHS representation of the process $X(t)$.

Proposition 3.5.4 (Loéve Theorem) *The Hilbert spaces $L^2(X, \mathbf{T})$ and $H(R)$ are isometrically isomorphic. Specifically, there exists a mapping S from $H(R)$ onto $L^2(X, \mathbf{T})$ that is one-to-one and preserves inner products.*

Corollary 3.5.1 *For every element $f \in H(R)$, there is a random variable $X_f = S(f) \in L^2(X, \mathbf{T})$ such that*

$$f(t) = \mathbb{E}[X_f \cdot X(t)].$$

Indeed, using the reproducing property of $R(t, s)$, we have

$$f(t) = (f, R(\cdot, t)) = (S(f), S(R(\cdot, t)))_{H(R)} = (S(f), X(t))_{L^2(\mathbb{P})} = \mathbb{E}[X_f \cdot X(t)].$$

The next result provides an RKHS representation of the process $X(t)$ when the covariance function $R(t, s)$ belongs to the Karhunen class.

Proposition 3.5.5 *Let $\{X(t), t \in T\}$ be a centered L^2-process with a covariance function $R(t, s)$ that belongs to the Karhunen class. Then, there exists an RKHS $H(R)$ that provides an RKHS representation of the process $X(t)$.*

It can be shown that the space $H(R)$ consists of all functions $g(t)$, $t \in \mathbf{T}$, which can be represented in the following form (see Parzen [239], p. 297):

$$g(t) = \int_E f(\lambda, t)h(\lambda)d\mu(\lambda)$$

for some (necessarily unique) function $h \in \overline{L}(f(t), t \in \mathbf{T})$. The corresponding norm is given by

$$\|g\|_{H(R)} = \|h\|_{L^2(\mu)}.$$

3.5.5 Revisiting the Spectral Representation of Stationary Processes

In this section, we present an extended version of the spectral representation theorem for stationary processes, known as the Cramér-Kolmogorov theorem (see Sect. 3.1.4). This theorem is a special case of the Karhunen integral representation theorem and is fundamental to the theory of stationary processes.

Let $\{X(t), t \in \mathbb{U}\}$ be a centered, discrete, or continuous time stationary L^2-process with covariance function $R(t, s) = r(t - s)$, where $\mathbb{U} = \mathbb{Z}$ in the discrete time case and $\mathbb{U} = \mathbb{R}$ in the continuous time case.

By the Herglotz theorem in the discrete time case, and the Bochner-Khintchine theorem in the continuous time case (see Sect. 3.1.4), there exists a unique finite Lebesgue-Stieltjes measure μ on the measure space $(\Lambda, \mathcal{B}(\Lambda))$, where $\Lambda = \mathbb{R}$ in the continuous time case, and $\Lambda = [-\pi.\pi)$ in the discrete time case, and $\mathcal{B}(\Lambda)$ is the Borel σ-algebra of subsets of Λ, such that for any $t, s \in \mathbb{U}$, the covariance function $r(u)$ of a stationary process $X(u)$ admits the following *spectral representation*:

$$R(t, s) = r(t - s) = \int_\Lambda e^{i\lambda(t-s)} d\mu(\lambda). \tag{3.5.16}$$

Observe that expression (3.5.16) is a special case of (3.5.8) with $f(t, \lambda) = \exp\{i\lambda t\}$. Thus, the covariance function of a stationary process belongs to the Karhunen class. Therefore, we can apply Karhunen's theorem to obtain the following result, known as the spectral representation theorem for stationary processes.

Proposition 3.5.6 (The Cramér-Kolmogorov Theorem) *Let $\{X(t),\ t \in \mathbb{U}\}$ be a centered, L^2-continuous stationary process with covariance function $r(t)$ and spectral measure $\mu(\cdot)$. The following assertions hold:*

(a) There is a unique orthogonal stochastic measure $\{Z(B), B \in \mathcal{B}(\Lambda)\}$ with values in $L^2(X)$ such that for each $t \in \mathbb{U}$

$$X(t) = \int_\Lambda e^{it\lambda} dZ(\lambda). \tag{3.5.17}$$

(b) The control measure of Z is μ; that is, for any $B_1, B_2 \in \mathcal{B}(\Lambda)$,

$$\mathbb{E}[Z(B_1)\overline{Z(B_2)}] = \mu(B_1 \cap B_2).$$

(c) There is a unique congruence V between the time domain $L^2(X)$ and the frequency domain $L^2(\mu)$ of the process $X(t)$ such that for any $u \in \mathbb{U}$,

$$V[e^{iu\lambda}] = X(u). \tag{3.5.18}$$

In particular, we have the following:

1. *The random variable Y belongs to $L^2(X)$ if and only if*

$$Y = \int_\Lambda \varphi(\lambda) dZ(\lambda) \quad \text{for some} \quad \varphi \in L^2(\mu). \tag{3.5.19}$$

2. *For any $Y_i \in L^2(X)$ and $\varphi_i(\lambda) = V[Y_i] \in L^2(\mu)$, $i = 1, 2$,*

$$(Y_1, Y_2)_{L^2(\mathbb{P})} = (\varphi_1, \varphi_2)_{L^2(\mu)}.$$

Remark 3.5.1 If $\{X(t), t \in \mathbb{R}\}$ is a stationary continuous time process with covariance function r and spectral density f, then by applying Proposition 3.5.5, we conclude that the RKHS $H(r)$ consists of all functions $g(t),\ t \in \mathbb{R}$, of the form

$$g(t) = \int_{-\infty}^{\infty} e^{i\lambda t} h(\lambda) f(\lambda) d\lambda$$

for some unique function $h \in L^2(f)$ for which

$$\|g\|_{H(r)}^2 = \|h\|_{L^2(f)}^2 < \infty.$$

3.6 Generalized Second-Order Stationary Processes

In this section, we introduce the concept of generalized stationary processes and briefly discuss some of their properties. For more information, we refer to the books by Gel'fand and Vilenkin [108], Major [222], and the papers by Gel'fand and Yaglom [109], and Yaglom [317].

From a physical perspective, the concept of an ordinary stochastic process $Y(t)$, $t \in \mathbb{R}$, refers to the measurement of random quantities at specific moments in time, without taking into account the values at other moments. However, in many cases, it is impossible to localize measurements to a specific point in time. As noted by Gel'fand and Vilenkin (see [108], p. 243), each actual measurement is performed using a device that possesses a certain degree of inertia (memory). Consequently, rather than measuring the ordinary process $Y(t)$ at a single moment, the device records a certain *averaged value* $X(\varphi)$:

$$X(\varphi) = \int_{\mathbb{R}} \varphi(t) Y(t) dt, \tag{3.6.1}$$

where $\varphi(t)$ is a function characterizing the device. Moreover, small changes in φ result in small changes in $X(\varphi)$. Consequently, we obtain a continuous linear functional, which leads to the concept of a *generalized stochastic process*.

Let $D = D(\mathbb{R})$ be the space of infinitely differentiable real-valued functions $\varphi(t)$ $(t \in \mathbb{R})$ with finite support. This means that the function φ vanishes outside some closed interval, referred to as the support of φ, denoted by $\text{supp}\{\varphi\}$). The topology in D is defined as follows: we say that a sequence of functions $\varphi_n \in D$ converges to a function $\varphi \in D$ as $n \to \infty$, and we write $\varphi_n \Rightarrow \varphi$, if $\text{supp}\{\varphi_n\} \subset [a, b]$ for all $n \in \mathbb{N}$ and $\varphi_n^{(k)}(t) \to \varphi^{(k)}(t)$ as $n \to \infty$, uniformly in $t \in [a, b]$ for all $k = 0, 1, \ldots$.

Definition 3.6.1 A generalized second-order stationary process $\{X(\varphi),\ \varphi \in D\}$ defined on a probability space $(\Omega, \mathcal{F}, \mathbb{P})$ is a random linear functional such that $\mathbb{E}[|X(\varphi)|^2] < \infty$ and for the mean functional $m(\varphi)$ and the covariance functional

$R(\varphi, \psi)$ the stationarity conditions hold:

$$m(\varphi) := (X(\varphi), 1) = m(\tau_t\varphi), \quad \varphi \in D,$$
$$R(\varphi, \psi) := (X(\varphi), X(\psi)) = R(\tau_t\varphi, \tau_t\psi), \quad \varphi, \psi \in D, \tag{3.6.2}$$

where τ_t is the shift operator: $[\tau_t\varphi](s) = \varphi(s+t)$, and $(\cdot, \cdot)$ is the inner product in the space $L^2(\mathbb{P}) := \{\xi = \xi(\omega) : \mathbb{E}[|\xi|^2] < \infty\}$.

We assume that the process $X(\varphi)$ is L^2-continuous, meaning that

$$\mathbb{E}[|X(\varphi_n) - X(\varphi)|^2] \to 0 \quad \text{as} \quad \varphi_n \Rightarrow \varphi,$$

and that it possesses a spectral density function $f(\lambda)$, $\lambda \in \mathbb{R}$.

Thus, the covariance functional (3.6.2) is continuous in each of its arguments and admits the spectral representation (see, e.g., Gel'fand and Vilenkin [108], p. 264):

$$R(\varphi, \psi) = \int_{\mathbb{R}} \hat{\varphi}(\lambda)\overline{\hat{\psi}(\lambda)} f(\lambda)\, d\lambda, \quad \varphi, \psi \in D,$$

where $\hat{\varphi}$ and $\hat{\psi}$ are the Fourier transforms of the functions φ and ψ, respectively. Furthermore, for some $p \geq 0$, we have

$$\int_{\mathbb{R}} \frac{f(\lambda)}{(1+\lambda^2)^p}\, d\lambda < \infty. \tag{3.6.3}$$

Similar to ordinary L^2-continuous stationary processes, a generalized stationary process $\{X(\varphi),\ \varphi \in D\}$ admits the spectral representation:

$$X(\varphi) = \int_{\mathbb{R}} \hat{\varphi}(\lambda) dZ_f(\lambda), \quad \varphi \in D,$$

where Z_f is the orthogonal stochastic measure corresponding to the spectral density function f (see Sect. 3.5.5).

Definition 3.6.2 A real-valued generalized process $\{X(\varphi),\ \varphi \in D\}$ is called Gaussian with mean functional $m(\varphi)$ and covariance functional $R(\varphi, \psi)$ if its characteristic functional $\Phi(\varphi) := \mathbb{E}[\exp\{iX(\varphi)\}]$ has the following form:

$$\Phi(\varphi) = \exp\left\{im(\varphi) - \frac{1}{2}R(\varphi, \varphi)\right\}, \quad \varphi \in D.$$

A real-valued stationary Gaussian generalized process $\varepsilon(\varphi)$ with zero mean and a covariance function equal to the delta function (defined as $\delta(\varphi) = \varphi(0)$) is called white noise.

The white noise is the derivative of the Wiener process, and its characteristic functional is given by (see Gel'fand and Vilenkin [108], p. 261)

$$\Phi_{\varepsilon}(\varphi) = \exp\left\{-\frac{1}{2}\int_0^{\infty}\varphi^2(t)dt\right\}, \qquad \varphi \in D.$$

Since the delta function is the Fourier transform of the Lebesgue measure, the spectral density of white noise is $f(\lambda) = 1$.

The class of generalized processes clearly includes all ordinary continuous time stationary processes. Additionally, it contains non-ordinary processes that are not defined in the classical sense. An example of such a process is the above-defined white noise.

Remark 3.6.1 If $Y(t)$ is a real-valued ordinary L^2-continuous stationary process such that $\mathbb{E}[|Y(t)|^2] \leq p(t)$ for some polynomial $p(t)$, then the integral in (3.6.1) is well-defined and determines a real-valued generalized stationary process $X(\varphi)$ (Gel'fand and Vilenkin [108], p. 257). On the other hand, the generalized stationary process $X(\varphi)$ defined by (3.6.1) uniquely determines the ordinary process $Y(t)$. Specifically, if $Y_1(t)$ and $Y_2(t)$ are two ordinary L^2-continuous stationary processes generating by (3.6.1) the same generalized stationary process $X(\varphi)$, then $Y_1(t) = Y_2(t)$ with probability 1 for each $t \in \mathbb{R}$. In this case, we will refer to the generalized stationary process $X(\varphi)$ as an ordinary process and will identify the processes $X(\varphi)$ and $Y(t)$. Additionally, note that for ordinary stationary processes, the condition (3.6.3) is satisfied with $p = 0$. Conversely, if the condition (3.6.3) is satisfied with $p = 0$, then the process $X(\varphi)$ must be expressible in the form (3.6.1), where $Y(t)$ is an ordinary process (see Yaglom [317]).

3.7 Bibliographical Notes

Section 3.1 Second-order (weakly) stationary processes were introduced by Khinchine in 1934, who provided a harmonic decomposition of their covariance functions. A.N. Kolmogorov conducted a comprehensive study of stationary sequences using Hilbert space methods in his pioneering papers [200, 201]. Since then, many researchers have made significant contributions to the development of this field.

Key concepts and basic results related to second-order stationary processes can be found in various texts. Notable references include works by Ash and Gardner [12], Brockwell and Davis [57], Cramér and Leadbetter [69], Doob [85], Gihman and Skorokhod [110], Grenander and Szegő [162], Ibragimov and Linnik [188], Rozanov [263], Shiryaev [277], and Yaglom [318], among others.

For Kolmogorov's isometric isomorphism theorem, see, for example, Grenander and Szegő [162], Section 10.5, and Ibragimov and Linnik [188], Section 16.6.

Section 3.2 More information about linear transformations of stationary processes can be found in the books by Gihman and Skorokhod [110] and Rozanov [263]. The proofs of Theorems 3.2.1 and 3.2.2 on the existence of spectral density functions, which can also be viewed as characterization theorems for linear processes, are provided in Ibragimov and Linnik [188], Section 16.7. For a detailed account of the relevant properties of Lévy processes, see, for example, the book by Protter [251].

Section 3.3 As mentioned in the text, the memory structure has traditionally been defined in two ways: in the time domain, by the decay rates of autocorrelations, and in the frequency domain, by the rates of explosion of low-frequency spectra (see, e.g., Beran [31], Beran et al. [33], Giraitis et al. [147], Guégan [163], Pipiras and Taqqu [246], Robinson [256], and references therein). Our definition of the memory structure, based on the spectral density function, differs slightly from the traditional approach.

More information about fractional Brownian motion (fBm) and fractional Gaussian noise (fGn) can be found in the books by Pipiras and Taqqu [246] and Samorodnitsky and Taqqu [269]. The fractional Riesz-Bessel motion (fRBm) was introduced in the paper by Anh et al. [5] and has since been extensively discussed in several papers, including those by Anh et al. [6, 9, 10], Gao [103], and Gao et al. [105, 106], among others. For the CARFIMA(p, H, q) model, we refer to the papers by Brockwell [56] and Tsai and Chan [304].

The Muckenhoupt condition (A_2) was first introduced in Muckenhoupt's paper [225]. More information about the Muckenhoupt condition (A_p) for any $p > 1$ and its applications in harmonic analysis can be found in the books by Böttcher and Karlovich [51], Garnett [107], and Stein [288]. The Muckenhoupt class of processes is discussed in the papers by Ginovyan [112, 114], Solev [281, 282], and Solev and Zerbet [284].

Section 3.4 A comprehensive treatment of single and multiple stochastic integrals with respect to orthogonal random measures can be found in several books, including Ash and Gardner [12], Gihman and Skorokhod [110], Ibragimov and Linnik [188], Major [222], Nualart [230], Peccati and Taqqu [240], and Shiryaev [277]. For more information about the multiple off-diagonal (Itô-type) and with-diagonal (Stratonovich-type) stochastic integrals with respect to Gaussian and Lévy processes, we refer to Farré et al. [97], Itô [193], Janson [196], Major [222], Nualart [230], Nualart and Schoutens [231], Peccati and Taqqu [240], and Pipiras and Taqqu [246] (see also Bai et al. [22]).

For the representation of the normalized quadratic functional as a double Wiener-Itô stochastic integral, see the papers by Doukhan et al. [86] and Bai et al. [22].

Section 3.5 The proof of the Karhunen-Loéve theorem, along with additional details, can be found in Ash and Gardner [12], Section 1.4, and Loéve [219], Section 37.5 (see also Parzen [238, 239]). For the general integral representations and the proof of Karhunen's theorem, we refer to Gihman and Skorokhod [110] (Section 4.5) and Parzen [239]. The representation via RKHSs and the proof of the Loéve theorem can be found in Hida and Hidsuda [175] (see also Parzen [239]). Spectral representations of stationary processes and the proof of the Cramér-Kolmogorov theorem can be found in many books (see, e.g., Ash and Gardner

[12] (Chapter 1), Cramér and Leadbetter [69] (Chapter 7), Doob [85] (Chapters 10 and 11), Ibragimov and Linnik [188] (Chapter 16), and Shiryaev [277] (Chapter 6)).

Section 3.6 A comprehensive treatment of generalized stationary processes is provided in the book by Gel'fand and Vilenkin [108]. We also refer to the book by Major [222] and the papers by Gel'fand and Yaglom [109], and Yaglom [317].

Chapter 4
The Trace Problem for Toeplitz Matrices and Operators

In this chapter, we present a unified approach to the trace approximation problem for Toeplitz and tapered Toeplitz matrices, as well as for operators generated by integrable real symmetric functions defined on the unit circle and the real line. We discuss the estimation of the corresponding approximation errors and provide applications to both discrete and continuous time stationary models with various memory structures. Additionally, we present estimates of Schatten norms for Toeplitz and tapered Toeplitz matrices, as well as for operators.

In Sect. 4.1, we briefly outline the significance of the trace approximation problem in the spectral and statistical analysis of stationary processes. In Sect. 4.2, we define the main objects discussed in this chapter—Toeplitz matrices and operators—and explore their various realizations. In Sect. 4.3, we state the trace approximation problem. In Sect. 4.4, we discuss the trace problem for Toeplitz matrices. Section 4.5 addresses the same problem for Toeplitz operators. The results presented in Sects. 4.4 and 4.5 are proved in Sect. 4.6. Section 4.7 is devoted to the applications of the trace approximation problem to discrete and continuous time stationary processes, with the results proved within the section. Section 4.8 discusses the trace approximation problem for tapered Toeplitz matrices and operators. Finally, in Sect. 4.9, we present estimates of Schatten norms for both Toeplitz and tapered Toeplitz matrices and operators.

4.1 Introduction

Toeplitz matrices and operators are of significant independent interest and have a wide range of applications across various fields of science, including economics, engineering, finance, hydrology, physics, and signal processing. They arise naturally in the context of stationary processes. This is because the covariance matrix of a discrete time stationary process is a truncated Toeplitz matrix generated by the

M. S. Ginovyan, *Random Toeplitz Functionals and Their Applications*,
Frontiers in Probability and the Statistical Sciences,
https://doi.org/10.1007/978-3-032-00961-6_4

spectral density of that process. Conversely, any non-negative summable function generates a Toeplitz matrix, which can be viewed as the spectral density of some discrete time stationary process. Consequently, the corresponding truncated Toeplitz matrix serves as the covariance matrix of that process. In the continuous time case, the Toeplitz matrix is replaced by a Toeplitz operator.

Truncated Toeplitz matrices and operators are of particular importance and serve as tools for studying many topics in the spectral and statistical analysis of discrete and continuous time stationary processes. These topics include central and non-central limit theorems, large deviations of Toeplitz-type random quadratic forms and functionals, estimation of spectral parameters and functionals, asymptotic expansions of estimators, hypothesis testing about the spectrum, and prediction of future values based on the observed past of the process.

The objective is to approximate the trace of the product of Toeplitz matrices (or operators) using the trace of a *single* Toeplitz matrix (or operator) generated by the product of their respective generating functions. It is important to note that the product of Toeplitz matrices (or operators) is not necessarily a Toeplitz matrix (or operator). The following two questions arise naturally:

How accurate is the approximation?

What is the rate of convergence of the approximation error?

These questions are known as the two trace approximation problems, referred to as Problem (A) and Problem (B). Similar issues also arise in the context of tapered Toeplitz matrices and operators.

The trace approximation problem and its applications in the statistical analysis and prediction of discrete time stationary processes date back to the classical monograph by Grenander and Szegő [162]. Since then, this problem has been explored by many authors across various classes of generating functions (symbols).

It is important to note that the trace approximation problem is particularly significant in the cases where the symbols of the underlying Toeplitz matrices and operators have singularities. Such cases frequently arise in problems involving the statistical analysis of long memory processes (where the spectral density is unbounded) and antipersistent processes (where the spectral density has zeros) in both discrete and continuous time.

4.2 Toeplitz Matrices and Operators

We begin by defining the main objects of study in this chapter: truncated Toeplitz matrices and operators generated by integrable real symmetric functions.

Toeplitz operators can be defined in several ways and have different realizations. This variety is important in applications, as it enables us to choose the most suitable realization for each specific problem.

An infinite matrix B is called *a Toeplitz matrix* if it has the following form:

$$B = ||b_{k-j}||_{k,j=0,1,\dots} = \begin{pmatrix} b_0 & b_{-1} & b_{-2} & \dots \\ b_1 & b_0 & b_{-1} & \dots \\ b_2 & b_1 & b_0 & \dots \\ \dots & \dots & \dots & \dots \end{pmatrix}, \tag{4.2.1}$$

where $b := \{b_k;\, k \in \mathbb{N}_0\}$ is a sequence of complex numbers. In other words, Toeplitz matrices have entries that depend on the differences between their indices.

An important and interesting case for our purposes occurs when the numbers $\{b_k;\, k \in \mathbb{Z}\}$ are the Fourier coefficients of an integrable real even function $\psi(\lambda)$ defined on $\mathbb{T} = [-\pi, \pi)$. In this case, we will say that the Toeplitz matrix B is generated by the function ψ and will use the notation:

$$B = B(\psi) := ||\widehat{\psi}(k-j)||_{k,j=0,1,\dots}, \tag{4.2.2}$$

where

$$b_k := \widehat{\psi}(k) = \int_{\mathbb{T}} e^{i\lambda k}\, \psi(\lambda)\, d\lambda, \quad k \in \mathbb{Z}. \tag{4.2.3}$$

The function ψ is referred to as the *generating function* or the *symbol* of the matrix $B(\psi)$. Recall that if $\psi(\lambda)$ is an even function ($\psi(-\lambda) = \psi(\lambda)$), then the Fourier coefficients $\widehat{\psi}(k)$ are real numbers.

To a Toeplitz matrix $B(\psi)$, we can associate a Toeplitz operator, which we also denote by $B(\psi)$ as follows. Let $H^{2+}(\mathbb{T})$ and $H^{2-}(\mathbb{T})$ be the Hardy spaces, defined by

$$H^{2+}(\mathbb{T}) := \left\{g(\lambda) \in L^2(\mathbb{T}) : \widehat{g}(k) = 0 \quad \text{for} \quad k < 0\right\},$$

$$H^{2-}(\mathbb{T}) := \left\{g(\lambda) \in L^2(\mathbb{T}) : \widehat{g}(k) = 0 \quad \text{for} \quad k \geq 0\right\}.$$

We have

$$L^2(\mathbb{T}) = H^{2+}(\mathbb{T}) \oplus H^{2-}(\mathbb{T}).$$

Denote by $P_+(\mathbb{T})$ and $P_-(\mathbb{T})$ the orthogonal projections of $L^2(\mathbb{T})$ onto $H^{2+}(\mathbb{T})$ and $H^{2-}(\mathbb{T})$, respectively. For a function $\psi(\lambda) \in L^1(\mathbb{T})$, the operator $B(\psi)$: $H^{2+}(\mathbb{T}) \to H^{2+}(\mathbb{T})$, defined by

$$B(\psi)\varphi = P_+(\mathbb{T})\boldsymbol{\psi}\varphi, \quad \varphi \in H^{2+}(\mathbb{T}),$$

where $\boldsymbol{\psi}$ denotes the multiplication operator by the function $\psi(\lambda)$, is referred to as a *Toeplitz operator with symbol* ψ. The operator $B(\psi)$ can also be considered in the space $L^2(\mathbb{T})$:

$$B(\psi)\varphi = P_+(\mathbb{T})\boldsymbol{\psi} P_+(\mathbb{T})\varphi, \quad \varphi \in L^2(\mathbb{T}). \tag{4.2.4}$$

Observe that the Toeplitz operator $B(\psi)$ has a Toeplitz matrix $B(\psi)$ in the basis $\{e^{i\lambda k},\ k \geq 0\}$ of the space $H^{2+}(\mathbb{T})$: for $k,\ j \geq 0$

$$\left(B(\psi)e^{ik\lambda}, e^{ij\lambda}\right) = \widehat{\psi}(k-j),$$

where $(\cdot, \cdot)$ is the inner products in the space $L^2(\mathbb{T})$.

A closely related to a Toeplitz matrix is a Hankel matrix. A Hankel matrix $H(\psi)$ generated by a function ψ is defined by

$$H(\psi) := ||\widehat{\psi}(k+j+1)||_{k,j=0,1,\ldots}, \tag{4.2.5}$$

where $\widehat{\psi}(k)$ are the Fourier coefficients of ψ given by (4.2.3). The corresponding Hankel operator is defined by

$$H(\psi)\varphi = P_-(\mathbb{T})\boldsymbol{\psi} P_+(\mathbb{T})\varphi, \quad \varphi \in L^2(\mathbb{T}). \tag{4.2.6}$$

There is a simple yet important identity that relates Toeplitz and Hankel operators:

$$B(fg) - B(f)B(g) = H(f)H(\widetilde{g}), \tag{4.2.7}$$

where $\widetilde{g}(\lambda) = g(1/\lambda)$, $\lambda \in \mathbb{T}$.

In this text, we will focus on truncated Toeplitz matrices and operators.

Let $\psi(\lambda)$ be an integrable real symmetric function defined on $\mathbb{T} = [-\pi, \pi)$. For $T = 1, 2, \ldots$, the $(T \times T)$*-truncated Toeplitz matrix* generated by ψ, denoted by $B_T(\psi)$, is defined by the following equation:

$$\begin{aligned} B_T(\psi) &:= \|\widehat{\psi}(t-s)\|_{t,s=1,2\ldots,T} \\ &= \begin{pmatrix} \widehat{\psi}(0) & \widehat{\psi}(-1) & \cdots \widehat{\psi}(1-T) \\ \widehat{\psi}(1) & \widehat{\psi}(0) & \cdots \widehat{\psi}(2-T) \\ \cdots & \cdots & \cdots\cdots \\ \widehat{\psi}(T-1) & \widehat{\psi}(T-2) & \cdots \widehat{\psi}(0) \end{pmatrix}, \end{aligned} \tag{4.2.8}$$

where $\widehat{\psi}(t)$ $(t \in \mathbb{Z})$ are the Fourier coefficients of ψ (see formula (4.2.3)).

Denote by $P_T(\mathbb{T})$ the orthogonal projection of $L^2(\mathbb{T})$ onto the space $\mathcal{P}_T(\mathbb{T})$ of trigonometric polynomials of degree at most T. Similar to Eq. (4.2.4), we can consider the truncated Toeplitz operator:

$$B_T(\psi)\varphi = P_T(\mathbb{T})\boldsymbol{\psi} P_T(\mathbb{T})\varphi, \ \varphi \in L^2(\mathbb{T}). \tag{4.2.9}$$

It turns out that there is an analogue of the identity (4.2.7) for truncated Toeplitz operators. This is expressed in what is known as *Widom's formula* (see Widom [315]):

$$B_T(fg) - B_T(f)B_T(g) = P_T H(f)H(\widetilde{g})P_T + R_T H(\widetilde{f})H(g)R_T, \qquad (4.2.10)$$

where the operators P_T and R_T are defined by

$$P_T(x_0, x_1, \ldots, x_T, \ldots) = (x_0, x_1, \ldots, x_T, 0, \ldots),$$
$$R_T(x_0, x_1, \ldots, x_T, \ldots) = (x_T, x_{T-1}, \ldots, x_0, 0, \ldots).$$

Now we define the truncated Toeplitz operators on $\mathbb{R} := (-\infty, \infty)$. Given a real number $T > 0$ and an integrable real symmetric function $\psi(\lambda)$ defined on $\mathbb{R}$, the *T-truncated Toeplitz operator* (also called the *Wiener-Hopf operator*) generated by ψ, denoted by $W_T(\psi)$, is defined by the following equation:

$$[W_T(\psi)u](t) = \int_0^T \hat{\psi}(t-s)u(s)ds, \quad u(s) \in L^2[0, T], \qquad (4.2.11)$$

where

$$\widehat{\psi}(t) = \int_{\mathbb{R}} e^{i\lambda t}\, \psi(\lambda)\, d\lambda, \quad t \in \mathbb{R} \qquad (4.2.12)$$

is the Fourier transform of $\psi(\lambda)$. Thus, $W_T(\psi)$ is an integral operator on $L^2[0, T]$ with kernel $\widehat{\psi}$.

Denote by $P_T(\mathbb{R})$ the orthogonal projection of $L^2(\mathbb{R})$ onto the space $\mathcal{E}_T(\mathbb{R})$ of entire functions of exponential type at most T. Similar to Eq. (4.2.9), we can consider the following realization of the operator $W_T(\psi)$:

$$W_T(\psi)\varphi = P_T(\mathbb{R})\boldsymbol{\psi} P_T(\mathbb{R})\varphi, \ \varphi \in L^2(\mathbb{R}), \qquad (4.2.13)$$

where $\boldsymbol{\psi}$ denotes the multiplication operator by the function $\psi(\lambda)$.

4.3 The Trace Approximation Problem

The problem of approximating the traces of products of truncated Toeplitz matrices and operators can be stated as follows.

Let $\mathcal{H} = \{h_1, h_2, \ldots, h_m\}$ be a collection of integrable real symmetric functions defined on the domain Λ, where $\Lambda = \mathbb{R}$ or $\Lambda = \mathbb{T}$. For a given number $T > 0$, let $A_T(h_k)$ denote either the $(T \times T)$-truncated Toeplitz matrix $(B_T(h_k))$ or the T-truncated Toeplitz operator $(W_T(h_k))$ generated by the function h_k, defined respectively by (4.2.8) and (4.2.11) with h_k instead of ψ. Further, let

$$\tau := \{\tau_k : \tau_k \in \{-1, 1\},\ k = 1, 2, \ldots, m\} \tag{4.3.1}$$

be a given sequence of $\pm$ 1's. Define

$$S_{A,\mathcal{H},\tau}(T) := \frac{1}{T} \operatorname{tr}\left[\prod_{k=1}^{m} \{A_T(h_k)\}^{\tau_k}\right], \tag{4.3.2}$$

where tr$[A]$ stands for the trace of A,

$$M_{\Lambda,\mathcal{H},\tau} := (2\pi)^{m-1} \int_\Lambda \prod_{k=1}^{m} [h_k(\lambda)]^{\tau_k} d\lambda, \tag{4.3.3}$$

and

$$\Delta_{A,\Lambda,\mathcal{H},\tau}(T) := |S_{A,\mathcal{H},\tau}(T) - M_{\Lambda,\mathcal{H},\tau}|. \tag{4.3.4}$$

The problem of interest is to approximate $S_{A,\mathcal{H},\tau}(T)$ by $M_{\Lambda,\mathcal{H},\tau}$ and to estimate the error rate for $\Delta_{A,\Lambda,\mathcal{H},\tau}(T)$ as $T \to \infty$. More precisely, for a given sequence $\tau = \{\tau_k \in \{-1, 1\},\ k = 1, 2, \ldots, m\}$, we aim to find conditions on the functions $\{h_k(\lambda),\ k = 1, 2, \ldots, m\}$ such that

$$\text{Problem (A):} \quad \Delta_{A,\Lambda,\mathcal{H},\tau}(T) = o(1) \quad \text{as} \quad T \to \infty, \quad \text{or}$$

$$\text{Problem (B):} \quad \Delta_{A,\Lambda,\mathcal{H},\tau}(T) = O(T^{-\gamma}), \quad \gamma > 0, \quad \text{as} \quad T \to \infty.$$

The trace approximation problem dates back to the classical monograph by Grenander and Szegő [162] and has been extensively studied in the literature (see, e.g., the survey paper by Ginovyan et al. [145] and references therein).

We will focus on the following special case, which is important from an application standpoint and is commonly discussed in the literature: $m = 2\nu$, $\tau_k = 1$ or $\tau_k = (-1)^k$, where $k = 1, 2, \ldots, m$, and

$$\begin{aligned} h_1(\lambda) &= h_3(\lambda) = \cdots = h_{2\nu-1}(\lambda) := f(\lambda) \\ h_2(\lambda) &= h_4(\lambda) = \cdots = h_{2\nu}(\lambda) := g(\lambda). \end{aligned} \tag{4.3.5}$$

This case is significant from an application standpoint, as it relates to various problems in the statistical analysis of stationary processes. These include asymptotic distributions and large deviations of Toeplitz-type quadratic functionals, estimation of spectral parameters and functionals, asymptotic expansions of estimators, and hypothesis testing concerning the spectrum. Some of these applications are discussed in Chaps. 5–10.

4.4 The Trace Problem for Toeplitz Matrices

Let $f(\lambda)$ be an integrable real symmetric function defined on $\mathbb{T} = [-\pi, \pi)$. For $T = 1, 2, \ldots$, let $B_T(f)$ be the $(T \times T)$ Toeplitz matrix generated by the function f:

$$B_T(f) := \|\widehat{f}(s - t)\|_{s,t=1,2,\ldots,T}, \tag{4.4.1}$$

where

$$\widehat{f}(t) = \int_{\mathbb{T}} e^{-i\lambda t} f(\lambda)\, d\lambda, \quad t \in \mathbb{Z}, \tag{4.4.2}$$

are the Fourier coefficients of f.

Observe that

$$\frac{1}{T}\mathrm{tr}\,[B_T(f)] = \frac{1}{T} \cdot T\widehat{f}(0) = \int_{-\pi}^{\pi} f(\lambda)d\lambda. \tag{4.4.3}$$

What happens to the relation (4.4.3) when the matrix $B_T(f)$ is replaced by a product of Toeplitz matrices? Note that the product of Toeplitz matrices is not necessarily a Toeplitz matrix.

The goal is to approximate the trace of the product of Toeplitz matrices by the trace of a single Toeplitz matrix generated by the product of their respective generating functions. Specifically, let $\mathcal{H} = \{h_1, h_2, \ldots, h_m\}$ be a collection of integrable real symmetric functions defined on $\mathbb{T}$. Define

$$S_{B,\mathcal{H}}(T) := \frac{1}{T}\mathrm{tr}\left[\prod_{i=1}^{m} B_T(h_i)\right], \tag{4.4.4}$$

$$M_{\mathbb{T},\mathcal{H}} := (2\pi)^{m-1} \int_{-\pi}^{\pi} \left[\prod_{i=1}^{m} h_i(\lambda)\right] d\lambda, \tag{4.4.5}$$

and let

$$\Delta(T) := \Delta_{B,\mathbb{T},\mathcal{H}}(T) = |S_{B,\mathcal{H}}(T) - M_{\mathbb{T},\mathcal{H}}|. \tag{4.4.6}$$

Observe that by (4.4.3)

$$M_{\mathbb{T},\mathcal{H}} = (2\pi)^{m-1} \int_{-\pi}^{\pi} \left[\prod_{i=1}^{m} h_i(\lambda)\right] d\lambda = \frac{1}{T}\mathrm{tr}\left[B_T\left(\prod_{i=1}^{m} h_i(\lambda)\right)\right]. \tag{4.4.7}$$

How well is $S_{B,\mathcal{H}}(T)$ approximated by $M_{\mathbb{T},\mathcal{H}}$? What is the rate of convergence to zero of the approximation error $\Delta_{B,\mathbb{T},\mathcal{H}}(T)$ as $T \to \infty$? These are Problems (A) and (B) for Toeplitz matrices, as stated in Sect. 4.3.

4.4.1 Problem (A) for Toeplitz Matrices

Recall that Problem (A) involves finding conditions on the functions $h_1(\lambda), h_2(\lambda), \ldots, h_m(\lambda)$ in (4.4.4)–(4.4.5) such that $\Delta_{B,\mathbb{T},\mathcal{H}}(T) = o(1)$ as $T \to \infty$.

In Theorems 4.4.1–4.4.4, we summarize the results concerning Problem (A) for Toeplitz matrices in the case where the exponents τ_k, $k = 1, 2, \ldots, m$, are all equal to 1 (see (4.3.1)).

Theorem 4.4.1 *Let $\Delta(T) := \Delta_{B,\mathbb{T},\mathcal{H}}(T)$ be as defined in (4.4.6). Each of the following conditions is sufficient for*

$$\Delta(T) = o(1) \quad \text{as} \quad T \to \infty. \tag{4.4.8}$$

(A1) *The function $\varphi(\mathbf{u})$ given by*

$$\varphi(\mathbf{u}) := \int_{-\pi}^{\pi} h_1(\lambda)h_2(\lambda - u_1)h_3(\lambda - u_2)\cdots h_m(\lambda - u_{m-1})\, d\lambda, \tag{4.4.9}$$

where $\mathbf{u} = (u_1, u_2, \ldots, u_{m-1}) \in \mathbb{R}^{m-1}$, belongs to $L^{m-2}(\mathbb{T}^{m-1})$ and is continuous at the point $\mathbf{0} = (0, 0, \ldots, 0)$.

(A2) *$h_i \in L^{p_i}(\mathbb{T})$, $1 \le p_i \le \infty$, $i = 1, 2, \ldots, m$, with $1/p_1 + \cdots + 1/p_m \le 1$.*

Theorem 4.4.2 *Let $m = 2\nu$, $h_1(\lambda) = h_3(\lambda) = \cdots = h_{2\nu-1}(\lambda) := f(\lambda)$ and $h_2(\lambda) = h_4(\lambda) = \cdots = h_{2\nu}(\lambda) := g(\lambda)$. Assume that there exist numbers $\alpha < 1$ and $\beta < 1$ such that for each $\delta > 0$,*

$$|f(\lambda)| = O\left(|\lambda|^{-\alpha-\delta}\right) \quad \text{and} \quad |g(\lambda)| = O\left(|\lambda|^{-\beta-\delta}\right) \quad \text{as} \quad \lambda \to 0.$$

Then the following assertions hold:

(a) If $\nu(\alpha + \beta) < 1$, then

$$\lim_{T\to\infty} \frac{1}{T} tr\,[B_T(f)B_T(g)]^\nu = (2\pi)^{2\nu-1} \int_{-\pi}^{\pi} [f(\lambda)g(\lambda)]^\nu d\lambda.$$

(b) If $\nu(\alpha + \beta) \ge 1$, then for every $\varepsilon > 0$,

$$tr\,[B_T(f)B_T(g)]^\nu = o\left(T^{\nu(\alpha+\beta)+\varepsilon}\right) \quad \text{as} \quad T \to \infty.$$

Theorem 4.4.3 *Assume that* $m = 4$, $h_1 = h_3 := f$, *and* $h_2 = h_4 := g$. *The following conditions are also sufficient for (4.4.8) with* $m = 4$*: the functions* f *and* g *satisfy* $f \in L^2(\mathbb{T})$, $g \in L^2(\mathbb{T})$, $fg \in L^2(\mathbb{T})$, *and*

$$\int_{-\pi}^{\pi} f^2(\lambda)g^2(\lambda-\mu)\,d\lambda \longrightarrow \int_{-\pi}^{\pi} f^2(\lambda)g^2(\lambda)\,d\lambda \quad \text{as} \quad \mu \to 0.$$

To state the next theorem, we recall the class $SV_0(\mathbb{R})$ of slowly varying at zero functions (see Definition 2.5.1).

Theorem 4.4.4 *Assume that* $m = 4$, $h_1 = h_3 := f$, *and* $h_2 = h_4 := g$. *The following conditions are also sufficient for (4.4.8) with* $m = 4$*: the functions* f *and* g *satisfy*

$$f(\lambda) \le |\lambda|^{-\alpha}L_1(\lambda), \quad |g(\lambda)| \le |\lambda|^{-\beta}L_2(\lambda), \quad \lambda \in \mathbb{T}, \tag{4.4.10}$$

for some $\alpha < 1$, $\beta < 1$ *with* $\alpha + \beta \le 1/2$, *where* L_1 *and* L_2 *are slowly varying at zero functions that satisfy*

$$L_i \in SV_0(\mathbb{R}), \;\; \lambda^{-(\alpha+\beta)}L_i(\lambda) \in L^2(\mathbb{T}), \;\; i = 1, 2. \tag{4.4.11}$$

The following result presents time domain sufficient conditions for (4.4.8) with $m = 4$. Specifically, the conditions are in terms of the Fourier transforms of the functions f and g (see Giraitis et al. [147], Lemma 6.3.4).

Theorem 4.4.5 *Let* $\widehat{f}(t)$ *and* $\widehat{g}(t)$ *be the Fourier transforms of functions* f *and* g, *respectively. If* $\widehat{f}$ *and* $\widehat{g}$ *satisfy the following conditions:*

$$\widehat{f} \in l^p\ (p \ge 1) \quad \text{and} \quad \widehat{g} \in l^q\ (q \ge 1) \quad \text{with} \quad 1/p + 1/q \ge 3/2,$$

then the relation (4.4.8) holds with $m = 4$.

We now consider the case in which the product in (4.3.2) also involves *inverse matrices*, specifically $\tau_k = (-1)^k$, $k = 1, 2, \ldots, m$. We assume that $m = 2\nu$, $\nu \in \mathbb{N}$. The functions from the collection $\mathcal{H} = \{h_1, h_2, \ldots, h_m\}$ that involve Toeplitz matrices are denoted by g_i, for $i = 1, 2, \ldots, \nu$, while those involving inverse Toeplitz matrices are denoted by f_i, for $i = 1, 2, \ldots, \nu$. We set

$$SI_{B,\mathcal{H}}(T) := \frac{1}{T}\mathrm{tr}\left[\prod_{i=1}^{\nu}[B_T(f_i)]^{-1}B_T(g_i)\right], \tag{4.4.12}$$

$$MI_{\mathbb{T},\mathcal{H}} := \frac{1}{2\pi}\int_{-\pi}^{\pi}\left[\prod_{i=1}^{\nu}\frac{g_i(\lambda)}{f_i(\lambda)}\right]d\lambda, \tag{4.4.13}$$

$$\Delta I_{B,\mathbb{T},\mathcal{H}}(T) := |SI_{B,\mathcal{H}}(T) - MI_{\mathbb{T},\mathcal{H}}|. \tag{4.4.14}$$

Theorem 4.4.6 *Let $\nu \in \mathbb{N}$, and $\alpha, \beta \in \mathbb{R}$ with $0 < \alpha, \beta < 1$ and $\nu(\beta - \alpha) < 1/2$. Suppose $f_i(\lambda)$ and $g_i(\lambda)$, for $i = 1, 2, \dots, \nu$, are symmetric real-valued functions satisfying the following conditions:*

(C1) *$f_i(\lambda)$, $i = 1, 2, \dots, \nu$, are non-negative and continuous at all $\lambda \in \mathbb{T} \setminus \{0\}$, $f_i^{-1}(\lambda)$ are continuous at all $\lambda \in \mathbb{T}\setminus\{0\}$, and for all $\delta > 0$ and $i = 1, 2, \dots, \nu$,*

$$f_i(\lambda) = O\left(|\lambda|^{-\alpha-\delta}\right) \quad \text{as} \quad \lambda \to 0;$$

(C2) *$\partial/(\partial\lambda) f_i^{-1}(\lambda)$ and $\partial^2/(\partial\lambda)^2 f_i^{-1}(\lambda)$ are continuous at all $\lambda \in \mathbb{T} \setminus \{0\}$, and for all $\delta > 0$ and $i = 1, 2, \dots, \nu$,*

$$\partial^k/(\partial\lambda)^k f_i^{-1}(\lambda) = O\left(|\lambda|^{-\alpha-k-\delta}\right) \quad \text{for} \quad k = 0, 1, 2;$$

(C3) *$g_i(\lambda)$ are continuous at all $\lambda \in \mathbb{T} \setminus \{0\}$, and for all $\delta > 0$ and $i = 1, 2, \dots, \nu$,*

$$g_i(\lambda) = O\left(|\lambda|^{-\beta-\delta}\right) \quad \text{as} \quad \lambda \to 0.$$

Then

$$\Delta I_{B,\mathbb{T},\mathcal{H}}(T) = o(1) \quad \text{as} \quad T \to \infty. \tag{4.4.15}$$

The proof of the next result can be found in Dzhaparidze [91], pages 75-76 (see also Davies [80]).

Theorem 4.4.7 *Let $f(\lambda)$ be a summable function on $\mathbb{T}$ such that $c \le f(\lambda) \le C$ with some constants $c > 0$ and $C < \infty$, and let $g(\lambda)$ be a function satisfying the condition: $g/f \in L^2(\mathbb{T})$. Then*

$$\lim_{T\to\infty} \frac{1}{T} tr\left[B_T^{-1}(f) B_T(g)\right]^2 = \frac{1}{2\pi} \int_{-\pi}^{\pi} \frac{g^2(\lambda)}{f^2(\lambda)} d\lambda. \tag{4.4.16}$$

4.4.2 Problem (B) for Toeplitz Matrices

Recall that Problem (B) involves finding conditions on the functions $h_1(\lambda)$, $h_2(\lambda), \dots, h_m(\lambda)$ in (4.4.4) such that for some $\gamma > 0$,

$$\Delta_{B,\mathbb{T},\mathcal{H}}(T) = O(T^{-\gamma}) \quad \text{as} \quad T \to \infty.$$

In Theorem 4.4.8, we summarize the results concerning Problem (B) for Toeplitz matrices, specifically for the case where $\tau_k = 1$, $k = 1, 2, \dots, m$.

First, we introduce some classes of functions (see, e.g., Taniguchi [292] and Taniguchi and Kakizawa [295]). Recall that $\mathbb{T} = [-\pi, \pi)$ and define

$$\mathcal{F}_1(\mathbb{T}) := \left\{ f \in L^1(\mathbb{T}) : \sum_{k=-\infty}^{\infty} |k||\widehat{f}(k)| < \infty; \right\}, \tag{4.4.17}$$

where $\widehat{f}(k) = \int_{\mathbb{T}} e^{-i\lambda k} f(\lambda)\, d\lambda$, $\widehat{f}(-k) = \widehat{f}(k)$, and

$$\mathcal{F}_2(\mathbb{T}) = \mathcal{F}_{\text{ARMA}}(\mathbb{T}) := \left\{ f : f(\lambda) = \frac{\sigma^2}{2\pi} \frac{|\theta_q(e^{-i\lambda})|^2}{|\psi_p(e^{-i\lambda})|^2} \right\}, \tag{4.4.18}$$

where $0 < \sigma^2 < \infty$, the polynomials $\theta_q(z) := \sum_{k=0}^{q} a_k z^k$ and $\psi_p(z) := \sum_{k=0}^{p} b_k z^k$ $(p, q \in \mathbb{N})$ are both bounded away from zero for $|z| \leq 1$.

Remark 4.4.1 The following implications were established in Taniguchi [292]:

(a) If $f_1, f_2 \in \mathcal{F}_1(\mathbb{T})$, then $f_1 f_2 \in \mathcal{F}_1(\mathbb{T})$.
(b) If $f \in \mathcal{F}_2(\mathbb{T})$, then $f \in \mathcal{F}_1(\mathbb{T})$ and $f^{-1} \in \mathcal{F}_2(\mathbb{T})$.

Theorem 4.4.8 *Assume that $\tau_k = 1$, $k = 1, 2, \ldots, m$, $\mathcal{H} = \{h_1, h_2, \ldots, h_m\}$, and let $\Delta(T) := \Delta_{B,\mathbb{T},\mathcal{H}}(T)$ be defined as in (4.4.6). The following assertions hold:*

(B1) *If $h_i \in \mathcal{F}_1(\mathbb{T})$, $i = 1, 2, \ldots, m$, then*

$$\Delta(T) = O(T^{-1}) \quad \text{as } T \to \infty.$$

(B2) *If the functions $h_i(\lambda)$, $i = 1, 2, \ldots, m$, have uniformly bounded derivatives on $\mathbb{T}$, then for any $\epsilon > 0$*

$$\Delta(T) = O(T^{-1+\epsilon}) \quad \text{as } T \to \infty.$$

(B3) *Assume that the function $\varphi(\mathbf{u})$ defined by (4.4.9) with some constants $C > 0$ and $\gamma \in (0, 1]$ satisfies*

$$|\varphi(\mathbf{u}) - \varphi(\mathbf{0})| \leq C|\mathbf{u}|^{\gamma}, \quad \mathbf{u} = (u_1, u_2, \ldots, u_{m-1}) \in \mathbb{T}^{m-1},$$

where $\mathbf{0} = (0, 0, \ldots, 0)$ and $|\mathbf{u}| = |u_1| + |u_2| + \cdots + |u_{m-1}|$. Then, for any $\varepsilon > 0$,

$$\Delta(T) = O\left(T^{-\gamma+\varepsilon}\right) \quad \text{as} \quad T \to \infty. \tag{4.4.19}$$

(B4) *Let $h_i(\lambda) \in \text{Lip}(\mathbb{T}; p_i, \gamma)$ with $p_i > 1$, $i = 1, 2, \ldots, m$, $1/p_1 + \cdots + 1/p_m \leq 1$, and $\gamma \in (0, 1]$. Then (4.4.19) holds for any $\varepsilon > 0$.*

(B5) *Let $h_i(\lambda)$, $i = 1, 2, \ldots, m$, be differentiable functions defined on $\mathbb{T} \setminus \{0\}$, such that for some constants $C_{1i} > 0$, $C_{2i} > 0$, and α_i, $i = 1, 2, \ldots, m$, satisfying*

$0 < \alpha_i < 1$, $\alpha := \sum_{i=1}^m \alpha_i < 1$, *the following inequalities hold:*

$$|h_i(\lambda)| \le C_{1i}|\lambda|^{-\alpha_i}, \quad |h_i'(\lambda)| \le C_{2i}|\lambda|^{-(\alpha_i+1)}, \quad \lambda \in \mathbb{T} \setminus \{0\}, \quad i = 1, \dots, m.$$

Then (4.4.19) holds for any $\varepsilon > 0$ with

$$\gamma = \frac{1}{m}(1-\alpha). \tag{4.4.20}$$

Remark 4.4.2 It is easy to see that under the conditions of **(B2)**, we have $h_i \in \mathrm{Lip}(\mathbb{T}; p, 1)$ for any $i = 1, 2, \dots, m$ and $p \ge 1$. Therefore, **(B4)** implies **(B2)**.

Example 4.4.1 Let $h_i(\lambda) = |\lambda|^{-\alpha_i}$, $\lambda \in \mathbb{T}$, $i = 1, 2, \dots, m$, with $0 < \alpha_i < 1$ and $\alpha := \sum_{i=1}^m \alpha_i < 1$. It is easy to see that the conditions of **(B5)** are satisfied, and hence we have (4.4.19) with γ as defined in (4.4.20).

The following results show that for the special case $m = 2$, the rates in Theorem 4.4.8 **(B4)** and **(B5)** can be significantly improved. Denote

$$\Delta_{2,B}(T) := \left| \frac{1}{T}\mathrm{tr}[B_T(h_1)B_T(h_2)] - 2\pi \int_{\mathbb{T}} h_1(\lambda)h_2(\lambda)\,d\lambda \right|. \tag{4.4.21}$$

Theorem 4.4.9 *Let $h_i(\lambda) \in \mathrm{Lip}(\mathbb{T}; p_i, \gamma_i)$ with $p_i > 1$, $1/p_1 + 1/p_2 = 1$, and $\gamma_i \in (0, 1]$, $i = 1, 2$, and let $\Delta_{2,B}(T)$ be defined as in* (4.4.21). *Then*

$$\Delta_{2,B}(T) = \begin{cases} O(T^{-(\gamma_1+\gamma_2)}), & \text{if } \gamma_1 + \gamma_2 < 1 \\ O(T^{-1}\ln T), & \text{if } \gamma_1 + \gamma_2 = 1 \\ O(T^{-1}), & \text{if } \gamma_1 + \gamma_2 > 1. \end{cases}$$

Theorem 4.4.10 *Assume that the functions $h_i(\lambda)$, $i = 1, 2$, satisfy the conditions of Theorem 4.4.8* **(B5)** *with $m = 2$, and that $\Delta_{2,B}(T)$ is defined as in* (4.4.21). *Then*

$$\Delta_{2,B}(T) = O\left(T^{-1+(\alpha_1+\alpha_2)}\right) \quad \text{as} \quad T \to \infty. \tag{4.4.22}$$

The following result, due to Taniguchi [292] (see also Taniguchi and Kakizawa [295], page 173), addresses the case when the product in (4.3.2) includes inverse matrices, specifically when $\tau_k = (-1)^k$ for $k = 1, 2, \dots, m$.

Theorem 4.4.11 *Let $SI_{B,\mathcal{H}}(T)$, $MI_{\mathbb{T},\mathcal{H}}$, and $\Delta I_{B,\mathbb{T},\mathcal{H}}(T)$ be defined as in (4.4.12)–(4.4.14), and let $\mathcal{F}_1(\mathbb{T})$ and $\mathcal{F}_2(\mathbb{T})$ be defined as in (4.4.17) and (4.4.18), respectively. If $f_i \in \mathcal{F}_2(\mathbb{T})$ and $g_i \in \mathcal{F}_1(\mathbb{T})$ for $i = 1, 2, \dots, \nu$, then*

$$\Delta I_{B,\mathbb{T},\mathcal{H}}(T) = O(T^{-1}) \quad \text{as } T \to \infty. \tag{4.4.23}$$

4.5 The Trace Problem for Toeplitz Operators

In this section, we examine Problems (A) and (B) concerning Toeplitz operators, specifically in the case where the generating functions are defined on the real line. Problem (A) involves $o(1)$ approximation, while Problem (B) deals with $O(T^{-\gamma})$ approximation with $\gamma > 0$.

Let $f(\lambda)$ be an integrable real symmetric function defined on $\mathbb{R}$ and let $W_T(f)$ be the *T-truncated Toeplitz operator* generated by $f(\lambda)$ (see 4.2.11):

$$[W_T(f)u](t) = \int_0^T \hat{f}(t-s)u(s)ds, \quad u(s) \in L^2[0,T], \tag{4.5.1}$$

where $\widehat{f}$ is the Fourier transform of f.

It follows from (4.5.1) and the formula for the traces of integral operators (see Proposition 2.7.7) that

$$\mathrm{tr}\,[W_T(f)] = \int_0^T \hat{f}(t-t)dt = T\hat{f}(0) = T\int_{-\infty}^{+\infty} f(\lambda)d\lambda. \tag{4.5.2}$$

We pose the same question as in the case of Toeplitz matrices: What happens to the relation (4.5.2) when we replace the single operator $W_T(f)$ with a product of such operators? It is important to note that the product of Toeplitz operators is not necessarily a Toeplitz operator.

The approach is similar to that used for Toeplitz matrices; we approximate the trace of the product of Toeplitz operators by the trace of a single Toeplitz operator generated by the product of their respective generating functions.

Specifically, let $\mathcal{H} = \{h_1, h_2, \ldots, h_m\}$ be a collection of integrable real symmetric functions defined on $\mathbb{R}$. Define

$$S_{W,\mathcal{H}}(T) := \frac{1}{T}\mathrm{tr}\left[\prod_{i=1}^m W_T(h_i)\right], \tag{4.5.3}$$

$$M_{\mathbb{R},\mathcal{H}} := (2\pi)^{m-1}\int_{-\infty}^{\infty}\left[\prod_{i=1}^m h_i(\lambda)\right]d\lambda, \tag{4.5.4}$$

and let

$$\Delta(T) := \Delta_{W,\mathbb{R},\mathcal{H}}(T) = |S_{W,\mathcal{H}}(T) - M_{\mathbb{R},\mathcal{H}}|. \tag{4.5.5}$$

Observe that, according to Eq. (4.5.2), we have

$$M_{\mathbb{R},\mathcal{H}} = (2\pi)^{m-1}\int_{-\infty}^{\infty}\left[\prod_{i=1}^m h_i(\lambda)\right]d\lambda = \frac{1}{T}\mathrm{tr}\left[W_T\left(\prod_{i=1}^m h_i(\lambda)\right)\right]. \tag{4.5.6}$$

How well is $S_{W,\mathcal{H}}(T)$ approximated by $M_{\mathbb{R},\mathcal{H}}$? What is the rate of convergence to zero of the approximation error $\Delta_{W,\mathbb{R},\mathcal{H}}(T)$ as $T \to \infty$? These are Problems (A) and (B) in this context.

4.5.1 Problem (A) for Toeplitz Operators

In Theorems 4.5.1–4.5.3, we summarize the results regarding Problem (A) for Toeplitz operators in the case where $\tau_k = 1$ for $k = 1, 2, \ldots, m$.

Theorem 4.5.1 *Let $\Delta(T) := \Delta_{W,\mathbb{R},\mathcal{H}}(T)$ be as defined in (4.5.5). Each of the following conditions is sufficient for*

$$\Delta(T) = o(1) \quad \text{as} \quad T \to \infty. \tag{4.5.7}$$

(A1) *The function $\varphi(\mathbf{u})$ defined by*

$$\varphi(\mathbf{u}) := \int_{-\infty}^{+\infty} h_1(\lambda)h_2(\lambda - u_1)h_3(\lambda - u_2)\cdots h_m(\lambda - u_{m-1})\, d\lambda, \tag{4.5.8}$$

where $\mathbf{u} = (u_1, u_2, \ldots, u_{m-1}) \in \mathbb{R}^{m-1}$, belongs to $L^{m-2}(\mathbb{R}^{m-1})$ and is continuous at $\mathbf{0} = (0, 0, \ldots, 0) \in \mathbb{R}^{m-1}$.

(A2) *$h_i \in L^1(\mathbb{R}) \cap L^{p_i}(\mathbb{R})$, $p_i > 1$, $i = 1, 2, \ldots, m$, with $1/p_1 + \cdots + 1/p_m \le 1$.*

Theorem 4.5.2 *Assume that $m = 4$, $h_1 = h_3 := f$, and $h_2 = h_4 := g$. The following conditions are also sufficient for (4.5.7) with $m = 4$: the functions f and g satisfy $f \in L^2(\mathbb{R})$, $g \in L^2(\mathbb{R})$, $fg \in L^2(\mathbb{R})$, and*

$$\int_{-\infty}^{+\infty} f^2(\lambda)g^2(\lambda - \mu)\, d\lambda \longrightarrow \int_{-\infty}^{+\infty} f^2(\lambda)g^2(\lambda)\, d\lambda \quad \text{as} \quad \mu \to 0. \tag{4.5.9}$$

Theorem 4.5.3 *Assume that $m = 4$, $h_1 = h_3 := f$, and $h_2 = h_4 := g$. The following conditions are also sufficient for (4.5.7) with $m = 4$: the functions f and g are integrable on $\mathbb{R}$, bounded on $\mathbb{R} \setminus (-a, a)$ with some $a > 0$, and satisfy*

$$f(\lambda) \le |\lambda|^{-\alpha} L_1(\lambda), \quad |g(\lambda)| \le |\lambda|^{-\beta} L_2(\lambda), \quad \lambda \in [-a, a], \tag{4.5.10}$$

for some $\alpha < 1$, $\beta < 1$ with $\alpha + \beta \le 1/2$, where L_1 and L_2 are slowly varying at zero functions that satisfy

$$L_i \in SV_0(\mathbb{R}), \ \lambda^{-(\alpha+\beta)} L_i(\lambda) \in L^2[-a, a],\ i = 1, 2. \tag{4.5.11}$$

4.5.2 Problem (B) for Toeplitz Operators

In Theorem 4.5.4, we summarize the results regarding Problem (B) for Toeplitz operators in the case where $\tau_k = 1$ for $k = 1, 2, \ldots, m$. Define the class of functions:

$$\mathcal{F}_1(\mathbb{R}) := \left\{ f \in L^1(\mathbb{R}) : \int_{-\infty}^{\infty} |t||\widehat{f}(t)|dt < \infty \right\}, \tag{4.5.12}$$

where $\widehat{f}(t) = \int_{\mathbb{R}} e^{-i\lambda t} f(\lambda)\, d\lambda$, $\widehat{f}(-t) = \widehat{f}(t)$.

Theorem 4.5.4 serves as the continuous version of Theorem 4.4.8.

Theorem 4.5.4 *Let $\mathcal{H} = \{h_1, h_2, \ldots, h_m\}$, and let $\Delta(T) := \Delta_{W,\mathbb{R},\mathcal{H}}(T)$ and $\varphi(\mathbf{u})$ be defined as in (4.5.5) and (4.5.8), respectively. The following assertions hold:*

(B1) *If $h_i \in \mathcal{F}_1(\mathbb{R})$, $i = 1, 2, \ldots, m$, then*

$$\Delta(T) = O(T^{-1}) \quad \text{as} \quad T \to \infty. \tag{4.5.13}$$

(B2) *Assume that $\varphi(\mathbf{u}) \in L^\infty(\mathbb{R}^{m-1})$ and with some constants $C > 0$ and $\gamma \in (0, 1]$,*

$$|\varphi(\mathbf{u}) - \varphi(\mathbf{0})| \le C|\mathbf{u}|^\gamma, \quad \mathbf{u} = (u_1, u_2, \ldots, u_{m-1}) \in \mathbb{R}^{m-1}, \tag{4.5.14}$$

where $\mathbf{0} = (0, 0, \ldots, 0)$ and $|\mathbf{u}| = |u_1| + |u_2| + \cdots + |u_{m-1}|$. Then for any $\varepsilon > 0$

$$\Delta(T) = O\left(T^{-\gamma+\varepsilon}\right) \quad \text{as} \quad T \to \infty. \tag{4.5.15}$$

(B3) *Let $h_i(\lambda) \in \mathrm{Lip}(\mathbb{R}; p_i, \gamma)$ with $p_i > 1$, $i = 1, 2, \ldots, m$, $1/p_1 + \cdots + 1/p_m \le 1$ and $\gamma \in (0, 1]$. Then (4.5.15) holds for any $\varepsilon > 0$.*

(B4) *Let $h_i(\lambda)$, $i = 1, 2, \ldots, m$, be differentiable functions defined on $\mathbb{R} \setminus \{0\}$, such that for some constants $C_i > 0$, $\sigma_i > 0$, and $\delta_i > 1$, $i = 1, 2, \ldots, m$, with $\sigma := \sum_{i=1}^m \sigma_i < 1$, the following inequalities hold:*

$$|h_i(\lambda)| \le \begin{cases} C_i|\lambda|^{-\sigma_i} & \text{if} \quad |\lambda| \le 1 \\ C_i|\lambda|^{-\delta_i} & \text{if} \quad |\lambda| > 1 \end{cases}, \tag{4.5.16}$$

$$|h_i'(\lambda)| \le \begin{cases} C_i|\lambda|^{-\sigma_i-1} & \text{if} \quad |\lambda| \le 1 \\ C_i|\lambda|^{-\delta_i-1} & \text{if} \quad |\lambda| > 1 \end{cases} \tag{4.5.17}$$

for all $i = 1, 2, \ldots, m$. *Then for any* $\varepsilon > 0$,

$$\Delta(T) = O\left(T^{-\gamma+\varepsilon}\right) \quad as \quad T \to \infty \tag{4.5.18}$$

with

$$\gamma = \frac{1}{m}(1-\sigma). \tag{4.5.19}$$

The two theorems presented below are continuous versions of Theorems 4.4.9 and 4.4.10. They demonstrate that for the special case where $m = 2$, the rates of convergence in Theorem 4.5.4 **(B3)** and **(B4)** can be significantly improved.

Denote

$$\Delta_{2,W}(T) := \left| \frac{1}{T}\mathrm{tr}[W_T(h_1)W_T(h_2)] - 2\pi \int_{\mathbb{R}} h_1(\lambda)h_2(\lambda)\, d\lambda \right|. \tag{4.5.20}$$

Theorem 4.5.5 *Let* $h_i(\lambda) \in \mathrm{Lip}(\mathbb{R}; p_i, \gamma_i)$ *with* $p_i > 1$, $1/p_1 + 1/p_2 = 1$, *and* $\gamma_i \in (0, 1]$, $i = 1, 2$, *and let* $\Delta_{2,W}(T)$ *be defined as in* (4.5.20). *Then*

$$\Delta_{2,W}(T) = \begin{cases} O(T^{-(\gamma_1+\gamma_2)}), & if \ \gamma_1 + \gamma_2 < 1 \\ O(T^{-1}\ln T), & if \ \gamma_1 + \gamma_2 = 1 \\ O(T^{-1}), & if \ \gamma_1 + \gamma_2 > 1. \end{cases}$$

Theorem 4.5.6 *Assume that the functions* $h_i(\lambda)$, $i = 1, 2$, *satisfy the conditions of Theorem 4.5.4* **(B4)** *with* $m = 2$, *and that* $\Delta_{2,W}(T)$ *is defined as in* (4.5.20). *Then*

$$\Delta_{2,W}(T) = O\left(T^{-1+(\sigma_1+\sigma_2)}\right) \quad as \quad T \to \infty. \tag{4.5.21}$$

4.6 Proofs of Theorems of Sects. 4.4 and 4.5

4.6.1 Proofs of Theorems of Sect. 4.4

The proofs of Theorems 4.4.1, 4.4.8, 4.4.9, and 4.4.10 are similar to their corresponding results for Toeplitz operators (Theorems 4.5.1–4.5.5), which we will prove in Sect. 4.6.2. For $m = 4$, the proofs of Theorems 4.4.1 **(A2)** and 4.4.2 (a) can be found in Giraitis et al. [147] (p. 68, Theorem 6.3.2). Theorems 4.4.3 and 4.4.5 were proved in Giraitis et al. [147] (p. 171, Lemma 6.3.4). The proof of Theorem 4.4.2 is provided in Fox and Taqqu [101], Theorem 1. Here, we will only prove Theorems 4.4.4 and 4.4.6.

4.6.1.1 Proof of Theorem 4.4.4

Proof of Theorem 4.4.4 For $f, g \in L^1(\mathbb{T})$ and $\mathbf{x} = (x_1, x_2, x_3, x_4)$, we define

$$F(\mathbf{x}) := f(x_1)f(x_2)g(x_3)g(x_4),$$

and

$$H_T(\mathbf{x}) := G_T(x_1 - x_3)G_T(x_2 - x_3)G_T(x_4 - x_1)G_T(x_4 - x_2),$$

where

$$G_T(u) = \sum_{k=1}^{T} e^{iku} = e^{iu(T+1)/2} \cdot D_T(u), \tag{4.6.1}$$

and $D_T(u)$ is the discrete Dirichlet kernel defined by (2.2.1). Using Proposition 2.7.7, it is easy to verify that

$$\operatorname{tr}(B_T(f)B_T(g))^2 = \int_{\mathbb{T}^4} F(\mathbf{x})H_T(\mathbf{x})d\mathbf{x}. \tag{4.6.2}$$

Thus, the result of the theorem for the case $\alpha + \beta < \frac{1}{2}$ follows from Theorem 4.4.2 (a) (see also Giraitis et al. [147], Theorem 6.3.2). Therefore, we prove the theorem in the critical case where $\alpha + \beta = \frac{1}{2}$. Specifically, we need to prove that

$$\lim_{T\to\infty} \frac{1}{T} \int_{\mathbb{T}^4} F(\mathbf{x})H_T(\mathbf{x})d\mathbf{x} = 8\pi^3 \int_{\mathbb{T}} f^2(x)g^2(x)dx, \tag{4.6.3}$$

provided that

$$f(x) \le |x|^{-\alpha}L(x), \ |g(x)| \le |x|^{-\beta}L(x), \quad x \in \mathbb{T}, \tag{4.6.4}$$

where $L = L_1 + L_2 \in SV_0(\mathbb{R})$ and

$$\alpha < 1, \ \beta < 1, \ \alpha + \beta = \frac{1}{2}, \quad \int_{\mathbb{T}} x^{-1}L^2(x)dx < \infty. \tag{4.6.5}$$

If $\alpha, \beta \ge 0$, then (4.6.4) implies $f \in L^{1/\alpha}(\mathbb{T})$ and $g \in L^{1/\beta}(\mathbb{T})$. Therefore, the result follows from Theorem 4.4.1 **(A2)**.

Assuming that $\beta < 0$, from (4.6.5), we have

$$\frac{1}{2} < \alpha < 1, \quad -\frac{1}{2} < \beta < 0. \tag{4.6.6}$$

For $\varepsilon \in (0, 1)$, we define

$$f_\varepsilon(x) := \begin{cases} 0, & \text{if } |x| < \varepsilon, \\ f(x), & \text{if } \varepsilon \le |x| \le \pi. \end{cases}$$

and

$$\mathbb{T}_{i,\varepsilon} := \left\{ \mathbf{x} \in \mathbb{T}^4 : |x_i| < \varepsilon \right\}, \quad i = 1, 2.$$

Then, we have

$$\frac{1}{T} \int_{\mathbb{T}^4} F(\mathbf{x}) H_T(\mathbf{x}) d\mathbf{x} =: J_T^1 + J_T^2,$$

where

$$J_T^1 := \frac{1}{T} \int_{\mathbb{T}^4} f_\varepsilon(x_1) f_\varepsilon(x_2) g(x_3) g(x_4) H_T(\mathbf{x}) d\mathbf{x}$$

and

$$|J_T^2| \le \frac{1}{T} \int_{\mathbb{T}_{1,\varepsilon}} |F(\mathbf{x}) H_T(\mathbf{x})| d\mathbf{x} + \frac{1}{T} \int_{\mathbb{T}_{2,\varepsilon}} |F(\mathbf{x}) H_T(\mathbf{x})| d\mathbf{x} =: I_{T,\varepsilon}^1 + I_{T,\varepsilon}^2.$$

Since $f_\varepsilon, g \in L^\infty(\mathbb{T})$, we have

$$\lim_{T\to\infty} J_T^1 = 8\pi^3 \int_{\mathbb{T}} f_\varepsilon^2(x) g^2(x) dx \to \int_{\mathbb{T}} f^2(x) g^2(x) dx \quad \text{as} \quad \varepsilon \to 0.$$

Therefore, to prove (4.6.3), it remains to show that

$$\lim_{\varepsilon\to 0,\ T\to\infty} \left(I_{T,\varepsilon}^1 + I_{T,\varepsilon}^2 \right) = 0. \tag{4.6.7}$$

We start by proving that

$$\lim_{\varepsilon\to 0,\ T\to\infty} I_{T,\varepsilon}^1 = \lim_{\varepsilon\to 0,\ T\to\infty} \frac{1}{T} \int_{\mathbb{T}_{1,\varepsilon}} |F(\mathbf{x}) H_T(\mathbf{x})| d\mathbf{x} = 0. \tag{4.6.8}$$

Define

$$B_{i,j} = \left\{ \mathbf{x} \in \mathbb{T}^4 : |x_i| \le \frac{|x_j|}{2} \right\}, \quad i = 1, 2, \ j = 3, 4,$$

$$B = \left\{ \mathbf{x} \in \mathbb{T}^4 : |x_1| < \varepsilon, \ |x_i| > \frac{|x_j|}{2}, \ i = 1, 2, \ j = 3, 4 \right\}.$$

Then, we can write

$$I^1_{T,\varepsilon} \le \frac{1}{T}\sum_{i=1}^{2}\sum_{j=3}^{4}\int_{B_{i,j}} |F(\mathbf{x})H_T(\mathbf{x})|d\mathbf{x} + \frac{1}{T}\int_B |F(\mathbf{x})H_T(\mathbf{x})|d\mathbf{x}$$

$$=: \sum_{i=1}^{2}\sum_{j=3}^{4} A_{i,j} + A. \tag{4.6.9}$$

Let $w \in SV_0(\mathbb{R})$ be a function satisfying

$$\int_{\mathbb{T}} x^{-1}L^2(x)w^{-4}(x)dx < \infty. \tag{4.6.10}$$

For $\mathbf{x} \in B_{1,3}$, we have $|x_3|/2 < |x_1 - x_3| < 2|x_3|$. Therefore, in view of (4.6.4) and Proposition 2.5.5, we can write

$$\begin{aligned}
A_{1,3} :&= \frac{1}{T}\int_{B_{1,3}} |F(\mathbf{x})H_T(\mathbf{x})|d\mathbf{x} \\
&\le Cw^4\left(\frac{1}{T}\right)\int_{B_{1,3}} |x_1|^{-\alpha}|x_2|^{-\alpha}|x_3|^{-\beta}|x_4|^{-\beta}L(x_1)L(x_2)L(x_3)L(x_4) \\
&\times |x_1 - x_3|^{-3/4}|x_2 - x_3|^{-3/4}|x_1 - x_4|^{-3/4}|x_2 - x_4|^{-3/4} \\
&\times w^{-1}(x_1 - x_3)w^{-1}(x_2 - x_3)w^{-1}(x_1 - x_4)w^{-1}(x_2 - x_4)d\mathbf{x} \\
&\le Cw^4\left(\frac{1}{T}\right)\int_{\mathbb{T}^2} |x_2|^{-\alpha}|x_4|^{-\beta}|x_2 - x_4|^{-3/4}L(x_2)L(x_4)w^{-1}(x_2 - x_4)dx_2 \\
&\times \int_{\mathbb{T}} |x_1|^{-\alpha}|x_1 - x_4|^{-3/4}L(x_1)w^{-1}(x_1 - x_4)dx_1 \\
&\times \int_{\mathbb{T}} |x_3|^{-\beta-3/4}|x_2 - x_3|^{-3/4}L(x_3)w^{-1}(x_3)w^{-1}(x_2 - x_3)dx_3dx_2dx_4.
\end{aligned}$$

Applying Proposition 2.5.3, followed by Proposition 2.5.4, we obtain

$$\begin{aligned}
A_{1,3} &\le Cw^4\left(\frac{1}{T}\right)\int_{\mathbb{T}^2} |x_2|^{-\alpha}|x_4|^{-\beta}|x_2 - x_4|^{-3/4}L(x_2)L(x_4)w^{-1}(x_2 - x_4) \\
&\times |x_4|^{-\alpha+1/4}L(x_4)w^{-1}(x_4)|x_2|^{-\beta-1/2}L(x_2)w^{-2}(x_2)dx_2dx_4 \\
&= Cw^4\left(\frac{1}{T}\right)\int_{\mathbb{T}} |x_4|^{-1/4}L^2(x_4)w^{-1}(x_4)
\end{aligned}$$

$$\times \int_{\mathbb{T}} |x_2|^{-1}|x_2 - x_4|^{-3/4} L^2(x_2) w^{-2}(x_2) w^{-1}(x_2 - x_4) dx_2 dx_4$$

$$\leq C w^4 \left(\frac{1}{T}\right) \int_{\mathbb{T}} |x_4|^{-1} L^2(x_4) w^{-4}(x_4) dx_4 = o(1), \tag{4.6.11}$$

as $T \to \infty$. The last relation in (4.6.11) follows from condition (4.6.10). Similarly, we can show that all the integrals in the first sum of (4.6.9) tend to zero as $T \to \infty$.

To estimate the last integral in (4.6.9), we use the inequalities in (4.6.4) and Proposition 2.2.2 to obtain

$$\begin{aligned}
A := &\frac{1}{T} \int_B |F(\mathbf{x}) H_T(\mathbf{x})| d\mathbf{x} \\
&\leq C T^3 \int_B |x_1|^{-\alpha} |x_2|^{-\alpha} |x_3|^{-\beta} |x_4|^{-\beta} L(x_1) L(x_2) L(x_3) L(x_4) \\
&\times \psi_T(x_1 - x_3) \psi_T(x_2 - x_3) \psi_T(x_1 - x_4) \psi_T(x_2 - x_4) d\mathbf{x} \\
&\leq C T^3 \int\limits_{(-2\varepsilon, 2\varepsilon)^2} |x_3|^{-1/2} |x_4|^{-1/2} L(x_3) L(x_4) \\
&\times \int_{\mathbb{T}} \psi_T(x_1 - x_3) \psi_T(x_1 - x_4) L(x_1) dx_1 \\
&\times \int_{\mathbb{T}} \psi_T(x_2 - x_3) \psi_T(x_2 - x_4) L(x_2) dx_2 dx_3 dx_4 \\
&\leq C T \int\limits_{(-2\varepsilon, 2\varepsilon)} |x_3|^{-1/2} L(x_3) \int_{\mathbb{T}} |x_4|^{-1/2} \psi_T^{3/2}(x_3 - x_4) L(x_4) dx_4 dx_3 \\
&\leq C \int\limits_{-2T\varepsilon}^{2T\varepsilon} |y|^{-1/2} L\left(\frac{y}{T}\right) \int_{-\infty}^{\infty} \frac{|x|^{-1/2}}{(1 + |x - y|)^{3/2}} L\left(\frac{x}{T}\right) dx dy.
\end{aligned} \tag{4.6.12}$$

Next, we show that

$$\int_{-\infty}^{\infty} \frac{|x|^{-1/2}}{(1 + |x - y|)^{3/2}} L\left(\frac{x}{T}\right) dx \leq C y^{-1/2} L\left(\frac{y}{T}\right), \quad y \in \mathbb{T}. \tag{4.6.13}$$

Indeed, for $y \in \mathbb{T}$, we have

$$\begin{aligned}
\int\limits_{|x| \leq |y|} \tfrac{|x|^{-1/2}}{(1+|x-y|)^{3/2}} L\left(\tfrac{x}{T}\right) dx &\leq C L\left(\tfrac{y}{T}\right) \textstyle\int_{\mathbb{T}} |x|^{-1/2} dx \\
&\leq C L\left(\tfrac{y}{T}\right) \leq C y^{-1/2} L\left(\tfrac{y}{T}\right).
\end{aligned} \tag{4.6.14}$$

According to Proposition 2.5.1, the function $t^{-1/2}L(t)$ is decreasing on the interval $(0, \delta)$ for some $\delta > 0$. Thus, assuming without loss of generality that $T > \pi/\delta$, we find that for $|x| > |y|$,

$$|x|^{-1/2}L\left(\frac{x}{T}\right) = T^{-1/2}\left(\frac{|x|}{T}\right)^{-1/2} L\left(\frac{x}{T}\right)$$
$$\leq T^{-1/2}\left(\frac{|y|}{T}\right)^{-1/2} L\left(\frac{y}{T}\right) = |y|^{-1/2}L\left(\frac{y}{T}\right).$$

Therefore, we have

$$\int\limits_{|x|>|y|} \frac{|x|^{-1/2}}{(1+|x-y|)^{3/2}} L\left(\frac{x}{T}\right) dx$$
$$\leq C|y|^{-1/2}L\left(\frac{y}{T}\right)\int_{-\infty}^{\infty} \frac{1}{(1+|x|)^{3/2}}dx \leq C|y|^{-1/2}L\left(\frac{y}{T}\right). \tag{4.6.15}$$

From (4.6.14) and (4.6.15), we derive (4.6.13).

Next, considering (4.6.12), (4.6.13), and (4.6.5), we conclude that as $\varepsilon \to 0$,

$$A \leq C \int\limits_{-2T\varepsilon}^{2T\varepsilon} |y|^{-1}L^2\left(\frac{y}{T}\right)dy = C\int\limits_{-2\varepsilon}^{2\varepsilon} |t|^{-1}L^2(t)dt = o(\varepsilon). \tag{4.6.16}$$

A combination of (4.6.9), (4.6.11) and (4.6.16) yields the desired relation (4.6.8). Similar arguments show that

$$\lim_{\varepsilon\to 0,\ T\to\infty} I^2_{T,\varepsilon} = \lim_{\varepsilon\to 0,\ T\to\infty} \frac{1}{T}\int_{\mathbb{T}_{2,\varepsilon}} |F(\mathbf{x})H_T(\mathbf{x})|d\mathbf{x} = 0.$$

Thus, the relation (4.6.7) and, consequently, Theorem 4.4.4 are proven. □

4.6.1.2 Proof of Theorem 4.4.6

In our proof of the theorem, we adopt the approach described in Dahlhaus's paper [74] and start by proving two lemmas.

The following lemma approximates the inverse matrix $B_T^{-1}(f)$ using the Toeplitz matrix $B_T\left(\{4\pi^2 f\}^{-1}\right)$ (see Sect. 9.1 for motivation).

Lemma 4.6.1 *Let $0 < \alpha < 1$ and let f be a function satisfying the conditions (C1) and (C2) (for $k = 0, 1$) of the theorem. Then, for each $\delta \in (0, \alpha/4)$, we have*

$$J_T := \|I - B_T^{1/2}(f)B_T\left(\{4\pi^2 f\}^{-1}\right)B_T^{1/2}(f)\|_2 = O(T^\delta), \tag{4.6.17}$$

where $\|A\|_2$ denotes the Euclidean norm of an operator A.

Proof Let $G_T(u)$ be defined as in (4.6.1). Taking into account the equality

$$\int_{-\pi}^{\pi} G_T(x-y)G_T(y-z)dy = 2\pi G_T(x-z),$$

for J_T in (4.6.17), we have

$$J_T^2 = (2\pi)^{-4}\int_{\mathbb{T}^4}\left(\frac{f(x_1)}{f(y_1)}-1\right)\left(\frac{f(x_2)}{f(y_2)}-1\right)G_T(x_1-y_1)G_T(y_1-x_2) \\ \times G_T(x_2-y_2)G_T(y_2-x_1)dx_1dx_2dy_1dy_2. \tag{4.6.18}$$

Let $L_T(x)$ be the periodic extension (with 2π period) of the function:

$$L_T(x) = \begin{cases} T & \text{for } |x| \le 1/T \\ 1/|x| & \text{for } \pi \ge |x| > 1/T. \end{cases} \tag{4.6.19}$$

Summation by parts with a constant C yields the following inequality:

$$|G_T(x)| \le CL_T(x). \tag{4.6.20}$$

Next, by condition (C2), for each $\delta \in (0,\alpha)$ and a constant C, we have

$$\left|\frac{f(x)}{f(y)}-1\right| \le C\frac{|y-x|^{1-3\delta}}{|x|^{1-\delta}}. \tag{4.6.21}$$

This may be derived by considering the cases $0 < y \le x/2$, $0 < x/2 < y \le x$, $0 < x < y \le (3x)/2$, and $0 < (3x)/2 < y$ separately (for details see Dahlhaus [74]).

A combination of (4.6.18), (4.6.20), and (4.6.21) with a constant C yields

$$J_T^2 \le C\int_{\mathbb{T}^4}|x_1|^{-1+\delta}|x_2|^{-1+\delta}L_T(x_1-y_1)^{3\delta}L_T(y_1-x_2) \\ \times L_T(x_2-y_2)^{3\delta}L_T(y_2-x_1)dx_1dx_2dy_1dy_2 \le CT^{6\delta}\ln^2 T, \tag{4.6.22}$$

and the result follows. □

Lemma 4.6.2 *Let $f(x)$ and $g(x)$ be positive symmetric functions such that there exist numbers α and β, $0 < \alpha, \beta < 1$, with $f^{-1}(x) = O(|x|^{\alpha})$ and $g(x) = O(|x|^{-\beta})$ as $x \to 0$. Then*

$$\|B_T^{-1/2}(f)B_T^{1/2}(g)\|_\infty = \|B_T^{1/2}(g)B_T^{-1/2}(f)\|_\infty = O\left(T^{\max\{(\beta-\alpha)/2),0\}}\right), \tag{4.6.23}$$

where $\|A\|_\infty$ denotes for the spectral norm of an operator A.

Proof Since the matrix $B_T(f)$ has rank T, we have

$$\begin{aligned}\|B_T^{1/2}(g)B_T^{-1/2}(f)\|_\infty^2 &= \sup_{|x|=1}\frac{x'B_T(g)x}{x'B_T(f)x}\\ &\le C\sup_{|x|=1}\frac{\int_{\mathbb{T}}|\lambda|^{-\beta}\left|\sum_{n=1}^T x_n e^{-i\lambda n}\right|^2 d\lambda}{\int_{\mathbb{T}}|\lambda|^{-\alpha}\left|\sum_{n=1}^T x_n e^{-i\lambda n}\right|^2 d\lambda}\\ &\le C\sup_{h\in\mathcal{P}_T}\frac{\int_{\mathbb{T}}|\lambda|^{-\beta}h(\lambda)d\lambda}{\int_{\mathbb{T}}|\lambda|^{-\alpha}h(\lambda)d\lambda},\end{aligned}$$

where $\mathcal{P}_T$ denotes the set of probability densities $h(\lambda)$ on $\mathbb{T}$ with $h(\lambda)\le T$. Notice that if $\beta\le\alpha$, the above expression is bounded. If $\beta>\alpha$, the supremum is attained by the function $h = T\mathbf{1}_{\{|\lambda|<1/(2T)\}}$. In this case, we can explicitly evaluate the bound, resulting in $O(T^{\beta-\alpha})$. □

Proof of Theorem 4.4.6 Let $n\le 2\nu$ and $j_1,\ldots,j_n\in\{1,\ldots,\nu\}$. By considering all combinations $\sum_{l=1}^n s_l f_{j_l}^{-1}$ and $\sum_{l=1}^n r_l g_{j_l}$ with $s_l, r_l\in\{0,1\}$, we obtain from Theorem 4.4.2 (a) that

$$\lim_{T\to\infty}\frac{1}{T}\mathrm{tr}\left[\prod_{l=1}^n B_T\left(\{4\pi^2 f_{j_l}\}^{-1}\right)B_T(g_{j_l})\right] = \frac{1}{2\pi}\int_{-\pi}^{\pi}\left[\prod_{l=1}^n\frac{g_{j_l}(\lambda)}{f_{j_l}(\lambda)}\right]d\lambda. \tag{4.6.24}$$

Since $B_T(g_j) = B_T(g_j^+) - B_T(g_j^-)$ with $g_j^+, g_j^-\ge 0$ we can assume, without loss of generality, that all g_j are non-negative. We define for $j=1,\ldots,\nu$,

$$\begin{aligned}A_j &:= B_T^{1/2}(g_j)B_T^{-1}(f_j)B_T^{1/2}(g_{j+1})\\ B_j &:= B_T^{1/2}(g_j)B_T\left(\{4\pi^2 f_j\}^{-1}\right)B_T^{1/2}(g_{j+1}),\end{aligned}$$

where we set $g_{\nu+1}=g_1$. With this notation, we have (see 4.4.12)

$$\frac{1}{T}\mathrm{tr}\left[\prod_{j=1}^{\nu}B_T^{-1}(f_j)B_T(g_j)\right] = \frac{1}{T}\mathrm{tr}\left[\prod_{j=1}^{\nu}A_j\right]. \tag{4.6.25}$$

We show that

$$D_T := \left|\frac{1}{T}\mathrm{tr}\left[\prod_{j=1}^{\nu}A_j - \prod_{j=1}^{\nu}B_j\right]\right| \to 0 \quad\text{as}\quad T\to\infty. \tag{4.6.26}$$

To this end, we write

$$
\begin{aligned}
D_T &= \left| \sum_{n=1}^{\nu} \frac{1}{T} \mathrm{tr} \left[\left(\prod_{j=1}^{n-1} B_j \right) (A_n - B_n) \left(\prod_{j=n+1}^{\nu} A_j \right) \right] \right| \\
&\leq \sum_{n=1}^{\nu} \frac{1}{T} \left\| \prod_{j=1}^{n-1} B_j \right\|_2 \|A_n - B_n\|_2 \prod_{j=n+1}^{\nu} \|A_j\|_\infty .
\end{aligned}
\tag{4.6.27}
$$

The relation (4.6.24) implies that

$$
\left\| \prod_{j=1}^{n-1} B_j \right\|_2 = O(T^{1/2}). \tag{4.6.28}
$$

Therefore, by applying Lemma 4.6.2, we obtain for all j $(j = 1, \dots, \nu)$ and $\delta > 0$

$$
\|A_j\|_\infty = O\left(T^{\max\{\beta - \alpha + 2\delta, 0\}} \right). \tag{4.6.29}
$$

Next, by Lemmas 4.6.1 and 4.6.2, we obtain for all n and $\delta > 0$ that

$$
\begin{aligned}
\|A_n - B_n\|_2 &\leq \|B_T^{1/2}(g_j) B_T^{-1/2}(f_j)\|_\infty \\
&\quad \times \|I - B_T^{1/2}(f_j) B_T\left(\{4\pi^2 f_j\}^{-1} \right) B_T^{1/2}(f_j)\|_2 \\
&\quad \times \|B_T^{-1/2}(f_j) B_T^{1/2}(g_{j+1})\|_\infty \\
&= O\left(T^{\beta - \alpha + 3\delta} \right).
\end{aligned}
\tag{4.6.30}
$$

From (4.6.27) to (4.6.30), we obtain (4.6.26). Finally, by combining (4.6.24)–(4.6.26), we conclude the proof of Theorem 4.4.6. □

4.6.2 *Proof of Theorems of Sect. 4.5*

4.6.2.1 Preliminaries: Technical Lemmas

We first present several technical lemmas. The proof of the next lemma can be found in Giraitis et al. [147], Lemma 6.3.2.

Lemma 4.6.3 *Let* $0 < \beta < 1$, $0 < \alpha < 1$, *and* $\alpha + \beta > 1$. *Then for any* $y \in \mathbb{R}$, $y \neq 0$,

$$\int_{\mathbb{R}} \frac{1}{|x|^{\alpha}|x+y|^{\beta}}dx = \frac{C}{|y|^{\alpha+\beta-1}}, \tag{4.6.31}$$

where C *is a constant depending on* α *and* β.

Denote $E_1 := \{(u_1, u_2, \ldots, u_n) \in \mathbb{R}^n : |u_i| \leq 1,\ i = 1, 2, \ldots, n\}$ and let $E_1^c := \mathbb{R}^n \setminus E_1$.

Lemma 4.6.4 *Let* $0 < \alpha \leq 1$ *and* $\frac{n}{n+1} < \beta < \frac{n+\alpha}{n+1}$. *Then, for* $i = 1, \ldots, n$, *we have*

$$B_i := \int_{E_1} \frac{|u_i|^{\alpha}}{|u_1 \cdots u_n(u_1 + \cdots + u_n)|^{\beta}} du_1 \cdots du_n < \infty. \tag{4.6.32}$$

Proof Using Lemma 4.6.3 and the notation $d\mathbf{u} = du_n du_{n-1} \cdots du_1$, we can write

$$\begin{aligned}
B_1 &\leq \int\limits_{\{|u_1|\leq 1\}} \frac{1}{|u_1|^{\beta-\alpha}} \int\limits_{\mathbb{R}^{n-2}} \frac{1}{|u_2 \cdots u_{n-1}|^{\beta}} \int\limits_{\mathbb{R}} \frac{1}{|u_n(u_1 + \cdots + u_n)|^{\beta}} d\mathbf{u} \\
&\leq C \cdot \int\limits_{\{|u_1|\leq 1\}} \frac{1}{|u_1|^{\beta-\alpha}} \int\limits_{\mathbb{R}^{n-2}} \frac{1}{|u_2 \cdots u_{n-1}|^{\beta}|u_1 + \cdots + u_{n-1}|^{2\beta-1}} du_{n-1} \cdots du_1 \\
&\leq C^2 \int\limits_{\{|u_1|\leq 1\}} \frac{1}{|u_1|^{\beta-\alpha}} \int\limits_{\mathbb{R}^{n-3}} \frac{1}{|u_2 \cdots u_{n-2}|^{\beta}|u_1 + \cdots + u_{n-2}|^{3\beta-2}} du_{n-2} \cdots du_1 \\
&\leq \cdots\cdots\cdots\cdots\cdots\cdots\cdots\cdots\cdots\cdots\cdots\cdots \\
&\leq C^{n-2} \int\limits_{\{|u_1|\leq 1\}} \frac{1}{|u_1|^{\beta-\alpha}} \int\limits_{\mathbb{R}} \frac{1}{|u_2|^{\beta}|u_1 + u_2|^{(n-1)\beta-n+2}} du_2 du_1 \\
&\leq C^{n-1} \int\limits_{\{|u_1|\leq 1\}} \frac{1}{|u_1|^{(n+1)\beta-\alpha-n+1}} du_1 < \infty,
\end{aligned}$$

yielding (4.6.32) for $i = 1$. The quantities $B_2, \ldots, B_n$ can be estimated similarly. □

Lemma 4.6.5 *Let* $\frac{n}{n+1} < \beta < 1$. *Then*

$$I := \int_{E_1^c} \frac{1}{|u_1 \cdots u_n(u_1 + \cdots + u_n)|^{\beta}} du_1 \cdots du_n < \infty. \tag{4.6.33}$$

Proof We have

$$
\begin{aligned}
I &\le \int\limits_{|u_1|>1} + \cdots + \int\limits_{|u_n|>1} \frac{1}{|u_1 \cdots u_n (u_1 + \cdots + u_n)|^\beta}\, du_1 \cdots du_n \\
&=: I_1 + \cdots + I_n.
\end{aligned} \tag{4.6.34}
$$

We estimate the integral I_1 in (4.6.34). Using Lemma 4.6.3, we can write

$$
\begin{aligned}
I_1 &\le \int\limits_{|u_1|>1} \frac{1}{|u_1|^\beta} \int\limits_{\mathbb{R}^{n-2}} \frac{1}{|u_2 \cdots u_{n-1}|^\beta} \int\limits_{\mathbb{R}} \frac{1}{|u_n(u_1 + \cdots + u_n)|^\beta}\, du_n \cdots du_1 \\
&\le C \int\limits_{|u_1|>1} \frac{1}{|u_1|^\beta} \int\limits_{\mathbb{R}^{n-2}} \frac{1}{|u_2 \cdots u_{n-1}|^\beta |u_1 + \cdots + u_{n-1}|^{2\beta-1}}\, du_{n-1} \cdots du_1 \\
&\le \cdots\cdots\cdots\cdots\cdots\cdots\cdots\cdots\cdots\cdots\cdots\cdots \\
&\le C^{n-1} \int\limits_{|u_1|>1} \frac{1}{|u_1|^{(n+1)\beta-n+1}}\, du_1 < \infty.
\end{aligned}
$$

The integrals $I_2, \ldots, I_n$ in (4.6.34) can be estimated similarly, leading to the desired result. □

Let $D_T(u)$ be the continuous Dirichlet kernel defined by (2.2.1). We define

$$
G_T(u) := \int_0^T e^{itu} dt = e^{iTu/2} D_T(u), \quad u \in \mathbb{R}, \tag{4.6.35}
$$

$$
\Phi_T(\mathbf{u}) := \frac{1}{(2\pi)^{m-1} T} \cdot D_T(u_1) \cdots D_T(u_{m-1}) D_T(u_1 + \cdots + u_{m-1}), \tag{4.6.36}
$$

$$
\Psi(\mathbf{u}) := \varphi(u_1, u_1 + u_2, \ldots, u_1 + \cdots + u_{m-1}), \tag{4.6.37}
$$

where $\mathbf{u} := (u_1, \ldots, u_{m-1}) \in \mathbb{R}^{m-1}$ and the function $\varphi(\mathbf{u})$, corresponding to the collection $\mathcal{H} = \{h_1, h_2, \ldots, h_m\}$, is defined by (4.5.8).

The next lemma includes several formulas for the trace of the product of Toeplitz operators.

Lemma 4.6.6 *Let $\mathcal{H} = \{h_1, h_2, \ldots, h_m\}$ be a collection of integrable real symmetric functions defined on $\mathbb{R}$, $\hat{h}_k$ be the Fourier transform of the function h_k*

($k = 1, \dots, m$), and $S_m(T) := S_{W,\mathcal{H}}(T)$ be defined as in (4.5.3). The following equalities hold:

$$\text{(a) } S_m(T) = \frac{1}{T}\int_0^T \cdots \int_0^T \hat{h}_1(x_1 - x_2)\hat{h}_2(x_2 - x_3)\cdots \hat{h}_m(x_m - x_1) \times dx_1 \cdots dx_m. \tag{4.6.38}$$

$$\text{(b) } S_m(T) = \frac{1}{T}\int_{\mathbb{R}^m} h_1(t_1)\dots h_m(t_m) G_T(t_1 - t_2) G_T(t_2 - t_3)\cdots \times G_T(t_m - t_1) dt_1 \dots dt_m. \tag{4.6.39}$$

$$\text{(c) } S_m(T) = (2\pi)^{m-1}\int_{\mathbb{R}^{m-1}} \Psi(\mathbf{u})\Phi_T(\mathbf{u})d\mathbf{u}. \tag{4.6.40}$$

Proof Observe that in view of Eq. (4.2.11) and Proposition 2.7.7, the product $\prod_{i=1}^m W_T(h_i)$ is an integral operator in $L^2[0, T]$ with kernel

$$K(t, s) = \int_{[0,T]^{m-1}} \hat{h}_1(t - v_1)\hat{h}_2(v_1 - v_2)\cdots \hat{h}_m(v_{m-1} - s)dv_1 \dots dv_{m-1}.$$

Therefore, the equality (4.6.38) follows from the formula for the traces of integral operators (see Proposition 2.7.7).

Next, by using the Fourier transforms of the functions h_k ($k = 1, \dots, m$) and formula (4.6.35), we can derive (4.6.39) from (4.6.38).

Finally, by making the change of variables:

$$t_1 = \lambda, \quad t_1 - t_2 = u_1, \quad t_2 - t_3 = u_2, \quad t_3 - t_4 = u_3, \dots, t_{m-1} - t_m = u_{m-1},$$

and observing that $u_1 + \cdots + u_{m-1} = t_1 - t_m$, in view of (4.6.35)–(4.6.37) and the equality

$$G_T(u_1)G_T(u_2)\cdots G_T\left(-\sum_{i=1}^{m-1} u_i\right) = D_T(u_1)D_T(u_2)\cdots D_T\left(\sum_{i=1}^{m-1} u_i\right),$$

from (4.6.39) we infer (4.6.40). □

Observe that for $m = 2$, we have

$$S_2(T) := 2\pi\int_{\mathbb{R}^2} F_T(\lambda - \mu)h_1(\lambda)h_2(\mu)\, d\lambda d\mu = \int_{-T}^T \left(1 - \frac{|t|}{T}\right)\hat{h}_1(t)\hat{h}_2(t)\, dt, \tag{4.6.41}$$

where $F_T(\cdot)$ is the Fejér kernel given by Eq. (2.2.2).

For numbers $m \in \mathbb{N}$ ($m \geq 4$) and $\delta > 0$, we define

$$\mathbb{E}_\delta := \{(u_1, \ldots, u_{m-1}) \in \mathbb{R}^{m-1} : |u_i| \leq \delta,\ i = 1, \ldots, m-1\},$$
$$\mathbb{E}_\delta^c := \mathbb{R}^{m-1} \setminus \mathbb{E}_\delta,$$

and

$$p(m) = \frac{m-2}{m-3}.$$

In the next lemma, we outline several properties of the kernel $\Phi_T(\mathbf{u})$, as defined in (4.6.36), demonstrating that $\Phi_T(\mathbf{u})$, similar to the classical Fejér kernel $F_T(u)$ (see Proposition 2.2.3), serves as an approximate identity.

Lemma 4.6.7 *The kernel $\Phi_T(\mathbf{u})$, $\mathbf{u} \in \mathbb{R}^{m-1}$, $m \geq 4$, possesses the following properties:*

$$(a) \quad \sup_T \int_{\mathbb{R}^{m-1}} |\Phi_T(\mathbf{u})|\, d\mathbf{u} = C_1 < \infty;$$

$$(b) \quad \int_{\mathbb{R}^{m-1}} \Phi_T(\mathbf{u})\, d\mathbf{u} = 1;$$

$$(c) \quad \lim_{T\to\infty} \int_{\mathbb{E}_\delta^c} |\Phi_T(\mathbf{u})|\, d\mathbf{u} = 0 \quad \textit{for any} \quad \delta > 0;$$

(d) *for any $\delta > 0$ there exists a constant $C_\delta > 0$ such that*

$$\int_{\mathbb{E}_\delta^c} |\Phi_T(\mathbf{u})|^{p(m)}\, d\mathbf{u} \leq C_\delta \quad \textit{for} \quad T > 0, \tag{4.6.42}$$

Proof The proofs of properties (a)–(c) can be found in Bentkus [26], Lemma 3.2 (see also the proof of Lemma 4.8.5). To prove property (d), first observe that for $T > 0$,

$$\int_{\mathbb{R}} |D_T(u)|^{p(m)}\, du \leq C \cdot T^{p(m)-1} \quad \text{and} \quad |D_T(u)| \leq C_\delta \quad \text{for} \quad |u| > \delta. \tag{4.6.43}$$

For $\mathbf{u} = (u_1, \ldots, u_{m-1}) \in \mathbb{R}^{m-1}$, we have

$$\begin{aligned}\int_{\mathbb{E}_\delta^c} \Phi_T^{p(m)}(\mathbf{u}) d\mathbf{u} &\leq \int\limits_{|u_1|>\delta} \Phi_T^{p(m)}(\mathbf{u}) d\mathbf{u} + \int\limits_{|u_2|>\delta} \Phi_T^{p(m)}(\mathbf{u}) d\mathbf{u} \\ &\quad + \ldots + \int\limits_{|u_{m-1}|>\delta} \Phi_T^{p(m)}(\mathbf{u}) d\mathbf{u} \\ &=: I_1 + I_2 + \cdots + I_{m-1}.\end{aligned} \tag{4.6.44}$$

It is sufficient to estimate the integral I_1 (the integrals $I_2, \dots, I_{m-1}$ can be estimated in a similar manner). We have

$$I_1 \le \int\limits_{|u_1|>\delta,\ |u_2|>\delta/(m-1)} \Phi_T^{p(m)}(\mathbf{u})d\mathbf{u} + \dots + \int\limits_{|u_1|>\delta,\ |u_{m-1}|>\delta/(m-1)} \Phi_T^{p(m)}(\mathbf{u})d\mathbf{u}$$

$$+ \int\limits_{|u_1|>\delta,\ |u_2|\le\delta/(m-1),\dots,|u_{m-1}|\le\delta/(m-1)} \Phi_T^{p(m)}(\mathbf{u})d\mathbf{u}$$

$$=: I_1^{(1)} + \dots + I_1^{(m-2)} + I_1^{(m-1)}. \tag{4.6.45}$$

According to (4.6.36) and (4.6.43), we have

$$I_1^{(1)} \le C_\delta \cdot \frac{1}{T^{p(m)}} \cdot \int\limits_{|u_2|>\delta/(m-1)} |D_T(u_2)|^{p(m)} \dots |D_T(u_{m-1})|^{p(m)}$$

$$\times |D_T(u_1 + \dots + u_{m-1})|^{p(m)}\, du_1 du_{m-1} \dots du_2$$

$$\le C_\delta \cdot \frac{1}{T^{p(m)}} \cdot T^{(p(m)-1)(m-2)} \int\limits_{|u_2|>\delta/(m-1)} \frac{1}{|u_2|^{p(m)}}\, du_2 \le C_\delta. \tag{4.6.46}$$

Likewise, we have

$$I_1^{(j)} \le C_\delta, \qquad j = 2, \dots, m-2. \tag{4.6.47}$$

Next, observe that in the integral $I_1^{(m-1)}$, we have $|u_1 + \dots + u_{m-1}| > \delta/(m-1)$. Hence, by (4.6.43),

$$I_1^{(m-1)} \le C_\delta \cdot \frac{1}{T^{p(m)}} \int\limits_{|u_1|>\delta} D_T^{p(m)}(u_1) \cdots D_n^{p(m)}(u_{m-1}) du_2 \dots du_{m-1} du_1$$

$$\le C_\delta \int\limits_{|u_1|>\delta} \frac{1}{|u_1|^{p(m)}}\, du_1 \le C_\delta. \tag{4.6.48}$$

From (4.6.44) to (4.6.48), we obtain (4.6.42). Lemma 4.6.7 is proved. □

Lemma 4.6.8 *If the function $\Psi(\mathbf{u})$ belongs to $L^1(\mathbb{R}^{m-1}) \cap L^{m-2}(\mathbb{R}^{m-1})$ and is continuous at the point $\mathbf{0} = (0, \dots, 0) \in \mathbb{R}^{m-1}$, then*

$$\lim_{T\to\infty} \int_{\mathbb{R}^{m-1}} \Psi(\mathbf{u})\Phi_T(\mathbf{u})d\mathbf{u} = \Psi(\mathbf{0}), \tag{4.6.49}$$

where $\mathbf{u} = (u_1, u_2, \dots, u_{m-1}) \in \mathbb{R}^{m-1}$ and $\Phi_T(\mathbf{u})$ is defined by (4.6.36).

Proof By Lemmas 4.6.6 (3) and 4.6.7 (b), we have

$$R_T := \int_{\mathbb{R}^{m-1}} \Psi(\mathbf{u})\Phi_T(\mathbf{u})d\mathbf{u} - \Psi(\mathbf{0}) = \int_{\mathbb{R}^{m-1}} [\Psi(\mathbf{u}) - \Psi(\mathbf{0})]\Phi_T(\mathbf{u})d\mathbf{u}. \tag{4.6.50}$$

For any $\varepsilon > 0$, we can find $\delta > 0$ to satisfy

$$|\Psi(\mathbf{u}) - \Psi(\mathbf{0})| < \frac{\varepsilon}{C_1}, \quad \mathbf{u} \in \mathbb{E}_\delta, \tag{4.6.51}$$

where C_1 is the constant from Lemma 4.6.7 (a). Consider the decomposition $\Psi = \Psi_1 + \Psi_2$ such that

$$\|\Psi_1\|_{(m-2)} \le \frac{\varepsilon}{\sqrt{C_\delta}} \quad \text{and} \quad \|\Psi_2\|_\infty < \infty, \tag{4.6.52}$$

where C_δ is as in Lemma 4.6.7 (d).

Observe that $\frac{1}{m-2} + \frac{1}{p(m)} = 1$, where $p(m) = \frac{m-2}{m-3}$. Therefore, by applying Lemma 4.6.7, Hölder's inequality, and utilizing formulas (4.6.51) and (4.6.52), for sufficiently large T, we obtain

$$\begin{aligned} |R(T)| &\le \int_{\mathbb{E}_\delta} |\Psi(\mathbf{u}) - \Psi(\mathbf{0})||\Phi_T(\mathbf{u})|d\mathbf{u} + \int_{\mathbb{E}_\delta^c} |\Psi_1(\mathbf{u})||\Phi_T(\mathbf{u})|d\mathbf{u} \\ &+ \int_{\mathbb{E}_\delta^c} |\Psi_2(\mathbf{u}) - \Psi(\mathbf{0})||\Phi_T(\mathbf{u})|d\mathbf{u} \le \frac{\varepsilon}{C_1}\int_{\mathbb{E}_\delta} |\Phi_T(\mathbf{u})|d\mathbf{u} \\ &+ \|\Psi_1\|_{(m-2)}\left[\int_{\mathbb{E}_\delta^c} \Phi_T^{p(m)}(\mathbf{u})d\mathbf{u}\right]^{1/p(m)} + C_2\int_{\mathbb{E}_\delta^c} |\Phi_T(\mathbf{u})|d\mathbf{u} \le 3\,\varepsilon. \end{aligned}$$

This, combined with (4.6.50), yields (4.6.49). Lemma 4.6.8 is proved. □

We define

$$\mu_T(A) := \frac{1}{T}\int_A G_T(u_1 - u_2)G_T(u_2 - u_3)\cdots G_T(u_m - u_1)du_1 \ldots du_m, \tag{4.6.53}$$

where $G_T(\cdot)$ is defined as in (4.6.35). Note that μ_T is a signed measure on the Borel sets of $\mathbb{R}^m$.

Let $\mathbb{C}_{loc}(\mathbb{R}^m)$ be the space of continuous functions on $\mathbb{R}^m$ with bounded support.

Lemma 4.6.9 *If $f \in \mathbb{C}_{loc}(\mathbb{R}^m)$, then*

$$\lim_{T\to\infty} \int_{\mathbb{R}^m} f(\mathbf{u})d\mu_T = (2\pi)^{m-1}\int_{\mathbb{R}} f(t, \ldots, t)dt. \tag{4.6.54}$$

Proof Making the change of variables:

$$u_1 = t, \quad u_1 - u_2 = t_1, \quad u_2 - u_3 = t_2, \quad u_3 - u_4 = t_3, \dots, u_{m-1} - u_m = t_{m-1},$$

and observing that $t_1 + \cdots + t_{m-1} = u_1 - u_m$, in view of (4.6.53) and Lemma 4.6.6 (b) and (c), we have

$$\int_{\mathbb{R}^m} f(\mathbf{u}) d\mu_T = (2\pi)^{m-1} \int_{\mathbb{R}^{m-1}} \Psi(\mathbf{t}) \Phi_T(\mathbf{t}) d\mathbf{t}, \tag{4.6.55}$$

where

$$\Psi(\mathbf{t}) = \int_{\mathbb{R}} f\left(t, t - t_1, t - t_1 - t_2, \dots t - \sum_{i=1}^{m-1} t_i\right) dt.$$

It is not difficult to verify that the function $\Psi((\mathbf{t})$ ($\mathbf{t} \in \mathbb{R}^{m-1}$) satisfies the conditions of Lemma 4.6.8, and

$$\lim_{\mathbf{t} \to \mathbf{0}} \Psi(\mathbf{t}) = \int_{\mathbb{R}} f(t, \dots, t) dt. \tag{4.6.56}$$

Therefore, by applying Lemma 4.6.8 along with Eqs. (4.6.55) and (4.6.56), we obtain (4.6.54). Lemma 4.6.9 is proved. □

Lemma 4.6.10 *If $f \in L^1(\mathbb{R})$, then the following inequalities hold:*

$$\int_{\mathbb{R}^m} |f(x_i)| d|\mu_T| \le C_1 \|f\|_{L^1(\mathbb{R})}, \quad i = 1, \dots, m, \tag{4.6.57}$$

$$\int_{\mathbb{R}^m} |f(x_i) f(x_j)| d|\mu_T| \le C_2 \|f\|^2_{L^2(\mathbb{R})}, \quad i, j = 1, \dots, m, \ i \ne j. \tag{4.6.58}$$

where C_1 and C_2 are absolute constants, and μ_T is defined as in (4.6.53).

Proof For simplicity, we prove the lemma in the common special case where $m = 4$. In the general case, the result can be derived using an induction argument. By Proposition 2.2.1 (b) and formula (4.6.35), for $T > 0$, we have

$$|G_T(x)| \le CT \psi_T(x), \quad \text{where} \quad \psi_T(x) = \frac{1}{1 + T|x|}, \quad x \in \mathbb{R}. \tag{4.6.59}$$

We first show that for $u, v \in \mathbb{R}$,

$$T \int_{\mathbb{R}} \psi_T(x + u) \psi_T(x + v) dx \le C_\delta \psi_T^{1-\delta}(u - v), \ \ \delta > 0. \tag{4.6.60}$$

Indeed, note that the inequality (4.6.60) is equivalent to the following:

$$\int_{\mathbb{R}} \frac{1}{(1+|x|)(1+|x+t|)}\,dx \le C_\delta \frac{1}{(1+|t|)^{1-\delta}}, \qquad t\in\mathbb{R}. \tag{4.6.61}$$

Denoting by $J(t)$ the left-hand side of (4.6.61), we can write

$$\begin{aligned} J(t) &= \int_{|x|<2|t|} \frac{1}{(1+|x|)(1+|x+t|)}\,dx + \int_{|x|\ge 2|t|} \frac{1}{(1+|x|)(1+|x+t|)}\,dx \\ &\le \frac{C}{1+|t|}\int_{|x|<2|t|}\left[\frac{1}{1+|x|}+\frac{1}{1+|x+t|}\right]dx + C\int_{|x|\ge 2|t|}\frac{1}{(1+|x|)^2}\,dx \\ &\le C\frac{\ln(1+|t|)}{1+|t|} + C\frac{1}{1+|t|} \le C_\delta\frac{1}{(1+|t|)^{1-\delta}}, \qquad \delta>0, \end{aligned}$$

yielding (4.6.61).

To prove (4.6.57) for $i=1$ (say), we apply (4.6.59) and the inequality (4.6.60) with $\delta=1/4$ to obtain

$$\begin{aligned} \int_{\mathbb{R}^4} |f(x_1)|d|\mu_T| &\le CT^3\int_{\mathbb{R}^4} |f(x_1)|\psi_T(x_1-x_3)\psi_T(x_3-x_2) \\ &\quad\times \psi_T(x_4-x_1)\psi_T(x_2-x_4)dx_1dx_2dx_3dx_4 \\ &\le CT\int_{\mathbb{R}}|f(x_1)|\int_{\mathbb{R}}\psi_T^{3/2}(x_1-x_2)dx_2dx_1 \le C_1\|f\|_{L^1(\mathbb{R})}. \end{aligned}$$

This proves (4.6.57).

To prove (4.6.58) for $i=1, j=2$ (say), we use (4.6.59), (4.6.60), and the Cauchy-Schwarz inequality to obtain

$$\begin{aligned} &\int_{\mathbb{R}^4} |f(x_1)f(x_2)|d|\mu_T| \\ &\quad\le CT^3\int_{\mathbb{R}^4}|f(x_1)f(x_2)|\psi_T(x_1-x_3)\psi_T(x_3-x_2) \\ &\quad\times \psi_T(x_4-x_1)\psi_T(x_2-x_4)dx_1dx_2dx_3dx_4 \\ &\quad\le CT\int_{\mathbb{R}^2}|f(x_1)f(x_2)|\psi_T^{3/2}(x_1-x_2)dx_1dx_2 \\ &\quad\le C\left\{T\int_{\mathbb{R}^2} f^2(x_1)\psi_T^{3/2}(x_1-x_2)dx_1dx_2\right\}^{1/2} \\ &\quad+\left\{T\int_{\mathbb{R}^2} f^2(x_2)\psi_T^{3/2}(x_1-x_2)dx_1dx_2\right\}^{1/2} \le C_2\int_{\mathbb{R}} f^2(x)dx. \end{aligned}$$

Lemma 4.6.10 is proved. □

Lemma 4.6.11 *Let* $f(\mathbf{u}) = \prod_{i=1}^{m} f_i(u_i)$ *with* $\mathbf{u} = (u_1, u_2, \dots, u_m)$*, where* $f_i \in L^1(\mathbb{R}) \cap L^\infty(\mathbb{R})$ *for* $i = 1, 2, \dots, m$*. Then the relation (4.6.54) holds.*

Proof Suppose $\|f_i\|_\infty \le M < \infty$, $i = 1, 2, \dots, m$. Using Proposition 2.1.4, for any $\varepsilon > 0$, we can find functions h_i and g_i, $i = 1, 2, \dots, m$, that satisfy

$$f_i = g_i + h_i, \quad g_i \in \mathbb{C}_{loc}(\mathbb{R}), \quad \|h_i\|_{L^1(\mathbb{R})} \le \varepsilon, \quad \|g_i\|_{\mathbb{C}} \le M. \tag{4.6.62}$$

Therefore, we have

$$\int_{\mathbb{R}^m} f d\mu_T = \int_{\mathbb{R}^m} \prod_{i=1}^{m} (g_i + h_i) d\mu_T = \int_{\mathbb{R}^m} \prod_{i=1}^{m} g_i d\mu_T + I_T. \tag{4.6.63}$$

By (4.6.62) and Lemma 4.6.10, we obtain

$$\begin{aligned} |I_T| &\le \sum_{j=1}^{m} \int_{\mathbb{R}^m} |h_j| \prod_{i=1, i\ne j}^{m} (|g_i| + |h_i|) d|\mu_T| \\ &\le C_M \sum_{j=1}^{m} \int_{\mathbb{R}^m} |h_j| d|\mu_T| \le C_M \sum_{j=1}^{m} \|h_j\|_{L^1(\mathbb{R})} \le C_M \varepsilon. \end{aligned} \tag{4.6.64}$$

Next, by Lemma 4.6.9, we have

$$\begin{aligned} \lim_{T\to\infty} \int_{\mathbb{R}^m} \prod_{i=1}^{m} g_i(u_i) d\mu_T &= \int_{\mathbb{R}} \prod_{i=1}^{m} g_i(u) du \\ &= \int_{\mathbb{R}} \prod_{i=1}^{m} [f_i(u) - h_i(u)]\, du = \int_{\mathbb{R}} f(u, \dots, u) du + J, \end{aligned} \tag{4.6.65}$$

where

$$|J| \le \sum_{j=1}^{m} \int_{\mathbb{R}} |h_j(u)| \prod_{i=1, i\ne j}^{m} (|f_i(u)| + |g_i(u)|) du \le C_M \varepsilon. \tag{4.6.66}$$

From (4.6.63) to (4.6.66), we derive (4.6.54). Lemma 4.6.11 is proved. □

4.6.2.2 Proofs of Theorems 4.5.1–4.5.5

Proof of Theorem 4.5.1 We begin with assertion **(A1)** of the theorem. By Lemmas 4.6.6 (3) and 4.6.7 (b), we have

$$\Delta(T) = (2\pi)^{m-1}|R(T)|,$$

where

$$R(T) = \int_{\mathbb{R}^{m-1}} [\Psi(\mathbf{u}) - \Psi(\mathbf{0})]\Phi_T(\mathbf{u})d\mathbf{u}.$$

It follows from Eq. (4.6.37) that the function $\Psi(\mathbf{u})$ belongs to $L^{m-2}(\mathbb{R}^{m-1})$ and is continuous at the point $\mathbf{0} = (0, \ldots, 0) \in \mathbb{R}^{m-1}$. Therefore, the result follows from Lemma 4.6.8. □

Proof of (A2) According to part **(A1)** of the theorem, it is sufficient to prove that the function

$$\varphi(\mathbf{u}) := \int_{-\infty}^{+\infty} h_1(\lambda)h_2(\lambda - u_1)h_3(\lambda - u_2) \cdots h_m(\lambda - u_{m-1})\, d\lambda, \qquad (4.6.67)$$

where $\mathbf{u} = (u_1, \ldots, u_{m-1}) \in \mathbb{R}^{m-1}$, belongs to $L^{m-2}(\mathbb{R}^{m-1})$ and is continuous at the point $\mathbf{0} = (0, \ldots, 0) \in \mathbb{R}^{m-1}$, provided that

$$h_i \in L^1(\mathbb{R}) \cap L^{p_i}(\mathbb{R}), \quad 1 \le p_i \le \infty, \quad i = 1, \ldots, m, \quad \sum_{i=1}^{m} \frac{1}{p_i} \le 1. \qquad (4.6.68)$$

It follows from Hölder's inequality, and formulas (4.6.67) and (4.6.68), that

$$|\varphi(\mathbf{u})| \le \prod_{i=1}^{m} ||f_i||_{L^{p_i}(\mathbb{R})} < \infty, \quad \mathbf{u} \in \mathbb{R}^{m-1}.$$

Therefore, $\varphi \in L^\infty(\mathbb{R}^{m-1})$. On the other hand, the condition $h_i \in L^1(\mathbb{R})$ and Eq. (4.6.67) imply that $\varphi \in L^1(\mathbb{R}^{m-1})$. Thus, by the L^p-interpolation theorem (Proposition 2.1.1), we conclude that $\varphi \in L^{m-2}(\mathbb{R}^{m-1})$.

To prove the continuity of the function $\varphi(\mathbf{u})$ at the point $\mathbf{0} = (0, \ldots, 0) \in \mathbb{R}^{m-1}$, we will consider three cases.

Case 1 $p_i < \infty$, $i = 1, \ldots, m$.
For an arbitrary $\varepsilon > 0$, we can find $\delta > 0$ that satisfies the condition (see (4.6.68)):

$$\|h_i(\lambda - u) - h_i(\lambda)\|_{L^{p_i}(\mathbb{R})} \le \varepsilon, \ i = 2, \ldots, m \quad \text{if} \quad |u| \le \delta. \qquad (4.6.69)$$

We fix $\mathbf{u} = (u_1, \dots, u_{m-1})$ with $|\mathbf{u}| < \delta$ and denote

$$\overline{h}_i(\lambda) := h_i(\lambda - u_{i-1}) - h_i(\lambda), \quad i = 2, \dots, m.$$

Then, in view of (4.6.67), we have

$$\varphi(\mathbf{u}) = \int_{\mathbb{R}} h_1(\lambda) \prod_{i=2}^{m-1} \left(\overline{h}_i(\lambda) + h_i(\lambda)\right) d\lambda = \varphi(\mathbf{0}) + W.$$

It follows from (4.6.69) that $\|\overline{h}_i\|_{p_i} \le \varepsilon$, $i = 2, \dots, m$. Observe that each integral comprising W contains at least one function $\overline{h}_i$ and can be estimated as follows:

$$\left| \int_{\mathbb{R}} h_1(u)\overline{h}_2(\lambda)h_3(\lambda)\dots h_{m-1}(\lambda)d\lambda \right|$$
$$\le \|h_1\|_{L^{p_1}} \|\overline{h}_2\|_{L^{p_2}} \|h_3\|_{L^{p_3}} \dots \|h_m\|_{L^{p_m}} \le C\varepsilon.$$

Case 2 $p_i \le \infty$, $i = 1, \dots, m$ and $\sum_{i=1}^{m} \frac{1}{p_i} < 1$.
There exist finite numbers $p_i' < p_i$, $i = 1, \dots, m$, such that $\sum_{i=1}^{m} 1/p_i' \le 1$. Hence, according to (4.6.68), we have $h_i \in L^{p_i'}$ $i = 1, \dots, m$, and φ is continuous at the point $\mathbf{0}$, as in Case 1.

Case 3 $p_i \le \infty$, $i = 1, \dots, m$ and $\sum_{i=1}^{m} \frac{1}{p_i} = 1$.
Note that at least one of the numbers p_i is finite. Without loss of generality, assume that $p_1 < \infty$. Then, for any $\varepsilon > 0$, we can find functions h_1', h_1'' such that

$$h_1 = h_1' + h_1'', \quad h_1' \in L^\infty(\mathbb{R}), \quad \|h_1''\|_{L^{p_1}} < \varepsilon. \tag{4.6.70}$$

Therefore,

$$\varphi(\mathbf{u}) = \varphi'(\mathbf{u}) + \varphi''(\mathbf{u}),$$

where the functions φ' and φ'' are defined by Eq. (4.6.67), with h_1 replaced by h_1' and h_1'', respectively. It follows from (4.6.70) that the function φ' is continuous at the point $\mathbf{0}$ (see Case 2), while by Hölder's inequality $|\varphi''(\mathbf{u})| \le C \cdot \varepsilon$. Therefore, for sufficiently small $|\mathbf{u}|$, we obtain

$$|\varphi(\mathbf{u}) - \varphi(\mathbf{0})| \le |\varphi'(\mathbf{u}) - \varphi'(\mathbf{0})| + |\varphi''(\mathbf{u}) - \varphi''(\mathbf{0})| \le (C+1)\varepsilon,$$

and the result follows. Theorem 4.5.1 is proved.

□

Remark 4.6.1 Observe that we have proven that, under the conditions stated in (4.6.68), the following relation holds:

$$\lim_{T\to\infty}\int_{\mathbb{R}^m}\left[\prod_{i=1}^m h_i(u_i)\right]d\mu_T=(2\pi)^{m-1}\int_{\mathbb{R}}\left[\prod_{i=1}^m h_i(u)\right]du,$$

where μ_T is defined as in (4.6.53).

Proof of Theorem 4.5.2 According to Theorem 4.5.1 (**A2**), it is sufficient to prove that for $\mathbf{t}=(t_1,t_2,t_3)\in\mathbb{R}^3$, the function

$$\varphi(\mathbf{t}):=\int_{\mathbb{R}}f(u)g(u-t_1)f(u-t_2)g(u-t_3)du \tag{4.6.71}$$

belongs to $L^2(\mathbb{R}^3)$ and is continuous at the point $\mathbf{0}=(0,0,0)$, provided that f and g satisfy the conditions of the theorem, specifically $f\in L^2(\mathbb{R})$, $g\in L^2(\mathbb{R})$, $fg\in L^2(\mathbb{R})$, and the relation (4.5.9) holds. Since

$$\varphi^2(\mathbf{t})=\int_{\mathbb{R}^2}f(u)f(v)g(u-t_1)g(v-t_1)f(u-t_2)f(v-t_2)g(u-t_3)g(v-t_3)dudv,$$

we have

$$\begin{aligned}\int_{\mathbb{R}^3}\varphi^2(\mathbf{t})\,d\mathbf{t}&=\int_{\mathbb{R}^2}\left[\int_{\mathbb{R}}g(u-t_1)g(v-t_1)\,dt_1\int_{\mathbb{R}}f(u-t_2)f(v-t_2)\,dt_2\right.\\&\quad\times\left.\int_{\mathbb{R}}g(u-t_3)g(v-t_3)\,dt_3\right]f(u)f(v)dudv\\&\le\|f\|^2_{L^2(\mathbb{R})}\|g\|^4_{L^2(\mathbb{R})}\int_{\mathbb{R}^2}f(u)f(v)dudv<\infty.\end{aligned}$$

Now we will prove the continuity of $\varphi(\mathbf{t})$ at the point $(0,0,0)$. Let ε be an arbitrary positive number. Denote

$$E_K=\{u\in\mathbb{R}:|f(u)|\le K\},\quad f_1(u)=\mathbf{1}_{E_K}(u)f(u),\quad f_2(u)=f(u)-f_1(u),$$

where $\mathbf{1}_{E_K}(u)$ is the indicator function of the set E_K, and the number $K>0$ is chosen to satisfy

$$\int_{\mathbb{R}\setminus E_k}f^2(u)g^2(u)du\le\varepsilon.$$

Then, we have

$$f=f_1+f_2,\quad\|f_1\|_\infty\le K,\quad\int_{\mathbb{R}}f_2^2(u)g^2(u)du\le\varepsilon. \tag{4.6.72}$$

The function $\varphi(\mathbf{t})$ in Eq. (4.6.71) can be represented as follows:

$$\begin{aligned}\varphi(\mathbf{t}) &= \int_{\mathbb{R}} f_1(u)g(u-t_1)f_1(u-t_2)g(u-t_3)du \\ &+ \int_{\mathbb{R}} f_2(u)g(u-t_1)f(u-t_2)g(u-t_3)du \\ &+ \int_{\mathbb{R}} f_1(u)g(u-t_1)f_2(u-t_2)g(u-t_3)du \\ &=: \varphi_1(\mathbf{t}) + \varphi_2(\mathbf{t}) + \varphi_3(\mathbf{t}).\end{aligned} \tag{4.6.73}$$

Now we estimate the functions $\varphi_k(\mathbf{t})$ for $k = 1, 2, 3$ separately. First, we have

$$\begin{aligned}\varphi_1(\mathbf{t}) &= \int_{\mathbb{R}} f_1(u)g(u-t_1)f_1(u-t_2)\left[g(u-t_3) - g(u)\right]du \\ &+ \int_{\mathbb{R}} f_1(u)g(u)f_1(u-t_2)\left[g(u-t_1) - g(u)\right]du \\ &+ \int_{\mathbb{R}} f_1(u)g^2(u)f_1(u-t_2)du =: I_1 + I_2 + I_3.\end{aligned} \tag{4.6.74}$$

Using Hölder's inequality, we can derive from (4.6.72) that

$$|I_1| \le K^2\|g\|_2 \cdot \|g(u+t_3) - g(u)\|_2 = o(1) \quad \text{as} \quad t_3 \to 0. \tag{4.6.75}$$

Similarly, we obtain

$$|I_2| = o(1) \quad \text{as} \quad t_1 \to 0. \tag{4.6.76}$$

Next, from (4.6.72), we have

$$\begin{aligned}|I_3 - \varphi(0,0,0)| &= \left|\int_{\mathbb{R}} f_1(u+t_2)g^2(u+t_2)f_1(u)du - \int_{\mathbb{R}} f_1^2(u)g^2(u)du\right| \\ &+ \left|\int_{\mathbb{R}} f_2^2(u)g^2(u)du\right| \\ &\le K\left\| f_1(u+t_2)g^2(u+t_2) - f_1(u)g_1^2(u)\right\|_1 + \varepsilon = o(1) + \varepsilon\end{aligned} \tag{4.6.77}$$

as $t_2 \to 0$. From (4.6.74) to (4.6.77), for sufficiently small $|\mathbf{t}|$, we obtain

$$|\varphi_1(\mathbf{t}) - \varphi(0,0,0)| \le 2\varepsilon. \tag{4.6.78}$$

Now we estimate the function $\varphi_2(\mathbf{t})$. We have

$$\begin{aligned}|\varphi_2(\mathbf{t})|^2 &\le \int_{\mathbb{R}} f_2^2(u)g^2(u-t_1)du \int_{\mathbb{R}} f_2^2(u-t_2)g^2(u-t_3)du \\ &= \left|\int_{\mathbb{R}} f^2(u)g^2(u-t_1)du - \int_{\mathbb{R}} f_1^2(u)g^2(u-t_1)du\right| \\ &\times \int_{\mathbb{R}} f^2(u)g^2(u+t_2-t_3)du \\ &\to \left|\int_{\mathbb{R}} f^2(u)g^2(u)du - \int_{\mathbb{R}} f_1^2(u)g^2(u)du\right| \int_{\mathbb{R}} f^2(u)g^2(u)du\end{aligned}$$

as $|\mathbf{t}| \to 0$. Therefore, in view of (4.6.72), for sufficiently small values of $|\mathbf{t}|$,

$$|\varphi_2(\mathbf{t})|^2 \le \varepsilon \int_{\mathbb{R}} f^2(u)g^2(u)du. \tag{4.6.79}$$

Similarly, we can prove that for sufficiently small values of $|\mathbf{t}|$,

$$|\varphi_3(\mathbf{t})|^2 \le \varepsilon \int_{\mathbb{R}} f^2(u)g^2(u)du. \tag{4.6.80}$$

A combination of (4.6.73) and (4.6.78)–(4.6.80) yields

$$\lim_{\mathbf{t}\to\mathbf{0}} \varphi(\mathbf{t}) = \varphi(0, 0, 0).$$

This completes the proof of the theorem. □

Proof of Theorem 4.5.3 In light of (4.6.39) and (4.6.53), we need to prove that the conditions in (4.5.10) and (4.5.11) imply

$$\lim_{T\to\infty} \frac{1}{T} \int_{\mathbb{R}^4} f(x_1)f(x_2)g(x_3)g(x_4)d\mu_T = 8\pi^3 \int_{\mathbb{R}} f^2(x)g^2(x)dx. \tag{4.6.81}$$

First, if $\alpha, \beta \ge 0$, then (4.5.10) and (4.5.11) imply that $f \in L^{1/\alpha}(\mathbb{R})$ and $g \in L^{1/\beta}(\mathbb{R})$. Therefore, the result follows from Theorem 4.5.1 **(A2)**. Assuming that $\beta < 0$, from (4.5.10), we have $g \in L^{\infty}(\mathbb{R})$. We define

$$\overline{f}(x) := \begin{cases} 0, & \text{if } x \in [-a/2, a/2] \\ f(x), & \text{otherwise} \end{cases}, \qquad \overline{g}(x) := \begin{cases} 0, & \text{if } x \in [-a, a] \\ g(x), & \text{otherwise} \end{cases},$$

where a is an arbitrary positive number. Let $\underline{f} := f - \overline{f}$ and $\underline{g} := g - \overline{g}$.

Then, we have

$$\begin{aligned}\frac{1}{T}\int_{\mathbb{R}^4} f(x_1)f(x_2)g(x_3)g(x_4)d\mu_T &= \frac{1}{T}\int_{\mathbb{R}^4} \overline{f}(x_1)f(x_2)g(x_3)g(x_4)d\mu_T\\ &+\frac{1}{T}\int_{\mathbb{R}^4} \underline{f}(x_1)\overline{f}(x_2)g(x_3)g(x_4)d\mu_T\\ &+\frac{1}{T}\int_{\mathbb{R}^4} \underline{f}(x_1)\underline{f}(x_2)g(x_3)g(x_4)d\mu_T\\ &=: I_T^1+I_T^2+I_T^3. \end{aligned} \tag{4.6.82}$$

Since $\overline{f},\ g \in L^\infty(\mathbb{R})$ and $f \in L^1(\mathbb{R})$ (see also Remark 4.6.1) for the integrals I_T^1 and I_T^2, we obtain

$$\begin{aligned}\lim_{T\to\infty} I_T^1 &= 8\pi^3\int_{\mathbb{R}} \overline{f}(x)f(x)g^2(x)dx = 8\pi^3\int_{|x|>a/2} f^2(x)g^2(x)dx,\\ \lim_{T\to\infty} I_T^2 &= 8\pi^3\int_{\mathbb{R}} \underline{f}(x)\overline{f}(x)g^2(x)dx = 0. \end{aligned} \tag{4.6.83}$$

Next, for the integral I_T^3, we have

$$\begin{aligned} I_T^3 &= \frac{1}{T}\int_{\mathbb{R}^4} \underline{f}(x_1)\underline{f}(x_2)\underline{g}(x_3)\underline{g}(x_4)d\mu_T + \frac{1}{T}\int_{\mathbb{R}^4} \underline{f}(x_1)\underline{f}(x_2)\underline{g}(x_3)\overline{g}(x_4)d\mu_T\\ &+\frac{1}{T}\int_{\mathbb{R}^4} \underline{f}(x_1)\underline{f}(x_2)\overline{g}(x_3)\underline{g}(x_4)d\mu_T\\ &+\frac{1}{T}\int_{\mathbb{R}^4} \underline{f}(x_1)\underline{f}(x_2)\overline{g}(x_3)\overline{g}(x_4)d\mu_T =: \sum_{i=1}^4 J_T^i. \end{aligned} \tag{4.6.84}$$

Now we examine the integrals J_T^i, $i = 1, 2, 3, 4$. First, let us consider J_T^1:

$$J_T^1 = \frac{1}{T}\int_{[-a,a]^4} \underline{f}(x_1)\underline{f}(x_2)\underline{g}(x_3)\underline{g}(x_4)d\mu_T.$$

Arguments similar to those that led to equality (4.6.3) can be used to prove that

$$\lim_{T\to\infty} J_T^1 = 8\pi^3\int_{-a/2}^{a/2} f^2(x)g^2(x)dx. \tag{4.6.85}$$

Since $f(x_1)f(x_2) \in L^1(\mathbb{R}^2)$, for any $\varepsilon > 0$, we can find $\delta > 0$ that satisfies

$$\int_{|x_1-x_2|<\delta} |f(x_1)f(x_2)|dx_1dx_2 < \varepsilon.$$

Because $g \in L^\infty(\mathbb{R})$, in view of (4.6.59) and (4.6.60), for sufficiently large T, we obtain

$$
\begin{aligned}
|J_T^2| &\le C \cdot T \int_{-a/2}^{a/2} \int_{-a/2}^{a/2} |f(x_1)f(x_2)| \int_{-a/2}^{a/2} \psi_T(x_1 - x_3)\psi_T(x_2 - x_3) \\
&\times \int_{|x_4|>a} x_4^{-2} dx_4 dx_3 dx_1 dx_2 \\
&\le C \int_{-a/2}^{a/2} \int_{-a/2}^{a/2} |f(x_1)f(x_2)|(1 + T|x_1 - x_2|)^{-1/2} dx_1 dx_2 \\
&\le C \int_{|x_1 - x_2|<\delta} |f(x_1)f(x_2)| dx_1 dx_2 \\
&+ (1 + T\delta)^{-1/2} \int_{|x_1 - x_2|\ge\delta} |f(x_1)f(x_2)| dx_1 dx_2 \le 2\varepsilon.
\end{aligned}
$$

This means that

$$
\lim_{T\to\infty} J_T^2 = 0. \tag{4.6.86}
$$

Likewise, it can be shown that

$$
\lim_{T\to\infty} J_T^3 = 0. \tag{4.6.87}
$$

To estimate the integral J_T^4 in (4.6.84), note that in this case, $|x_i - x_j| > a/2$ for $i = 1, 2$ and $j = 3, 4$. Therefore, we have

$$
\begin{aligned}
|J_T^4| &\le \frac{C}{T} \int_{\mathbb{R}^4} \underline{f}(x_1)\underline{f}(x_2)\overline{g}(x_3)\overline{g}(x_4) dx_1 dx_2 dx_3 dx_4 \\
&\le \frac{C}{T} \|f\|_{L^1(\mathbb{R})}^2 \|g\|_{L^1(\mathbb{R})}^2 \to 0 \quad \text{as} \quad T \to \infty.
\end{aligned} \tag{4.6.88}
$$

From (4.6.84) to (4.6.88), we obtain

$$
\lim_{T\to\infty} I_T^3 = 8\pi^3 \int_{-a/2}^{a/2} f^2(x)g^2(x)dx. \tag{4.6.89}
$$

A combination of (4.6.82), (4.6.83), and (4.6.89) yields (4.6.81). This completes the proof of the theorem. □

Proof of Theorem 4.5.4 We begin with assertion **(B1)** of the theorem. First, the condition $h_i \in \mathcal{F}_1$ implies that there exists a constant $C > 0$ such that

$$\hat{h}_i \in L^1(\mathbb{R}) \quad \text{and} \quad |\hat{h}_i(t)| \le C, \quad t \in \mathbb{R}, \quad i = 1, 2, \dots, m. \tag{4.6.90}$$

By Lemma 4.6.6, we have

$$T \cdot S(T) = \int_0^T \dots \int_0^T \hat{h}_1(u_1 - u_2)\hat{h}_2(u_2 - u_3) \dots \hat{h}_m(u_m - u_1) du_1 \dots du_m.$$

Making the change of variables:

$$u_1 - u_2 = t_1, \quad u_2 - u_3 = t_2, \dots, u_{m-1} - u_m = t_{m-1},$$

and observing that $t_1 + \cdots + t_{m-1} = u_1 - u_m$, we obtain

$$T \cdot S(T) = \int_0^T \int_{u_m - t_1 - \cdots - t_{m-1} - T}^{u_m - t_1 - \cdots - t_{m-1}} \int_{u_m - t_1 - \cdots - t_{m-2} - T}^{u_m - t_1 - \cdots - t_{m-2}} \cdots \tag{4.6.91}$$

$$\cdots \int_{u_m - t_1 - T}^{u_m - t_1} \hat{h}_1(t_1) \cdots \hat{h}_{m-1}(t_{m-1})\hat{h}_m(-t_1 - \cdots - t_{m-1}) d\mathbf{t}_{m-1} du_m$$

$$= \int_{-T}^{T} \cdots \int_{-T}^{T} \hat{h}_1(t_1) \cdots \hat{h}_{m-1}(t_{m-1})\hat{h}_m(-t_1 - \cdots - t_{m-1}) \left[T - l(\mathbf{t}_{m-1})\right] d\mathbf{t}_{m-1},$$

where we used the notation: $\mathbf{t}_{m-1} := (t_1, \dots, t_{m-1})$ and $d\mathbf{t}_{m-1} := dt_1 \cdots dt_{m-1})$. The function $l(\mathbf{t}_{m-1}) := l(t_1, \dots, t_{m-1})$ is chosen appropriately and satisfies

$$|l(\mathbf{t}_{m-1})| \le 2\left(|t_1| + \cdots + |t_{m-1}|\right). \tag{4.6.92}$$

On the other hand, by (4.5.4) and Parseval-Plancherel's equality, we have

$$M := M_{\mathbb{R},\mathcal{H}} = (2\pi)^{m-1} \int_{-\infty}^{\infty} \left[\prod_{i=1}^{m} h_i(\lambda)\right] d\lambda \tag{4.6.93}$$

$$= \int_{-\infty}^{\infty} \cdots \int_{-\infty}^{\infty} \hat{h}_1(t_1) \cdots \hat{h}_{m-1}(t_{m-1})\hat{h}_m(-t_1 - \cdots - t_{m-1}) d\mathbf{t}_{m-1}.$$

It follows from (4.5.4), (4.6.91), and (4.6.93) that

$$\begin{aligned}S(T)-M &:= S_{W,\mathcal{H}}(T)-M_{\mathbb{R},\mathcal{H}}\\ &= -\frac{1}{T}\int_{[-T,\,T]^{m-1}} \hat{h}_1(t_1)\cdots\hat{h}_{m-1}(t_{m-1})\hat{h}_m(-t_1-\cdots-t_{m-1})l(\mathbf{t}_{m-1})d\mathbf{t}_{m-1}\\ &+\int_{\mathbb{R}^{m-1}\setminus[-T,\,T]^{m-1}} \hat{h}_1(t_1)\cdots\hat{h}_{m-1}(t_{m-1})\hat{h}_m(-t_1-\cdots-t_{m-1})d\mathbf{t}_{m-1}\\ &=: \Delta_T^1+\Delta_T^2. \end{aligned}\tag{4.6.94}$$

Since $h_i \in \mathcal{F}_1$, $i=1,2,\ldots,m$, by (4.6.90), (4.6.92), and (4.6.94), we have

$$|T\cdot\Delta_T^1| \le 2C\sum_{i=1}^{m-1}\int_{\mathbb{R}^{m-1}} \left|\hat{h}_1(t_1)\cdots\hat{h}_{m-1}(t_{m-1})t_i\right|d\mathbf{t}_{m-1} =: C_1<\infty. \tag{4.6.95}$$

Further, observe that

$$\mathbb{R}^{m-1}\setminus[-T,\,T]^{m-1}\subset\bigcup_{i=1}^{m}\left\{(t_1,\ldots,t_{m-1})\in\mathbb{R}^{m-1}:|t_i|>T\right\} =: \bigcup_{i=1}^{m} E_i.$$

Therefore, by (4.6.90) and (4.6.94), we have

$$|T\cdot\Delta_T^2| \le 2C\sum_{i=1}^{m-1}\int_{E_i} \left|\hat{h}_1(t_1)\ldots\hat{h}_{m-1}(t_{m-1})t_i\right|d\mathbf{t}_{m-1} =: C_2<\infty. \tag{4.6.96}$$

Finally, from (4.6.94) to (4.6.96), we obtain assertion **(B1)**. □

***Proof of Assertion* (B2)** By Lemmas 4.6.6 (3) and 4.6.7 (b), and Eq. (4.5.5), we can conclude that

$$\Delta(T)=(2\pi)^{m-1}\left|\int_{\mathbb{R}^{m-1}}[\Psi(\mathbf{u})-\Psi(\mathbf{0})]\Phi_T(\mathbf{u})d\mathbf{u}\right|,\quad \mathbf{0}=(0,\ldots,0). \tag{4.6.97}$$

It follows from (4.5.14) and (4.6.37) that for $\mathbf{u}=(u_1,\ldots,u_{m-1})\in\mathbb{R}^{m-1}$,

$$|\Psi(\mathbf{u})-\Psi(\mathbf{0})| \le (m-1)C\left(|u_1|^\gamma+\cdots+|u_{m-1}|^\gamma\right). \tag{4.6.98}$$

Let $\varepsilon \in (0, \gamma)$. By applying Proposition 2.2.1 (c) with $\delta = (1 + \varepsilon - \gamma)/m$, and using Eqs. (4.6.97) and (4.6.98), we can write

$$\begin{aligned}\Delta(T) &\leq C_m \int_E |\Psi(\mathbf{u}) - \Psi(\mathbf{0})||\Phi_T(\mathbf{u})|d\mathbf{u} \\ &+ C_m \int_{E^c} |\Psi(\mathbf{u}) - \Psi(\mathbf{0})||\Phi_T(\mathbf{u})|d\mathbf{u} \\ &\leq \frac{C_m}{T^{1-m\delta}} \sum_{i=1}^{m-1} \int_E \frac{|u_i|^\gamma}{|u_1 \cdots u_{m-1}(u_1 + \cdots + u_{m-1})|^{1-\delta}} du_1 \cdots du_{m-1} \\ &+ 2\|\varphi\|_\infty \frac{C_m}{T^{1-m\delta}} \int_{E^c} \frac{1}{|u_1 \cdots u_{m-1}(u_1 + \cdots + u_{m-1})|^{1-\delta}} du_1 \cdots du_{m-1},\end{aligned} \tag{4.6.99}$$

where $E := \{(u_1, u_2, \ldots, u_{m-1}) \in \mathbb{R}^{m-1} : |u_i| \leq 1, \ i = 1, 2, \ldots, m-1\}$ and $E^c := \mathbb{R}^{m-1} \setminus E$. Since

$$\frac{m-1}{m} < 1 - \delta < \frac{m-1+\gamma}{m},$$

we can apply Lemmas 4.6.4 and 4.6.5 with $\alpha = \gamma$, $n = m - 1$, and $\beta = 1 - \delta$ to conclude that all the integrals in (4.6.99) are finite. Since $1 - m\delta = \gamma - \varepsilon$, the assertion **(B2)** of the theorem follows from (4.6.99). □

Proof of Assertion (B3) According to assertion **(B2)**, it is sufficient to prove that the function

$$\varphi(\mathbf{u}) := \int_{\mathbb{R}} h_1(\lambda)h_2(\lambda - u_1) \cdots h_m(\lambda - u_{m-1})d\mathbf{u}, \tag{4.6.100}$$

where $\mathbf{u} = (u_1, \ldots, u_{m-1}) \in \mathbb{R}^{m-1}$, belongs to $L^\infty(\mathbb{R}^{m-1})$, and with some positive constant C

$$|\varphi(\mathbf{u}) - \varphi(\mathbf{0})| \leq C|\mathbf{u}|^\gamma, \quad \mathbf{u} \in \mathbb{R}^{m-1}, \tag{4.6.101}$$

provided that

$$h_i \in \text{Lip}(\mathbb{R}; p_i, \gamma), \ 1 \leq p_i \leq \infty, \ i = 1, 2, \ldots, m, \ \text{and} \ \sum_{i=1}^m \frac{1}{p_i} \leq 1. \tag{4.6.102}$$

It follows from Hölder's inequality and (4.6.102) that

$$|\varphi(\mathbf{u})| \le \prod_{i=1}^{m} ||h_i||_{p_i} < \infty, \quad \mathbf{u} \in \mathbb{R}^{m-1}.$$

Therefore, $\varphi \in L^\infty(\mathbb{R}^{m-1})$.

To prove the inequality (4.6.101), we fix $\mathbf{u} = (u_1, \ldots, u_{m-1}) \in \mathbb{R}^{m-1}$ and denote

$$\overline{h}_i(\lambda) := h_i(\lambda - u_{i-1}) - h_i(\lambda), \quad \lambda \in \mathbb{R}, \quad i = 2, \ldots, m. \tag{4.6.103}$$

Since $h_i \in \mathrm{Lip}(\mathbb{R}; p_i, \gamma)$, we have

$$\|\overline{h}_i\|_{p_i} \le C_i |\mathbf{u}|^\gamma, \quad i = 2, \ldots, m. \tag{4.6.104}$$

By (4.6.100) and (4.6.103), we have

$$\varphi(\mathbf{u}) = \int_{\mathbb{R}} h_1(\lambda) \prod_{i=2}^{m} \left(\overline{h}_i(\lambda) + h_i(\lambda)\right) d\lambda = \varphi(\mathbf{0}) + W.$$

Each of the $(2^{m-1} - 1)$ integrals comprising W contains at least one function $\overline{h}_i$. In view of (4.6.104), these integrals can be estimated as follows:

$$\left| \int_{\mathbb{R}} h_1(\lambda)\overline{h}_2(\lambda)h_3(\lambda) \cdots h_m(\lambda) d\lambda \right|$$
$$\le \|h_1\|_{p_1} \|\overline{h}_2\|_{p_2} \|h_3(u)\|_{p_2} \cdots \|h_m\|_{p_m} \le C|\mathbf{u}|^\gamma.$$

This completes the proof of assertion **(B3)**. □

To prove assertion **(B4)** of Theorem 4.5.4 and Theorem 4.5.6, we will use the following lemma.

Lemma 4.6.12 *Let $p > 1$, $0 < \sigma < 1/p$, and let $f(\lambda)$ be a differentiable function defined on $\mathbb{R} \setminus \{0\}$ such that for some constant $C > 0$,*

$$|f(\lambda)| \le \begin{cases} C|\lambda|^{-\sigma} & \text{if } |\lambda| \le 1 \\ C|\lambda|^{-\delta} & \text{if } |\lambda| > 1 \end{cases}, \quad |f'(\lambda)| \le \begin{cases} C|\lambda|^{-\sigma-1} & \text{if } |\lambda| \le 1 \\ C|\lambda|^{-\delta-1} & \text{if } |\lambda| > 1 \end{cases}.$$

Then $f \in \mathrm{Lip}(p, 1/p - \sigma)$.

Proof Let $b \in (0, 1/2)$ be fixed. Then, we have

$$\int_{|\lambda| \le 2b} |f(\lambda + b) - f(\lambda)|^p d\lambda \le (2C)^p \int_0^{3b} |\lambda|^{-p\sigma} d\lambda \le C_1 b^{1-p\alpha}. \tag{4.6.105}$$

Next, for $|\lambda| > 2b$, we have $f(\lambda+b) - f(\lambda) = f'(\xi) \cdot b$ for some $\xi \in (\lambda, \lambda+b)$. Therefore, we can write

$$\int_{2b<|\lambda|<1/2} |f(\lambda+b) - f(\lambda)|^p d\lambda \le C^p b^p \int_b^{1/2} |\lambda|^{-p(\sigma+1)} d\lambda \le C b^{1-p\sigma} \tag{4.6.106}$$

and

$$\begin{aligned} &\int_{1/2<|\lambda|<\infty} |f(\lambda+b) - f(\lambda)|^p d\lambda \\ &\le C^p b^p \int_{1/2}^1 |\lambda|^{-p(\sigma+1)} d\lambda + C^p b^p \int_1^\infty |\lambda|^{-p(\delta+1)} d\lambda \\ &\le C b^p \le C b^{1-p\sigma}. \end{aligned} \tag{4.6.107}$$

From (4.6.105) to (4.6.107), we obtain

$$\|f(\lambda+b) - f(\lambda)\|_p \le C b^{1/p-\sigma},$$

implying that $f \in \text{Lip}(p, 1/p - \sigma)$. □

Proof of Assertion* (B4) *of Theorem 4.5.4 We set

$$\frac{1}{p_i} := \sigma_i + \frac{1}{m}\left[1 - (\sigma_1 + \cdots + \sigma_m)\right], \quad i = 1, 2, \ldots, m.$$

Then, we have

$$\sum_{i=1}^m \frac{1}{p_i} = 1; \ \ \frac{1}{p_i} - \sigma_i + \frac{1}{m}\left[1 - (\sigma_1 + \cdots + \sigma_m)\right] = \gamma > 0, \quad i = 1, 2, \ldots, m,$$

and

$$0 < \sigma_i < \frac{1}{p_i} < 1 < \delta_i, \qquad i = 1, 2, \ldots, m.$$

Therefore, according to Lemma 4.6.12, we have $h_i \in \text{Lip}(p_i, \gamma)$, $i = 1, 2, \ldots, m$. By applying assertion **(B3)**, we obtain (4.5.18) with γ as defined in (4.5.19). □

Proof of Theorem 4.5.5 Observe first that by Eq. (4.6.41), we have

$$\begin{aligned} S_{2,W}(T) &:= \frac{1}{T}\mathrm{tr}[W_T(h_1)W_T(h_2)] \\ &= 2\pi \int_{\mathbb{R}}\int_{\mathbb{R}} F_T(s-t)h_1(s)h_2(t)\,dt\,ds, \end{aligned} \tag{4.6.108}$$

where $F_T(u)$ is the Fejér kernel (see 2.2.2).

Below we use the properties of the Fejér kernel $F_T(u)$ stated in Proposition 2.2.3. Since the function $F_T(u)$ is even, in view of (4.6.108), we can write

$$S_{2,W}(T) = \pi \int_{\mathbb{R}}\int_{\mathbb{R}} F_T(u)\,[h_1(u+t)h_2(t) + h_1(t)h_2(u+t)]\,dudt. \tag{4.6.109}$$

Consequently, taking into account (2.2.3) and the equality

$$\int_{\mathbb{R}} h_1(t)h_2(t)dt = \int_{\mathbb{R}} h_1(u+t)h_2(u+t)dt,$$

by (4.6.109), we obtain

$$\begin{aligned} \Delta_{2,W}(T) &:= \left|\frac{1}{T}\mathrm{tr}[W_T(h_1)W_T(h_2)] - 2\pi\int_{\mathbb{R}} h_1(t)h_2(t)\,dt\right| \\ &= \left|\pi\int_{\mathbb{R}} F_T(u)\int_{\mathbb{R}}(h_1(t)-h_1(u+t))(h_2(u+t)-h_2(t))dtdu\right|. \end{aligned} \tag{4.6.110}$$

Next, using Hölder's inequality, we find from (4.6.110) that

$$\Delta_{2,W}(T) \le \pi\int_{\mathbb{R}} F_T(u)||h_1(u+\cdot)-h_1(\cdot)||_p||h_2(u+\cdot)-h_2(\cdot)||_q du.$$

Thus, we can write

$$\Delta_{2,W}(T) \le C_1\int_0^1 F_T(u)|u|^{\gamma_1+\gamma_2}du + C_2||h_1||_p||h_2||_q\int_{u>1} F_T(u)du.$$

Therefore, the result follows from the properties of the Fejér kernel stated in Eqs. (2.2.6) and (2.2.7). □

Proof of Theorem 4.5.6 It follows from Lemma 4.6.12 that under the assumptions of the theorem, $h_i \in \mathrm{Lip}(p_i, 1/p_i - \sigma_i)$, $i = 1, 2$. Hence, by applying Theorem 4.5.5 with $\gamma_i = \frac{1}{p_i} - \sigma_i$, we obtain (4.5.21). □

4.7 Applications to Stationary Processes

In this section, we apply the results from Sects. 4.4 and 4.5 to important special cases where the generating functions of the Toeplitz matrices and operators represent the spectral densities of discrete time ARFIMA$(0, d, 0)$ stationary processes or continuous time stationary fractional Riesz-Bessel motions. Further applications are discussed in Chap. 5.

We use the following notation: $m = 2\nu$;

$$h_1(\lambda) = h_3(\lambda) = \cdots = h_{2\nu-1}(\lambda) := f_1(\lambda)$$
$$h_2(\lambda) = h_4(\lambda) = \cdots = h_{2\nu}(\lambda) := f_2(\lambda);$$

and

$$S_{\nu,A}(T) = \frac{1}{T}\operatorname{tr}[A_T(f_1)A_T(f_2)]^\nu, \tag{4.7.1}$$

$$\Delta_{\nu,A}(T) := \left| S_{\nu,A}(T) - (2\pi)^{2\nu-1} \int_\Lambda [f_1(\lambda) f_2(\lambda)]^\nu \, d\lambda \right|, \tag{4.7.2}$$

where either $A_T(f_i) = B_T(f_i)$ and $\Lambda = \mathbb{T}$ or $A_T(f_i) = W_T(f_i)$ and $\Lambda = \mathbb{R}$, $i = 1, 2$.

4.7.1 Applications to ARFIMA Time Series

The following theorem provides an error bound for the approximation $\Delta_{\nu,B}(T)$ when the associated Toeplitz matrices are generated by the spectral densities of two discrete time ARFIMA$(0, d, 0)$ stationary processes.

Theorem 4.7.1 *Let $f_i(\lambda)$, $i = 1, 2$, be the spectral densities of two ARFIMA$(0, d, 0)$ stationary processes defined as (see Sect. 3.3.3):*

$$f_i(\lambda) = \frac{\sigma_i^2}{2\pi} \left|1 - e^{-i\lambda}\right|^{-2d_i}, \quad i = 1, 2 \tag{4.7.3}$$

with $0 < \sigma_i^2 < \infty$ and $0 < d_i < 1/2$. Then, under the condition $d := d_1 + d_2 < 1/(2\nu)$, $\nu \in \mathbb{N}$, for any $\varepsilon > 0$, we have

$$\Delta_{\nu,B}(T) = O\left(T^{-\gamma+\varepsilon}\right) \quad \text{as} \quad T \to \infty \tag{4.7.4}$$

with

$$\gamma = \frac{1}{2\nu} - (d_1 + d_2). \tag{4.7.5}$$

Proof Assuming that $\lambda \in (0, \pi]$ (the case $\lambda \in [-\pi, 0)$ is treated similarly) and taking into account that $|1 - e^{-i\lambda}| = 2\sin(\lambda/2)$, we have for $i = 1, 2$

$$f_i(\lambda) = \frac{\sigma_i^2}{2\pi} \cdot 2^{-2d_i} \left[\sin\frac{\lambda}{2}\right]^{-2d_i},$$

$$f_i'(\lambda) = \frac{\sigma_i^2}{2\pi} \cdot \left[-2d_i 2^{-2d_i-1}\left(\sin\frac{\lambda}{2}\right)^{-2d_i-1} \cos\frac{\lambda}{2}\right]. \tag{4.7.6}$$

The conditions of Theorem 4.4.8 **(B5)** are clearly satisfied for the functions mentioned above, with $\alpha_i = 2d_i$ and $C_{1i} = C_{2i} = \sigma_i^2$ for $i = 1, 2$. Therefore, the result follows from Theorem 4.4.8 **(B5)** with γ defined in (4.7.5). □

The next theorem offers an explicit second-order asymptotic expansion for the trace $S_{1,B}(T)$, as defined in (4.7.1), in the case where the Toeplitz matrices are generated by the spectral densities given in (4.7.3). It shows that, in this special case, the second-order asymptotic expansion effectively removes the singularity and provides a substantially better approximation for $S_{1,B}(T)$.

Theorem 4.7.2 *Let $f_i(\lambda)$, $i = 1, 2$, be the spectral densities of two ARFIMA$(0, d, 0)$ stationary processes defined by (4.7.3) with $0 < \sigma_i^2 < \infty$ and $0 < d_i < 1/2$, $i = 1, 2$. Then, under the condition $d := d_1 + d_2 < 1/2$, the following asymptotic relation holds:*

$$\begin{aligned} S_{1,B}(T) &:= \frac{1}{T}\operatorname{tr}[B_T(f_1)B_T(f_2)] \\ &= 2\pi \int_{-\pi}^{\pi} f_1(\lambda) f_2(\lambda)\, d\lambda - \frac{C(d_1, d_2)}{T^{1-2d}} + o\left(\frac{1}{T^{1-2d}}\right) \end{aligned} \tag{4.7.7}$$

as $T \to \infty$, where

$$C(d_1, d_2) = \frac{2\sigma_1^2\sigma_2^2\pi^2}{\cos(\pi d_1)\cos(\pi d_2)\Gamma(2d_1)\Gamma(2d_2)} \cdot \frac{1}{2d(1-2d)}. \tag{4.7.8}$$

Proof It follows from (4.7.6) that the functions f_i ($i = 1, 2$) given in (4.7.3) satisfy the conditions of Theorem 4.4.10, with $\alpha_i = 2d_i$ ($i = 1, 2$). Consequently, relation (4.7.7) is a special case of relation (4.4.22) in Theorem 4.4.10. □

Remark 4.7.1 The asymptotic relation (4.7.7) was first established by Lieberman and Phillips [214] through direct calculations using the explicit forms of the functions f_i given in (4.7.3).

4.7.2 Applications to Fractional Riesz-Bessel Motions

We now assume that the underlying model is a continuous time stationary process, specified by fractional Riesz-Bessel motion, with a spectral density given by (see 3.3.20)

$$f(\lambda) = c\,|\lambda|^{-2\alpha}(1+\lambda^2)^{-\beta}, \quad \lambda \in \mathbb{R},\ 0 < c < \infty,\ 0 < \alpha < 1,\ \beta \geq 0. \tag{4.7.9}$$

Let $\Delta_{\nu,W}(T)$ be defined as in (4.7.2) with $A_T(f_i) = W_T(f_i)$, $i = 1, 2$, and $\Lambda = \mathbb{R}$. The following result is an immediate consequence of Theorem 4.5.1 **(A1)**.

Theorem 4.7.3 *Let $f_1(\lambda) = f(\lambda)$ be the spectral density of a fractional Riesz-Bessel motion defined by (4.7.9), and let $f_2(\lambda) = g(\lambda)$ be an integrable real symmetric function on $\mathbb{R}$. If for some $p, q \geq 1$ with $1/p + 1/q \leq 1/\nu$ ($\nu \in \mathbb{N}$), we have $g(\lambda) \in L^q(\mathbb{R})$ and $0 < \alpha < 1/(2p)$, $\alpha + \beta > 1/2$, then*

$$\Delta_{\nu,W}(T) = o(1) \quad \text{as} \quad T \to \infty.$$

The following result is a consequence of Theorem 4.5.4 **(B3)** and Lemma 4.6.12.

Theorem 4.7.4 *Let $f_1(\lambda) = f(\lambda)$ be as in (4.7.9) with $0 < \alpha < 1/(2p)$ and $\alpha + \beta > 1/2$, and let $f_2(\lambda) = g(\lambda)$ be an integrable real symmetric function from the class* Lip$(q, 1/p - 2\alpha)$ *with $1/p + 1/q \leq 1/\nu$, $\nu \in \mathbb{N}$. Then, for any $\varepsilon > 0$,*

$$\Delta_{\nu,W}(T) = O\left(T^{-\gamma+\varepsilon}\right) \quad \text{as} \quad T \to \infty. \tag{4.7.10}$$

with $\gamma = 1/p - 2\alpha$.

Proof By (4.7.9), we have

$$f'(\lambda) = -\frac{2\alpha + 2(\alpha+\beta)\lambda^2}{\lambda^{2\alpha+1}(1+\lambda)^{\beta+1}}. \tag{4.7.11}$$

It follows from Eqs. (4.7.9) and (4.7.11) that the functions f and f' satisfy the conditions of Lemma 4.6.12 with $\sigma = 2\alpha$ and $\delta = 2\alpha + 2\beta$. Hence, by Lemma 4.6.12, we have $f(\lambda) \in \text{Lip}(p, 1/p - 2\alpha)$. Therefore, the result follows from Theorem 4.5.4 **(B3)**. □

Theorem 4.7.5 *Let $f_i(\lambda)$, $i = 1, 2$, be the spectral density functions of two fractional Riesz-Bessel motions defined as*

$$f_i(\lambda) = \frac{C_i}{|\lambda|^{2\alpha_i}(1+\lambda^2)^{\beta_i}}, \quad 0 < \alpha_i < 1/2, \quad \alpha_i + \beta_i > 1/2, \quad i = 1, 2. \tag{4.7.12}$$

If $\alpha_1 + \alpha_2 < 1/(2\nu)$, *then (4.7.10) holds for any* $\varepsilon > 0$ *with*

$$\gamma = \frac{1}{2\nu} - (\alpha_1 + \alpha_2). \tag{4.7.13}$$

Proof It follows from (4.7.11) and (4.7.12) that the functions f_1 and f_2 satisfy the conditions of Theorem 4.5.4 **(B4)** with $\sigma_i = 2\alpha_i$ and $\delta_i = 2\alpha_i + 2\beta_i$, $i = 1, 2$. Therefore, the result follows from Theorem 4.5.4 **(B4)** with γ defined in (4.7.13). □

The next theorem, a continuous version of Theorem 4.7.2, provides an explicit second-order asymptotic expansion for the trace $S_{1,W}(T)$ in the case where the Toeplitz operators are generated by the spectral densities given by (4.7.12). It shows that, similar to the discrete time case, in this special case, a second-order asymptotic expansion effectively eliminates the singularity and delivers a significantly improved approximation for $S_{1,W}(T)$.

Theorem 4.7.6 *Let* $f_i(\lambda)$, $i = 1, 2$, *be defined as in (4.7.12). Then, under the condition* $\alpha := \alpha_1 + \alpha_2 < 1/2$, *the following asymptotic relation holds:*

$$\begin{aligned} S_{1,W}(T) &:= \frac{1}{T}\mathrm{tr}[W_T(f_1)W_T(f_2)] \\ &= 2\pi \int_{-\infty}^{+\infty} f_1(\lambda) f_2(\lambda)\, d\lambda - \frac{C(\alpha_1, \alpha_2)}{T^{1-2\alpha}} + o\left(\frac{1}{T^{1-2\alpha}}\right) \end{aligned} \tag{4.7.14}$$

as $T \to \infty$, *where*

$$C(\alpha_1, \alpha_2) = \frac{2C_1C_2\pi^2}{\cos(\pi\alpha_1)\cos(\pi\alpha_2)\Gamma(2\alpha_1)\Gamma(2\alpha_2)} \cdot \frac{1}{2\alpha(1-2\alpha)}. \tag{4.7.15}$$

The proof relies on the following lemma, which presents an asymptotic formula for the covariance function of an fRBm process.

Lemma 4.7.1 *Let* $f(\lambda)$ *be defined as in (4.7.9) with* $0 < \alpha < 1/2$ *and* $\beta > 1/2$, *and let* $r(t) := \hat{f}(t)$ *be the Fourier transform of* $f(\lambda)$. *Then*

$$r(t) = t^{2\alpha-1} \cdot \frac{\pi C}{\cos(\pi\alpha)\Gamma(2\alpha)} \cdot (1 + o(1)) \quad \text{as} \quad t \to \infty. \tag{4.7.16}$$

Remark 4.7.2 Taking into account the reflection formula for the gamma function $\Gamma(\cdot)$,

$$\Gamma(2\alpha)\Gamma(1-2\alpha) = \pi/\sin(2\pi\alpha),$$

the asymptotic relation (4.7.16) can be written in the following form

$$r(t) = Ct^{2\alpha-1}\sin(\pi\alpha)\Gamma(1-2\alpha) \cdot (1 + o(1)) \quad \text{as} \quad t \to \infty. \tag{4.7.17}$$

Proof of Lemma 4.7.1 We use the technique applied in the paper by Inoue and Kasahara [192], where the relation (4.7.16) was proved for $\beta = 1$.

Since the underlying process $X(t)$ is real-valued, we have for $t > 0$,

$$r(t) := \int_{-\infty}^{+\infty} e^{it\lambda} f(\lambda)\, d\lambda = 2\int_0^\infty \frac{C}{\lambda^{2\alpha}(1+\lambda^2)^\beta} \cos(t\lambda)\, d\lambda. \tag{4.7.18}$$

Using the change of variable $t\lambda = 1/u$, $\lambda = 1/(tu)$, $d\lambda = -du/(tu^2)$, we obtain from (4.7.18) that

$$\begin{aligned} r(t) &= 2C\int_0^\infty \frac{1}{((1/(tu))^{2\alpha}(1+1/(tu)^2))^\beta} \cos(1/u)\, \frac{du}{tu^2} \\ &= 2C\cdot t^{2\alpha-1}\int_0^\infty \left(\frac{(tu)^2}{1+(tu)^2}\right)^\beta u^{2\alpha-2}\cos(1/u)\, du \\ &= 2C\cdot t^{2\alpha-1}\int_0^\infty L(tu)k(u)\, du, \end{aligned} \tag{4.7.19}$$

where

$$L(u) = \left(\frac{u^2}{1+u^2}\right)^\beta \quad \text{and} \quad k(u) = u^{2\alpha-2}\cos(1/u). \tag{4.7.20}$$

Choose $\delta > 0$ such that $\delta < \min(1-2\alpha, 2\alpha)$. Then the improper integrals

$$\int_{0+}^1 u^{-\delta} k(u)\, du \quad \text{and} \quad \int_1^{\infty-} u^\delta k(u)\, du$$

exist. Taking into account that the function $L(u)$ in (4.7.20) is quasi-monotone (see Sect. 2.5), we can apply the Bojanic-Karamata theorem (see Bingham et al. [43], Theorem 4.1.5) to conclude that

$$\int_0^\infty k(u)L(tu)\, du \longrightarrow \int_{0+}^{\infty-} k(u)\, du \quad \text{as} \quad t\to\infty. \tag{4.7.21}$$

Next, using the formula (see, e.g., Dwight [87])

$$\int_0^{\infty-} x^{-p}\cos(mx)\, dx = \frac{\pi m^{p-1}}{2\cos(p\pi/2)\Gamma(p)}, \qquad 0 < p < 1, \quad m > 0,$$

with $p = 2\alpha$ and $m = 1$, and the change of variable $1/u = v$, we obtain

$$\int_{0+}^{\infty-} k(u)\,du = \int_{0}^{\infty-} u^{2\alpha-2}\cos(1/u)\,du = \int_{0}^{\infty-} v^{-2\alpha}\cos(v)\,dv$$
$$= \frac{\pi}{2\cos(\pi\alpha)\Gamma(2\alpha)}. \tag{4.7.22}$$

From (4.7.18) to (4.7.22), we obtain the desired relation (4.7.16). □

Proof of Theorem 4.7.6 In view of Eq. (4.6.41) and Parseval-Plancherel's equality, we have

$$S_{1,W}(T) = \int_{-T}^{T}\left(1 - \frac{|t|}{T}\right) r_1(t) r_2(t)\,dt \tag{4.7.23}$$
$$= 2\pi \int_{-\infty}^{+\infty} f_1(\lambda) f_2(\lambda)\,d\lambda - I_1 - I_2,$$

where

$$I_1 = \int_{|t|>T} r_1(t) r_2(t)\,dt \tag{4.7.24}$$

and

$$I_2 = \frac{1}{T}\int_{-T}^{T} |t| r_1(t) r_2(t)\,dt. \tag{4.7.25}$$

Therefore, for $\alpha := \alpha_1 + \alpha_2 < 1/2$, from Lemma 4.7.1 and Eq. (4.7.24), we have as $T \to \infty$

$$I_1 = \int_{|t|>T} r_1(t) r_2(t)\,dt = 2\int_{t>T} r_1(t) r_2(t)\,dt$$
$$= \frac{2C_1C_2\pi^2}{\cos(\pi\alpha_1)\cos(\pi\alpha_2)\Gamma(2\alpha_1)\Gamma(2\alpha_2)} \cdot \int_{t>T} t^{2(\alpha-1)}\,(1 + o(1))$$
$$= \frac{2C_1C_2\pi^2}{\cos(\pi\alpha_1)\cos(\pi\alpha_2)\Gamma(2\alpha_1)\Gamma(2\alpha_2)} \cdot \frac{1}{1-2\alpha} \cdot T^{2\alpha-1}\,(1 + o(1)). \tag{4.7.26}$$

For the integral I_2, from Lemma 4.7.1 and Eq. (4.7.25), we have as $T \to \infty$ and $\alpha := \alpha_1 + \alpha_2 < 1/2$

$$
\begin{aligned}
I_2 &= \frac{1}{T}\int_{-T}^{T} |t| r_1(t) r_2(t)\, dt = \frac{2}{T}\int_0^T t\, r_1(t) r_2(t)\, dt \\
&= \frac{2}{T} \cdot \frac{C_1 C_2 \pi^2}{\cos(\pi\alpha_1)\cos(\pi\alpha_2)\Gamma(2\alpha_1)\Gamma(2\alpha_2)} \int_0^T t^{2\alpha-1}\, dt\, (1+o(1)) \\
&= \frac{2C_1 C_2 \pi^2}{\cos(\pi\alpha_1)\cos(\pi\alpha_2)\Gamma(2\alpha_1)\Gamma(2\alpha_2)} \cdot \frac{1}{2\alpha} \cdot T^{2\alpha-1}\, (1+o(1)) .
\end{aligned} \tag{4.7.27}
$$

A combination of relations (4.7.23), (4.7.26), and (4.7.27) concludes the proof.

□

Remark 4.7.3

(a) Similar to the ARFIMA case (see the proof of Theorem 4.7.2), the asymptotic relation in (4.7.14) can be deduced from Theorem 4.5.6.

(b) We analyze the behavior of the approximations as $\alpha_1, \alpha_2 \to 1/4$. First, observe that the first-order asymptotic formula has a pole when $\alpha := \alpha_1 + \alpha_2 = 1/2$. Specifically, by denoting $\beta := \beta_1 + \beta_2$, and using (4.7.12) along with the change of variable $\lambda^2 = u$, we have

$$
\begin{aligned}
2\pi \int_{-\infty}^{+\infty} f_1(\lambda) f_2(\lambda)\, d\lambda &= 2\pi C_1 C_2 \int_{-\infty}^{+\infty} \frac{1}{|\lambda|^{2\alpha}(1+\lambda^2)^{\beta}} d\lambda \\
&= 2\pi C_1 C_2 \int_0^{\infty} \frac{u^{-1/2-\alpha}}{(1+u)^{\beta}} du.
\end{aligned} \tag{4.7.28}
$$

Applying the formula (see, e.g., Dwight [87])

$$
\int_0^{\infty} \frac{u^{m-1}}{(1+u)^{m+n}} du = \frac{\Gamma(m)\Gamma(n)}{\Gamma(m+n)}, \quad m > 0,\ n > 0, \tag{4.7.29}
$$

with $m = 1/2 - \alpha$ and $n = \beta - m = \alpha + \beta - 1/2$, from (4.7.28), we find that

$$
\int_{-\infty}^{+\infty} f_1(\lambda) f_2(\lambda)\, d\lambda = \frac{\Gamma(1/2-\alpha)\Gamma(\alpha+\beta-1/2)}{\Gamma(\beta)}. \tag{4.7.30}
$$

Then, using the Laurent expansion of the gamma function $\Gamma(1/2-\alpha)$ around the pole $\alpha = 1/2$, we obtain from (4.7.30) that as $\alpha = \alpha_1 + \alpha_2 \to 1/2$,

$$
2\pi \int_{-\infty}^{+\infty} f_1(\lambda) f_2(\lambda)\, d\lambda = \frac{4\pi C_1 C_2}{1-2\alpha} + O(1). \tag{4.7.31}
$$

However, as $\alpha_1, \alpha_2 \to 1/4$, the asymptotic behavior of the second-order term $C(\alpha_1, \alpha_2)$ (see (4.7.15)) is readily observed to be

$$\begin{aligned} C(\alpha_1, \alpha_2) &= \frac{2\pi^2 C_1 C_2}{\cos(\pi\alpha_1)\cos(\pi\alpha_2)\Gamma(2\alpha_1)\Gamma(2\alpha_2)} \cdot \frac{1}{2\alpha(1-2\alpha)} \\ &= \frac{2\pi^2 C_1 C_2}{\cos^2(\pi/4)[\Gamma(1/2)]^2} \cdot \frac{1}{1-2\alpha} + O(1) \\ &= \frac{4\pi C_1 C_2}{1-2\alpha} + O(1). \end{aligned} \tag{4.7.32}$$

Thus, the pole in the first-order approximation is eliminated by the second-order approximation, so that the approximation

$$2\pi \int_{-\infty}^{+\infty} f_1(\lambda) f_2(\lambda)\, d\lambda - \frac{C(\alpha_1, \alpha_2)}{T^{1-2(\alpha_1+\alpha_2)}} \tag{4.7.33}$$

is bounded as $\alpha_1, \alpha_2 \to 1/4$. This good behavior explains why the second-order approximation produces a good approximation that does uniformly well over $\alpha_1, \alpha_2 \in [0, 1/4]$, including the limits of the domain.

(c) The second-order equivalence holds along an arbitrary ray for which $\alpha = \alpha_1 + \alpha_2 \to 1/2$.
Indeed, let $\alpha_1^0 \in [0, 1/2]$ be any fixed number such that $\alpha_1 \to \alpha_1^0$ and $\alpha = \alpha_1 + \alpha_2 \to 1/2$; then the representation (4.7.31) for the first-order asymptotic term continues to apply. On the other hand, using the reflection formula

$$\Gamma(z)\Gamma(1-z) = \pi/\sin(\pi z)$$

with $z = 2\alpha_1^0$, we have from (4.7.15) as $\alpha_1 \to \alpha_1^0$ and $\alpha = \alpha_1 + \alpha_2 \to 1/2$,

$$\begin{aligned} C(\alpha_1, \alpha_2) &= \frac{2\pi^2 C_1 C_2}{\cos(\alpha_1\pi)\cos(\alpha_2\pi)\Gamma(2\alpha_1)\Gamma(2\alpha_2)} \cdot \frac{1}{2\alpha(1-2\alpha)} \\ &= \frac{2\pi^2 C_1 C_2}{\cos(\alpha_1^0\pi)\cos[(1-2\alpha_1^0)\pi/2]\Gamma(2\alpha_1^0)\Gamma(1-2\alpha_1^0)} \cdot \frac{1}{1-2\alpha} + O(1) \\ &= \frac{2\pi^2 C_1 C_2}{\cos(\alpha_1^0\pi)\sin(\alpha_1^0\pi)} \cdot \frac{\sin(2\alpha_1^0\pi)}{\pi} \cdot \frac{1}{1-2\alpha} + O(1) \\ &= \frac{4\pi C_1 C_2}{1-2\alpha} + O(1). \end{aligned}$$

This shows that the second-order equivalence remains valid along any arbitrary ray as $\alpha = \alpha_1 + \alpha_2$ approaches 1/2.

Thus, in this special case, the second-order asymptotic expansion removes the singularity present in the first-order approximation and provides a substantially improved approximation of the original functional.

(d) A similar analysis of the behavior of approximations for ARFIMA models was conducted by Lieberman and Phillips [214].

4.8 The Trace Problem for Tapered Toeplitz Matrices and Operators

4.8.1 Tapering: Motivation

Tapering is a technique designed to enhance the accuracy of statistical estimators by applying specific preliminary transformations to the original data before conducting the estimation procedure. In the statistical analysis of stationary processes, data are often tapered before calculating the statistic of interest. Consequently, the statistical inference procedure, instead of the original data,

$$\mathbf{X}_T := \{X(t),\ t \in D_T\},$$

relies on the *tapered data*:

$$\mathbf{X}_T^h := \{h_T(t)X(t),\ t \in D_T\}, \tag{4.8.1}$$

where $D_T := [0, T]$ in the continuous time case and $D_T := \{1, \dots, T\}$ in the discrete time case, and

$$h_T(t) := h(t/T) \tag{4.8.2}$$

with $h(t)$, $t \in \mathbb{R}$, being a *taper function* to be specified below.

To analyze the properties of the tapered statistical estimators, we need limit theorems for the tapered quadratic functionals of the observed processes. This requires addressing the trace approximation problem for tapered Toeplitz matrices and operators.

For $k \in \mathbb{N}$, denote by $H_{k,T}(\lambda)$ the *tapered Dirichlet-type kernel*, defined by

$$H_{k,T}(\lambda) := \begin{cases} \sum_{t=1}^{T} h_T^k(t) e^{-i\lambda t} & \text{in the d.t. case,} \\ \int_0^T h_T^k(t) e^{-i\lambda t} dt & \text{in the c.t. case,} \end{cases} \tag{4.8.3}$$

and put

$$H_{k,T} := H_{k,T}(0). \tag{4.8.4}$$

Throughout the book, we will assume that the taper function $h(\cdot)$ satisfies the following assumption unless stated otherwise.

Assumption 4.8.1 The taper $h : \mathbb{R} \to \mathbb{R}$ is a continuous non-negative function of bounded variation and of bounded support [0, 1], such that $H_k \neq 0$, where

$$H_k := \lim_{T\to\infty} (1/T)H_{k,T}, \quad \text{and } H_{k,T} \text{ is defined as in (4.8.4).} \tag{4.8.5}$$

Observe that in the continuous time case, we have

$$H_k = \int_0^1 h^k(t)dt. \tag{4.8.6}$$

Note The case $h(t) = \mathbf{1}_{[0,1]}(t)$, where $\mathbf{1}_{[0,1]}(\cdot)$ denotes the indicator function of the interval [0, 1], will be referred to as the *non-tapered case*.

Remark 4.8.1 The taper function $h(t)$ normally has a maximum at $t = 1/2$ and decreases smoothly to zero as t tends to 0 or 1. For the discrete time case, an example of a taper function $h(t)$ satisfying Assumption 4.8.1 is the Tukey-Hanning taper function $h(t) = 0.5(1 - \cos(\pi t))$ for $t \in [0, 1]$. For the continuous time case, a simple example of a taper function $h(t)$ satisfying Assumption 4.8.1 is the function $h(t) = 1 - |t|$ for $t \in [0, 1]$.

For more details about tapering, including its motivations and benefits in statistical inference problems, see Appendix A.

4.8.2 *Tapered Toeplitz Matrices and Operators*

Let $\psi(\lambda)$ be an integrable real symmetric function defined on $[-\pi, \pi)$, and let $h(t)$, $t \in [0, 1]$, be a taper function. For $T = 1, 2, \ldots$, the *$(T \times T)$-truncated tapered Toeplitz matrix* generated by ψ and h, denoted by $B_T^h(\psi)$, is defined by the following equation (see formula (4.2.8) for the non-tapered case):

$$B_T^h(\psi) := \|\widehat{\psi}(t-s)h_T(t)h_T(s)\|_{t,s=1,2\ldots,T}, \tag{4.8.7}$$

where $\widehat{\psi}(t)$ $(t \in \mathbb{Z})$ are the Fourier coefficients of the function $\psi(\cdot)$, and $h_T(\cdot)$ is defined as in (4.8.2).

Given a real number $T > 0$ and an integrable real symmetric function $\psi(\lambda)$ defined on $\mathbb{R}$, the *T-truncated tapered Toeplitz operator* (also known as the *tapered Wiener-Hopf operator*) generated by ψ and a taper function h, denoted by $W_T^h(\psi)$, is defined as follows (see formula (4.2.11) for the non-tapered case):

$$[W_T^h(\psi)u](t) = \int_0^T \hat{\psi}(t-s)u(s)h_T(s)ds, \quad u(s) \in L^2([0, T]; h_T), \tag{4.8.8}$$

where $\hat{\psi}(\cdot)$ is the Fourier transform of $\psi(\cdot)$, and $L^2([0,T]; h_T)$ denotes the weighted L^2-space with respect to the measure $h_T(t)dt$.

Similar to the non-tapered case (see Sect. 4.3), the trace approximation problem for tapered Toeplitz matrices and operators can be stated as follows.

Given a collection of integrable real symmetric functions $\mathcal{H} = \{\psi_1, \psi_2, \ldots, \psi_k\}$ defined on Λ, where $\Lambda = [-\pi.\pi)$ or $\Lambda = \mathbb{R}$, and a taper function h. Let $A_T^h(\psi_i)$ denote either the $T \times T$ tapered Toeplitz matrix $B_T^h(\psi_i)$, or the T-truncated tapered Toeplitz operator $W_T^h(\psi_i)$ generated by the functions ψ_i and h, defined respectively by (4.8.7) and (4.8.8) with ψ_i used in place of ψ. Define

$$S_{A,\mathcal{H},h}(T) := \frac{1}{T}\operatorname{tr}\left[\prod_{i=1}^{k} A_T^h(\psi_i)\right], \tag{4.8.9}$$

$$M_{\Lambda,\mathcal{H},h} := (2\pi)^{k-1} H_k \int_\Lambda \left[\prod_{i=1}^{k} \psi_i(\lambda)\right] d\lambda, \tag{4.8.10}$$

where H_k is defined as in (4.8.5), and let

$$\Delta_h(T) := \Delta_{A,\Lambda,\mathcal{H},h}(T) = |S_{A,\mathcal{H},h}(T) - M_{\Lambda,\mathcal{H},h}|. \tag{4.8.11}$$

The problem is to approximate $S_{A,\mathcal{H},h}(T)$ by $M_{\Lambda,\mathcal{H},h}$ and estimate the error rate for $\Delta_h(T)$ as $T \to \infty$. Here, we consider only Problem (A), which involves finding conditions on the functions $\{\psi_k(\lambda),\ k = 1, 2, \ldots, m\}$ such that

$$\Delta_h(T) = o(1) \quad \text{as} \quad T \to \infty. \tag{4.8.12}$$

The following theorem presents the tapered versions of the results stated in Theorems 4.4.1 and 4.5.1.

Theorem 4.8.1 *Let $\Delta_h(T)$ be defined as in (4.8.11) with a taper function h satisfying Assumption 4.8.1. Then each of the following conditions is sufficient for* (4.8.12).

(C1) *The function $\varphi(\mathbf{u})$ defined by*

$$\varphi(\mathbf{u}) := \int_\Lambda \psi_1(\lambda)\psi_2(\lambda - u_1)\psi_3(\lambda - u_2)\cdots\psi_k(\lambda - u_{k-1})\, d\lambda, \tag{4.8.13}$$

where $\mathbf{u} = (u_1, u_2, \ldots, u_{k-1}) \in \Lambda^{k-1}$, belongs to $L^{k-2}(\Lambda^{k-1})$ and is continuous at the point $\mathbf{0} = (0, 0, \ldots, 0) \in \Lambda^{k-1}$.

(C2) $\psi_i \in L^1(\Lambda) \cap L^{p_i}(\Lambda)$, $p_i > 1$, $i = 1, 2, \ldots, k$, *with* $1/p_1 + \cdots + 1/p_k \le 1$.

The next theorem presents the tapered versions of the results stated in Theorems 4.4.3, 4.4.4, 4.5.2, and 4.5.3.

Theorem 4.8.2 *Let $\Delta_h(T)$ be defined as in (4.8.11) with a taper function h satisfying Assumption 4.8.1. Assume that $k = 4$, $\psi_1 = \psi_3 := f$ and $\psi_2 = \psi_4 := g$. Then each of the following conditions is sufficient for* (4.8.12) *with $m = 4$.*

(C3) *The functions f and g satisfy $f \in L^2(\Lambda)$, $g \in L^2(\Lambda)$, $fg \in L^2(\Lambda)$, and*

$$\int_\Lambda f^2(\lambda)g^2(\lambda-\mu)\,d\lambda \longrightarrow \int_\Lambda f^2(\lambda)g^2(\lambda)\,d\lambda \quad \text{as} \quad \mu \to 0.$$

(C4) *The functions f and g are integrable on $\mathbb{R}$ and bounded outside any neighborhood of the origin and satisfy for some $a > 0$*

$$f(\lambda) \le |\lambda|^{-\alpha} L_1(\lambda), \quad |g(\lambda)| \le |\lambda|^{-\beta} L_2(\lambda), \quad \lambda \in [-a, a], \tag{4.8.14}$$

for some $\alpha < 1$, $\beta < 1$ with $\alpha + \beta \le 1/2$, where L_1 and L_2 are slowly varying at zero functions satisfying

$$L_i \in SV_0(\mathbb{R}), \;\; \lambda^{-(\alpha+\beta)} L_i(\lambda) \in L^2[-a, a], \; i = 1, 2. \tag{4.8.15}$$

4.8.3 *Proofs of Theorems 4.8.1 and 4.8.2*

We will focus exclusively on proving the results related to Toeplitz operators. The proofs for the corresponding results involving Toeplitz matrices are similar. First, we will present several technical lemmas.

4.8.3.1 Preliminaries: Technical Lemmas

In this section, we prove several auxiliary lemmas regarding the properties of continuous time tapered Dirichlet- and Fejér-type kernels. Some of these properties for the discrete time tapered case were established by Dahlhaus [70]. The non-tapered analogs of these properties are demonstrated in Sects. 2.2 and 4.6.2.

An important role in our analysis of the aforementioned properties is played by the function $L_T(\cdot) : \mathbb{R} \to \mathbb{R}$, $T \in \mathbb{R}^+$, defined by (cf. 4.6.19)

$$L_T(t) = \begin{cases} T & \text{for } |t| \le 1/T \\ 1/|t| & \text{for } |t| > 1/T. \end{cases} \tag{4.8.16}$$

The function $L_T(t)$ has several notable properties, as discussed in the papers by Dahlhaus [70] for the discrete time case, and by Eichler [95], as well as Ginovyan and Sahakyan [140] for the continuous time case. In our proofs of Theorems 4.8.1 and 4.8.2, we will use the following property of the function $L_T(t)$.

Lemma 4.8.1 *Let* $m = 0, 1, \ldots,$ *and*

$$L(t) := L_1(t) = \begin{cases} 1 & \text{for } |t| \le 1 \\ 1/|t| & \text{for } |t| > 1, \end{cases} \qquad t \in \mathbb{R}. \tag{4.8.17}$$

Then for $u \in \mathbb{R}$, *we have*

$$Q(u) := \int_{\mathbb{R}} L(t)L(t+u)\ln^m(2+|t|)dt \le M \ln^{m+1}(2+|u|)L(u), \tag{4.8.18}$$

where M *is a constant depending on* m.

Proof For $|u| \le 2$, we have

$$\begin{aligned} Q(u) &\le \ln^m 3 \int_{|t|\le 3} dt + \int_{|t|>3} \frac{1}{|t(t+u)|} \ln^m(2+|t|)dt \\ &\le M + M \int_1^\infty \frac{1}{t^2} \ln^m(2+|t|)dt \le M \ln^{m+1}(2+|u|)L(u). \end{aligned} \tag{4.8.19}$$

In the case when $|u| > 2$, we can write

$$\begin{aligned} Q(u) &= \int_{|t|\le 1} \frac{\ln^m(2+|t|)}{|t+u|} dt + \int_{|t+u|\le 1} \frac{\ln^m(2+|t|)}{|t|} dt \\ &+ \int_{|t|>1,\ |t+u|>1} \frac{\ln^m(2+|t|)}{|t(t+u)|} dt =: Q_1(u) + Q_2(u) + Q_3(u). \end{aligned} \tag{4.8.20}$$

Now we estimate the functions $Q_i(u)$, $i = 1, 2, 3$. Assuming, without loss of generality, that $u > 2$, for $Q_1(u)$, we have

$$Q_1(u) = \int_{-1}^{1} \frac{\ln^m(2+|t|)}{t+u} dt \le \frac{M}{u-1} \le M \ln^{m+1}(2+|u|)L(u). \tag{4.8.21}$$

Similarly, for $Q_2(u)$, we obtain the estimate

$$Q_2(u) \le M \ln^{m+1}(2+|u|)L(u). \tag{4.8.22}$$

To estimate $Q_3(u)$, we first represent it in the following form:

$$\begin{aligned} Q_3(u) &= \int_{-\infty}^{-u-1} + \int_{-u+1}^{-1} + \int_1^\infty \frac{\ln^m(2+|t|)}{|t(t+u)|} dt \\ &=: Q_{31}(u) + Q_{32}(u) + Q_{33}(u). \end{aligned} \tag{4.8.23}$$

We now estimate the functions $Q_{3i}(u)$, $i = 1, 2, 3$. We have

$$
\begin{aligned}
Q_{31}(u) &\le \int_1^\infty \frac{\ln^m(t+u+2)}{t(t+u)}dt \le \frac{\ln^m(3u)}{u}\int_1^u \frac{1}{t}dt + M\int_u^\infty \frac{\ln^m(t+u)}{(t+u)^2}dt \\
&\le M\frac{\ln^{m+1} u}{u} + M\int_{2u}^\infty \frac{\ln^m t}{t^2}dt \\
&\le M\frac{\ln^{m+1} u}{u} \le M\ln^{m+1}(2+|u|)L(u).
\end{aligned} \tag{4.8.24}
$$

Similarly, for $Q_{33}(u)$, we obtain

$$
Q_{33}(u) \le M\ln^{m+1}(2+|u|)L(u). \tag{4.8.25}
$$

Finally, for $Q_{32}(u)$, we have

$$
Q_{32}(u) \le \int_1^{u+1} \frac{\ln^m t}{t(t-u)}dt \le \frac{\ln^{m+1} u}{u} \le M\ln^{m+1}(2+|u|)L(u). \tag{4.8.26}
$$

By combining the inequalities (4.8.19)–(4.8.26), we arrive at (4.8.18).

Lemma 4.8.1 is proved. □

The following two lemmas present properties of the Dirichlet-type kernel $H_{k,T}(\lambda)$ as defined in (4.8.3).

Lemma 4.8.2 *The function $H_{k,T}(\lambda)$ defined by (4.8.3) satisfies the following identity:*

$$
\int_\Lambda H_{k,T}(\lambda-u)H_{j,T}(u-\mu)du = 2\pi H_{k+j,T}(\lambda-\mu), \quad \lambda,\mu\in\Lambda. \tag{4.8.27}
$$

Proof The lemma we prove for $\Lambda = \mathbb{R}$. The proof for $\Lambda = \mathbb{T}$ is similar (see also Dahlhaus [70]). Without loss of generality, we assume that $\mu = 0$ and observe that $H_{k,T} \in L^2(\mathbb{R})$. Hence, by (4.8.3), we have

$$
\widehat{H_{k,T}}(\zeta) = 2\pi h_T^k(-\zeta) \quad \text{and} \quad \widehat{H_{k,T}(\lambda-\cdot)}(\zeta) = 2\pi h_T^k(\zeta)e^{-i\lambda\zeta},
$$

where $\hat{g}$ is the Fourier transform of g. By using Parseval-Plancherel's identity, we obtain

$$
\begin{aligned}
\int_\mathbb{R} H_{k,T}(\lambda-u)H_{j,T}(u)du &= \frac{1}{2\pi}(2\pi)^2\int_\mathbb{R} h_T^k(\zeta)h_T^j(\zeta)e^{-i\lambda\zeta}d\zeta \\
&= 2\pi\int_\mathbb{R} h_T^{k+j}(\zeta)e^{-i\lambda\zeta}d\zeta = 2\pi H_{k+j,T}(\lambda),
\end{aligned}
$$

and the result follows. □

Lemma 4.8.3 *Let the functions $H_{k,T}(\cdot)$ and $L_T(\cdot)$ be defined by (4.8.3) and (4.8.16), respectively. Then for all $k \in \mathbb{N}$ and a constant C_k independent of T, the following inequality holds:*

$$|H_{k,T}(\lambda)| \le C_k\, L_T(\lambda). \tag{4.8.28}$$

Proof Since the taper function h is assumed to have bounded variation, denoting by $V(h)$ the total variation of h, we can write

$$\int_{\mathbb{R}} \left| \prod_{i=1}^{k} h(t+u_i) - h^k(t) \right| dt \le ||h||_\infty^{k-1} V(h) \sum_{i=1}^{k} |u_i|. \tag{4.8.29}$$

Therefore, for the Fourier transform of $h_T(t) = h(t/T)$, we obtain

$$|H_{k,T}(\lambda)| \le \frac{1}{2} \int_{\mathbb{R}} \left| h^k(t) - h^k\left(t - \frac{\pi}{\lambda}\right) \right| dt \le \frac{1}{2} ||h||_\infty^{k-1} V(h) k\pi |\lambda|^{-1}. \tag{4.8.30}$$

On the other hand, we have

$$|H_{k,T}(\lambda)| \le ||h||_\infty^{k} T. \tag{4.8.31}$$

A combination of (4.8.30) and (4.8.31), in view of (4.8.16), yields (4.8.28). □

For a number k ($k = 2, 3, \dots$) and a taper function h satisfying Assumption 4.8.1, consider the following tapered Fejér-type kernel function:

$$\Phi_{k,T}^h(\mathbf{u}) := \frac{H_T(\mathbf{u})}{(2\pi)^{k-1} H_{k,T}(0)}, \quad \mathbf{u} = (u_1, \dots, u_{k-1}) \in \mathbb{R}^{k-1}, \tag{4.8.32}$$

where

$$H_T(\mathbf{u}) := H_{1,T}(u_1) \cdots H_{1,T}(u_{k-1}) H_{1,T}\left(-\sum_{j=1}^{k-1} u_j \right). \tag{4.8.33}$$

The function $H_{k,T}(\cdot)$ is defined by (4.8.3) with $H_{k,T}(0) = T \cdot H_k \neq 0$ (see 4.8.5).

For $\mathbf{u} = (u_1, \dots, u_{k-1}) \in \mathbb{R}^{k-1}$, denote

$$\Psi(\mathbf{u}) := \varphi(u_1, u_1 + u_2, \dots, u_1 + \cdots + u_{k-1}), \tag{4.8.34}$$

where the function $\varphi(\mathbf{u})$, corresponding to the collection $\mathcal{H} = \{\psi_1, \psi_2, \dots, \psi_k\}$, is defined by (4.8.13).

The next lemma is a tapered version of Lemma 4.6.6 and provides formulas for the trace of the product of tapered Toeplitz operators.

Lemma 4.8.4 *Let $\mathcal{H} = \{\psi_1, \psi_2, \ldots, \psi_k\}$ be a collection of integrable real symmetric functions defined on $\mathbb{R}$, and let $S_k^h(T) := S_{W,\mathcal{H},h}(T)$ be defined as in (4.8.9). The following equalities hold:*

$$
\begin{aligned}
S_k^h(T) &= \frac{1}{T}\int_{\mathbb{R}^k} \psi_1(x_1)\ldots\psi_k(x_k)H_{1,T}(x_1-x_2)H_{1,T}(x_2-x_3)\cdots \\
&\quad\times H_{1,T}(x_{k-1}-x_k)H_{1,T}(x_k-x_1)dx_1\ldots dx_k \\
&= (2\pi)^{k-1}H_k\int_{\mathbb{R}^{k-1}} \Psi(\mathbf{u})\Phi_{k,T}^h(\mathbf{u})d\mathbf{u},
\end{aligned}
\tag{4.8.35}
$$

where $\mathbf{u} = (u_1, \ldots, u_{k-1}) \in \mathbb{R}^{k-1}$, $d\mathbf{u} = du_1\cdots du_{k-1}$, and H_k is defined in (4.8.5).

Proof The result follows from the formulas (4.8.3), (4.8.6), (4.8.8), (4.8.9), (4.8.13), (4.8.32)–(4.8.34), a formula for the traces of integral operators (see Proposition 2.7.7), and the change of variables:

$$u_1 = x_1 - x_2,\ u_2 = x_2 - x_3,\ \ldots,\ u_{k-1} = x_{k-1} - x_k.$$

□

The following lemma, a tapered version of Lemma 4.6.7, shows that similar to the non-tapered case, the tapered kernel $\Phi_{k,T}^h(\mathbf{u})$, defined by (4.8.32), serves as an approximate identity.

Lemma 4.8.5 *For any $k = 2, 3, \ldots$ and a taper function h satisfying Assumption 4.8.1, the kernel $\Phi_{k,T}^h(\mathbf{u})$, where $\mathbf{u} = (u_1, \ldots, u_{k-1})$, has the following properties:*

(a) $\sup_{T>0}\int_{\mathbb{R}^{k-1}}\left|\Phi_{k,T}^h(\mathbf{u})\right| d\mathbf{u} = C_1 < \infty$.
(b) $\int_{\mathbb{R}^{k-1}}\Phi_{k,T}^h(\mathbf{u})\,d\mathbf{u} = 1$.
(c) $\lim_{T\to\infty}\int_{\mathbb{E}_\delta^c}\left|\Phi_{k,T}^h(\mathbf{u})\right| d\mathbf{u} = 0$ *for any* $\delta > 0$.
(d) If $k > 2$, for any $\delta > 0$, there exists a constant $M_\delta > 0$ such that for $T > 0$

$$\left\|\Phi_{k,T}^h\right\|_{L^{p_k}(\mathbb{E}_\delta^c)} \le M_\delta, \tag{4.8.36}$$

where $p_k = \frac{k-2}{k-3}$ for $k > 3$, $p_3 = \infty$, $\mathbb{E}_\delta^c = \mathbb{R}^{k-1}\setminus\mathbb{E}_\delta$, and

$$\mathbb{E}_\delta = \{\mathbf{u} = (u_1, \ldots, u_{k-1}) \in \mathbb{R}^{k-1} : |u_i| \le \delta,\ i = 1, \ldots, k-1\}.$$

Proof We begin with assertion (a). Note that by (4.8.3), we have

$$H_{1,T}(\lambda) := \int_{\mathbb{R}} h\left(\frac{t}{T}\right)e^{-i\lambda t}dt = T\int_{\mathbb{R}} h(t)e^{-iT\lambda t}dt = T\hat{h}(T\lambda), \tag{4.8.37}$$

where $\hat{h}$ is the Fourier transform of h. Hence, in view of (4.8.33), we can write

$$\begin{aligned}
&\int_{\mathbb{R}^{k-1}} |H_T(\mathbf{u})|\, d\mathbf{u} \\
&= T^k \int_{\mathbb{R}^{k-1}} \left|\hat{h}(Tu_1)\cdots\hat{h}(Tu_{k-1})\hat{h}(Tu_1+\cdots+Tu_{k-1})\right| du_1\cdots du_{k-1} \\
&= T \int_{\mathbb{R}^{k-1}} \left|\hat{h}(u_1)\cdots\hat{h}(u_{k-1})\hat{h}(u_1+\cdots+u_{k-1})\right| du_1\cdots du_{k-1}. \qquad (4.8.38)
\end{aligned}$$

Since, by assumption, h is a function of bounded variation with support on [0, 1], we have

$$\left|\hat{h}(u)\right| \le M\cdot L(u) \quad \text{for} \quad u\in\mathbb{R},$$

where the function L is defined as in (4.8.17), and M is a constant that depends on h. Taking into account that $H_{k,T}(0) = T\cdot H_k$, from Eqs. (4.8.18), (4.8.32), and (4.8.38), we obtain

$$\begin{aligned}
&\int_{\mathbb{R}^{k-1}} \left|\Phi^h_{k,T}(\mathbf{u})\right| d\mathbf{u} \\
&\le M \int_{\mathbb{R}^{k-1}} |L(u_1)\cdots L(u_{k-1})L(u_1+\cdots+u_{k-1})|\, du_1\cdots du_{k-1} \\
&\le M \int_{\mathbb{R}^{k-2}} |L(u_1)\cdots L(u_{k-2})L(u_1+\cdots u_{k-2})| \\
&\times \ln(2+|u_1+\cdots+u_{k-2}|)du_1\cdots du_{k-2} \\
&\le M \int_{\mathbb{R}^{k-3}} |L(u_1)\cdots L(u_{k-3})L(u_1+\cdots u_{k-3})| \\
&\times \ln^2(2+|u_1+\cdots+u_{k-3}|)du_1\cdots du_{k-3} \\
&\le \cdots\cdots\cdots\cdots\cdots\cdots\cdots\cdots\cdots\cdots \\
&\le M \int_{\mathbb{R}} \left|L^2(u_1)\right| \ln^{k-2}(2+|u_1|)du_1 < \infty, \qquad (4.8.39)
\end{aligned}$$

and assertion (a) follows.

The statement in part (b) follows directly from Eqs. (4.8.27), (4.8.32), and (4.8.33).

To prove assertion (c), we write

$$\begin{aligned}
\int_{\mathbb{E}^c_\delta} |\Phi^h_{k,T}(\mathbf{u})| &\le \int\limits_{|u_1|>\delta} + \int\limits_{|u_2|>\delta} + \cdots + \int\limits_{|u_{k-1}|>\delta} |\Phi^h_{k,T}(\mathbf{u})| d\mathbf{u} \\
&=: J_1 + J_2 + \cdots + J_{k-1}.
\end{aligned}$$

Using the same arguments as in (4.8.39), we obtain for $s = 1, 2, \ldots, k-1$

$$J_s \le M \int_{|u_s|>T\delta} \left|L^2(u_s)\right| \ln^{k-2}(2+|u_s|)du_s \to 0 \quad \text{as} \quad T \to \infty,$$

and the result follows.

To prove assertion (d), we first note that by (4.8.28), for $T > 0$

$$\int_{\mathbb{R}} H_{1,T}^{p_k}(u)du \le C \cdot T^{p_k-1} \quad \text{and} \quad \left|H_{1,T}(u)\right| \le C_\delta \quad \text{for} \quad |u| > \delta. \tag{4.8.40}$$

For $k = 3$, by (4.8.28), we have

$$\left|\Phi_{3,T}^h(u_1, u_2)\right| = \left|\frac{H_{1,T}(u_1)H_{1,T}(u_2)H_{1,T}(u_1+u_2)}{(2\pi)^2 T H_3}\right| \le M_\delta, \quad (u_1, u_2) \in \mathbb{E}_\delta^c,$$

since for any $(u_1, u_2) \in \mathbb{E}_\delta^c$, at least two of the three numbers $|u_1|$, $|u_2|$, $|u_1 + u_2|$ are greater than $\delta/3$.

In the case when $k \ge 4$, we have

$$\begin{aligned}\int_{\mathbb{E}_\delta^c} \left|\Phi_{k,T}^h(\mathbf{u})\right|^{p_k} d\mathbf{u} &\le \int_{|u_1|>\delta} \left|\Phi_{k,T}^h(\mathbf{u})\right|^{p_k} d\mathbf{u} + \cdots \\ &+ \int_{|u_{k-1}|>\delta} \left|\Phi_{k,T}^h(\mathbf{u})\right|^{p_k} d\mathbf{u} =: I_1 + \cdots + I_{k-1}.\end{aligned} \tag{4.8.41}$$

We estimate I_1, and the integrals $I_2, \ldots, I_{k-1}$ can be estimated similarly. We have

$$\begin{aligned}I_1 \le &\int_{|u_1|>\delta,\ |u_2|>\delta/k} \left|\Phi_{k,T}^h(\mathbf{u})\right|^{p_k} d\mathbf{u} + \cdots + \int_{|u_1|>\delta,\ |u_{k-1}|>\delta/k} \left|\Phi_{k,T}^h(\mathbf{u})\right|^{p_k} d\mathbf{u} \\ + &\int_{|u_1|>\delta,\ |u_2|\le\delta/k,\ldots,|u_{k-1}|\le\delta/k} \left|\Phi_{k,T}^h(\mathbf{u})\right|^{p_k} d\mathbf{u} =: I_{12} + I_{13} + \cdots + I_{1k}.\end{aligned} \tag{4.8.42}$$

According to (4.8.32) and (4.8.40), we have

$$\begin{aligned}I_{12} &\le C_\delta \cdot \frac{1}{T^{p_k}} \int_{|u_2|>\delta/k} H_{1,T}^{p_k}(u_2) \cdots H_{1,T}^{p_k}(u_{k-1}) H_{1,T}^{p_k}(u_1 + \cdots + u_{k-1}) \\ &\quad \times du_1 du_3 \cdots du_{k-1} du_2 \\ &\le C_\delta T^{(k-2)(p_k-1)-p_k} \int_{|u_2|>\delta/k} \frac{1}{u_2^2} du_2 \le M_\delta.\end{aligned} \tag{4.8.43}$$

Likewise, we obtain

$$I_{1s} \leq M_\delta, \qquad s = 3, 4, \ldots, k-1. \tag{4.8.44}$$

Note that in the integral I_{1k}, we have $|u_1 + \cdots + u_{k-1}| > \delta - \delta(k-2)/k > \delta/k$. Hence, by (4.8.40), we obtain

$$\begin{aligned} I_{1k} &\leq C_{\delta/k} \cdot \frac{1}{T^{p_k}} \int\limits_{|u_1|>\delta} H_{1,T}^{p_k}(u_1) H_{1,T}^{p_k}(u_2) \cdots H_{1,T}^{p_k}(u_{k-1}) du_2 \cdots du_{k-1} du_1 \\ &\leq M_\delta \int\limits_{|u_1|>\delta} \frac{1}{u_1^2} du_1 \leq M_\delta. \end{aligned} \tag{4.8.45}$$

From (4.8.41) to (4.8.45), we obtain (4.8.36). Lemma 4.8.5 is proved. □

Lemma 4.8.6 *Assume that the function Ψ belongs to $L^1(\mathbb{R}^{k-1}) \bigcap L^{k-2}(\mathbb{R}^{k-1})$ and is continuous at the point $\mathbf{v} = (v_1, \ldots, v_{k-1})$ (L^0 is the space of measurable functions). Then*

$$\lim_{T\to\infty} \int_{\mathbb{R}^{k-1}} \Psi(\mathbf{u}+\mathbf{v}) \Phi_{k,T}^h(\mathbf{u}) d\mathbf{u} = \Psi(\mathbf{v}). \tag{4.8.46}$$

Proof The assertion we prove for $\mathbf{v} = \mathbf{0} := (0, \ldots, 0)$. In the general case, one can consider the function $\overline{\Psi}(\mathbf{u}) = \Psi(\mathbf{u}+\mathbf{v})$ with $\overline{\Psi}(0) = \Psi(\mathbf{v})$.

By Lemma 4.8.5 (b), we have

$$R_{k,T}^h := \int_{\mathbb{R}^{k-1}} \Psi(\mathbf{u}) \Phi_{k,T}^h(\mathbf{u}) d\mathbf{u} - \Psi(\mathbf{0}) = \int_{\mathbb{R}^{k-1}} [\Psi(\mathbf{u}) - \Psi(\mathbf{0})] \Phi_{k,T}^h(\mathbf{u}) d\mathbf{u}. \tag{4.8.47}$$

Next, for any $\varepsilon > 0$, we can find $\delta > 0$ to satisfy

$$|\Psi(\mathbf{u}) - \Psi(\mathbf{0})| < \frac{\varepsilon}{C_1} \quad \text{for} \quad \mathbf{u} \in \mathbb{E}_\delta, \tag{4.8.48}$$

where C_1 is the constant from Lemma 4.8.5 (a).

For $k = 2$, we have

$$\left|\Phi_{2,T}^h(u)\right| = \left|\frac{H_{1,T}(u) H_{1,T}(-u)}{(2\pi) T H_2}\right| \leq \frac{M_\delta}{T}, \quad |u| > \delta.$$

Therefore, by Lemma 4.8.5 (a) and (c), and Eq. (4.8.47), for sufficiently large T, we obtain

$$
\begin{aligned}
|R^h_{2,T}| &\le \int_{|u|\le\delta} + \int_{|u|>\delta} |\Psi(u)-\Psi(0)| \left|\Phi^h_{2,T}(u)\right| du \\
&\le \frac{\varepsilon}{C_1}\int_{\mathbb{R}} \left|\Phi^h_{2,T}(u)\right| du + \frac{M_\delta}{T}\int_{\mathbb{R}} |\Psi(u)| du \\
&+ |\Psi(0)| \int_{|u|>\delta} \left|\Phi^h_{2,T}(u)\right| du \le 3\varepsilon.
\end{aligned}
$$

For $k > 2$, we consider the decomposition $\Psi = \Psi_1 + \Psi_2$ such that

$$
\|\Psi_1\|_{k-2} \le \frac{\varepsilon}{M_\delta} \quad \text{and} \quad \|\Psi_2\|_\infty < \infty, \tag{4.8.49}
$$

where M_δ is defined as in Lemma 4.8.5 (d). By applying Lemma 4.8.5 (a)–(d) and using formulas (4.8.47)–(4.8.49), for sufficiently large T, we obtain

$$
\begin{aligned}
|R^h_{k,T}| &\le \int_{\mathbb{E}_\delta} |\Psi(\mathbf{u})-\Psi(\mathbf{0})| \left|\Phi^h_{k,T}(\mathbf{u})\right| d\mathbf{u} + \int_{\mathbb{E}^c_\delta} |\Psi_1(\mathbf{u})| \left|\Phi^h_{k,T}(\mathbf{u})\right| d\mathbf{u} \\
&+ \int_{\mathbb{E}^c_\delta} |\Psi_2(\mathbf{u})-\Psi(\mathbf{0})| \left|\Phi^h_{k,T}(\mathbf{u})\right| d\mathbf{u} \le \frac{\varepsilon}{C_1}\int_{\mathbb{E}_\delta} \left|\Phi^h_{k,T}(\mathbf{u})\right| d\mathbf{u} \\
&+ \|\Psi_1\|_{k-2} \left\|\Phi^h_{k,T}\right\|_{L^{p_k}(\mathbb{E}^c_\delta)} + \left(\|\Psi_2\|_\infty + |\Psi(0)|\right) \int_{\mathbb{E}^c_\delta} \left|\Phi^h_{k,T}(\mathbf{u})\right| d\mathbf{u} \le 3\,\varepsilon.
\end{aligned}
$$

This, combined with (4.8.47), yields (4.8.46) for $\mathbf{v} = \mathbf{0}$. Lemma 4.8.6 is proved. □

We define

$$
\mu^h_T(A) := \frac{1}{T}\int_A H_T(\mathbf{u})d\mathbf{u}, \tag{4.8.50}
$$

where $\mathbf{u} = (u_1, \ldots, u_k) \in \mathbb{R}^k$, and $H_T(\mathbf{u})$ is defined as in (4.8.33).

Let $\mathbb{C}_{loc}(\mathbb{R}^k)$ be the space of continuous functions on $\mathbb{R}^k$ with bounded support. The following lemma is a tapered version of Lemma 4.6.9.

Lemma 4.8.7 *If $f \in \mathbb{C}_{loc}(\mathbb{R}^k)$, then*

$$
\lim_{T\to\infty}\int_{\mathbb{R}^k} f(\mathbf{x})\, d\mu^h_T = (2\pi)^{k-1} H_k \int_{\mathbb{R}} f(u,\ldots,u)du, \tag{4.8.51}
$$

where $\mathbf{x} = (x_1, \ldots, x_k) \in \mathbb{R}^k$, $\mu^h_T(A)$ is defined as in (4.8.50), *and H_k is defined as in* (4.8.5).

Proof The proof is similar to that of Lemma 4.6.9. Making a change of variables

$$x_1 = u, \quad x_1 - x_2 = u_1, \quad x_2 - x_3 = u_2, \dots, x_{k-1} - x_k = u_{k-1},$$

in view of Eqs. (4.8.32), (4.8.33), (4.8.50), and Lemma 4.8.4, we can write

$$\int_{\mathbb{R}^k} f(\mathbf{x}) d\mu_T^h = (2\pi)^{k-1} H_k \int_{\mathbb{R}^{k-1}} \Psi(\mathbf{u}) \Phi_{k,T}^h(\mathbf{u})) d\mathbf{u}, \tag{4.8.52}$$

where

$$\Psi(\mathbf{u}) = \int_{\mathbb{R}} f\left(u, u - u_1, u - u_1 - u_2, \dots u - \sum_{i=1}^{k-1} u_i\right) du.$$

It is straightforward to verify that the function Ψ satisfies the conditions of Lemma 4.8.6, and

$$\lim_{\mathbf{u} \to \mathbf{0}} \Psi(\mathbf{u}) = \int_{\mathbb{R}} f(u, \cdots, u) du. \tag{4.8.53}$$

Therefore, by applying Lemma 4.8.6 from (4.8.52) and (4.8.53), we obtain (4.8.51).
The next lemma is the tapered version of Lemma 4.6.10. □

Lemma 4.8.8 *If $f \in L^1(\mathbb{R})$, then the following inequalities hold:*

$$\int_{\mathbb{R}^k} |f(x_i)| d|\mu_T^h| \le C_1 \|f\|_{L^1(\mathbb{R})}, \quad i = 1, \dots, k, \tag{4.8.54}$$

$$\int_{\mathbb{R}^k} |f(x_i) f(x_j)| d|\mu_T^h| \le C_2 \|f\|_{L^2(\mathbb{R})}^2, \quad i, j = 1, \dots, k, \; i \ne j. \tag{4.8.55}$$

where C_1 and C_2 are absolute constants, and μ_T^h is defined as in (4.8.50).

Proof Since h is a function of bounded variation with support on [0, 1], in view of (4.8.3), for $T > 0$, we have

$$|H_{1,T}(x)| \le C_h T \psi_T(x), \quad \text{where} \quad \psi_T(x) = \frac{1}{1 + T|x|}, \quad x \in \mathbb{R}. \tag{4.8.56}$$

The remainder of the proof follows the arguments used in the proof of Lemma 4.6.10, employing inequality (4.8.56) instead of (4.6.59). □

The next lemma is the tapered version of Lemma 4.6.11.

Lemma 4.8.9 *Let $\phi(\mathbf{u}) = \prod_{i=1}^{k} \phi_i(u_i)$ with $\phi_i \in L^1(\mathbb{R}) \bigcap L^\infty(\mathbb{R})$, $i = 1, \dots, k$. Then the asymptotic relation (4.8.51) holds.*

Proof The proof reiterates the arguments presented in the proof of Lemma 4.6.11, utilizing Lemma 4.8.8 in place of Lemma 4.6.10. □

4.8.3.2 Proofs of Theorems 4.8.1 and 4.8.2

Proof of Theorem 4.8.1 We begin with assertion **(C1)** of the theorem. By Eq. (4.8.9) and Lemma 4.8.4, we have

$$S_{W,\mathcal{H},h}(T) = (2\pi)^{k-1} H_k \int_{\mathbb{R}^{k-1}} \Psi(\mathbf{u}) \Phi^h_{k,T}(\mathbf{u}) d\mathbf{u}.$$

By Eqs. (4.8.10), (4.8.13), and (4.8.34), we obtain

$$M_{\mathbb{R},\mathcal{H},h} := (2\pi)^{k-1} H_k \int_{\mathbb{R}} \left[\prod_{i=1}^{k} \psi_i(\lambda) \right] d\lambda = (2\pi)^{k-1} H_k \Psi(\mathbf{0}). \qquad (4.8.57)$$

Next, from Eqs. (4.8.11), (4.8.3.2), (4.8.57), and Lemma 4.8.5 (b), we obtain

$$\begin{aligned} \Delta_h(T) &= |S_{W,\mathcal{H},h}(T) - M_{\mathbb{R},\mathcal{H},h}| = \\ &= (2\pi)^{k-1} H_k \left| \int_{\mathbb{R}^{k-1}} [\Psi(\mathbf{u}) - \Psi(\mathbf{0})] \Phi^h_{k,T}(\mathbf{u}) d\mathbf{u} \right|. \end{aligned} \qquad (4.8.58)$$

Finally, by applying Lemma 4.8.6 with $\mathbf{v} = \mathbf{0}$, we can conclude the proof of part **(C1)** of the theorem.

To prove part **(C2)** of the theorem, we observe that in the proof of Theorem 4.5.1 **(A2)**, it was established that the condition $\psi_i \in L^1(\mathbb{R}) \cap L^{p_i}(\mathbb{R})$, $p_i > 1$, $i = 1, 2, \ldots, k$, with $1/p_1 + \ldots + 1/p_k \leq 1$ implies that the function $\Psi(\mathbf{u})$ defined by (4.8.34) belongs to $L^1(\mathbb{R}^{k-1}) \bigcap L^{k-2}(\mathbb{R}^{k-1})$ and is continuous at the point $\mathbf{0} = (0, \ldots, 0) \in \Lambda^{k-1}$ $(k = 2, 3, \ldots)$. Therefore, the result follows from Lemma 4.8.6 with $\mathbf{v} = \mathbf{0}$. Theorem 4.8.1 is proved. □

Proof of Theorem 4.8.2 The proof of part **(C3)** of the theorem is similar to that of Theorem 4.5.2, while the proof of part **(C4)** is analogous to that of Theorem 4.5.3, employing the tapered versions of the corresponding lemmas. Therefore, we omit the details. □

4.9 Estimates for Schatten Norms

In this section, we present various estimates of Schatten norms for both Toeplitz and tapered Toeplitz matrices and operators. For the definitions and properties of the Schatten classes and norms, see Sect. 2.7.2.

4.9.1 Estimates for Schatten Norms: The Non-tapered Case

The following theorem provides an upper bound for the Schatten norms of truncated Toeplitz matrices and operators.

Theorem 4.9.1 *Let $\psi(u) \in L^1(\Lambda) \cap L^p(\Lambda)$, with $1 \le p \le \infty$, and let $A_T(\psi)$ denote either the $(T \times T)$-truncated Toeplitz matrix $B_T(\psi)$ or the T-truncated Toeplitz operator $W_T(\psi)$ generated by the function ψ, defined respectively by Eqs. (4.2.8) and (4.2.11). Then, for the Schatten norm of $A_T(\psi)$, the following inequality holds:*

$$||A_T(\psi)||_p \le T^{1/p}||\psi||_p. \tag{4.9.1}$$

Proof The theorem we prove for Toeplitz operators: $A_T(\psi) = W_T(\psi)$. The proof for matrices is similar. It is known (see Gohberg and Krein [155], Theorem 13.1, or Zhu [320], Theorem 2.6.) that the Riesz-Thorin interpolation theorem applies to the Schatten classes $\mathcal{S}_p$. Therefore, it suffices to establish the inequality in (4.9.1) for $p = 1$ and $p = \infty$.

Case $p = \infty$. Taking into account the definition of the operator $W_T(\psi)$ (see 4.2.11), in view of Proposition 2.7.6 (d), we have

$$\begin{aligned} ||W_T(\psi)||_\infty = ||W_T(\psi)|| &= \sup_{u \in L^2(\mathbb{R}),\ ||u||_2=1} |(W_T(\psi)u, u)| \\ &= \sup_{u \in L^2(\mathbb{R}),\ ||u||_2=1} \left| \int_0^T \int_0^T \widehat{\psi}(t-s)u(t)u(s)\,dt\,ds \right| \qquad (4.9.2) \\ &= \sup_{u \in L^2(\mathbb{R}),\ ||u||_2=1} \left| \int_0^T \int_0^T \left[\int_{-\infty}^{+\infty} e^{i\lambda(t-s)} \psi(\lambda)\,d\lambda \right] u(t)u(s)\,dt\,ds \right|. \end{aligned}$$

A square integrable function $u(t)$ is also integrable on $[0, T]$. Hence, by switching the order of integration in (4.9.2), we obtain

$$\begin{aligned} ||W_T(\psi)||_\infty &= \sup_{u \in L^2(\mathbb{R}),\ ||u||_2=1} \left| \int_{-\infty}^{+\infty} \psi(\lambda) \left[\int_0^T u(t)e^{it\lambda}\,dt \int_0^T u(s)e^{-is\lambda}\,ds \right] d\lambda \right| \\ &\le \sup_{u \in L^2(\mathbb{R}),\ ||u||_2=1} \int_{-\infty}^{+\infty} |\psi(\lambda)| \left| \int_0^T u(t)\,e^{i\lambda t} dt \right|^2 d\lambda. \qquad (4.9.3) \end{aligned}$$

Next, using the Parseval-Plancherel theorem, we obtain

$$||W_T(\psi)||_\infty \le ||\psi(\lambda)||_\infty \sup_{||u||_2=1} \int_{-\infty}^{+\infty} \left| \int_0^T u(t)\,e^{i\lambda t} dt \right|^2 d\lambda = ||\psi||_\infty,$$

yielding (4.9.1) for $p = \infty$.
Case $p = 1$. We need to prove that

$$||W_T(\psi)||_1 \leq T \int_{-\infty}^{+\infty} |\psi(\lambda)| d\lambda. \tag{4.9.4}$$

First, suppose that $\psi(\lambda) \geq 0$. Then, $W_T(\psi) \geq 0$. Hence, using the formula for the trace of an integral operator (see Proposition 2.7.7), we have in view of Proposition 2.7.6 (a) that

$$||W_T(\psi)||_1 = \text{tr } W_T(\psi) = T \int_{-\infty}^{+\infty} |\psi(\lambda)| d\lambda,$$

yielding (4.9.4) for $\psi(\lambda) \geq 0$. It is easy to see that the inequality (4.9.4) remains valid when the function $\psi(\lambda)$ preserves its sign. Therefore, setting $\psi_+(\lambda) = \max(0, \psi(\lambda))$ and $\psi_-(\lambda) = \min(0, \psi(\lambda))$, we can write

$$\begin{aligned} ||W_T(\psi)||_1 &= ||W_T(\psi_+) + W_T(\psi_-)||_1 \\ &\leq ||W_T(\psi_+)||_1 + ||W_T(\psi_-)||_1 = T \int_{-\infty}^{+\infty} |\psi(\lambda)| d\lambda. \end{aligned}$$

This completes the proof. □

Remark 4.9.1 The proof of the theorem for the case $p = \infty$ can be simplified by considering the realization of the operator $W_T(\psi)$ in terms of a multiplication operator and an orthogonal projector. Specifically, let P_T denote the orthogonal projection from $L^2(\mathbb{R})$ onto the space $\mathcal{E}_T(\mathbb{R})$ of entire functions of exponential type at most T, and let $\boldsymbol{\psi}$ be the multiplication operator by the function ψ. Then, we have (see 4.2.13)

$$W_T(\psi)\varphi = P_T \boldsymbol{\psi} P_T \varphi, \quad \varphi \in L^2(\mathbb{R}). \tag{4.9.5}$$

Therefore,

$$||W_T(\psi)||_\infty = ||P_T \boldsymbol{\psi} P_T||_\infty \leq ||\psi||_\infty.$$

Theorem 4.9.2 *Let* $\psi(u) \in L^1(\Lambda) \cap L^p(\Lambda)$, *where* $1 < p < \infty$. *Then*

$$||A_T(\psi)||_\infty = o\big(T^{1/p}\big) \quad as \quad T \to \infty. \tag{4.9.6}$$

Proof The theorem we prove for Toeplitz operators: $A_T(\psi) = W_T(\psi)$. Let N_T be a positive function of T, which we will specify later. We set

$$M_T = \{\lambda \in \mathbb{R};\ |\psi(\lambda)| > N_T\}. \tag{4.9.7}$$

Then, by (4.9.3), we have

$$\mu_T := ||W_T(\psi)||_\infty \le \sup_{u\in L^2(\mathbb{R}),\ ||u||_2=1} \int_{-\infty}^{+\infty} |\psi(\lambda)| \left| \int_0^T u(t)\, e^{i\lambda t} dt \right|^2 d\lambda. \tag{4.9.8}$$

For $u(t) \in L^2[0, T]$ with $||u||_2 = 1$, we have $\left|\int_0^T u(t)\, e^{i\lambda t} dt\right|^2 \le T$. Hence, by applying the Parseval-Plancherel theorem, we can derive from (4.9.7) and (4.9.8) that

$$\mu_T \le 2\pi N_T + T \int_{M_T} |\psi(\lambda)|\, d\lambda. \tag{4.9.9}$$

Next, we show that for every p $(1 < p < \infty)$

$$\int_{M_T} |\psi(\lambda)|\, d\lambda \le \gamma_T^p\, N_T^{(1-p)}, \tag{4.9.10}$$

where

$$\gamma_T = \left(\int_{M_T} |\psi(\lambda)|^p\, d\lambda \right)^{1/p}. \tag{4.9.11}$$

Indeed, by Hölder's inequality, we have

$$\int_{M_T} |\psi(\lambda)|\, d\lambda \le \gamma_T \left[m(M_T)\right]^{(p-1)/p}, \tag{4.9.12}$$

where $m(M_T)$ is the Lebesgue measure of the set M_T.

Next, using Chebyshev's inequality, we find that

$$m(M_T) \le \gamma_T^p\, N_T^{-p}. \tag{4.9.13}$$

A combination of (4.9.12) and (4.9.13) yields (4.9.10).

Now, from (4.9.9) and (4.9.10), we obtain

$$\mu_T \le 2\pi N_T + T \gamma_T^p\, N_T^{(1-p)}. \tag{4.9.14}$$

If $\psi \in L^\infty(\mathbb{R})$, then by setting $N_T = \|\psi\|_\infty$ for all $T > 0$, we find that $\gamma_T = 0$. Consequently, (4.9.14) implies that $\mu_T = O(1)$ as $T \to \infty$.

Now, suppose $\psi \notin L^\infty(\mathbb{R})$ and for a fixed $T > 0$, consider the function:

$$F(N) := N - T^{1/p} \left(\int_{\{\lambda : |\psi(\lambda)| > N\}} |\psi(\lambda)|^p\, d\lambda \right)^{1/p}, \quad N \in [0, \infty).$$

Since $F(0) < 0$ and $\lim_{N\to\infty} F(N) = +\infty$, there exists a positive number $N = N_T$ with $F(N_T) = 0$. Thus, we have

$$N_T = T^{1/p}\left(\int_{\{\lambda:|\psi(\lambda)|>N_T\}} |\psi(\lambda)|^p\right) = T^{1/p}\gamma_T. \tag{4.9.15}$$

For $\psi \notin L^\infty(\mathbb{R})$, the equality (4.9.15) implies that $\lim_{T\to\infty} N_T = \infty$. Therefore, $\gamma_T = o(1)$, and from (4.9.14) and (4.9.15), we obtain

$$\mu_T < CT^{1/p}\gamma_T = o(T^{1/p}) \quad \text{as} \quad T \to \infty.$$

Theorem 4.9.2 is proved. □

Theorem 4.9.3 *Let $\psi(u) \in L^1(\Lambda) \cap L^2(\Lambda)$. Then*

$$\frac{1}{T}||A_T(\psi)||_2^2 \longrightarrow 2\pi||\psi||_2^2 < \infty \quad as \quad T \to \infty. \tag{4.9.16}$$

Proof Using (4.2.11) and the formula for the Hilbert-Schmidt norm of integral operators (see Sect. 2.7.3, Eq. (2.7.10)), we can write

$$\begin{aligned}\frac{1}{T}||W_T(\psi)||_2^2 &= \frac{1}{T}\int_0^T\int_0^T |\widehat{\psi}(t-s)|^2\,dt\,ds\\ &= \frac{1}{T}\int_{-T}^T (T-|t|)|\widehat{\psi}(t)|^2\,dt = \int_{-T}^T\left(1-\frac{|t|}{T}\right)|\widehat{\psi}(t)|^2\,dt.\end{aligned} \tag{4.9.17}$$

Passing to the limit as $T \to \infty$ and applying Parseval-Plancherel's identity, from (4.9.17) we obtain (4.9.16). Theorem 4.9.3 is proved. □

Proposition 4.9.1 *Let $A_T(f)$ and $A_T(g)$ denote either the $(T \times T)$-truncated Toeplitz matrices $B_T(f)$ and $B_T(g)$ or the T-truncated Toeplitz operators $W_T(f)$ and $W_T(g)$ generated by the functions f and g, respectively. If the pair (f, g) satisfies condition $(\mathcal{H})$ (see Definition 2.1.1), then*

$$\lim_{T\to\infty} T^{-1/2}\|A_T(f)\,A_T(g)\|_\infty = 0. \tag{4.9.18}$$

Proof By Proposition 2.1.9, under the condition $(\mathcal{H})$, there exist numbers p_1 ($p_1 > p$) and q_1 ($q_1 > q$), such that $H_p(\beta_1) \subset L_{p_1}$, $H_q(\beta_2) \subset L_{q_1}$, and $1/p_1 + 1/q_1 \le 1/2$. Hence, applying Theorem 4.9.2, we have

$$T^{-1/2}\|A_T(f)\,A_T(g)\|_\infty = o\big(T^{1/p_1+1/q_1-1/2}\big) \quad \text{as} \quad T \to \infty. \tag{4.9.19}$$

Taking into account that $1/p_1 + 1/q_1 - 1/2 \le 0$, from (4.9.19) we obtain (4.9.18). □

As an immediate consequence of Proposition 2.7.6 (c) and Theorem 4.9.1, we have the following result.

Proposition 4.9.2 *Let $A_T(f)$ and $A_T(g)$ be as in Proposition 4.9.1 with f and g satisfying $f \in L^1(\Lambda) \cap L^p(\Lambda)$ and $g \in L^1(\Lambda) \cap L^q(\Lambda)$ and $1/p + 1/q = 1/2$. Then*

$$||A_T(f)A_T(g)||_2 \le T^{1/2}||f||_p||g||_q. \tag{4.9.20}$$

4.9.2 Estimates for Schatten Norms: The Tapered Case

Let $\psi(u) \in L^1(\Lambda) \cap L^p(\Lambda)$, with $1 \le p \le \infty$, and let $A_T^h(\psi)$ denote either the $(T \times T)$-truncated tapered Toeplitz matrix $(B_T^h(\psi))$ or the T-truncated tapered Toeplitz operator $(W_T^h(\psi))$ generated by function ψ and a taper function h satisfying Assumption 4.8.1, defined respectively by (4.8.7) and (4.8.8). The following theorem provides estimates for the Schatten norms of tapered truncated Toeplitz matrices and operators.

Theorem 4.9.4 *The following assertions hold:*

(a) For $\psi(u) \in L^1(\Lambda) \cap L^p(\Lambda)$ with $1 \le p \le \infty$, we have

$$||A_T^h(\psi)||_p \le T^{1/p}||\psi||_p. \tag{4.9.21}$$

(b) For $\psi(u) \in L^1(\Lambda) \cap L^p(\Lambda)$ with $1 < p < \infty$, we have

$$\lambda_T := ||A_T^h(\psi)||_\infty = o\big(T^{1/p}\big) \quad as \quad T \to \infty. \tag{4.9.22}$$

(c) For $\psi(u) \in L^1(\Lambda) \cap L^2(\Lambda)$, we have

$$\frac{1}{T}||A_T^h(\psi)||_2^2 \longrightarrow 2\pi H_4||\psi||_2^2 \quad as \quad T \to \infty, \tag{4.9.23}$$

where H_4 is defined as in (4.8.5).

Proof The proofs of assertions (a) and (b) closely resemble those for the non-tapered case, requiring only minor modifications (see the proofs of Theorems 4.9.1 and 4.9.2). Therefore, we will focus on proving assertion (c) for Toeplitz operators $W_T(\psi)$. The proof for matrices follows a similar approach.

Using the formula for the Hilbert-Schmidt norm of integral operators (see Sect. 2.7.3, Eq. 2.7.10), we have

$$||W_T^h(\psi)||_2^2 = \int_0^T \int_0^T |\widehat{\psi}(t-s)|^2 |h_T(t)h_T(s)|^2 \, dt \, ds. \tag{4.9.24}$$

Making the change of variables $t - s = u$ and noting that the taper function h is supported on $[0, 1]$, we obtain from (4.9.24) that

$$||W_T^h(\psi)||_2^2 = \int_{\mathbb{R}}\int_{\mathbb{R}} |\widehat{\psi}(u)|^2 |h_T(s+u)h_T(s)|^2 \, du \, ds. \tag{4.9.25}$$

Therefore, taking into account that $h_T(t) = h(t/T)$ and using the change of variables $s/T = v$, we can write

$$\begin{aligned} \frac{1}{T}||W_T^h(\psi)||_2^2 &= \frac{1}{T}\int_{\mathbb{R}} |\widehat{\psi}(u)|^2 \left[\int_0^T |h(s/T + u/T)h(s/T)|^2 \, ds\right] du \\ &= \int_{\mathbb{R}} |\widehat{\psi}(u)|^2 \left[\int_0^1 |h(v + u/T)h(v)|^2 \, dv\right] du. \end{aligned} \tag{4.9.26}$$

For the inside integral on the right-hand side of (4.9.26), in view of (4.8.6), we have

$$\lim_{T\to\infty} \int_0^1 |h(v + u/T)h(v)|^2 \, dv = \int_0^1 h^4(v) \, dv = H_4. \tag{4.9.27}$$

Finally, using Parseval-Plancherel's equality from (4.9.26) and (4.9.27), we obtain the desired relation (4.9.23). □

Remark 4.9.2 The results of Propositions 4.9.1 and 4.9.2 hold true for the tapered Toeplitz operators $A_T^h(f)$ and $A_T^h(g)$.

4.9.3 Estimates for Schatten Norms with Fisher-Hartwig-Type Singularities

In this section, we examine Toeplitz matrices and show that for a specific important class of generating functions f exhibiting *Fisher-Hartwig type singularities*, the estimate in (4.9.1) for the Toeplitz matrix $B_T(f)$ can be significantly improved.

An archetypal example of a generating function with a Fisher-Hartwig-type singularity is the following:

$$\omega_\alpha^+(\lambda) := \begin{cases} \lambda^{-\alpha} & \text{for } \lambda \in (0, \pi), \\ 0 & \text{for } \lambda \in (-\pi, 0), \end{cases} \qquad 0 < \alpha < 1. \tag{4.9.28}$$

For ω_α^+ given by (4.9.28), one can compute the Frobenius norm of the Toeplitz matrix $B_T(\omega_\alpha^+)$, which is just the Schatten norm for $p = 2$, to obtain that

$$||B_T(\omega_\alpha^+)||_2 \le \begin{cases} C_2(\alpha)T^{1/2} & \text{if } \alpha < 1/2, \\ C_2(\alpha)(T\ln T)^{1/2} & \text{if } \alpha = 1/2, \\ C_2(\alpha)T^{\alpha} & \text{if } \alpha > 1/2, \end{cases} \tag{4.9.29}$$

where $C_2(\alpha) \in (0, \infty)$ is a constant that depends on α (see, e.g., Böttcher [47], and Vong and Jin [311]).

Define $\omega_\alpha^-(\lambda) := \omega_\alpha^+(-\lambda)$, and consider the class $\mathcal{F}(\beta, \gamma)$ of generating functions f of the form

$$f(\lambda) := \omega_\beta^+(\lambda)u(\lambda) + \omega_\gamma^-(\lambda)v(\lambda), \tag{4.9.30}$$

where $0 < \beta < 1$, $0 < \gamma < 1$, $u \in L^\infty$, and $v \in L^\infty$.

We will say that a function f has a *single Fisher-Hartwig singularity of degree* α if f is of the form (4.9.30) with $\beta = \gamma = \alpha$.

Examples of functions belonging to the class $\mathcal{F}(\beta, \gamma)$ are the following:

$$f(\lambda) = |\lambda|^{-\alpha} = \omega_\alpha^+(\lambda) + \omega_\alpha^-(\lambda), \tag{4.9.31}$$

$$f(\lambda) = |e^{i\lambda} - 1|^{-\alpha} = |\omega_\alpha^+(\lambda) + \omega_\alpha^-(\lambda)| \frac{|\lambda|^{-\alpha}}{(2|\sin(\lambda/2)|)^\alpha}, \tag{4.9.32}$$

$$f(\lambda) = \begin{cases} |e^{i\lambda} - 1|^{-\beta} \exp(i\delta_1(\lambda - \pi)) \text{ for } \lambda < 0, \\ |e^{i\lambda} - 1|^{-\gamma} \exp(i\delta_2(\pi - \lambda)) \text{ for } \lambda > 0, \end{cases} \tag{4.9.33}$$

where $\alpha, \beta, \gamma \in (0, 1)$ and $\delta_1, \delta_2 \in \mathbf{C}$.

Note that the class $\mathcal{F}(\beta, \gamma)$ specifically includes all functions that have a single Fisher-Hartwig singularity. This means it includes all functions of the form (4.9.33) where $\beta = \gamma$ (see, e.g., Böttcher and Silbermann [52]). The following theorem was proved by Böttcher [47].

Theorem 4.9.5 *Let $f \in \mathcal{F}(\beta, \gamma)$, that is, f is of the form* (4.9.30) *with $\beta, \gamma \in (0, 1)$ and $u, v \in L^\infty$, and let $\alpha = \max(\beta, \gamma)$. Let $B_T(f)$ be the Toeplitz matrix generated by the function f. If $1/\alpha \le p < \infty$, then there exists a constant $C = C_p(f) \in (0, \infty)$ that depends on p and α such that for all $T \ge 1$, the following inequality holds:*

$$||B_T(f)||_p \le \begin{cases} CT^\alpha (\ln T)^{1+\alpha} & \text{if } p = 1/\alpha, \\ CT^\alpha \ln T & \text{if } 1/\alpha < p < \infty, \\ CT^\alpha & \text{if } p = \infty. \end{cases} \tag{4.9.34}$$

Proof The proof involves the following steps. First, we observe that $B_T(f) = B_T(s_T f)$, where $s_T f$ represents the $(T-1)$st partial sum of the Fourier series of the function f. We then apply the inequality (4.9.1) with $A_T(f) = B_T(f)$ to obtain

$$||B_T(f)||_p \le T^{1/p} ||s_T f||_p. \tag{4.9.35}$$

Finally, the representation of the function $s_T f$ via the Dirichlet kernel is used to estimate the norm $||s_T f||_p$ (for details, see Böttcher [47]). □

Comparing the two estimates in (4.9.29) and in (4.9.34), Böttcher [47] observed that there is room for improvement in inequality (4.9.34). He conjectured that the term $T^{\alpha}(\ln T)^{1+\alpha}$ in (4.9.34) could be replaced by $(T \ln T)^{\alpha}$, and that the term $T^{\alpha} \ln T$ in (4.9.34) could be improved to T^{α}.

The proof of these conjectures was provided in the paper by Vong and Jin [311]. Specifically, by utilizing the theory of unitarily invariant norms, they obtained the following improvement of Theorem 4.9.5.

Theorem 4.9.6 *Let f be a function with a single Fisher-Hartwig singularity of degree α. Then, for $1 \le p < \infty$, there exists a positive constant C such that for all $T \ge 1$, the following inequality holds:*

$$||B_T(f)||_p \le \begin{cases} C\,(1-\alpha)^{-1}(1-\alpha p)^{-1/p}T^{1/p} & \text{if } \alpha p < 1, \\ C\,(1-\alpha)^{-1}(T \ln T)^{\alpha} & \text{if } \alpha p = 1, \\ C\,(1-\alpha)^{-1}(\alpha p - 1)^{-1/p}T^{\alpha} & \text{if } \alpha p > 1. \end{cases} \tag{4.9.36}$$

The estimate in (4.9.36) is clearly superior to that in (4.9.34) and aligns with the estimate in (4.9.29).

Remark 4.9.3 In Böttcher [47], lower bounds for the Schatten norm $||B_T(f)||_p$ were also obtained. Specifically, it was proved that if the generating function f is of the form (4.9.30) with $\beta, \gamma \in (0, 1)$ and $u, v \in L^{\infty}$, and if the functions u and v satisfy additional smoothness conditions, then for $1/\alpha \le p \le \infty$, there exists a constant $K(f) \in (0, \infty)$ that depends only on f, such that for all $T \ge 1$, the following inequality holds:

$$||B_T(f)||_p \ge K(f)T^{\alpha}.$$

Now, as an application of Theorems 4.9.5 and 4.9.6, we estimate the Schatten norm of the covariance matrix for a long memory ARFIMA(p, d, q) model.

Recall that an ARFIMA(p, d, q) model $X(t)$ is a stationary process with the spectral density (see Sect. 3.3):

$$f(\lambda) = |1 - e^{-i\lambda}|^{-2d} f_1(\lambda), \quad d < 1/2, \tag{4.9.37}$$

where $f_1(\lambda)$ is the spectral density of an ARMA(p, q) process, given by (3.3.9). For $0 < d < 1/2$, the model $X(t)$ specified by (4.9.37) exhibits long memory.

In view of (4.9.32) and (4.9.37), we have

$$f(\lambda) = |\omega_{2d}^{+}(\lambda) + \omega_{2d}^{-}(\lambda)| f_2(\lambda), \tag{4.9.38}$$

where

$$f_2(\lambda) = \frac{|\lambda|^{-2d}}{(2|\sin(\lambda/2)|)^{2d}} f_1(\lambda). \tag{4.9.39}$$

Taking into account the assumption $0 < 2d < 1$, it is easy to see that the assumptions of Theorem 4.9.5 are satisfied with $\beta = \gamma = \alpha = 2d$ and $u = v = f_2$, where f_2 is defined as in (4.9.39). Thus, we have the following corollary of Theorem 4.9.5 for a long memory ARFIMA(p, d, q) model.

Corollary 4.9.1 *Let $B_T(f)$ be the covariance matrix of a long memory ARFIMA(p, d, q) model, that is, $B_T(f)$ is the Toeplitz matrix generated by the function* (4.9.37) *with $0 < d < 1/2$. Then for the Schatten norm $||B_T(f)||_p$, the inequality in* (4.9.34) *is satisfied with $\alpha = 2d$ provided that $1/(2d) \le p < \infty$.*

Since the spectral density f of a long memory ARFIMA(p, d, q) model has a single Fisher-Hartwig singularity of degree $2d$ (as shown in (4.9.38)), we can derive the following corollary of Theorem 4.9.6.

Corollary 4.9.2 *For the covariance matrix $B_T(f)$ of a long memory ARFIMA (p, d, q) model, the estimate in* (4.9.36) *holds with $\alpha = 2d$.*

4.10 Bibliographical Notes and Comments

Section 4.2 For a detailed account of the theory and properties of Toeplitz matrices and operators, we refer to the monographs by Böttcher and Silbermann [52, 53]. Here, we provide an overview of the essential concepts needed for our purposes.

Section 4.3 The trace approximation problem for Toeplitz matrices and operators, along with its applications in the statistical analysis and prediction of stationary processes, can be traced back to the classical monograph by Grenander and Szegő [162]. Subsequently, many authors have explored this problem for various classes of generating functions (see, e.g., Avram [13], Dahlhaus [74, 75], Fox and Taqqu [100, 101], Ginovyan [117, 118, 120], Ginovyan and Sahakyan [135–138], Giraitis et al. [147], Giraitis and Surgailis [148], Hasminskii and Ibragimov [171], Ibragimov [180], Kac [198], Lieberman and Phillips [214], Rosenblatt [259, 260], Taniguchi [292], Taniguchi and Kakizawa [295], Taqqu [298], and references therein). More references on this topic can be found in the survey paper by Ginovyan et al. [145].

Notice that, from an application viewpoint, the trace approximation problem is particularly important in cases where the symbols of Toeplitz matrices and operators exhibit singularities. Such cases arise in various problems of spectral and statistical analysis related to long memory and antipersistent stationary processes. Some of these applications are discussed in Chaps. 5–10.

Section 4.4 Assertion **(A1)** of Theorem 4.4.1 for $m = 4$, $h_1 = h_3 := f$ and $h_2 = h_4 := g$ was proved by Ginovyan and Sahakyan [135]. Assertion **(A2)** was established by Avram [13] (see also Theorem 6.3.2 in Giraitis et al. [147]). In the special case where $p_i = \infty$ for $i = 1, 2, \dots, m$, meaning that all h_i are bounded functions, this was first established by Grenander and Szegő [162], Section 7.4.

For $m = 4$, with $p_1 = p_3 = 2$ and $p_2 = p_4 = \infty$, assertion **(A2)** was first proved by Ibragimov [180] and by Rosenblatt [259]. In the paper by Avram [13], along with assertion **(A2)**, it was also proved that if $\bar{p} + \varepsilon > 1$ with $\varepsilon > 0$, where $\bar{p} := \sum_{i=1}^{m} 1/p_i$, then

$$\lim_{T\to\infty} \frac{1}{T^{\bar{p}+\varepsilon}} \mathrm{tr}\left[\prod_{i=1}^{m} B_T(h_i)\right] = 0.$$

The proof of Theorem 4.4.2 is provided in the paper by Fox and Taqqu [101] (Theorem 1). For $m = 4$ the proof of assertion (a) of Theorem 4.4.2 can also be found in the monograph by Giraitis et al. [147] (Theorem 6.3.2). Theorems 4.4.3 and 4.4.5 were proved in Giraitis et al. [147] (Lemma 6.3.4). Theorem 4.4.4 was proved by Ginovyan and Sahakyan [135]. Under the assumption $\alpha + \beta < 1/2$, the result of Theorem 4.4.4 was first obtained by Fox and Taqqu [101] (cf. Theorem 4.4.2). Theorem 4.4.6 was proved by Dahlhaus [74]. The proof of Theorem 4.4.7 can be found in the monograph by Dzhaparidze [91] (pp. 75–76); see also the paper by Davies [80].

Assertion **(B1)** of Theorem 4.4.8 was proved by Taniguchi [292] (see also the monograph by Taniguchi and Kakizawa [295], p. 173). Assertion **(B2)**, which is weaker than **(B1)** but holds under less stringent conditions, was proved by Lieberman and Phillips [214]. Assertions **(B3)**–**(B5)** for $m = 4$ were proved by Ginovyan and Sahakyan [138].

Concerning Assertion **(B2)** of Theorem 4.4.8, Lieberman and Phillips [214] claimed a more accurate rate of convergence for $\Delta(T)$. To present their result, we introduce the following notation. For $\alpha \in (-\infty, 1)$, we define the function spaces F_α and $F_\alpha^{(1)}$:

$$F_\alpha := \left\{ f \in L^1(\mathbb{T}) : \sup_{x\in\mathbb{T}_0} |x|^\alpha |f(x)| < \infty \right\},$$

$$F_\alpha^{(1)} := \left\{ f \in F_\alpha \cap \mathbb{C}_0^1(\mathbb{T}) : \sup_{x\in\mathbb{T}_0} |x|^{\alpha+1} \left|\frac{\mathrm{d}f}{\mathrm{d}x}(x)\right| < \infty \right\},$$

where $\mathbb{C}_0^1(\mathbb{T})$ denotes the set of functions on $\mathbb{T}$ that are continuously differentiable on $\mathbb{T}_0 := \mathbb{T} \setminus \{0\}$. Also, we write $x_+ := \max\{x, 0\}$ for $x \in \mathbb{R}$.

In Theorem 2 of Lieberman and Phillips [214], the following result was claimed: *Let $p \in \mathbb{N}$ and $\alpha, \beta \in (-\infty, 1)$ be such that $p(\alpha + \beta) < 1$. Then for even functions $f \in F_\alpha^{(1)}$ and $g \in F_\beta^{(1)}$ the following asymptotic relation holds for any $\varepsilon > 0$:*

$$\Delta(T) = O(T^{-1+p(\alpha+\beta)_+ +\epsilon}) \quad \text{as} \quad T \to \infty, \tag{4.10.1}$$

where $\Delta(T)$ is defined as in (4.4.6) *with $m = 4$, $h_1 = h_3 := f$, and $h_2 = h_4 := g$.*

Unfortunately, as noted by Ginovyan and Sahakyan [138], the proof of this result presented by Lieberman and Phillips [214] contains an inaccuracy related to the application of the power counting theorem developed by Fox and Taqqu [101] (see Remark 3 in Ginovyan and Sahakyan [138] for further details).

The gap in the proof of Theorem 2 from Lieberman and Phillips [214] was recently addressed by Takabatake [291].

Theorem 4.4.9 was proved by Ginovyan [121]. Theorem 4.4.10 was established by Ginovyan and Sahakyan [138]. Theorem 4.4.11, which addresses the case when the product in (4.3.2) also involves inverse matrices, is due to Taniguchi [292] (see also Taniguchi and Kakizawa [295], page 173). For $m = 2$, a similar result was obtained by Coursol and Dacunha-Castelle [67], without the ARMA assumption.

Section 4.5 For the special case $m = 4$, $h_1 = h_3 := f$, and $h_2 = h_4 := g$, Theorem 4.5.1 was proved by Ginovyan and Sahakyan [135]. For arbitrary m, the proof was presented in Ginovyan et al. [145]. Theorems 4.5.2 and 4.5.3 were proved by Ginovyan and Sahakyan [135]. Theorems 4.5.4–4.5.5 were proved in Ginovyan et al. [145] (see also Ginovyan and Sahakyan [137]).

It would be interesting to extend the results of Theorems 4.5.2 and 4.5.3 to arbitrary $m > 4$ and to establish the continuous analogs of Dahlhaus's theorem (Theorem 4.4.6) and Taniguchi's theorem (Theorem 4.4.11) for Toeplitz operators.

Section 4.7 Theorem 4.7.2 was first proved by Lieberman and Phillips [214], where the asymptotic relation (4.7.7) was established through direct calculations using the explicit forms of the functions f_i given in (4.7.3).

Theorems 4.7.1 and 4.7.3–4.7.6 were proved by Ginovyan and Sahakyan [137] (see also Ginovyan et al. [145]).

Section 4.8 Tapered Toeplitz matrices and operators naturally arise in the context of statistical inferences based on tapered data. For more information about tapering, including its motivations and benefits in statistical inference problems, see Appendix A. Tapered Toeplitz matrices have been considered by Dahlhaus [70], Doukhan et al. [86], Eichler [95], Ludeña [220], and others in the context of statistical inferences for discrete time stationary processes. Tapered Toeplitz operators were studied in the papers by Ginovyan and Sahakyan [140–143] in the context of statistical inferences for continuous time stationary processes. Theorems 4.8.1 and 4.8.2 were proved by Ginovyan and Sahakyan [140].

Section 4.9 Theorems 4.9.1–4.9.3 and Proposition 4.9.1 were proved by Ginovyan [120] (see also Ginovyan and Sahakyan [135, 136]). The proof of Theorem 4.9.4 can be found in Ginovyan and Sahakyan [141].

For more information about generating functions with Fisher-Hartwig-type singularities, we refer the reader to the book by Böttcher and Silbermann [52]. Theorem 4.9.5 was proved by Böttcher [47], while the proof of Theorem 4.9.6 is provided by Vong and Jin [311].

Chapter 5
Central Limit Theorems for Random Toeplitz Quadratic Functionals

In this chapter, we present results on central limit theorems for random Toeplitz-type quadratic functionals of stationary processes. The underlying processes are either Gaussian or linear processes with memory, defined in discrete or continuous time. As applications, we provide Berry-Esséen bounds, as well as large and moderate deviation results for these functionals.

In Sect. 5.1, we introduce the functionals under consideration and examine three sets of frequency domain conditions required for these functionals to satisfy the CLT: the (L^p, L^q) condition, the (α, β) condition, and the trace condition. In Sect. 5.2, we present results that include both frequency domain and time domain sufficient conditions for the CLT applicable to Gaussian and linear processes, with a primary focus on the frequency domain results. In Sect. 5.3, we discuss the most commonly used methods and tools for proving central limit theorems: the method of characteristic functions, the method of cumulants or moments, the approximation method, and several widely used convergence theorems, including the Cramér-Wold device. Section 5.4 contains proofs of the theorems stated in Sect. 5.2. In Sect. 5.5, we briefly discuss operator conditions for the CLT. Additionally, as applications of the limit theorems presented in Sect. 5.2, we establish Berry-Esséen bounds, along with large and moderate deviation results for random Toeplitz-type quadratic functionals.

5.1 Introduction

A significant part of large-sample statistical inference relies on the limit theorems of probability theory, which involve sums and quadratic functionals of stationary observations. Depending on the memory (or dependence) structure of the underlying processes, these functionals, once normalized, can have different limits, and the proofs of such limit theorems generally employ different methods. Here, we focus

M. S. Ginovyan, *Random Toeplitz Functionals and Their Applications*,
Frontiers in Probability and the Statistical Sciences,
https://doi.org/10.1007/978-3-032-00961-6_5

on quadratic functionals and prove central limit theorems for random Toeplitz-type quadratic functionals. The term "central limit theorem" (CLT) refers to the statement that a suitably standardized quadratic functional converges in distribution to a Gaussian random variable. Limit theorems in which a suitably standardized quadratic functional converges in distribution to a non-Gaussian random variable are referred to as "non-central limit theorems" (NCLT). Non-central limit theorems for Toeplitz-type quadratic functionals we prove in Chap. 6.

5.1.1 The Functionals Under Consideration

Let $\{X(u),\ u \in \mathbb{U}\}$ be a centered real-valued stationary process with spectral density (s.d.) $f(\lambda)$, $\lambda \in \Lambda$, and covariance function (c.f.) $r(t)$, $t \in \mathbb{U}$. We will consider both the continuous time (c.t.) and discrete time (d.t.) cases simultaneously. In the continuous time case, $\mathbb{U} = \mathbb{R} := (-\infty, \infty)$, while in the discrete time case, $\mathbb{U} = \mathbb{Z} := \{0, \pm 1, \pm 2, \ldots\}$. The domain Λ of the frequency variable λ is $\Lambda = \mathbb{R}$ in the continuous time case and $\Lambda := [-\pi.\pi)$ in the discrete time case.

We are interested in the asymptotic distribution (as $T \to \infty$) of the following random Toeplitz-type quadratic functionals of the process $X(u)$:

$$Q_T := \begin{cases} \sum_{t=1}^{T} \sum_{s=1}^{T} \widehat{g}(t-s) X(t) X(s) & \text{in the d.t. case,} \\ \\ \int_0^T \int_0^T \widehat{g}(t-s) X(t) X(s)\, dt\, ds & \text{in c.t. case,} \end{cases} \tag{5.1.1}$$

where

$$\widehat{g}(t) := \int_{\Lambda} e^{i\lambda t}\, g(\lambda)\, d\lambda, \quad t \in \mathbb{U} \tag{5.1.2}$$

is the Fourier transform of some real, even, integrable function $g(\lambda)$, $\lambda \in \Lambda$. We will refer to $g(\lambda)$ and its Fourier transform $\widehat{g}(t)$ as a *generating function* and *generating kernel* for the functional Q_T, respectively. In the discrete time case, the functions $f(\lambda)$ and $g(\lambda)$ are assumed to be 2π-periodic and are extended periodically to the entire real line $\mathbb{R}$. In the continuous time case, the process $X(u)$ is assumed to be measurable and mean square continuous (see 3.1.3).

The limiting distributions of the functionals in (5.1.1) are completely determined by the spectral density $f(\lambda)$ (or covariance function $r(t)$) and the generating function $g(\lambda)$ (or generating kernel $\widehat{g}(t)$). Depending on their properties, the limits can be either Gaussian (i.e., Q_T, with appropriate normalization, satisfies the central limit theorem) or non-Gaussian.

The following two questions arise naturally:

(a) Under what conditions on $f(\lambda)$ (resp. $r(t)$) and $g(\lambda)$ (resp. $\widehat{g}(t)$) will the limits be Gaussian? This is referred to as the *CLT problem*.
(b) Describe the limit distributions if they are non-Gaussian (the *NCLT problem*).

Quadratic functionals of the form (5.1.1) arise naturally in the context of both nonparametric and parametric statistical estimation of the spectrum of stationary processes. Their limiting distributions are essential for establishing the asymptotic properties of the corresponding estimators (see Chaps. 8 and 9).

The problem of describing the asymptotic distribution of random Toeplitz-type quadratic forms and functionals of stationary processes has a long history. It dates back to the classical monograph by Grenander and Szegő [162], which addressed the CLT problem for Gaussian processes as an application of the authors' theory on the asymptotic behavior of the trace of products of truncated Toeplitz matrices and operators.

Since then, many researchers have examined both the CLT and NCLT problems for various stationary models (see, e.g., the survey paper by Ginovyan and Taqqu [146] and the references therein).

In this chapter, we will explore the CLT problem for both discrete and continuous time stationary processes. The NCLT problem will be discussed in Chap. 6.

5.1.2 Conditions for CLT and NCLT

Conditions expressed in terms of the spectral density $f(\lambda)$ and the generating function $g(\lambda)$ are referred to as *frequency domain conditions*. In contrast, conditions expressed in terms of the covariance function $r(t)$ and the generating kernel $\widehat{g}(t)$ are called *time domain conditions*.

There are three sets of frequency domain conditions that functionals of the form (5.1.1) must satisfy in order to meet the central limit theorem (CLT). These conditions define the ranges of applicability for both the CLT and the NCLT:

(a) The (L^p, L^q) condition
(b) The (α, β) condition
(c) The trace condition

All three are "compensation" conditions, meaning that the good behavior of one function, say g, can offset the poor behavior of f and vice versa.

(a) The (L^p, L^q) condition. Let $f \in L^1(\Lambda) \cap L^p(\Lambda)$ $(p \geq 2)$ and $g \in L^1(\Lambda) \cap L^q(\Lambda)$ $(q \geq 2)$. If $1/p + 1/q \leq 1/2$, then the functional Q_T, with appropriate normalization, satisfies the CLT (see Theorem 5.2.1 **(C)**). On the other hand, when $1/p + 1/q > 1/2$, then Q_T generally does not satisfy the CLT. This condition originates from the classical works of Rosenblatt [258, 259], where it was shown that for a sample covariance $\widehat{r}(t)$ (a functional of the form (5.1.1) with $g(\lambda) = \cos(t\lambda)$) to satisfy the CLT, the spectral density f must meet the condition $f \in L^p(\mathbb{T})$ with $p \geq 2$. This represents a (L^p, L^q) condition with $p \geq 2$ and $q = \infty$.

(b) The (α, β) condition. If both the spectral density f and the generating function g are *regularly varying* at the origin with orders α and β, respectively (see Sect. 2.5), then it is the sum $\alpha + \beta$ that determines the limiting distribution of the functional

Q_T. Specifically, when $\alpha + \beta \leq 1/2$, then the limiting distribution of Q_T is Gaussian, i.e., the functional Q_T, with appropriate normalization, satisfies the CLT (see Theorem 5.2.2). In contrast, when $\alpha+\beta > 1/2$, the limiting distribution of Q_T is non-Gaussian (see Theorems 6.2.5 and 6.2.6).

(c) The trace condition. This condition, which is broader and encompasses both conditions (a) and (b), is essentially an implicit condition. It is expressed in terms of the traces of products of truncated Toeplitz matrices in the discrete time case and operators in the continuous time case.

The goal is to approximate the traces of products of Toeplitz matrices and operators—which are not necessarily Toeplitz—by the traces of a single Toeplitz matrix and a Toeplitz operator, respectively. Specifically, let $A_T(f)$ denote either the $T \times T$ Toeplitz matrix $B_T(f)$ or the T-truncated Toeplitz operator $W_T(f)$ generated by the spectral density f. Similarly, let $A_T(g)$ denote either the $T \times T$ Toeplitz matrix $B_T(g)$ or the T-truncated Toeplitz operator $W_T(g)$ generated by the function g (for definitions, see Sect. 4.3). The trace condition is as follows: if

$$fg \in L^1(\Lambda) \cap L^2(\Lambda)$$

and

$$T^{-1}\big[\mathrm{tr}\big((A_T(f)A_T(g))^2\big) - \mathrm{tr}\big(A_T(f^2g^2)\big)\big] \to 0 \quad \text{as} \quad T \to \infty,$$

then the quadratic functional Q_T in (5.1.1), with appropriate normalization, satisfies the CLT (see Theorems 5.2.1 **(A)** and 5.2.3).

We will also discuss the time domain counterparts of the (L^p, L^q) and (α, β) conditions.

5.2 CLT for Toeplitz-Type Quadratic Functionals for Gaussian and Linear Processes

In this section, we present sufficient conditions for quadratic forms and functionals of the form (5.1.1) to satisfy the CLT. The processes considered include both discrete and continuous time Gaussian or linear processes with memory. Both the matrix and the operator that characterize the quadratic form and functional are Toeplitz.

5.2.1 Frequency Domain Conditions

Let the functional $Q_T := Q_T(f, g)$ be defined as in (5.1.1). We include the function f in the notation $Q_T(f, g)$ to emphasize that the distribution of the functional also depends on the spectral density f. By $\widetilde{Q}_T$, we denote the standard normalized quadratic functional:

$$\widetilde{Q}_T := T^{-1/2}\left(Q_T - \mathbb{E}[Q_T]\right). \tag{5.2.1}$$

The notation

$$\widetilde{Q}_T \xrightarrow{d} \eta \sim N(0, \sigma^2) \quad \text{as} \quad T \to \infty \tag{5.2.2}$$

will mean that the distribution of the random variable $\widetilde{Q}_T$ converges (as $T \to \infty$) to a centered normal distribution with variance σ^2.

Let $A_T(f)$ and $A_T(g)$ denote either the $T \times T$ Toeplitz matrices (in the discrete time case) or the T-truncated Toeplitz operators (in the continuous time case) generated by the functions f and g, respectively (for definitions, see Sect. 4.2, formulas (4.2.8) and (4.2.11)). Note that $A_T(f)$ serves as the covariance matrix in the discrete time case and as the covariance operator in the continuous time case for the process $\{X(u),\ u \in \mathbb{U}\}$.

We assume below that $f, g \in L^1(\Lambda)$, and without loss of generality, that $g \geq 0$. We also define

$$\sigma_Q^2 := 16\pi^3 \int_\Lambda f^2(\lambda) g^2(\lambda)\, d\lambda. \tag{5.2.3}$$

As usual, $\Lambda = [-\pi, \pi)$ in the discrete time case, and $\Lambda = \mathbb{R}$ in the continuous time case.

The theorems presented below provide sufficient conditions, expressed in terms of functions f and g, to ensure central limit theorems for standard normalized quadratic functionals $\widetilde{Q}_T$ in both discrete and continuous time Gaussian processes. Certain assumptions imposed on f allow for long-range dependence ($f(0) = \infty$), while others accommodate discontinuities at different frequencies. Occasionally, the favorable behavior of one function, say g, can compensate for the unfavorable behavior of f and vice versa.

Theorem 5.2.1 *Let f, g, $A_T(f)$, $A_T(g)$, and $\widetilde{Q}_T$ be defined as above. Each of the following conditions is sufficient for the functional Q_T to satisfy the CLT:*

$$\widetilde{Q}_T \xrightarrow{d} \eta \sim N(0, \sigma_Q^2) \quad \text{as} \quad T \to \infty, \tag{5.2.4}$$

with variance σ_Q^2 given by (5.2.3).

(A) $f \cdot g \in L^1(\Lambda) \cap L^2(\Lambda)$ *and*

$$\chi_2(\widetilde{Q}_T) := \frac{2}{T} tr\left[A_T(f) A_T(g)\right]^2 \longrightarrow 16\pi^3 \int_\Lambda f^2(\lambda) g^2(\lambda) d\lambda \tag{5.2.5}$$

as $T \to \infty$, where tr$[A]$ stands for the trace of the operator (or the matrix) A, and $\chi_2(\widetilde{Q}_T)$ is the second cumulant of $\widetilde{Q}_T$.

(B) *The function*

$$\varphi(\mathbf{u}) := \varphi(u_1, u_2, u_3) = \int_\Lambda f(\lambda)g(\lambda - u_1)f(\lambda - u_2)g(\lambda - u_3)\, d\lambda \qquad (5.2.6)$$

belongs to $L^2(\Lambda^3)$ and is continuous at the point $\mathbf{0} = (0, 0, 0)$.

(C) *$f \in L^1(\Lambda) \cap L^p(\Lambda)$ $(p \geq 2)$ and $g \in L^1(\Lambda) \cap L^q(\Lambda)$ $(q \geq 2)$ with $1/p + 1/q \leq 1/2$.*

(D) *$f \in L^1(\Lambda) \cap L^2(\Lambda)$, $g \in L^1(\Lambda) \cap L^2(\Lambda)$, $fg \in L^2(\Lambda)$, and*

$$\int_\Lambda f^2(\lambda)g^2(\lambda - \mu)\, d\lambda \longrightarrow \int_\Lambda f^2(\lambda)g^2(\lambda)\, d\lambda \quad \text{as} \quad \mu \to 0.$$

Remark 5.2.1 Note that assertion **(A)** of the theorem implies assertions **(B)**–**(D)**, and assertion **(B)** implies assertions **(C)** and **(D)** (see Giraitis and Surgailis [148] and Ginovyan and Sahakyan [135, 136]).

To present the next theorem, which includes an (α, β) sufficient condition for the CLT, we recall the class $SV_0(\mathbb{R})$ of slowly varying at zero functions, as introduced in Sect. 2.5 (see Definition 2.5.1).

Theorem 5.2.2 *Assume that the functions f and g are integrable on $\mathbb{R}$, bounded outside any neighborhood of the origin, and satisfy the following conditions for some $a > 0$:*

$$f(\lambda) \leq |\lambda|^{-\alpha}L_1(\lambda), \quad |g(\lambda)| \leq |\lambda|^{-\beta}L_2(\lambda), \quad \lambda \in [-a, a], \qquad (5.2.7)$$

for some $\alpha < 1$, $\beta < 1$ with $\alpha + \beta \leq 1/2$, where $L_1(x)$ and $L_2(x)$ are slowly varying at zero functions satisfying

$$L_i \in SV_0(\mathbb{R}), \ \lambda^{-(\alpha+\beta)}L_i(\lambda) \in L^2[-a, a], \ i = 1, 2. \qquad (5.2.8)$$

Then $\widetilde{Q}_T \xrightarrow{d} \eta \sim N(0, \sigma_Q^2)$ as $T \to \infty$ with variance σ_Q^2 given by (5.2.3).

Remark 5.2.2 The conditions $\alpha < 1$ and $\beta < 1$ in Theorem 5.2.2 ensure that the Fourier transforms of the functions f and g are well-defined. It is important to note that when $\alpha > 0$, the process $X(t)$ may exhibit long-range dependence. We also allow here $\alpha + \beta$ to take the critical value 1/2. The assumptions $f \cdot g \in L^1(\mathbb{R})$, $f, g \in L^\infty(\mathbb{R} \setminus [-a, a])$, and the condition (5.2.8) imply that $f \cdot g \in L^2(\mathbb{R})$. Therefore, the variance σ_Q^2 in (5.2.3) is finite.

Remark 5.2.3 The functions

$$f(\lambda) = |\lambda|^{-\alpha}|\ln|\lambda||^{-\gamma} \quad \text{and} \quad g(\lambda) = |\lambda|^{-\beta}|\ln|\lambda||^{-\gamma}, \tag{5.2.9}$$

where $\alpha < 1$, $\beta < 1$, $\alpha + \beta \leq 1/2$, and $\gamma > 1/2$, provide examples of spectral density f and generating function g that satisfy the conditions of Theorem 5.2.2 (see Ginovyan and Sahakyan [135]).

Remark 5.2.4 The slowly varying functions L_1 and L_2 in (5.2.7) are important because they provide significant flexibility in the choice of spectral density f and generating function g. In Theorem 5.2.2, the assumption that the functions L_1 and L_2 belong to the class $SV_0(\mathbb{R})$ rather than simply being slowly varying at zero is made to address the critical case $\alpha + \beta = 1/2$. Specifically, Theorem 5.2.2 shows that in this critical case, the limiting distribution of the standard normalized quadratic form $\widetilde{Q}_T$, as given by (5.2.1), is Gaussian and depends on the slowly varying factors L_1 and L_2 through the functions f and g.

Remark 5.2.5 The functions f and g in Theorem 5.2.2 have singularities at the point $\lambda = 0$ and are bounded in any neighborhood of this point. It can be shown that the choice of the point $\lambda = 0$ is not essential; rather, it can be any point $\lambda_0 \in [-a, a]$. By utilizing the properties of the products of Toeplitz matrices and operators, it can be shown that Theorem 5.2.2 remains valid when the functions f and g exhibit singularities of the form (5.2.9) at the same finite number of points within the interval $[-a, a]$ (see Ginovyan and Sahakyan [135]).

Remark 5.2.6 Assertion **(A)** of Theorem 5.2.1 implies Theorem 5.2.2. On the other hand, for the functions $f(\lambda) = \lambda^{-3/4}$ and $g(\lambda) = \lambda^{3/4}$, which satisfy the conditions of Theorem 5.2.2, the function $\varphi(u_1, u_2, u_3)$ in (5.2.6) is not defined for $u_2 = 0$, $u_1 \neq 0$, $u_3 \neq 0$. This demonstrates that assertion **(B)** of Theorem 5.2.1 does not generally imply Theorem 5.2.2 (see Ginovyan and Sahakyan [135]).

Remark 5.2.7 The critical case $\alpha + \beta = 1/2$ was partially investigated by Terrin and Taqqu [303]. They first proved that if $\alpha, \beta < 1$ and $\alpha + \beta > 1/2$, then the random variable

$$\widehat{Q}_n = \frac{1}{n^{\alpha+\beta}L_1(1/n)L_2(1/n)}\,(Q_n - \mathbb{E}[Q_n])$$

converges in distribution to a non-Gaussian random variable $Y(\alpha, \beta)$, which can be represented as a double Wiener-Itô integral (see also Theorem 6.2.5 with $\tau = 1$). Then, by analyzing the distribution of the limiting random variable $Y(\alpha, \beta)$, which exists only when $\alpha+\beta > 1/2$, they showed that when $0 < \alpha, \beta < 1, \alpha+\beta > 1/2$, $\alpha_0, \beta_0 > 0$, and $\alpha_0+\beta_0 = 1/2$, the random variable $(\alpha+\beta-1/2)Y(\alpha, \beta)$ converges in distribution to a Gaussian random variable as $\alpha \to \alpha_0$ and $\beta \to \beta_0$.

Giraitis and Surgailis [148] conjectured that the relation (5.2.5), and consequently, the CLT for the functional Q_T, holds under the single condition that the integral on the right-hand side of (5.2.5) is finite. Ginovyan [119] demonstrated that the finiteness of this integral does not guarantee convergence in (5.2.5). He conjectured that both the positivity and finiteness of the integral in (5.2.5) are sufficient for Q_T to satisfy the central limit theorem.

The following proposition demonstrates that the positivity and finiteness of the integral in (5.2.5) are not sufficient for the functional Q_T to satisfy the CLT.

Proposition 5.2.1 *There exist a spectral density $f(\lambda)$ and a generating function $g(\lambda)$ such that*

$$0 < \int_{-\pi}^{\pi} f^2(\lambda)\, g^2(\lambda)\, d\lambda < \infty \tag{5.2.10}$$

and

$$\lim_{T\to\infty} \sup \chi_2(\widetilde{Q}_T) = \lim_{n\to\infty} \sup \frac{2}{T} tr\,(B_T(f)B_T(g))^2 = \infty. \tag{5.2.11}$$

Specifically, the condition (5.2.10) does not guarantee convergence in (5.2.5), and therefore, it is not sufficient for the functional Q_T to satisfy the CLT.

The following two theorems provide sufficient conditions for the CLT for discrete time linear processes.

Giraitis and Surgailis [148] proved that assertions **(A)**, **(C)**, and **(D)** of Theorem 5.2.1 remain valid for discrete time linear processes. In particular, they established the following theorem (see also Giraitis et al. [147], Theorem 6.3.1).

Theorem 5.2.3 *Let $\{X(u),\ u \in \mathbb{Z}\}$ be a stationary linear process of the form* (3.2.7) *with spectral density f, driven by IID(0,1) noise $\{\xi(u),\ u \in \mathbb{Z}\}$ that has a finite fourth moment. Let Q_T be a quadratic form generated by the function g given by* (5.1.1). *Additionally, let $B_T(f)$ and $B_T(g)$ be the $T \times T$ Toeplitz matrices generated by the functions f and g, respectively (see formula* (4.2.8)*). Assume that*

$$\frac{1}{T} tr\big[B_T(f)B_T(g)\big]^2 \longrightarrow 8\pi^3 \int_{-\pi}^{\pi} f^2(\lambda)g^2(\lambda)d\lambda < \infty, \tag{5.2.12}$$

Then the CLT holds for Q_T; that is,

$$\widetilde{Q}_T := T^{-1/2}\,(Q_T - \mathbb{E}[Q_T]) \xrightarrow{d} \eta \sim N(0, \sigma^2_{QL}) \quad \text{as} \quad T \to \infty, \tag{5.2.13}$$

where

$$\sigma^2_{QL} = 16\pi^3 \int_{-\pi}^{\pi} f^2(\lambda)g^2(\lambda)d\lambda + \chi_4 \left[2\pi \int_{-\pi}^{\pi} f(\lambda)g(\lambda)d\lambda\right]^2, \tag{5.2.14}$$

Here $\chi_4 = \kappa_4$ if the innovation variance $\sigma^2 = 1$ and $\chi_4 = \kappa_4/\sigma^4$ if $\sigma^2 \neq 1$, where κ_4 is the fourth cumulant of $\xi(0)$.

The next theorem states that the CLT for Q_T holds when either the spectral density f or the generating function g is sufficiently smooth (see Giraitis and Taqqu [149], Theorem 2.2).

Theorem 5.2.4 *Let $X(u)$, f, g, and Q_T be as defined in Theorem 5.2.3. Assume that either $f \in \mathrm{Lip}_\alpha(\mathbb{T})$ or $g \in \mathrm{Lip}_\alpha(\mathbb{T})$ with $\alpha > 1/2$, and $fg \in L^2(\mathbb{T})$. Then the relation* (5.2.13) *is satisfied with σ^2_{QL} given by (5.2.14).*

Remark 5.2.8 An analog of Theorem 5.2.3 for continuous time linear processes will be proved in Chap. 6 (see Theorem 6.4.1).

5.2.2 *Time Domain Conditions*

In this subsection, we present time domain sufficient conditions for a quadratic form Q_T given in (5.1.1) to satisfy the CLT. Specifically, these conditions are expressed in terms of the covariance function $r(t)$ and the generating kernel $\widehat{g}(t)$. The processes considered here will be discrete time Gaussian or linear processes with memory. It is important to note that the time domain conditions stated below are generally not equivalent to the frequency domain conditions specified in Theorems 5.2.1 and 5.2.2. Furthermore, the method of proof for establishing the CLT under time domain conditions differs from that used for frequency domain conditions. In this context, diagrams are utilized, and the method of moments is applied. This is because, under the frequency domain conditions, one can employ an approximation technique to replace the potentially unbounded spectral density with a bounded one. This approach allows us to approximate the bivariate quadratic forms by univariate sums of m-dependent random variables (see Sect. 5.3). However, such an approximation technique does not work with time domain conditions. Therefore, one must directly address the covariances, which tend to decrease slowly.

The following theorem provides sufficient conditions in terms of the covariance function $r(t)$ and the generating kernel $\widehat{g}(t)$ that ensure the central limit theorem for the quadratic form $\widetilde{Q}_T$, as defined in Eq. (5.1.1), for discrete time Gaussian processes (see Giraitis et al. [147], Lemma 6.3.4, and Giraitis and Taqqu [150]).

Theorem 5.2.5 *If the covariance function $r(t)$ and the generating kernel $\widehat{g}(t)$ satisfy the following conditions:*

$$r \in l^p \ (p \geq 1) \quad \text{and} \quad \widehat{g} \in l^q \ (q \geq 1) \quad \text{with} \quad 1/p + 1/q \geq 3/2,$$

then the CLT holds for the quadratic form Q_T with limiting variance σ^2_Q given by Eq. (5.2.3).

The next theorem, proved by Giraitis and Taqqu [150], demonstrates that under certain rather restrictive conditions on the covariance function $r(t)$ and the generating kernel $\widehat{g}(t)$, the long-range dependence of the process $X(t)$ can be compensated by the fast decay of the generating kernel $\widehat{g}(t)$, thereby ensuring that the CLT for Q_T holds. These conditions guarantee that the frequency domain sufficient conditions provided in Theorem 5.2.2 are satisfied. The theorem involves quasi-monotone sequences.

Recall that a sequence $\{b(t), t \in \mathbb{Z}\}$ is said to be *quasi-monotonically convergent* to 0 if $b(t) \to 0$ and $b(t+1) \le b(t)(1+c/t)$ as $t \to 0$ for some $c > 0$. Additionally, the sequence $b(t)$ has bounded variation if $\sum_{t=1}^{\infty} |b(t+1) - b(t)| < \infty$.

Theorem 5.2.6 *Assume that* $r(t) = |t|^{-\gamma_1} L_1(|t|)$ *and* $\widehat{g}(t) = |t|^{-\gamma_2} L_2(|t|)$ *with* $0 < \gamma_1, \gamma_2 < 3$, *and* $\gamma_1 + \gamma_2 > 3/2$. *Suppose, in addition, that both sequences* $\{r(t)\}$ *and* $\{\widehat{g}(t)\}$ *have bounded variation and are quasi-monotonically convergent to* 0;

if $1 < \gamma_1 < 3$, $r(t)$ *has the same sign for large t and satisfies* $\sum_{t \in \mathbb{Z}} r(t) = 0$;
if $1 < \gamma_2 < 3$, $\widehat{g}(t)$ *has the same sign for large t and satisfies* $\sum_{t \in \mathbb{Z}} \widehat{g}(t) = 0$.
Then, the CLT holds for the quadratic form Q_T. *The limiting variance is expressed by Eq. (5.2.3) provided that* $0 < \gamma_1, \gamma_2 < 1$.

5.3 Tools

The most commonly used methods to prove central limit theorems are:

(a) The method of characteristic functions
(b) The method of cumulants or moments
(c) The approximation method

To prove the central limit theorems stated in Sect. 5.2, we use either the method of cumulants or the approximation method, depending on the underlying model. If the model is Gaussian, we apply the method of cumulants; otherwise, we use the approximation method, which reduces the quadratic integral form to a single integral form (see Lemmas 5.4.2 and 5.4.3).

5.3.1 *The Characteristic Functions and Cumulant Criteria for the CLT*

The characteristic function criterion for the CLT is based on the fact that convergence in distribution is equivalent to the pointwise convergence of the corresponding characteristic functions. The general cumulant criterion for the CLT relies on the following well-known result (see, e.g., Giraitis et al. [147], Corollary 4.2.1).

Proposition 5.3.1 *Let the random variables* $\{X_T, T \in \mathbb{R}\}$ *have all moments finite, and let* $\mathbb{E}[X_T] \to 0$, $\mathrm{Var}[X_T] \to \sigma^2$, *and* $\chi_k(X_T) := \mathrm{Cum}_k(X_T) \to 0$ *as* $T \to \infty$ *for all* $k = 3, 4, \ldots$. *Then* $X_T \xrightarrow{d} X \sim N(0, \sigma^2)$ *as* $T \to \infty$.

The characteristic functions and cumulant criteria for the CLT regarding quadratic functionals of Gaussian models are based on the following general result (see Ibragimov [180]).

Let ξ be a Gaussian random variable taking values in a separable Hilbert space H. Specifically, ξ has the following characteristic functional:

$$\varphi(h) = \exp\left\{m(h) - \frac{1}{2}(Rh, h)\right\}, \quad h \in H, \tag{5.3.1}$$

where $m(h)$ is a continuous linear functional and the correlation operator R is a self-adjoint, completely continuous operator with finite trace. We assume, without loss of generality, that $m(h) \equiv 0$. Let A be any self-adjoint bounded linear operator.

Proposition 5.3.2 (Ibragimov [180]) *Let the operators R and A be defined as above. The quadratic form $(A\xi, \xi)$ has the same distribution as the sum $\sum_{k=1}^{\infty} \lambda_k \xi_k^2$, where $\{\xi_k, k \in \mathbb{N}\}$ are independent $N(0, 1)$ Gaussian random variables and $\{\lambda_k, k \in \mathbb{N}\}$ are the eigenvalues of the operator $T := RA$.*

Proof Without loss of generality, we can assume that the operator R is strictly positive, that is, $(R\varphi, \varphi) = 0$ iff $\varphi = 0$. Let $\{\psi_k\}$ be the complete orthonormal system in H consisting of the eigenvectors of the operator R and let $\{\rho_k\}$ be the corresponding eigenvalues. Let H_n denote the finite-dimensional subspace spanned by $\{\psi_1, \dots, \psi_n\}$, and let P_n be the orthoprojector in H onto H_n. Define the operators

$$A_n := P_n A P_n, \quad R_n := P_n R P_n, \quad T_n := P_n T P_n.$$

Let $\lambda_k^{(n)}$ denote the eigenvalues of the operator T_n indexed so that $|\lambda_1^{(n)}| \geq |\lambda_2^{(n)}| \geq \cdots |\lambda_n^{(n)}| \geq \cdots$. Since the operators A, R, and T are compact, we have (see, e.g., Riesz and Sz.-Nagy [254], Section 5.85)

$$||A - A_n|| \to 0, \ ||R - R_n|| \to 0, \ ||T - T_n|| \to 0 \quad \text{as} \quad n \to \infty. \tag{5.3.2}$$

Hence,

$$\mathbb{E}[|(A\xi, \xi) - (A_n\xi, \xi)|] \leq ||A - A_n|| \, \mathbb{E}[|\xi|^2] \to 0 \quad \text{as} \quad n \to \infty.$$

Therefore, for any $u \in \mathbb{R}$, we have

$$\lim_{n\to\infty} \mathbb{P}\{(A_n\xi, \xi) < u\} = \mathbb{P}\{(A\xi, \xi) < u\}.$$

Further, from the results of Sevastyanov [276], it follows that

$$\lambda_k^{(n)} \to \lambda_k \quad \text{as} \quad n \to \infty, \tag{5.3.3}$$

$$\lim_{n\to\infty}\left[\sum_{k=1}^{n}(\lambda_k^{(n)})^2-\sum_{k=1}^{\infty}\lambda_k^2\right]=\frac{\sigma^2}{2}\geq 0, \tag{5.3.4}$$

$$\varphi(t)=\exp\left\{-\frac{\sigma^2t^2}{2}\right\}\prod_{k=1}^{\infty}|1-2it\lambda_k|^{-1/2}, \tag{5.3.5}$$

where $\varphi(t)$ is the characteristic function of the random variable $(A\xi, \xi)$.

Therefore, it remains to show that all the numbers λ_k (and only these) are the eigenvalues of the operator T, and $\sigma = 0$.

Let $\lambda_k^{(n)} \to \lambda_k$ as $n \to \infty$, and let $\varphi_k^{(n)}$ be the corresponding sequence of eigenvectors of the operator T_n. By (5.3.2), we have as $n \to \infty$

$$||T\varphi_k^{(n)}-T_n\varphi_k^{(n)}||=||T\varphi_k^{(n)}-\lambda_k^{(n)}\varphi_k^{(n)}||\leq||T-T_n||\to 0. \tag{5.3.6}$$

Since T is a compact operator, we can assume that the sequence $T\varphi_k^{(n)}, n = 1, 2, \ldots$, is convergent. Then, from (5.3.3), it follows that the sequence $\varphi_k^{(n)}, n = 1, 2, \ldots$, converges to some φ_k satisfying $||\varphi_k|| = 1$ and $T\varphi_k = \lambda_k\varphi_k$.

Let now φ be any eigenvector and λ be the corresponding eigenvalue of the operator T. If $\varphi \neq \lim_{n\to\infty}\varphi_k^{(n)}$ for arbitrary k, then $|(\varphi, \varphi_k^{(n)})| \to 0$ as $n \to \infty$ for any k. Moreover, we have

$$\begin{aligned}||\varphi||^2&=\frac{1}{\lambda^2}\lim_{n\to\infty}(T_n\varphi, T_n\varphi)=\frac{1}{\lambda^2}\sum_{k=1}^{n}(\lambda_k^{(n)})^2|(\varphi,\varphi_k^{(n)})|^2\\&\leq\max_{1\leq k\leq p}|(\varphi,\varphi_k^{(n)})|\sum_{k=1}^{p}|\lambda_k^{(n)}|^2+\sum_{k=p+1}^{n}|\lambda_k^{(n)}|^2.\end{aligned} \tag{5.3.7}$$

We show that all the "series" $\sum_{k=1}^{n}|\lambda_k^{(n)}|^2$ are uniformly convergent. Since the operator $T = RA$ is nuclear (trace class), we have

$$\sup\sum_{k=1}^{\infty}|(Tu_k, u_k)|=C_0<\infty,$$

where the supremum is taken over all orthonormal bases $\{u_k\}$. Therefore, we have

$$\sum_{k=1}^{n}|\lambda_k^{(n)}|=\sum_{k=1}^{n}|(P_nTP_n\varphi_k^{(n)},\varphi_k^{(n)})|=\sum_{k=1}^{n}|(T\varphi_k^{(n)},\varphi_k^{(n)})|\leq C_0.$$

Next, since $\lim_{k\to\infty} |\lambda_k| = 0$, then $\lim_{k,n\to\infty} |\lambda_k^{(n)}| = 0$. Thus, for sufficiently large n and p, we have

$$\sum_{k=p+1}^{n} |\lambda_k^{(n)}|^2 \le \max_{k\ge p+1} |\lambda_k^{(n)}| C_0 < \varepsilon \tag{5.3.8}$$

Since ε can be arbitrarily small, the relations (5.3.7) and (5.3.8) imply that every eigenvalue λ of the operator T is of the form $\lim_{n\to\infty} \lambda_k^{(n)}$.

Finally, the uniform convergence of the "series" $\sum_{k=1}^{n} |\lambda_k^{(n)}|^2$ implies that

$$\lim_{n\to\infty} \sum_{k=1}^{n} (\lambda_k^{(n)})^2 = \sum_{k=1}^{\infty} \lambda_k^2,$$

and thus, in view of (5.3.4), we have $\sigma = 0$. This concludes the proof of the proposition. □

As a consequence of Proposition 5.3.2, we have the following result (see Ibragimov [180]).

Corollary 5.3.1 *If under the conditions of Proposition 5.3.2,* $||A||_\infty \cdot ||R||_\infty < \varepsilon$ *and* $\mathrm{Var}(A\xi, \xi) > 0$*, then for all sufficiently small* $\varepsilon > 0$ *and* $u \in \mathbb{R}$*, we have*

$$\left| \mathbb{P}\left\{ \frac{(A\xi, \xi) - \mathbb{E}[(A\xi, \xi)]}{\sqrt{\mathrm{Var}(A\xi, \xi)}} < u \right\} - \Phi(u) \right| \le C_1 \varepsilon, \tag{5.3.9}$$

where $\Phi(u)$ *is the standard normal distribution function.*

Proof It is easy to see that $\lambda_1 := \max_k \lambda_k < \varepsilon$, where $\{\lambda_k, k \in \mathbb{N}\}$ are the eigenvalues of the operator RA. Denoting by $\varphi(t)$ the characteristic function of the random variable

$$\frac{(A\xi, \xi) - \mathbb{E}[(A\xi, \xi)]}{\sqrt{\mathrm{Var}(A\xi, \xi)}},$$

we have

$$\log \varphi(t) = -\frac{t^2}{2} \cdot 2 \sum_k \mu_k^2 + \frac{2\theta |t|^3}{3} \cdot 2 \sum_k |\mu_k|^3 = -\frac{t^2}{2} + \varepsilon\theta |t|^3, \tag{5.3.10}$$

where $\mu_k = \lambda_k [\mathrm{Var}(A\xi, \xi)]^{-1/2}$, and $\theta \in [0, 1]$.

Applying Esséen's theorem (see, e.g., Petrov [242], Section 5.2) from (5.3.10), we obtain (5.3.9). □

Remark 5.3.1 It follows from Proposition 2.7.4 that the sets of nonzero eigenvalues of the operators RA, AR, and $R^{1/2}AR^{1/2}$ coincide, where $R^{1/2}$ is the positive definite square root of R.

Remark 5.3.2 A different proof of Proposition 5.3.2 was provided by Pipiras and Taqqu [246], employing the technique of Wiener-Itô multiple integrals of order two, along with the diagram formula (see Proposition 4.5.1 and Lemma 4.5.2 in [246]). Additionally, see Breuer and Major [54] and Saulis and Statulevicius [271], Proposition 5.2.

5.3.2 The Approximation Method

The cumulants method is useful for proving central limit theorems when all moments exist. An alternative approach that does not require the finiteness of higher-order moments is the *m-dependent approximation method.*

Recall that a process $\{X(t),\ t\in\mathbb{U}\}$ is said to be *m-dependent* if, for any $s\in\mathbb{U}$, the σ-fields $\sigma\{X(t), t\le s\}$ and $\sigma\{X(t), t\ge s+m+1\}$ are independent. This means that random variables that are sufficiently far apart in time are independent. The most common example of a stationary m-dependent process is the moving average of order m (MA(m)) model.

The m-dependent approximation method is based on the following propositions (see Ibragimov and Linnik [188], Theorem 19.2.1, and Beran et al. [33], Proposition 4.1).

Proposition 5.3.3 *Let $\{X_m(t),\ t\in\mathbb{N}\}$ be a stationary sequence of m-dependent random variables with* $\mathrm{Var}[X_m(0)]<\infty$, *and let $S_{T,m} := \sum_{t=1}^{T} X_m(t)$. Then*

$$T^{-1/2}S_{T,m}\xrightarrow{d}\sigma_m Z, \tag{5.3.11}$$

where $Z\sim N(0,1)$, and

$$\sigma_m^2 := \mathrm{Var}[X_m(0)] + 2\sum_{t=1}^{\infty}\mathrm{Cov}(X_m(0), X_m(t)). \tag{5.3.12}$$

Proposition 5.3.4 *Let $\{X(t),\ t\in\mathbb{N}\}$ be a stationary sequence, $\{X_m(t),\ t\in\mathbb{N}\}$ be a stationary sequence of m-dependent random variables, and $\{c_T,\ T\in\mathbb{N}\}$ be a sequence of constants. Define $S_T := \sum_{t=1}^{T} X(t)$ and $S_{T,m} := \sum_{t=1}^{T} X_m(t)$. Assume that the following conditions hold:*

(a) $c_T^{-1}S_{T,m}\xrightarrow{d} S_T$ *as* $T\to\infty$.

(b) $S_m\xrightarrow{P} S$ *as* $m\to\infty$.

(c) For each $\gamma > 0$,

$$\lim_{m\to\infty} \limsup_{T\to\infty} \mathbb{P}\left(c_T^{-1}|S_{T,m} - S_T| > \gamma\right) = 0.$$

Then

$$c_T^{-1} S_T \xrightarrow{d} S \text{ as } T \to \infty.$$

The proposition that follows presents the asymptotic distribution of the sum of the products of two stationary linear processes.

Let $\{\xi(t),\ t \in \mathbb{Z}\}$ be a sequence of centered i.i.d. noise with finite forth moments. Let $\sigma_\xi^2 := \mathrm{Var}(\xi_0)$ and $\kappa_4 := \mathrm{Cum}_4(\xi_0)$. Denote

$$S_T := \sum_{t=1}^{T} X_1(t)X_2(t), \qquad T \in \mathbb{N}, \tag{5.3.13}$$

where

$$X_i(t) := \sum_{s\in\mathbb{Z}} g_i(t-s)\xi_(s),\ \ t \in \mathbb{Z},\ i = 1, 2, \tag{5.3.14}$$

are two discrete time linear processes driven by the same i.i.d. noise $\{\xi(t),\ t \in \mathbb{Z}\}$), and $g_i(t)$, $(t \in \mathbb{Z},\ i = 1, 2)$ are two square-summable time-invariant filters.

Proposition 5.3.5 *Let S_T and $X_i(t)$ be defined as in* (5.3.13) *and* (5.3.14)*, respectively. Assume that the covariance functions $r_i(t)$ of $X_i(t)$ ($i = 1, 2$) satisfy the following condition:*

$$\sum_{t\in\mathbb{Z}} r_i^2(t) < \infty,\ \ i = 1, 2.$$

Then

$$T^{-1/2}\left(S_T - \mathbb{E}[S_T]\right) \xrightarrow{d} N(0, \sigma_S^2), \tag{5.3.15}$$

where

$$\sigma_S^2 := 2\sum_{t\in\mathbb{Z}} r_1(t)r_2(t) + (\kappa_4/\sigma_\xi^4)r_{12}^2, \tag{5.3.16}$$

and $r_{12} := \mathbb{E}[X_1(0)X_2(0)]$.

A similar result applies to continuous time processes. Let $X_i(t)$, $(t \in \mathbb{R},\ i = 1, 2)$ be two continuous time linear processes driven by the same OIP $\{\xi(t),\ t \in \mathbb{R}\}$)

satisfying $\mathbb{E}|d\,\xi(t)|^2 = \sigma^2 dt$ (see 3.2.10):

$$X_i(t) := \int_{\mathbb{R}} g_i(t-s)d\xi(s), \quad \int_{\mathbb{R}} |a(t)|^2 dt < \infty, \ t \in \mathbb{R}, \ i = 1, 2, \tag{5.3.17}$$

where $g_i(t)$, $(t \in \mathbb{R}, i = 1, 2)$ are two square-integrable time-invariant filters. Denote

$$S_T := \int_0^T X_1(t)X_2(t)dt, \quad T \in \mathbb{R}. \tag{5.3.18}$$

Proposition 5.3.6 *Let $X_i(t)$ and S_T be defined as in* (5.3.17) *and* (5.3.18), *respectively. Assume that the covariance functions $r_i(t)$ of $X_i(t)$ $(i = 1, 2)$ satisfy the following condition:*

$$\int_{\mathbb{R}} r_i^2(t)dt < \infty, \ i = 1, 2.$$

Then

$$T^{-1/2}\left(S_T - \mathbb{E}[S_T]\right) \xrightarrow{d} N(0, \sigma_S^2), \tag{5.3.19}$$

where

$$\sigma_S^2 := 2\int_{\mathbb{R}} r_1(t)r_2(t)dt + (\kappa_4/\sigma_\xi^4)r_{12}^2, \tag{5.3.20}$$

and $r_{12} := \mathbb{E}[X_1(0)X_2(0)]$.

For the proof of Proposition 5.3.5, see Giraitis et al. [147], Sect. 4.5, Theorem 4.5.2. The proof of Proposition 5.3.6 is similar.

5.3.3 Convergence Theorems

In this section, we present several convergence theorems that will be frequently used in our proofs of limit theorems for quadratic forms and functionals.

Proposition 5.3.7 (Slutsky's Theorem) *Let $\{X_n, Y_n, X, n \geq 1\}$ be random variables defined in the same probability space $(\Omega, \mathcal{F}, \mathbb{P})$, and let c be a finite constant. If $X_n \xrightarrow[n\to\infty]{d} X$ and $Y_n \xrightarrow[n\to\infty]{d} c$, then $X_n + Y_n \xrightarrow[n\to\infty]{d} X + c$ and $X_nY_n \xrightarrow[n\to\infty]{d} cX$.*

Slutsky's theorem is sometimes referred to as the *converging together lemma*. The following generalizations of Slutsky's theorem will be useful for analyzing time series models (see, e.g., Resnick [253]). Theorem 8.6.2).

Proposition 5.3.8 (Second Converging Together Lemma) *Let* $\{X_{k,n}, X_n, Y_n, X, n \geq 1, k \geq 1\}$ *be random variables defined in the same probability space* $(\Omega, \mathcal{F}, \mathbb{P})$. *Assume that for each* k, $X_{k,n} \xrightarrow{d} X_k$ *as* $n \to \infty$ *and* $X_k \xrightarrow{d} X$ *as* $k \to \infty$. *Suppose further that for any* $\varepsilon > 0$,

$$\lim_{k\to\infty} \limsup_{n\to\infty} \mathbb{P}[|X_{k,n} - Y_n| > \varepsilon] = 0.$$

Then we have $Y_n \xrightarrow{d} X$ *as* $n \to \infty$.

Proposition 5.3.9 (Third Converging Together Lemma) *Let* $\{X_n, Z_{n,\varepsilon}, Y_{n,\varepsilon}, n \geq 1, \varepsilon > 0\}$ *be random variables such that* $X_n = Z_{n,\varepsilon} + Y_{n,\varepsilon}$. *Assume that there are random variables* $\{Z, Z_\varepsilon, \varepsilon > 0\}$ *such that for any* $\varepsilon > 0$, $Z_{n,\varepsilon} \xrightarrow{d} Z_\varepsilon$ *as* $n \to \infty$ *and* $Z_\varepsilon \xrightarrow{d} Z$ *as* $\varepsilon \to 0$. *Suppose further that*

$$\lim_{\varepsilon\to 0} \limsup_{n\to\infty} \operatorname{Var}(Y_{n,\varepsilon}) = 0.$$

Then we have $X_n \xrightarrow{d} Z$ *as* $n \to \infty$.

In several proofs of asymptotic normality in this book, we employ the following *truncation argument*, which is an immediate consequence of Proposition 5.3.9 (see Giraitis et al. [147], Section 4.2).

Proposition 5.3.10 (The Truncation Argument) *Let* $\{X_n, X_{n,\varepsilon}, R_{n,\varepsilon}, n \geq 1, \varepsilon > 0\}$ *be random variables such that* $X_n = X_{n,\varepsilon} + R_{n,\varepsilon}$. *Assume that for any* $\varepsilon > 0$, $X_{n,\varepsilon} \xrightarrow{d} Z_\varepsilon \sim N(0, \sigma_\varepsilon^2)$, $\operatorname{Var}(Y_{n,\varepsilon}) \to \sigma_\varepsilon^2$ *as* $n \to \infty$. *Suppose further that* $\lim_{\varepsilon\to 0} \limsup_{n\to\infty} \operatorname{Var}(R_{n,\varepsilon}) = 0$ *and* $\sigma^2 := \lim_{n\to\infty} \operatorname{Var}(X_n)$ *exists. Then we have* $X_n \xrightarrow{d} Z \sim N(0, \sigma^2)$ *as* $n \to \infty$.

The next proposition presents the so-called discretization argument (see the proof of Theorem 18.7.1 in Ibragimov and Linnik [188]).

Proposition 5.3.11 (The Discretization Argument) *Let*

$$S_T := \sigma_T^{-1} \int_0^T X(t)dt, \quad \sigma_T^2 := \mathbb{E}\left(\int_0^T X(t)dt\right)^2 \to \infty \tag{5.3.21}$$

as $T \to \infty$, *and let*

$$S_{[T]} := \sigma_{[T]}^{-1} \int_0^{[T]} X(t)dt = \sigma_{[T]}^{-1} \sum_{j=0}^{[T]} \xi_j, \tag{5.3.22}$$

where

$$\xi_j := \int_{j-1}^{j} X(t)dt, \quad j = 0, 1, \ldots, [T].$$

Then the random variables S_T and $S_{[T]}$ have the same limiting distributions as $T \to \infty$.

In the next proposition, we present the *Cramér-Wold device*, which is a key step in the proofs of multivariate limit theorems (see, e.g., Billingsley [40], Theorem 29.4, and Bockwell and Davis [57], Proposition 6.3.1).

Proposition 5.3.12 (The Cramér-Wold Device) *For k-dimensional random vectors $\mathbf{X}_n := (X_{n1}, \ldots, X_{nk})$ and $\mathbf{X} = (X_1, \ldots, X_k)$, a necessary and sufficient condition for $\mathbf{X}_n \underset{n\to\infty}{\xrightarrow{d}} \mathbf{X}$ is that*

$$\sum_{i=1}^{k} t_i X_{ni} \underset{n\to\infty}{\xrightarrow{d}} \sum_{i=1}^{k} t_i X_i$$

for each $(t_1, \ldots, t_k) \in \mathbb{R}^k$. In other words, every fixed linear combination of the coordinates of $\mathbf{X}_n$ must converge in distribution to the corresponding linear combination of the coordinates of $\mathbf{X}$.

5.4 Proofs of Theorems of Sect. 5.2

5.4.1 Lemmas

Let $A_T(f)$ denote the covariance matrix (or operator) of a Gaussian process $\{X(u),\ u \in \mathbb{U}\}$. Specifically, $A_T(f)$ is either the $T \times T$ Toeplitz matrix $B_T(f)$ or the T-truncated Toeplitz operator $W_T(f)$ generated by the spectral density f. Similarly, $A_T(g)$ denotes either the $T \times T$ Toeplitz matrix or the T-truncated Toeplitz operator generated by the function g (for definitions, see formulas (4.2.8) and (4.2.11)).

As a consequence of Proposition 5.3.2, we obtain the following result, which establishes a link between the distribution of the quadratic functional Q_T in (5.1.1) and the trace problem for Toeplitz matrices and operators discussed in Chap. 4.

Lemma 5.4.1 *The following assertions hold:*

(a) The quadratic functional Q_T in (5.1.1) has the same distribution as the sum

$$\sum_{k=1}^{\infty} \lambda_{k,T}\xi_k^2 \quad \left(\sum_{k=1}^{T} \lambda_{k,T}\xi_k^2 \quad \text{in the discrete time case}\right),$$

where $\{\xi_k, k \geq 1\}$ *are independent* $N(0, 1)$ *Gaussian random variables and* $\{\lambda_{k,T}, k \geq 1\}$ *are the eigenvalues of the operator* $A_T(f)A_T(g)$.

(b) The characteristic function $\varphi_T(t)$ *of* Q_T *is given by*

$$\varphi_T(t) = \prod_{k=1}^{\infty} |1 - 2it\lambda_{k,T}|^{-1/2}. \tag{5.4.1}$$

(c) The k-th-order cumulant $\chi_k(\cdot)$ *of* Q_T *is given by*

$$\begin{aligned}\chi_k(Q_T) &= 2^{k-1}(k-1)! \sum_{j=1}^{\infty} \lambda_{j,T}^k \\ &= 2^{k-1}(k-1)!\, tr[A_T(f)A_T(g)]^k,\end{aligned} \tag{5.4.2}$$

where $tr[A]$ *stands for the trace of the operator* A.

The product in (5.4.1) *and the sum in* (5.4.2) *are over* $j = 1, \ldots, T$ *in the discrete time case.*

Proof Follows from Proposition 5.3.2 with $R = A_T(f)$ and $A = A_T(g)$. □

The next proposition provides estimates for the k-th-order cumulant $\chi_k(Q_T)$ of the quadratic functional Q_T, as defined in (5.1.1).

Proposition 5.4.1 *Let* $f \in L^1(\Lambda) \cap L^p(\Lambda)$ $(p \geq 1)$ *and* $g \in L^1(\Lambda) \cap L^q(\Lambda)$ $(q \geq 1)$. *The following assertions hold:*

(a) If $1/p + 1/q \leq 1/k$, *then for all* $k \in \mathbb{N}$ *as* $T \to \infty$

$$\chi_k(Q_T) \sim T \times 2^{k-1}(k-1)!\,(2\pi)^{2k-1} \int_{\Lambda} f^k(\lambda)g^k(\lambda)d\lambda. \tag{5.4.3}$$

(b) For all $k = 3, 4, \ldots,$

$$\chi_k(Q_T) \leq C\,\mathrm{Var}(Q_T) \times \left(||f||_p||g||_q T^{1/p+1/q}\right)^{k-2}. \tag{5.4.4}$$

(c) If $1/p + 1/q = 1/2$, *then for all* $k = 2, 3, 4, \ldots,$

$$\chi_k(Q_T) \leq C\, T^{k/2} \times ||f||_p^k ||g||_q^k. \tag{5.4.5}$$

Proof *Assertion (a) follows from* (5.4.2) *and Theorem 4.4.1* **(A2)** *in the discrete time case, and Theorem 4.5.1* **(A2)** *in the continuous time case.*

To prove assertion (b), observe first that

$$\mathrm{Var}(Q_T) = 2tr[A_T(f)A_T(g)]^2. \tag{5.4.6}$$

Hence, by (5.4.2)*, we can write*

$$\chi_k(Q_T) \le 2^{k-2}(k-1)!\mathrm{Var}(Q_T)\,[\mu_T(f,g)]^{k-2}, \tag{5.4.7}$$

where $\mu_T(f,g) := ||A_T(f)A_T(g)||_\infty$.

Using the arguments from the proof of Theorem 4.9.2 and Hölder's inequality (see also Bentkus and Rudzkis [28], Lemma 3.1), we infer that

$$||A_T(f)||_\infty \le (2\pi)^{1-1/p}||f||_p T^{1/p} \quad \text{and} \quad ||A_T(g)||_\infty \le (2\pi)^{1-1/q}||g||_q T^{1/q}.$$

Therefore, we have

$$\mu_T(f,g) := ||A_T(f)A_T(g)||_\infty \le (2\pi)^{1-1/p-1/q}||f||_p||g||_q T^{1/p+1/q}. \tag{5.4.8}$$

A combination of (5.4.6)–(5.4.8) *yields* (5.4.4).

Assertion (c) follows from assertion (b), formula (5.4.6)*, and Proposition 4.9.2.*

□

Let $\{Y(t),\ t \in \mathbb{R}\}$ be a real-valued, centered, separable stationary Gaussian process with the spectral density $f_Y(\lambda)$. Consider the quadratic functionals:

$$L_T := \int_0^T Y^2(t)\,dt \tag{5.4.9}$$

and

$$\widetilde{L}_T := T^{-1/2}\left(L_T - \mathbb{E}[L_T]\right). \tag{5.4.10}$$

Lemma 5.4.2 *Assume that* $f_Y(\lambda) \in L^1(\mathbb{R}) \cap L^2(\mathbb{R})$*. Then the distribution of the normalized quadratic functional* $\widetilde{L}_T$ *converges (as* $T \to \infty$*) to the centered normal distribution* $N(0, \sigma_Y^2)$ *with variance*

$$\sigma_Y^2 = 4\pi \int_{-\infty}^{+\infty} f_Y^2(\lambda)\,d\lambda. \tag{5.4.11}$$

Proof For $T > 0$ denote by $W_T(f_Y)$ the Toeplitz operator generated by the function f_Y (see Eq. (4.5.1)). Let $R(t) := \hat{f}_Y(t)$ be the generating kernel of $W_T(f_Y)$; thus, $R(t)$ serves as the covariance function of $Y(t)$. By $\mu_j = \mu_j(T)$, $j \ge 1$ and $e_j(t) = e_j(t,T) \in L_2[0.T]$, $j \ge 1$, we denote the eigenvalues and the corresponding orthonormal eigenfunctions of $W_T(f_Y)$. Therefore, we have

$$\int_0^T R(t-s)e_j(s)\,ds = \mu_j e_j(t), \quad j \ge 1. \tag{5.4.12}$$

Since by Mercer's theorem (see Proposition 2.7.9)

$$R(t-s) = \sum_{j=1}^{\infty} \mu_j e_j(t) e_j(s) \tag{5.4.13}$$

with positive and summable eigenvalues $\{\mu_j\}$, we have the Karhunen-Loéve expansion (see Proposition 3.5.1):

$$Y(t) = \sum_{j=1}^{\infty} \sqrt{\mu_j} \xi_j e_j(t), \tag{5.4.14}$$

where $\{\xi_j\}$ are independent $N(0, 1)$ random variables. Therefore, by (5.4.10), (5.4.14), and Lemma 5.4.1 (a), we have

$$\widetilde{L}_T = \frac{1}{\sqrt{T}} \sum_{j=1}^{\infty} \mu_j (\xi_j^2 - 1). \tag{5.4.15}$$

Denote by $\chi_k(\widetilde{L}_T)$ the k-th-order cumulant of $\widetilde{L}_T$. By (5.4.15) and Lemma 5.4.1c), we have

$$\chi_k(\widetilde{L}_T) = \begin{cases} 0, & \text{for } k = 1, \\ (k-1)!\, 2^{k-1} T^{-k/2} \operatorname{tr}[W_T(f_Y)]^k, & \text{for } k \geq 2. \end{cases} \tag{5.4.16}$$

By (5.4.16) with $k = 2$ and Theorem 4.9.3, we have as $T \to \infty$

$$\chi_2(\widetilde{L}_T) = \frac{2}{T} ||W_T(f_Y)||_2^2 \longrightarrow 4\pi \int_{-\infty}^{+\infty} f_Y^2(\lambda)\, d\lambda < \infty. \tag{5.4.17}$$

Next, by (5.4.16) for $k \geq 3$, we obtain

$$|\chi_k(\widetilde{L}_T)| \leq C \frac{1}{T} ||W_T(f_Y)||_2^2 \, T^{1-k/2} \mu_T^{k-2}, \tag{5.4.18}$$

where $\mu_T = ||W_T(f_Y)||_\infty$.

Finally, by (5.4.17) and Theorem 4.9.2, the right-hand side of (5.4.18) converges to zero as $T \to \infty$. This concludes the proof of the lemma. □

5.4.2 Proofs of Theorems

Proof of Theorem 5.2.1 First, note that assertions **(B)**, **(C)**, and **(D)** of the theorem follow from assertion **(A)** and the corresponding results from Chap. 4. Specifically, we have the following:

Proof of assertion ***(B)***. In the discrete time case, the result follows from assertion **(A)** and Theorem 4.4.1 **(A1)** with $m = 4$, $h_1 = h_3 := f$, and $h_2 = h_4 := g$. In the continuous time case, it follows from **(A)** and Theorem 4.5.1 **(A1)** with $m = 4$, $h_1 = h_3 := f$, and $h_2 = h_4 := g$.

Proof of assertion ***(C)***. In the discrete time case, the result follows from assertion **(A)** and Theorem 4.4.1 **(A2)** with $m = 4$, $h_1 = h_3 := f$, and $h_2 = h_4 := g$. In the continuous time case, it follows from **(A)** and Theorem 4.5.1 **(A2)** with $m = 4$, $h_1 = h_3 := f$, and $h_2 = h_4 := g$.

Proof of assertion ***(D)***. In the discrete time case, the result follows from assertion **(A)** and Theorem 4.4.3. In the continuous time case, it follows from **(A)** and Theorem 4.5.2.

Thus, we only need to prove assertion **(A)** of the theorem. The proof for the discrete time case can be found in the book by Giraitis et al. [147] (Theorem 6.3.1). Therefore, we will focus on proving assertion **(A)** for the continuous time case.

Proof of assertion ***(A)***. By Theorem 3.2.2, the underlying process $X(t)$ admits the moving average representation:

$$X(t) = \int_{-\infty}^{+\infty} \widehat{a}(t-s)\, d\xi(s), \tag{5.4.19}$$

where

$$\int_{-\infty}^{+\infty} |\widehat{a}(t)|^2\, dt < \infty, \tag{5.4.20}$$

while $\xi(s)$ is a process with orthogonal increments such that $\mathbb{E}[d\xi(s)] = 0$ and $\mathbb{E}|d\xi(s)|^2 = ds$. Moreover, the spectral density $f(\lambda)$ of $X(t)$ can be represented as

$$f(\lambda) = 2\pi\, |a(\lambda)|^2, \tag{5.4.21}$$

where $a(\lambda)$ is the inverse Fourier transform of the filter $\widehat{a}(t)$. We set

$$a_1(\lambda) = (2\pi)^{1/2} a(\lambda) \cdot [g(\lambda)]^{1/2}, \tag{5.4.22}$$

where $g(\lambda)$ is the generating function, which we assume to be non-negative without loss of generality.

Consider a process $Y(t)$ $(t \in \mathbb{R})$ defined by

$$Y(t) = \int_{-\infty}^{+\infty} \widehat{a_1}(t-s)\, d\xi(s), \tag{5.4.23}$$

in which $\widehat{a_1}(t)$ is the Fourier transform of $a_1(\lambda)$ and $\xi(s)$ is as defined in (5.4.19).

Since $fg \in L^1(\mathbb{R}) \cap L^2(\mathbb{R})$, we can apply Parseval-Plancherel's identity to conclude that

$$\int_{-\infty}^{+\infty} |\widehat{a_1}(t)|^2\, dt = 2\pi \int_{-\infty}^{+\infty} |a_1(\lambda)|^2\, d\lambda = 4\pi^2 \int_{-\infty}^{+\infty} f(\lambda)\, g(\lambda)\, d\lambda < \infty. \tag{5.4.24}$$

Thus, $Y(t)$ is a well-defined stationary process with spectral density:

$$\varphi(\lambda) := |a_1(\lambda)|^2 = 2\pi\, f(\lambda)\, g(\lambda). \tag{5.4.25}$$

Since $fg \in L^1(\mathbb{R}) \cap L^2(\mathbb{R})$, the process $Y(t)$ defined by Eq. (5.4.23) satisfies the conditions of Lemma 5.4.2. Therefore, the combination of Lemma 5.4.2 and the approximation Lemma 5.4.3 that follows completes the proof of Theorem 5.2.1 **(A)**. □

Lemma 5.4.3 *Let the functionals Q_T and L_T be as defined in (5.1.1) and (5.4.9), respectively. Then, under the assumptions of Theorem 5.2.1* **(A)***, the following relation holds:*

$$\mathrm{Var}(Q_T - L_T) = o(T) \quad \text{as} \quad T \to \infty. \tag{5.4.26}$$

Proof Observe first that, by (5.1.1) and (5.4.19), we have

$$Q_T = \int_{\mathbb{R}^2} \left[\int_0^T \int_0^T \widehat{g}(t-s)\widehat{a}(t-u_1)\widehat{a}(s-u_2)\, dt\, ds \right] d\xi(u_1)\, d\xi(u_2). \tag{5.4.27}$$

Similarly, by (5.4.9) and (5.4.23), we obtain

$$L_T = \int_{\mathbb{R}^2} \left[\int_0^T \widehat{a_1}(t-u_1)\widehat{a_1}(s-u_2)\, dt \right] d\xi(u_1)\, d\xi(u_2). \tag{5.4.28}$$

Define

$$d_{1T}(u_1, u_2) := \int_0^T \int_0^T \widehat{g}(t-s)\widehat{a}(t-u_1)\widehat{a}(s-u_2)\, dt\, ds \tag{5.4.29}$$

and

$$d_{2T}(u_1,u_2) := \int_0^T \int_0^T \delta(t-s)\widehat{a}_1(t-u_1)\widehat{a}_1(s-u_2)\,dt\,ds$$
$$= \int_0^T \widehat{a}_1(t-u_1)\widehat{a}_1(s-u_2)\,dt, \tag{5.4.30}$$

where $\delta(t)$ is the delta function. Then, from Eqs. (5.4.27)–(5.4.30), we have

$$Q_T - L_T = \int_{\mathbb{R}^2} \left[d_{1T}(u_1,u_2) - d_{2T}(u_1,u_2)\right] d\xi(u_1)\, d\xi(u_2). \tag{5.4.31}$$

Since the underlying process $X(t)$ is Gaussian, we obtain

$$\operatorname{Var}(Q_T - L_T) = 2\int_{\mathbb{R}^2} \left[d_{1T}(u_1,u_2) - d_{2T}(u_1,u_2)\right]^2 du_1\, du_2. \tag{5.4.32}$$

Define

$$p_1(\lambda_1,\lambda_2,\mu) := a(\lambda_1)a(\lambda_2)g(\mu), \tag{5.4.33}$$
$$p_2(\lambda_1,\lambda_2,\mu) := a_1(\lambda_1)a_1(\lambda_2)\hat{\delta}(\mu) = a(\lambda_1)a(\lambda_2)[g(\lambda_1)]^{1/2}[g(\lambda_2)]^{1/2}. \tag{5.4.34}$$

Then, by Parseval-Plancherel's identity, we have

$$\int_{\mathbb{R}^2} d_{iT}^2(u_1,u_2)\,du_1\,du_2$$
$$= (2\pi)^2 \int_{\mathbb{R}^2} \left|\int_{\mathbb{R}} G_T(\lambda_1-\mu)G_T(\mu-\lambda_2)p_i(\lambda_1,\lambda_2,\mu)\,d\mu\right| d\lambda_1\, d\lambda_2$$
$$= (2\pi)^2 T||p_i||_T^2, \quad i=1,2, \tag{5.4.35}$$

where $G_T(u)$ is given by (4.6.35), $||p||_T^2 = (p,p)_T$,

$$(p,p')_T = \int_{\mathbb{R}^4} p(\lambda_1,\lambda_2,\lambda_3)\overline{p'(\lambda_1,\lambda_2,\lambda_4)}\,d\mu_T, \tag{5.4.36}$$

and the measure μ_T is defined by (4.6.53).

In view of (5.4.32) and (5.4.35), we have

$$\operatorname{Var}(Q_T - L_T) = 8\pi^2 T\|p_1 - p_2\|_T^2 \tag{5.4.37}$$

For any $K > 0$, we consider the following sets:

$$E_1^K = \{u \in \mathbb{R} : |a(u)| < K\}, \quad E_2^K = \{u \in \mathbb{R} : g(u) < K\}, \tag{5.4.38}$$

and define the truncated functions:

$$\begin{aligned} p_1^K(\mathbf{u}) &:= p_1(\mathbf{u})\mathbf{1}_1^K(u_1)\mathbf{1}_1^K(u_2)\mathbf{1}_2^K(u_3), \\ p_2^K(\mathbf{u}) &:= p_2(\mathbf{u})\mathbf{1}_1^K(u_1)\mathbf{1}_1^K(u_2)\mathbf{1}_2^K(u_1)\mathbf{1}_2^K(u_2), \end{aligned} \tag{5.4.39}$$

where $\mathbf{u} = (u_1, u_2, u_3) \in \mathbb{R}^3$ and $\mathbf{1}_j^K(u)$ is the indicator function of the set E_j^K, $j = 1, 2$. Then, we have

$$\|p_1 - p_2\|_T^2 \le 3\left(\|p_1^K - p_2^K\|_T^2 + \|p_1 - p_1^K\|_T^2 + \|p_2 - p_2^K\|_T^2\right). \tag{5.4.40}$$

Now, the relations in (5.4.33), (5.4.34), and (5.4.39) imply that

$$\|p_1^K - p_2^K\|_T^2 = \int_{\mathbb{R}^4} \Gamma d\mu_T,$$

where $\Gamma = \Gamma(u_1, u_2, u_3, u_4)$ is a sum of four functions that satisfy the conditions of Lemma 4.6.11 with $m = 4$. Since $\Gamma(u, u, u, u) = 0$ for $u \in \mathbb{R}$, by applying Lemma 4.6.11, we obtain

$$\lim_{T\to\infty} \|p_1^K - p_2^K\|_T = \int_{\mathbb{R}} \Gamma(u, u, u, u)du = 0. \tag{5.4.41}$$

Next, according to (5.4.36), we have

$$\|p_1\|_T^2 = \|p_1^K + (p_1 - p_1^K)\|_T^2 = \|p_1\|_T^2 + 2(p_1^K, p_1 - p_1^K)_T + \|p_1 - p_1^K\|_T^2.$$

Therefore,

$$\|p_1 - p_1^K\|_T^2 \le \big|\|p_1\|_T^2 - \|p_1^K\|_T^2\big| + 2\big|(p_1^K, p_1 - p_1^K)_T\big|. \tag{5.4.42}$$

By (5.2.5), (5.4.35), and Lemma 4.6.6 (b), we have

$$\|p_1\|_T^2 = (2\pi)^{-2}\frac{1}{T}\mathrm{tr}\,[B_T(f)B_T(g)]^2 \to 2\pi \int_{\mathbb{R}} f^2(u)g^2(u)du, \tag{5.4.43}$$

while, according to Lemma 4.6.11 and (5.4.35), we obtain

$$\|p_1^K\|_T^2 \to 2\pi \int_{F_K} f^2(u)g^2(u)du, \tag{5.4.44}$$

where $F_K := \{u \in \mathbb{R} : f(u) < K, \ g(u) < K\}$.

From (5.4.43) and (5.4.44), we obtain as $K \to \infty$

$$\lim_{T\to\infty} \left(\|p_1\|_T^2 - \|p_1^K\|_T^2 \right) = \int_{\mathbb{R}\setminus F_K} f^2(u) g^2(u) du = o(1). \tag{5.4.45}$$

To estimate the last term on the right-hand side of (5.4.42), we denote

$$\Gamma_K(u_1, u_2, u_3, u_4) = p_1^K(u_1, u_2, u_3) \left[p_1(u_1, u_2, u_4) - p_1^K(u_1, u_2, u_4) \right].$$

From (5.4.38) and (5.4.39), for $\Gamma_K(u_1, u_2, u_3, u_4) \neq 0$, we have

$$|a(u_1)| < K, \ |a(u_2)| < K, \ g(u_3) < K, \ g(u_4) > K, \tag{5.4.46}$$

Next, for any $L > K$ and $\mathbf{u} = (u_1, u_2, u_3, u_4)$, we have

$$(p_1^K, p_1 - p_1^K)_T = \int_{\mathbb{R}^4} \Gamma_K(\mathbf{u}) d\mu_T = \int_{\mathbb{R}^4} \Gamma_K(\mathbf{u}) \mathbf{1}_2^L(u_4) d\mu_T + I, \tag{5.4.47}$$

where with some constant C_K depending on K

$$|I| \leq C_K \int_{\mathbb{R}^4} g(u_4) \left(1 - \mathbf{1}_2^L(u_4)\right) d|\mu_T|. \tag{5.4.48}$$

It follows from (5.4.33), (5.4.34), and (5.4.39) that $\Gamma_K(\mathbf{u})\mathbf{1}_2^L(u_4)$ is a linear combination of functions that satisfy the conditions of Lemma 4.6.11 with $m = 4$. Therefore, by applying Lemma 4.6.11 and noting that $\Gamma_K(u, u, u, u) = 0$ for $u \in \mathbb{R}$, we obtain

$$\lim_{T\to\infty} \int_{\mathbb{R}^4} \Gamma_K(\mathbf{u}) \mathbf{1}_2^L(u_4) d\mu_T = \int_{\mathbb{R}} \Gamma_K(u, u, u, u) \mathbf{1}_2^L(u) du = 0. \tag{5.4.49}$$

For a given $\varepsilon > 0$ and sufficiently large L by (4.6.57), we have

$$C_K \int_{\mathbb{R}^4} g(u) \left(1 - \mathbf{1}_2^L(u)\right) d|\mu_T| \leq C_K \int_{\{u:\, g(u) > L\}} g(u) du \leq \varepsilon. \tag{5.4.50}$$

From (5.4.47) to (5.4.50), we obtain

$$\lim_{T\to\infty} (p_1^K, p_1 - p_1^K)_T = 0.$$

This, combined with (5.4.42) and (5.4.45), yields

$$\lim_{T\to\infty} \|p_1 - p_1^K\|_T = 0. \tag{5.4.51}$$

Finally, we prove that

$$\lim_{T\to\infty} \|p_2 - p_2^K\|_T = 0. \tag{5.4.52}$$

Indeed, according to (5.4.34), (5.4.39), and (4.6.58), we have

$$\|p_2 - p_2^K\|_T \le \int_{\mathbb{R}^4} [1 - \mathbf{1}_1^K(u_1)] f(u_1)g(u_1)f(u_2)g(u_2)d|\mu_T|$$
$$+ \int_{\mathbb{R}^4} [1 - \mathbf{1}_1^K(u_2)] f(u_1)g(u_1)f(u_2)g(u_2)d|\mu_T|$$
$$\le \int_{\{u:|f(u)|>\sqrt{K}\}} f^2(u)g^2(u)du + \int_{\{u:|g(u)|>K\}} f^2(u)g^2(u)du = o(1),$$

when $K \to \infty$ (uniformly on T).

A combination of (5.4.37), (5.4.40), (5.4.41), (5.4.51), and (5.4.52) yields (5.4.26). Lemma 5.4.3 is proved. □

Proof of Theorem 5.2.2 In the discrete time case, the result follows from Theorems 5.2.1 **(A)** and 4.4.4. In the continuous time case, it follows from Theorems 5.2.1 **(A)** and 4.5.3. □

Proof of Proposition 5.2.1 We construct functions f and g that satisfy the conditions (5.2.10) and (5.2.11). Let $p \ge 2$ be a fixed number. We choose a number $q > 1$ such that $\frac{1}{p} + \frac{1}{q} > 1$. For such p and q, consider the functions $f_0(\lambda)$ and $g_0(\lambda)$ defined by

$$f_0(\lambda) = \begin{cases} \left(2^s/s^2\right)^{1/p} & \text{if } 2^{-s-1} \le \lambda \le 2^{-s},\ s = 2m \\ 0 & \text{if } 2^{-s-1} \le \lambda \le 2^{-s},\ s = 2m+1, \end{cases} \tag{5.4.53}$$

$$g_0(\lambda) = \begin{cases} \left(2^s/s^2\right)^{1/q} & \text{if } 2^{-s-1} \le \lambda \le 2^{-s},\ s = 2m+1 \\ 0 & \text{if } 2^{-s-1} \le \lambda \le 2^{-s},\ s = 2m, \end{cases} \tag{5.4.54}$$

where m is a positive integer.

It is easy to see that $f_0(\lambda) \in L^p(\mathbb{T})$, $g_0(\lambda) \in L^q(\mathbb{T})$, and $f_0(\lambda)\, g_0(\lambda) \in L^r(\mathbb{T})$ for every r. Additionally, since the functions $f_0(\lambda)$ and $g_0(\lambda)$ have disjoint supports, we have

$$\int_{-\pi}^{\pi} f_0^2(\lambda)g_0^2(\lambda)\, d\lambda = 0.$$

We first show that for $1/p + 1/q > 1$

$$\frac{1}{T}\operatorname{tr}\left(B_T(f_0)B_T(g_0)\right)^2 \longrightarrow \infty \quad \text{as} \quad T \to \infty, \tag{5.4.55}$$

where $B_T(f_0)$ and $B_T(g_0)$ are the truncated Toeplitz matrices generated by the functions f_0 and g_0, respectively.

Indeed, let λ_k, $k = 1, \ldots, n$, denote the eigenvalues of the matrix $B_T(f_0)B_T(g_0)$. We have

$$\sum_{k=1}^{T} \lambda_k = \operatorname{tr}[B_T(f_0)B_T(g_0)] = 2\int_0^{\pi}\int_0^{\pi} \frac{\sin^2[T(t-\lambda)/2]}{\sin^2[(t-\lambda)/2]} f_0(\lambda)\, g_0(t)\, d\lambda\, dt$$

$$= 2\int_0^{\pi} \frac{\sin^2[(Tx)/2]}{\sin^2[x/2]} h(x)\, dx, \tag{5.4.56}$$

where

$$h(x) = \int_0^{\pi} f_0(\lambda)\, g_0(\lambda + x)\, d\lambda.$$

We have $h(0) = 0$ and for $x \neq 0$ the segment $[2^{-s-1} + x, 2^{-s} + x]$ is contained in $[2^{-s}, 2^{-s+1}]$ as long as $x \in [2^{-s-1}, 2^{-s}]$.

Therefore, for $s = 2m$ and $\lambda, x \in [2^{-s-1}, 2^{-s}]$, we have

$$f_0(\lambda)\, g_0(\lambda + x) = \left(\frac{2^s}{s^2}\right)^{1/p} \left(\frac{2^{s-1}}{(s-1)^2}\right)^{1/q} \gtrsim c \cdot \left(\frac{2^s}{s^2}\right)^{1/p+1/q} \tag{5.4.57}$$

and

$$h(x) \gtrsim c \cdot x^{1-1/p-1/q}\, (\ln x)^{-2/p-2/q}. \tag{5.4.58}$$

Here and below, c stands for various positive constants. It is easy to see that the relations (5.4.57) and (5.4.58) remain valid in the case where $s = 2m + 1$.

Therefore, taking into account that for $0 < x < 1/T$,

$$\sin^2[(Tx)/2] \geq cT^2 \sin^2[x/2],$$

in view of (5.4.56) and (5.4.58), we obtain

$$\sum_{k=1}^{T} \lambda_k = 2T^2 \int_0^{\pi} \frac{\sin^2[(Tx)/2]}{T^2 \sin^2[x/2]} h(x)\, dx \geq c \cdot T^2 \int_0^{1/T} h(x)\, dx$$

$$\gtrsim cT^2 \int_0^{1/T} x^{1-1/p-1/q}\, (\ln x)^{-2/p-2/q}\, dx$$

$$\geq c \cdot T^{1/p+1/q}\, (\ln T)^{-2/p-2/q}.$$

Applying the Cauchy-Schwarz inequality, we find that

$$\begin{aligned}\frac{1}{T}\mathrm{tr}\,(B_T(f_0)B_T(g_0))^2 &= \frac{1}{T}\sum_{k=1}^{T}\lambda_k^2 \geq \frac{1}{T^2}\left(\sum_{k=1}^{T}\lambda_k\right)^2\\ &\geq c\cdot T^{2/p+2/q-2}\,(\ln T)^{-4/p-4/q}\,.\end{aligned}$$

For $1/p+1/q > 1$, it is clear that the last expression tends to infinity as $T\to\infty$, which implies the relation in (5.4.55).

Next, for an arbitrary finite positive constant C, we set $g_{\pm}(\lambda) = g_0(\lambda) \pm C$. Since the functions $f_0(\lambda)$ and $g_0(\lambda)$ have disjoint supports, we have

$$\int_{-\pi}^{\pi} f_0^2(\lambda)\, g_{\pm}^2(\lambda)\, d\lambda = \int_{-\pi}^{\pi} f_0^2(\lambda)\,(g_0(\lambda)\pm C)^2\, d\lambda = C^2\int_{-\pi}^{\pi} f_0^2(\lambda)\, d\lambda < \infty,$$

implying that (5.2.10) is fulfilled. Next, by (5.4.55) and by Theorem 4.4.1 **(A2)** with $p_1 = p \geq 2$ and $p_2 = \infty$, we have as $T\to\infty$,

$$\frac{1}{T}\, C^2\mathrm{tr}\left(B_T^2(f_0)\right) \longrightarrow 8\pi^3C^2\int_{-\pi}^{\pi} f_0^2(\lambda)\, d\lambda < \infty. \tag{5.4.59}$$

On the other hand, we have

$$\begin{aligned}&\mathrm{tr}\,(B_T(f_0)B_T(g_{\pm}))^2 = \mathrm{tr}\,(B_T(f_0)B_T(g_0\pm C))^2\\ &= \mathrm{tr}\,(B_T(f_0)B_T(g_0))^2 \pm 2C\,\mathrm{tr}\left(B_T^2(f_0)B_T(g_0)\right) + C^2\mathrm{tr}\left(B_T^2(f_0)\right),\end{aligned}$$

which, combined with (5.4.55) and (5.4.59), implies

$$\begin{aligned}&\frac{1}{T}\mathrm{tr}\,(B_T(f_0)B_T(g_+))^2 + \frac{1}{T}\mathrm{tr}\,(B_T(f_0)B_T(g_-))^2\\ &= \frac{2}{T}\mathrm{tr}\,(B_T(f_0)B_T(g_0))^2 + \frac{2}{T}C^2\mathrm{tr}\left(B_T^2(f_0)\right) \to \infty \quad \text{as} \quad T\to\infty.\end{aligned}$$

Therefore, either

$$\lim_{T\to\infty}\sup\frac{1}{T}\mathrm{tr}\,(B_T(f_0)B_T(g_+))^2 = \infty,$$

or

$$\lim_{T\to\infty}\sup\frac{1}{T}\mathrm{tr}\,(B_T(f_0)B_T(g_-))^2 = \infty.$$

Thus, we obtain

$$\lim_{T\to\infty} \sup \chi_2(\widetilde{Q}_T) = \lim_{T\to\infty} \sup \frac{2}{T} \operatorname{tr}(B_T(f)B_T(g))^2 = \infty$$

with $f = f_0$ and $g = g_+$ or $g = g_-$. This completes the proof. □

5.5 Complements and Applications

5.5.1 Operator Conditions for CLT

In this subsection, we assume that the model $X(t)$ is a discrete time Gaussian process defined on a probability space $(\Omega, \mathcal{F}, \mathbb{P})$. We are interested in the asymptotic normality of the quadratic form $Q_T := Q_T(f, g)$, as given by Eq. (5.1.1) in the discrete time case. Recall the notation (see formula (5.2.1)):

$$\widetilde{Q}_T(f, g) := T^{-1/2} \left(Q_T(f, g) - \mathbb{E}[Q_T(f, g)]\right). \tag{5.5.1}$$

and (see formula (5.2.3)):

$$\sigma^2(f, g) := \sigma_Q^2 = 16\pi^3 \int_{-\pi}^{\pi} f^2(\lambda)g^2(\lambda)\, d\lambda. \tag{5.5.2}$$

As before, let $L^2(\mathbb{P})$ be the L^2-space constructed from a probability measure $\mathbb{P}$ such that $X(t) \in L^2(\mathbb{P})$, $t \in \mathbb{Z}$.

Solev and Gerville-Reache [283] observed that for a fixed spectral density f and a number $T \in \mathbb{N}$, the quadratic form $\widetilde{Q}_T(f, g)$ in (5.5.1) can be interpreted as the value of a linear operator defined by

$$\widetilde{\mathbf{Q}}_T : g \mapsto \widetilde{Q}_T(f, g).$$

It turns out that to study the asymptotic normality of the quadratic form $\widetilde{Q}_T(f, g)$, it is sufficient, for a small enough number $\varepsilon > 0$, to identify the sets B_ε of functions g (which may depend on f) for which the following relation holds:

$$\limsup_{T\to\infty} \sup_{g\in B_\varepsilon} ||\widetilde{Q}_T(f, g)||_{L^2(\mathbb{P})} \le \varepsilon, \tag{5.5.3}$$

where $||\cdot||_{L^2(\mathbb{P})}$ is the norm in the space $L^2(\mathbb{P})$.

It can be shown that $||\widetilde{Q}_T(f, g)||_{L^2(\mathbb{P})}$ possesses the following symmetry property when considered as a function of the pair (f, g): if g is a non-negative function, then

$$||\widetilde{Q}_T(f, g)||^2_{L^2(\mathbb{P})} = ||\widetilde{Q}_T(g, f)||^2_{L^2(\mathbb{P})}. \tag{5.5.4}$$

The following theorem provides "operator sufficient conditions" for the asymptotic normality of the quadratic form $\widetilde{Q}_T(f, g)$ (see Solev and Gerville-Reache [283]).

Theorem 5.5.1 *Let $\mathcal{G}_0$ and $\mathcal{G}$ ($\mathcal{G}_0 \subset \mathcal{G}$) be linear subsets in the space $L^1(\mathbb{T})$. Assume that the following conditions hold:*

(a) For every $g_0 \in \mathcal{G}_0$, the limit

$$\lim_{T\to\infty} ||\widetilde{Q}_T(f, g_0)||^2_{L^2(\mathbb{P})} = \sigma^2(f, g_0) < \infty$$

exists, and the following relation holds:

$$\widetilde{Q}_T(f, g_0) \xrightarrow{d} \eta_0 \sim N(0, \sigma^2(f, g_0)) \quad \text{as} \quad T \to \infty;$$

(b) For every $g \in \mathcal{G}$ and for a number $\varepsilon > 0$, there exists a function $g_0 \in \mathcal{G}_0$ (depending on g and ε) such that

$$\limsup_{T\to\infty} ||\widetilde{Q}_T(f, g - g_0)||_{L^2(\mathbb{P})} \le \varepsilon, \quad ||g - g_0)||_{f^2} \le \varepsilon,$$

where $||\cdot||_{f^2}$ is the norm in the space $L^2(\mathbb{T}, f^2)$.

Then $\sigma^2(f, g) < \infty$, the limit

$$\lim_{T\to\infty} ||\widetilde{Q}_T(f, g)||^2_{L^2(\mathbb{P})} = \sigma^2(f, g)$$

exists, and the quadratic form $Q_T(f, g)$ satisfies the CLT:

$$\widetilde{Q}_T(f, g) \xrightarrow{d} \eta \sim N(0, \sigma^2(f, g)) \quad \text{as} \quad T \to \infty. \tag{5.5.5}$$

The theorem states that if the CLT for the quadratic form $Q_T(f, g)$ holds for every $g_0 \in \mathcal{G}_0$ (part (a)), then it also holds for every $g \in \mathcal{G}$ that is close to g_0 in the metric of the space $L^2(\mathbb{T}, f^2)$ (part (b)).

In Solev and Gerville-Reache [283], constructive sufficient conditions were also derived for the quadratic form $Q_T(f, g)$ to satisfy the CLT.

5.5.2 Berry-Esséen Bounds for Random Toeplitz Quadratic Functionals

In this section, we briefly discuss the Berry-Esséen bounds in the CLT for random Toeplitz quadratic functionals, applicable to both continuous and discrete time Gaussian stationary processes (see Nourdin and Peccati [229], Taniguchi [293], and the references therein).

Let the quadratic functionals Q_T and $\widetilde{Q}_T$ be defined as in (5.1.1) and (5.2.1), respectively. Denote

$$\widehat{Q}_T := \widetilde{Q}_T/\sqrt{\mathrm{Var}(\widetilde{Q}_T)}, \tag{5.5.6}$$

and let Z be the standard normal random variable: $Z \sim N(0, 1)$.

The CLT for Q_T presented in Sect. 5.2 states that $\widehat{Q}_T \overset{d}{\to} Z$ as $T \to \infty$.

The next logical step is to examine the proximity between the distribution of $\widehat{Q}_T$ and the standard normal distribution, which entails determining the convergence rate in the CLT. Such results are known as *Berry-Esséen bounds (or asymptotics)*.

In the discrete time case, for a special quadratic functional, Berry-Esséen bounds were established in Tanoguchi [293]. For the continuous time case, Berry-Esséen-type bounds were obtained by Nourdin and Peccati [229]. The next theorem, which can be viewed as an extension of Theorem 5.2.1 **(C)**, encompasses both discrete and continuous time cases.

Theorem 5.5.2 *Let the functional $\widehat{Q}_T$ be defined as in (5.5.6), and let $\Phi(z) = P(Z \le z)$, where $Z \sim N(0, 1)$. Assume that $f(\lambda) \in L^1(\Lambda) \cap L^p(\Lambda)$ $(p \ge 1)$ and $g(\lambda) \in L^1(\Lambda) \cap L^q(\Lambda)$ $(q \ge 1)$. The following assertions hold:*

1. *If $1/p + 1/q \le 1/4$, then there exists a constant $C = C(f, g) > 0$ such that, for all $T > 0$, we have*

$$\sup_{z\in\mathbb{R}} |P(\widehat{Q}_T \le z) - \Phi(z)| \le \frac{C}{\sqrt{T}}. \tag{5.5.7}$$

2. *If $1/p + 1/q \le 1/8$ and $\int_\Lambda f^3(\lambda)g^3(\lambda)\, d\lambda \ne 0$, then there exist a constant $c = c(f, g) > 0$ and a number $T_0 = T_0(f, g) > 0$ such that for $T > T_0$, we have*

$$\sup_{z\in\mathbb{R}} |P(\widehat{Q}_T \le z) - \Phi(z)| \ge \frac{c}{\sqrt{T}}. \tag{5.5.8}$$

More precisely, for any $z \in \mathbb{R}$, we have as $T \to \infty$,

$$\sqrt{T}|P(\widehat{Q}_T \le z) - \Phi(z)| \longrightarrow \sqrt{\frac{2}{3}}\frac{\int_\Lambda f^3(\lambda)g^3(\lambda)d\lambda}{\left(\int_\Lambda f^2(\lambda)g^2(\lambda)d\lambda\right)^{3/2}}(1 - z^2)e^{-z^2/2}. \tag{5.5.9}$$

Remark 5.5.1 In the continuous time case, Theorem 5.5.2 was proved by Nourdin and Peccati [229] using a general central limit theorem from Sect. 5.2 (Theorem 5.2.1 (**C**)), Proposition 5.4.1 (a), and Stein's method. The proof for the discrete time case is similar to that of the continuous time case.

5.5.3 *Large and Moderate Deviations for Random Toeplitz Quadratic Functionals*

In this section, we present results on large and moderate deviations for Toeplitz quadratic functionals, which are applicable to both continuous and discrete time Gaussian stationary processes. For more information on this topic, we refer to the works of Bercu et al. [35], Bryc and Dembo [58], Dembo and Zeitouni [81], Gamboa et al. [102], Ihara [190], Kakizawa [199], Sato et al. [270], Taniguchi and Kakizawa [295], and other references therein.

The large deviations theory examines the exponential decay of probabilities associated with certain random systems. To illustrate the concept behind large deviations theory, we begin with the classical example concerning the behavior of sample means of independent, identically distributed (i.i.d.) random variables.

Let $\{X_k, k = 1, \ldots, n\}$ be an i.i.d. random sample from the distribution of a random variable X with mean $m := \mathbb{E}[X]$. Consider the sample means:

$$\widehat{S}_n := \frac{1}{n}\sum_{k=1}^{n} X_k, \quad n \in \mathbb{N}.$$

The weak law of large numbers (LLN) states that for any $\varepsilon > 0$,

$$\mathbb{P}\{|\widehat{S}_n - m| \geq \varepsilon\} \to 0 \quad \text{as} \quad n \to \infty. \tag{5.5.10}$$

By the central limit theorem, the "typical" value of $\widehat{S}_n$ is of the order $n^{-1/2}$. However, it is often the case that the convergence to zero in (5.5.10) is *exponentially fast*. In particular, in 1938, H. Cramér showed that under appropriate conditions on the moment generating function $M(a) := \mathbb{E}[\exp(aX_1)]$, the rate of decay of the probabilities in (5.5.10) is indeed exponential. He also provided a formula for the exponential decay rate (see, e.g., Bucklew [59], p. 9, or Dembo and Zeitouni [81], Theorem 2.2.3).

This result, the first in the theory of large deviations, has been extensively generalized to cover various new situations. These include i.i.d. random vectors taking values in $\mathbb{R}^d$ and certain infinite-dimensional spaces, as well as Gaussian processes, Markov processes, and stochastic processes that arise in statistics, information theory, and statistical mechanics. Therefore, a general framework is needed to encompass these diverse scenarios.

Let $\mathfrak{X}$ be a complete separable metric space, and let $\{\xi_k, k \in \mathbb{N}\}$ be a sequence of random vectors taking values in $\mathfrak{X}$. Typically, the sequence ξ_n converges in probability to a point x_0 in $\mathfrak{X}$ as $n \to \infty$. Thus, for any Borel subset A of $\mathfrak{X}$ whose closure does not contain x_0, the sequence of probabilities

$$\mathbf{P}_n\{A\} := \mathbb{P}\{\xi_n \in A\} \tag{5.5.11}$$

converges to zero as $n \to \infty$.

The large deviations theory addresses the fundamental question of whether these probabilities converge to zero exponentially fast as $n \to \infty$. If they do, the theory aims to express the decay rate as a function of the set A.

The exponential decay of the probabilities in (5.5.11) is encapsulated in the key concept of the *large deviation principle (LDP)*.

For simplicity, we define the LDP in the case where $\mathfrak{X} = \mathbb{R}$ (see, e.g., Bryc and Dembo [58], Dembo and Zeitouni [81], and Taniguchi and Kakizawa [295]).

Definition 5.5.1 Let $\{\xi_n, n \in \mathbb{Z}\}$ be a sequence of real-valued random variables defined on the probability space $(\Omega, \mathcal{F}, \mathbb{P})$. We say that $\{\xi_n\}$ satisfies the LDP with speed $a_n \to \infty$ and rate function $I : \mathbb{R} \to [0, \infty]$ if the following three conditions hold.

(a) For each $b \in \mathbb{R}$, the level set $\{x \in \mathbb{R} : I(x) \leq b\}$ is compact.
(b) For each closed set $C \subset \mathbb{R}$, the upper large deviation bound holds:

$$\limsup_{n\to\infty} a_n^{-1} \log \mathbb{P}\{\xi_n \in C\} \leq - \inf_{x\in C} I(x). \tag{5.5.12}$$

(c) For each open set $G \subset \mathbb{R}$, the lower large deviation bound holds:

$$\liminf_{n\to\infty} a_n^{-1} \log \mathbb{P}\{\xi_n \in G\} \geq - \inf_{x\in G} I(x). \tag{5.5.13}$$

Remark 5.5.2 Condition (a) implies that the rate function $I(x)$ is lower semicontinuous. Specifically, if $x_n \to x$, then $\liminf_{n\to\infty} I(x_n) \geq I(x)$. This condition also guarantees that the rate function $I(x)$ is unique.

Remark 5.5.3 Conditions (b) and (c) show that the infimum of $I(x)$ over the entire space $\mathbb{R}$ is zero. Together with condition (a), this implies that $I(x)$ attains its infimum over any nonempty closed set $C \subset \mathbb{R}$, which is positive whenever C does not contain a point where $I(x) = 0$ (see Ellis [96], p. 34).

Remark 5.5.4 A natural question related to (b) and (c) is to find conditions on a nonempty set $E \subset \mathbb{R}$ under which the following large deviation relation holds:

$$\limsup_{n\to\infty} a_n^{-1} \log \mathbb{P}\{\xi_n \in E\} = - \inf_{x\in E} I(x). \tag{5.5.14}$$

It is straightforward to verify that if $E \subset \mathbb{R}$ is an I*—continuity set* in the sense that

$$\inf_{x \in E^o} I(x) = \inf_{x \in E^c} I(x),$$

where E^o and E^c denote the interior and closure of E, respectively, then by conditions (b) and (c), the large deviation relation (5.5.14) holds.

It is important to explore the LDP for random variables of the form $m_n \xi_n$, where m_n is a sequence of positive numbers that tends to infinity as $n \to \infty$ but at a slower rate than the standard normalizing factor $n^{1/2}$ in the central limit theorem for ξ_n; that is, $m_n = o(n^{1/2})$ as $n \to \infty$. This scenario is often referred to as the *moderate deviation principle (MDP)*, as it provides an intermediate estimate between the CLT and the LDP.

The key tool used to establish the LDP and MDP is the Gärtner-Ellis theorem (see, e.g., Dembo and Zeitouni [81], Theorem 2.3.6).

Proposition 5.5.1 (Gärtner-Ellis Theorem) *Let $\{\xi_n, n \in \mathbb{Z}\}$ be a sequence of real-valued random variables defined on a probability space $(\Omega, \mathcal{F}, \mathbb{P})$, and let a_n be a sequence of positive numbers that tends to infinity as $n \to \infty$. We set*

$$M_n(s) := a_n^{-1} \log \mathbb{E}[\exp(s a_n \xi_n)].$$

If the limit

$$\lim_{n \to \infty} M_n(s) := M(s) \in \mathbb{R}$$

exists for all $s \in \mathbb{R}$, and the limiting function $M(s)$ is differentiable for all $s \in \mathbb{R}$, then the sequence $\{\xi_n\}$ satisfies the LDP with speed a_n and the rate function $I(x)$, which is the Legendre-Fenchel transform of $M(s)$:

$$I(x) = \sup_{\lambda \in \mathbb{R}} \{sx - M(s)\}.$$

Now, let Q_T be the Toeplitz quadratic functional of a Gaussian stationary process $X(t)$, defined by (5.1.1), with spectral density $f(\lambda)$ and generating function $g(\lambda)$.

The next result provides sufficient conditions, in terms of the functions f and g, to ensure that the LDP for the normalized quadratic functionals $\{\frac{1}{T} Q_T\}$ holds for both discrete and continuous time processes.

Theorem 5.5.3 *Assume that $f(\lambda)g(\lambda) \in L^\infty(\Lambda)$. Then the random variable $\frac{1}{T} Q_T$ satisfies the LDP with speed $a_T = \frac{1}{T}$ and rate function $I(x)$:*

$$I(x) = \sup_{-\infty < y < 1/(2C)} \{xy - V(f, g; y)\},$$

where $C = \text{ess sup} f(\lambda)|g(\lambda)|$ *and for* $y < 1/(2C)$

$$V(f, g; y) = -\frac{1}{4\pi}\int_{\Lambda} \log(1 - 2yf(\lambda)g(\lambda))d\lambda.$$

Remark 5.5.5 In the special case where $g(\lambda) = 1$, Theorem 5.5.3 was proved by Bryc and Dembo [58]. For the general case, we refer to Bercu et al. [35].

Moderate deviation results for quadratic forms in Gaussian stationary processes have been obtained in the papers by Bryc and Dembo [58] and Kakizawa [199]. In particular, the following theorem was proved by Kakizawa [199].

Theorem 5.5.4 *Assume that* $f \in L^p(\mathbb{T})$ $(p \geq 2)$ *and* $g \in L^q(\mathbb{T})$ $(q \geq 2)$ *with* $1/p+1/q < 1/2$. *Let* $\{a_T, T \in \mathbb{N}\}$ *be a sequence of positive real numbers such that*

$$1/a_T + T^{2(1/p+1/q)-1} \to 0 \quad \text{as} \quad T \to \infty.$$

Then the random variable

$$a_T^{-1/2}\widetilde{Q}_T = (Ta_T)^{-1/2}[Q_T - \mathbb{E}[Q_T]]$$

satisfies the LDP with speed a_T *and rate function* $I(x)$*:*

$$I(x) = \sup_{y\in\mathbb{R}}\{xy - y^2V(f, g)/2\},$$

where

$$V(f, g) = 16\pi^3\int_{\mathbb{T}} f^2(\lambda)g^2(\lambda)\,d\lambda.$$

Remark 5.5.6 In the special case where $g(\lambda) = 1$, Theorem 5.5.4 was first proved by Bryc and Dembo [58]. Furthermore, they demonstrated that the result also holds for continuous time processes with unbounded spectral density f.

Remark 5.5.7 Using the results from Chaps. 4 and 5, it can be shown that the result of Theorem 5.5.4 remains valid for quadratic functionals in continuous time processes, as well as for tapered quadratic functionals (for more details, see Sect. 7.5).

The large deviation principle has been established for several important statistics, including the log-likelihood ratio and quadratic forms, in both short and long memory Gaussian processes across various papers. Here, we will highlight a few of them.

The paper by Sato et al. [270] presents large deviation theorems for quadratic forms and log-likelihood ratios in both short memory and long memory discrete time Gaussian processes. It demonstrates that when the underlying process is not centered, the asymptotic behavior in the long memory case differs significantly from that in the short memory case.

In the papers by Bercu et al. [34, 35], a large deviation principle is established for Toeplitz quadratic forms of discrete time centered stationary Gaussian processes. This is achieved by utilizing results on the asymptotic behavior of the eigenvalues of a product of two Toeplitz matrices, along with Widom's formula (see 4.2.10).

A functional large deviations principle for quadratic forms of Gaussian stationary processes, in both discrete and continuous time, was proved in the paper by Gamboa et al. [102].

Large deviation results for spectral estimates in discrete time short memory stationary Gaussian processes have been obtained in the papers by Bentkus and Rudzkis [28] and Kakizawa [199].

Moderate deviation results for the empirical spectral density (periodogram) and nonlinear functionals of discrete time stationary moving average processes with i.i.d. innovations have been obtained in the paper by Djellout et al. [82].

5.6 Bibliographical Notes

For the discrete time case, assertions **(A)** and **(D)** of Theorem 5.2.1 were proved by Giraitis and Surgailis [148] (see also the monograph by Giraitis et al. [147]). Assertion **(B)** was proved by Ginovyan and Sahakyan [135]. Assertion **(C)** for $p = q = \infty$ was first established by Grenander and Szegö [162], Section 11.7, while the case $p = 2$, $q = \infty$ was proved by Ibragimov [180] and Rosenblatt [259]. In the general discrete time case, assertion **(C)** was proved by Avram [13].

For the continuous time case, assertions **(A)**–**(D)** were proved by Ginovyan [120] and by Ginovyan and Sahakyan [136].

For the discrete time case, the result of Theorem 5.2.2 under the condition $\alpha + \beta < 1/2$ was first obtained by Fox and Taqqu [101]. In the general case, which includes the critical value $\alpha + \beta = 1/2$, it was proved by Ginovyan and Sahakyan [135]. In the continuous time case, the result was proved by Ginovyan and Sahakyan [136].

Proposition 5.2.1, in the special case where $\int_{\mathbb{T}} f^2(\lambda)g^2(\lambda)d\lambda = 0$, was proved by Ginovyan [119]. The general case was established by Ginovyan and Sahakyan [135]. It is important to identify the limiting distribution of the quadratic form Q_T for this special model.

Theorem 5.2.3 was proved by Giraitis and Surgailis [148] (see also Giraitis et al. [147], Theorem 6.3.1). Theorem 5.2.4 was established by Giraitis and Taqqu [149]. Theorems 5.2.5 and 5.2.6, which involve time domain conditions for the CLT, were proved by Giraitis and Taqqu [150] (see also Giraitis and Taqqu [151]).

Central limit theorems for functionals of Gaussian fields have been established in the papers by Anh et al. [7], Breuer and Major [54], Doukhan et al. [86], Ginovyan [125], and Ludeña [220].

Chapter 6
Functional Limit Theorems for Toeplitz Processes

In this chapter, we establish weak convergence in the space $\mathbb{C}[0, 1]$ for the continuous time case and in the space $\mathbb{D}[0, 1]$ for the discrete time case of normalized stochastic processes. These processes are generated by Toeplitz type quadratic functionals of Gaussian and Lévy-driven linear stationary processes that exhibit long-range dependence. We have established both central and non-central functional limit theorems.

Section 6.1 outlines the problem addressed in this chapter, introduces the concepts of weak convergence and tightness, and presents Prohorov's theorem. Section 6.2 discusses functional central and non-central limit theorems for Gaussian models, with proofs provided in Sect. 6.3. Functional central and non-central limit theorems for Lévy-driven linear models are presented in Sect. 6.4, with proofs provided in Sect. 6.5.

6.1 Introduction

6.1.1 The Problem

In this chapter, we focus on describing the limiting behavior of stochastic processes generated by Toeplitz type quadratic functionals of a stationary process $X(u), u \in \mathbb{U}$, with spectral density f and covariance function r. Specifically, our object of study is the following process:

$$Q_T(\tau) = \begin{cases} \sum_{t=1}^{[T\tau]} \sum_{s=1}^{[T\tau]} \widehat{g}(t-s)X(t)X(s) & \text{in the d.t. case,} \\ \int_0^{T\tau} \int_0^{T\tau} \widehat{g}(t-s)X(t)X(s)\,dt\,ds & \text{in the c.t. case,} \end{cases} \tag{6.1.1}$$

M. S. Ginovyan, *Random Toeplitz Functionals and Their Applications*,
Frontiers in Probability and the Statistical Sciences,
https://doi.org/10.1007/978-3-032-00961-6_6

where $\tau \in [0, 1]$, $\widehat{g}(t)$, $t \in \mathbb{U}$, is the Fourier transform of some integrable even function $g(\lambda)$, $\lambda \in \Lambda$, and the symbol $[\cdot]$ denotes the greatest integer.

More precisely, the goals are:

(a) To establish limit theorems of the form

$$d_T^{-1}\left(Q_T(\tau) - \mathbb{E}[Q_T(\tau)]\right) \xrightarrow{f.d.d.} L(\tau) \text{ as } T \to \infty, \tag{6.1.2}$$

where d_T is a normalizing factor, $L(\tau)$ is the limiting process, and the symbol $\xrightarrow{f.d.d.}$ denotes convergence of finite-dimensional distributions

(b) To extend the convergence of finite-dimensional distributions to weak convergence in the space $\mathbb{C}[0, 1]$ for the continuous time case and in the space $\mathbb{D}[0, 1]$ for the discrete time case

The limiting process $L(\tau)$ in (6.1.2) is entirely determined by the spectral density $f(\lambda)$ (or covariance function $r(t)$) and the generating function $g(\lambda)$ (or generating kernel $\widehat{g}(t)$). Depending on their properties, $L(\tau)$ can be Gaussian or non-Gaussian.

We say that the process $Q_T(\tau)$ exhibits a *functional non-central limit theorem* (FNCLT) if the limiting process $L(\tau)$ in (6.1.2) is non-Gaussian, or if it is Gaussian, it differs from Brownian motion. We say that a *functional central limit theorem* (FCLT) for $Q_T(\tau)$ holds if we have the following weak convergence in the space $\mathbb{C}[0, 1]$ for the continuous time case (and in the space $\mathbb{D}[0, 1]$ for the discrete time case):

$$\widetilde{Q}_T(\tau) \Longrightarrow \sigma B(\tau) \text{ as } T \to \infty, \quad \tau \in [0, 1], \tag{6.1.3}$$

where $\sigma > 0$, $B(\tau)$ is a standard Brownian motion, and $\widetilde{Q}_T(\tau)$ is the standard normalized version of the process $Q_T(\tau)$:

$$\widetilde{Q}_T(\tau) := T^{-1/2}\left(Q_T(\tau) - \mathbb{E}[Q_T(\tau)]\right), \quad \tau \in [0, 1]. \tag{6.1.4}$$

The proofs of functional central limit theorems and non-central limit theorems generally employ different methods. Specifically, the proofs of central limit theorems rely on an approximation approach that reduces the quadratic integral form to a single integral form (see Lemmas 5.4.2, 5.4.3, 6.3.1, and 6.3.2). In contrast, the proofs of functional non-central limit theorems utilize the spectral representation of the underlying process, the properties of Wiener-Itô integrals, and a method developed by Surgailis [289] to establish the convergence in distribution of quadratic functionals to double Wiener-Itô integrals (see also Giraitis et al. [147], Corollary 4.7.1 and Proposition 14.3.2).

It is important to note that if the underlying process $X(t)$ is not necessarily Gaussian, additional complications arise due to the contribution of the random diagonal term in the double stochastic integral with respect to Lévy noise. This term is absent in the case of Gaussian noise (see Remark 6.4.2).

6.1.2 Weak Convergence. Tightness. Prohorov's Theorems

To prove functional limit theorems for appropriately normalized stochastic processes, one must verify two key assertions:

(a) The convergence of finite-dimensional distributions
(b) Tightness

Regarding the first assertion, we typically prove one-dimensional convergence, as extensions to the multivariate case are usually straightforward. These convergence methods were discussed in Sects. 5.3 and 5.4.1 (see also Sect. 6.3.1).

Now, we turn to the second assertion—tightness—which plays a fundamental role in both the theory of weak convergence and its applications.

A detailed account of weak convergence in various function spaces and tightness can be found in the monograph by Billingsley [41] (see also the seminal paper by Prohorov [250]). Here, we provide a brief overview of the essential concepts needed for our purposes.

Definition 6.1.1 Let S be a metric space and let $\mathcal{B} := \mathcal{B}(S)$ be the Borel σ-field. Let $\{\mathbb{P}_n\}$ and $\mathbb{P}$ be probability measures defined on the measure space $(S, \mathcal{B}(S))$. We say that a sequence of probability measures $\mathbb{P}_n$ converges weakly to a probability measure $\mathbb{P}$ and write $\mathbb{P}_n \Longrightarrow \mathbb{P}$ if, for every real-valued, bounded, and continuous function g on S, we have

$$\lim_{n\to\infty} \int_S g(s)d\mathbb{P}_n(s) = \int_S g(s)d\mathbb{P}(s).$$

We say that a sequence of random variables $\{X_n\}$ taking values in S converges weakly to a random variable X taking values in S and write $X_n \Longrightarrow X$ if

$$\mathbb{E}[g(X_n)] \to \mathbb{E}[g(X)]$$

whenever g is a real-valued, bounded, and continuous function.

Saying that X_n converges weakly to X is equivalent to stating the laws of X_n converge weakly to the law of X. To see this, if $\mathbb{P}_n$ and $\mathbb{P}$ are the laws of X_n and X, respectively, that is, $\mathbb{P}_n(B) = \mathbb{P}(X_n \in B)$ and $\mathbb{P}(B) = \mathbb{P}(X \in B)$ for each Borel subset $B \in \mathcal{B}(S)$, then $\mathbb{E}[g(X_n)] = \int_S g(s)d\mathbb{P}_n(s)$ and $\mathbb{E}[g(X)] = \int_S g(s)d\mathbb{P}(s)$ (see, e.g., Bass [23], Section 30.1).

Definition 6.1.2 A family of probability measures $\Pi := \{\mathbb{P}_n\}$ defined on the measure space $(S, \mathcal{B}(S))$ is said to be tight if, for every $\varepsilon > 0$, there exists a number n_0 and a compact set $K \subset S$ such that $\mathbb{P}(K) > 1 - \varepsilon$ for all $n > n_0$ and for all $\mathbb{P}_n \in \Pi$.

Definition 6.1.3 A family of probability measures $\Pi := \{\mathbb{P}_n\}$ on $(S, \mathcal{B}(S))$ is called relatively compact if every sequence of elements of Π contains a weakly convergent subsequence. Specifically, for every sequence $\{\mathbb{P}_n\}$ in Π there exist a subsequence $\{\mathbb{P}_{n(k)}\}$ and a probability measure $\mathbb{P}$ (defined on $(S, \mathcal{B}(S))$ but not necessarily an element of Π) such that $\mathbb{P}_{n(k)} \Longrightarrow \mathbb{P}$ as $k \to \infty$.

The next remarkable result is due to Yu.V. Prohorov [250] (see also Billingsley [41], Theorems 5.1 and 5.2).

Proposition 6.1.1 (Prohorov's First Theorem) *Let $\Pi := \{\mathbb{P}_n\}$ be a family of probability measures on $(S, \mathcal{B}(S))$. The following assertions hold:*

(a) If $\Pi := \{\mathbb{P}_n\}$ is tight, then it is relatively compact.
(b) Suppose that S is separable and complete. If $\Pi := \{\mathbb{P}_n\}$ is relatively compact, then it is tight.

We first consider weak convergence and tightness in the space $\mathbb{C} := \mathbb{C}[0, 1]$ of continuous functions on the unit interval, equipped with the uniform topology: for $x, y \in \mathbb{C}[0, 1]$:

$$\rho(x, y) := \sup_{t\in[0,1]} |x(t) - y(t)|.$$

Let $\omega_x(\delta)$ be the modulus of continuity of a function $x(t)$ on $[0, 1]$, defined as (see Sect. 2.1):

$$\omega_x(\delta) := \sup_{|s-t|\le\delta} |x(s) - x(t)|.$$

Since the space $\mathbb{C}[0, 1]$ is separable and complete, in this case, by Prohorov's theorem (Proposition 6.1.1), *the tightness and relative compactness coincide.*

The next proposition provides a necessary and sufficient condition for the tightness (relative compactness) of a family of measures in the space $\mathbb{C}[0, 1]$ (see Billingsley [41], Theorem 7.3).

Proposition 6.1.2 *The family of measures $\Pi := \{\mathbb{P}_n\}$ in the space $\mathbb{C}[0, 1]$ is tight if and only if the following two conditions hold:*

(a) For each positive η, there exist numbers M and n_0 such that

$$\mathbb{P}_n\{x : |x(0)| \ge M\} \le \eta \quad \text{for} \quad n \ge n_0. \tag{6.1.5}$$

(b) For each positive ε and η, there exist numbers $\delta \in (0, 1)$ and n_0 such that

$$\mathbb{P}_n\{x : \omega_x(\delta) \ge \varepsilon\} \le \eta \quad \text{for} \quad n \ge n_0. \tag{6.1.6}$$

The next proposition provides a powerful method for proving weak convergence in the function space $\mathbb{C}[0, 1]$ (see Billingsley [41], Theorem 7.1).

Proposition 6.1.3 *Let* $\{\mathbb{P}_n, n \in \mathbb{N}\}$ *and* $\mathbb{P}$ *be probability measures on the measure space* $(\mathbb{C}[0,1], \mathcal{B}(\mathbb{C}[0,1]))$. *If the finite-dimensional distributions of* $\mathbb{P}_n$ *converge weakly to those of* $\mathbb{P}$, *and if* $\{\mathbb{P}_n, n \in \mathbb{N}\}$ *is tight, then* $\mathbb{P}_n \Longrightarrow \mathbb{P}$ *as* $n \to \infty$.

Now let $(\Omega, \mathcal{F}, \mathbb{P})$ be a probability space, and let $X_n(t) := X_n(t, \omega)$, $(t \in [0,1]$, $\omega \in \Omega$, $n \in \mathbb{N})$ be a sequence of stochastic processes defined on $(\Omega, \mathcal{F}, \mathbb{P})$ with state space $(\mathbb{C}[0,1], \mathcal{B}(\mathbb{C}[0,1])$.

As an immediate consequence of Proposition 6.1.2, we obtain the following result.

Proposition 6.1.4 (Prohorov's Second Theorem) *Let* $\{X_n := X_n(t)\}$ *be a sequence of continuous processes. Assume that for each positive* ε *and* η *there exist numbers* M, n_0 *and* $\delta \in (0,1)$ *(depending on* ε *and* η*) such that for* $n \geq n_0$, *the following conditions hold:*

$$1.\quad \mathbb{P}(|X_n(0)| \geq M) \leq \eta.$$

$$2.\quad \mathbb{P}(\omega_{X_n}(\delta) \geq \varepsilon) \leq \eta. \tag{6.1.7}$$

Then the sequence $\{X_n\}$ *is tight.*

Remark 6.1.1 Observe that the second condition in Proposition 6.1.4 is satisfied if the following inequality holds:

$$\mathbb{E}\left[|X_n(s) - X_n(t)|^a\right] \leq C\,|s-t|^{b+1} \tag{6.1.8}$$

for constants $a > 0$, $b > 0$, and $C > 0$ (see Lemma 5.1 in Ibragimov [180]).

Proposition 6.1.5 *Let* $\{X_n(t)\}$ *and* $X(t)$ *be continuous processes. If the finite-dimensional distributions of* $X_n(t)$ *converge weakly to those of* X, *and if* $\{X_n(t)\}$ *is tight, then* $X_n \Longrightarrow X$ *as* $n \to \infty$.

The analogs of the above results regarding weak convergence and tightness in the space $\mathbb{C}[0,1]$ also apply to the Skorokhod space $\mathbb{D}[0,1]$. This space consists of real-valued functions $x(t)$ on $[0,1]$ that are right-continuous and possess left-hand limits (càdlàg functions), equipped with the Skorokhod topology rather than the uniform topology. It is important to note that the Skorokhod topology, when restricted to $\mathbb{C}[0,1]$, coincides with the uniform topology on that space (for details, see Billingsley [41], Sections 12 and 13, and Bass [23], Chapter 34).

6.2 Functional Limit Theorems for Gaussian Models

We begin by presenting the functional central limit theorems, which state that the limit of the normalized process $\widetilde{Q}_T(\tau)$, as defined by Eq. (6.1.4), is Brownian motion.

6.2.1 Functional CLT for the Process $Q_T(\tau)$

In the discrete time case, the following result extends Theorem 5.2.5 and presents sufficient time domain conditions for the process $\widetilde{Q}_T(\tau)$ to satisfy a functional CLT (see Giraitis and Taqqu [153]).

Theorem 6.2.1 *If the covariance function $r(t)$ and the generating kernel $\widehat{g}(t)$ satisfy the following conditions:*

$$r(\cdot) \in l^p, \quad \widehat{g}(\cdot) \in l^q \quad \text{for some} \quad p, q \geq 1, \ \frac{1}{p} + \frac{1}{q} \geq \frac{3}{2}, \tag{6.2.1}$$

then the functional CLT holds for the process $Q_T(\tau)$. Specifically, the convergence in (6.1.3) *occurs in the space* $\mathbb{D}[0, 1]$ *with limiting variance σ^2 given by*

$$\sigma^2 := \sum_{t,s,v \in \mathbb{Z}} \widehat{g}(t)\widehat{g}(s)\text{Cov}\left(P(X(v), X(v+t)), P(X(0), X(s))\right), \tag{6.2.2}$$

where $P(X(t), X(s)) := X(t)X(s) - \mathbb{E}[X(t)X(s)]$.

Now we consider the continuous time case. The next result, an extension of Theorem 5.2.1 **(A)**, involves the convergence of the finite-dimensional distributions of the process $\widetilde{Q}_T(\tau)$ to those of a standard Brownian motion (see Bai et al. [21]).

Theorem 6.2.2 *Assume that the spectral density f and the generating function g satisfy the following conditions:*

$$f \cdot g \in L^1(\mathbb{R}) \cap L^2(\mathbb{R}) \tag{6.2.3}$$

and

$$\mathbb{E}[\widetilde{Q}_T^2(1)] \to 16\pi^3 \int_{\mathbb{R}} f^2(x) g^2(x) dx \ \text{ as } T \to \infty. \tag{6.2.4}$$

Then, we have the following convergence of the finite-dimensional distributions:

$$\widetilde{Q}_T(\tau) \stackrel{f.d.d.}{\longrightarrow} \sigma_Q B(\tau) \ \text{ as } T \to \infty,$$

where $\widetilde{Q}_T(\tau)$ is defined as in (6.1.4), $B(\tau)$ is a standard Brownian motion, and (see (5.2.3)*):*

$$\sigma_Q^2 := 16\pi^3 \int_{\mathbb{R}} f^2(x) g^2(x) dx. \tag{6.2.5}$$

To extend the convergence of finite-dimensional distributions in Theorem 6.2.2 to weak convergence in the space $\mathbb{C}[0, 1]$, we impose an additional condition on the underlying Gaussian process $X(t)$ and the generating function g. It is convenient to express this condition in the time domain, specifically regarding the covariance function $r := \hat{f}$ and the generating kernel $a := \hat{g}$. The following condition serves as a "continuous" analog to condition (6.2.1) in Theorem 6.2.1:

$$r(\cdot) \in L^p(\mathbb{R}), \quad a(\cdot) \in L^q(\mathbb{R}) \quad \text{for some} \quad p, q \geq 1, \ \frac{1}{p} + \frac{1}{q} \geq \frac{3}{2}. \tag{6.2.6}$$

Remark 6.2.1 In fact, under (6.2.3), the condition (6.2.6) is sufficient for the convergence in (6.2.4). Indeed, let $\bar{p} = p/(p-1)$ and $\bar{q} = q/(q-1)$ be the Hölder conjugates of p and q, respectively. Since $1 \leq p, q \leq 2$, one has by the Hausdorff-Young inequality (see Proposition 2.3.4) that $\|f\|_{\bar{p}} \leq c_p \|r\|_p$ and $\|g\|_{\bar{q}} \leq c_q \|a\|_q$. Hence, in view of (6.2.6), we have

$$f(\cdot) \in L^{\bar{p}}, \quad g(\cdot) \in L^{\bar{q}}, \quad \frac{1}{\bar{p}} + \frac{1}{\bar{q}} = 2 - \frac{1}{p} - \frac{1}{q} \leq 1/2. \tag{6.2.7}$$

Then, the convergence in (6.2.4) follows from Theorem 5.2.1 **(C)**.

Remark 6.2.2 Observe that condition (6.2.6) is fulfilled if the functions $r(\cdot)$ and $a(\cdot)$ satisfy the following conditions: there exist constants $C > 0$, α^*, and β^* such that

$$|r(t)| \leq C(1 \wedge |t|^{\alpha^*-1}), \qquad |a(t)| \leq C(1 \wedge |t|^{\beta^*-1}), \tag{6.2.8}$$

where $0 < \alpha^*, \beta^* < 1/2$, and $\alpha^* + \beta^* < 1/2$. Indeed, to see this, first note that $r(\cdot)$, $a(\cdot) \in L^\infty(\mathbb{R})$. We can then choose $p, q \geq 1$ such that $p(\alpha^* - 1) < -1$ and $q(\beta^* - 1) < -1$. This choice implies that $r(\cdot) \in L^p(\mathbb{R})$ and $a(\cdot) \in L^q(\mathbb{R})$. Additionally, since $1/p + 1/q < 2 - \alpha^* - \beta^*$ and $2 - \alpha^* - \beta^* > 3/2$, one can further choose p, q to satisfy $1/p + 1/q \geq 3/2$.

The following results, two functional central limit theorems, extend Theorems 5.2.1 **(A)** and 5.2.2 to the weak convergence in the space $\mathbb{C}[0, 1]$ of the stochastic process $\widetilde{Q}_T(\tau)$ to a standard Brownian motion.

Theorem 6.2.3 *Let the spectral density f and the generating function g satisfy condition (6.2.3). Let the covariance function $r(\cdot)$ and the generating kernel $a(\cdot)$ satisfy condition (6.2.6). Then, we have the following weak convergence in the space $\mathbb{C}[0, 1]$:*

$$\widetilde{Q}_T(\tau) \Longrightarrow \sigma_Q B(\tau) \ \text{as} \ T \to \infty,$$

where $\widetilde{Q}_T(\tau)$ is defined as in (6.1.4), σ_Q is defined as in (6.2.5), and $B(\tau)$ is a standard Brownian motion.

Recall the class $SV_0(\mathbb{R})$ of slowly varying at zero functions $u(x)$, $x \in \mathbb{R}$, that satisfy the following conditions: for some $a > 0$, $u(x)$ is bounded on $[-a, a]$,

$\lim_{x\to 0} u(x) = 0$, $u(x) = u(-x)$ and $0 < u(x) < u(y)$ for $0 < x < y < a$ (see Definition 2.5.1).

Theorem 6.2.4 *Assume that the functions f and g are integrable on $\mathbb{R}$ and bounded outside any neighborhood of the origin. Furthermore, for some $a > 0$, they satisfy the following conditions:*

$$f(x) \le |x|^{-\alpha} L_1(x), \quad |g(x)| \le |x|^{-\beta} L_2(x), \quad x \in [-a, a] \tag{6.2.9}$$

for some $\alpha < 1$, $\beta < 1$ with $\alpha + \beta \le 1/2$, where $L_1(x)$ and $L_2(x)$ are slowly varying at zero functions that satisfy

$$L_i \in SV_0(\mathbb{R}), \quad x^{-(\alpha+\beta)} L_i(x) \in L^2[-a, a], \quad i = 1, 2. \tag{6.2.10}$$

Additionally, let the covariance function $r(\cdot)$ and the generating kernel $a(\cdot)$ satisfy condition (6.2.6). Then, we have the following weak convergence in the space $\mathbb{C}[0, 1]$:

$$\widetilde{Q}_T(\tau) \Longrightarrow \sigma_Q B(\tau) \ \text{as}\ T \to \infty,$$

where $\widetilde{Q}_T(\tau)$ is defined as in (6.1.4), σ_Q is defined as in (6.2.5), and $B(\tau)$ is a standard Brownian motion.

6.2.2 Functional NCLT for the Process $Q_T(\tau)$

In this section, we present functional non-central limit theorems in which the limit of the appropriately normalized process $Q_T(t)$, as defined by Eq. (6.1.1), is not a Brownian motion.

6.2.2.1 Non-central Limit Theorems (Discrete Time)

For discrete time Gaussian processes, the problem of describing the limiting distribution of the quadratic form (cf. (5.1.1)):

$$Q_T := \sum_{t=1}^{T} \sum_{s=1}^{T} \widehat{g}(t-s) X(t) X(s), \quad T \in \mathbb{N}, \tag{6.2.11}$$

when it is non-Gaussian, dates back to the classical papers of Rosenblatt [258–260]. Since then, this problem has been studied in a series of papers by Fox and Taqqu, Taqqu, and Terrin and Taqqu (see, e.g., [99, 296, 300–302], and references therein).

Suppose that the spectral density f and the generating function g are regularly varying functions at the origin:

$$f(\lambda) = |\lambda|^{-\alpha} L_1(\lambda) \quad \text{and} \quad g(\lambda) = |\lambda|^{-\beta} L_2(\lambda), \quad \alpha < 1, \beta < 1, \tag{6.2.12}$$

where $L_1(\lambda)$ and $L_2(\lambda)$ are slowly varying at zero functions. The conditions $\alpha < 1$ and $\beta < 1$ ensure that the Fourier coefficients of f and g are well-defined. When $\alpha > 0$, the model $\{X(t), t \in \mathbb{Z}\}$ exhibits long memory.

It is the sum $\alpha + \beta$ that determines the asymptotic behavior of the quadratic form Q_T. If $\alpha + \beta \leq 1/2$, then by Theorem 5.2.2, the standard normalized quadratic form

$$T^{-1/2} \left(Q_T - \mathbb{E}[Q_T]\right)$$

converges in distribution to a Gaussian random variable. Conversely, if $\alpha + \beta > 1/2$, the convergence to a Gaussian distribution fails, making it important to describe the corresponding limiting distribution.

We consider a more general problem—describing the limiting distribution of the following normalized process:

$$\widetilde{Q}_T(\tau) := d_T^{-1} \left(Q_T(\tau) - \mathbb{E}[Q_T(\tau)]\right), \tag{6.2.13}$$

where d_T is a suitably chosen normalization factor and $\{Q_T(\tau), \tau \in [0, 1]\}$ is defined as in (6.1.1) (in the discrete time case).

In Rosenblatt [258, 260] it was demonstrated that if a discrete time centered Gaussian stationary process $X(t)$ has the covariance function:

$$r(t) = (1 + t^2)^{\alpha/2 - 1/2} \quad \text{with} \quad 1/2 < \alpha < 1,$$

then the random variable

$$\widetilde{Q}_T := T^{-\alpha} \sum_{t=1}^{T} \left[X^2(t) - 1\right] \tag{6.2.14}$$

has a non-Gaussian limiting distribution. Furthermore, the corresponding limiting distribution was explicitly described in terms of characteristic functions.

Observe that the random variable in Eq. (6.2.14) is a special case of (6.2.13) with $\tau = 1$, $1/2 < \alpha < 1$, $\beta = 0$, $d_T = T^{\alpha}$, and $\widehat{g}(t - s) = \delta_{t,s}$, where $\delta_{t,s}$ is the Kronecker delta.

In Taqqu [296, 300], Rosenblatt's result was extended by demonstrating that the stochastic process

$$\widetilde{Q}_T(\tau) := T^{-\alpha} \sum_{t=1}^{[T\tau]} \left[X^2(t) - 1\right]$$

converges weakly in the space $\mathbb{D}[0, 1]$ to a stochastic process $Q(\tau)$, which has the following double Wiener-Itô integral representation:

$$Q(\tau) := C_\alpha \int_{\mathbb{R}^2}'' \frac{e^{i\tau(x+y)} - 1}{i(x+y)} |x|^{-\alpha/2} |y|^{-\alpha/2} dZ(x)dZ(y), \qquad (6.2.15)$$

where $Z(\cdot)$ is a complex-valued Gaussian random measure defined on the Borel σ-algebra $\mathcal{B}(\mathbb{R})$, satisfying $\mathbb{E}[Z(B)] = 0$, $\mathbb{E}[|Z(B)|^2] = |B|$, and $\overline{Z(-B)} = Z(B)$ for any $B \in \mathcal{B}(\mathbb{R})$. The double prime in the integral (6.2.15) indicates that the integration excludes the diagonals $x = \pm y$ (for the definition and properties of the Wiener-Itô integrals, see Sect. 3.4).

The stochastic process $Q(\tau)$ defined by Eq. (6.2.15) is referred to as the *Rosenblatt process*. The distribution of the random variable $Q(1)$, known as the *Rosenblatt distribution*, is described in Veillette and Taqqu [307] (see also Pipiras and Taqqu [246], Section 4.7).

The following result, proved by Terrin and Taqqu [301] using power counting theorems (see Sect. 2.6), describes the non-Gaussian limiting distribution of the appropriately normalized process $Q_T(\tau)$.

Theorem 6.2.5 *Let the functions f and g be as defined in (6.2.12) with $\alpha < 1$, $\beta < 1$ and slowly varying at zero and bounded on bounded intervals factors L_1 and L_2. Let the process $Q_T(\tau)$ be as defined in* (6.1.1). *Then for $\alpha + \beta > 1/2$ the process:*

$$Z_T(\tau) := \frac{1}{T^{\alpha+\beta} L_1(1/T) L_2(1/T)} \left(Q_T(\tau) - \mathbb{E}[Q_T(\tau)]\right) \qquad (6.2.16)$$

converges (as $T \to \infty$) weakly in the space $\mathbb{D}[0, 1]$ *to the process:*

$$Z(\tau) := \int_{\mathbb{R}^2}'' K_\tau(x, y) dZ(x)dZ(y), \qquad (6.2.17)$$

where

$$K_\tau(x, y) = |xy|^{-\alpha/2} \int_{\mathbb{R}} \frac{e^{i\tau(x+u)} - 1}{i(x+u)} \cdot \frac{e^{i\tau(y-u)} - 1}{i(y-u)} |u|^{-\beta} du. \qquad (6.2.18)$$

The double prime in the integral (6.2.17) indicates that the integration excludes the diagonals $x = \pm y$.

Remark 6.2.3 The limiting process in (6.2.17) is real-valued and non-Gaussian and satisfies $\mathbb{E}[Z(\tau)] = 0$ and

$$\mathbb{E}[Z^2(\tau)] = \int_{\mathbb{R}^2} |K_\tau(x, y)|^2 dxdy.$$

This process is self-similar with parameter $H = \alpha + \beta \in (1/2, 2)$. Specifically, the processes $\{Z(a\tau), \tau \in [0, 1]\}$ and $\{a^H Z(\tau), \tau \in [0, 1]\}$ have the same finite-dimensional distributions for all $a > 0$.

6.2.2.2 Non-central Limit Theorems (Continuous Time)

Now we present a *non-central limit theorem* for the process $Q_T(\tau)$ in the continuous time case, specifically when $Q_T(\tau)$ in (6.1.1) is defined through integrals.

Let the spectral density f and the generating function g satisfy the following conditions:

$$f(x) = |x|^{-\alpha} L_1(x) \quad \text{and} \quad g(x) = |x|^{-\beta} L_2(x), \quad x \in \mathbb{R}, \tag{6.2.19}$$

where $\alpha < 1$, $\beta < 1$, and $L_1(x)$ and $L_2(x)$ are slowly varying at zero functions that satisfy

$$\int_{\mathbb{R}} |x|^{-\alpha} L_1(x) dx < \infty \quad \text{and} \quad \int_{\mathbb{R}} |x|^{-\beta} L_2(x) dx < \infty.$$

We additionally assume that the functions $L_1(x)$ and $L_2(x)$ satisfy Potter's bound (see Proposition 2.5.2). Specifically, for any $\epsilon > 0$, there exist constants $C_i > 0$ such that if T is sufficiently large, then

$$\frac{L_i(x/T)}{L_i(1/T)} \leq C_i(|x|^{\epsilon} + |x|^{-\epsilon}), \quad i = 1, 2. \tag{6.2.20}$$

Remark 6.2.4 Note that a sufficient condition for (6.2.20) to hold is that the functions $L_1(x)$ and $L_2(x)$ are bounded on the intervals $[a, \infty)$ for any $a > 0$ (see Proposition 2.5.2). This condition applies to the slowly varying functions in Theorem 6.2.4.

We are now interested in the limiting process of the following normalized version of the process $Q_T(\tau)$, as defined in Eq. (6.1.1), with the functions f and g specified in (6.2.19):

$$Z_T(\tau) := \frac{1}{T^{\alpha+\beta} L_1(1/T) L_2(1/T)} \left(Q_T(\tau) - \mathbb{E}[Q_T(\tau)] \right). \tag{6.2.21}$$

Theorem 6.2.6 *Let the functions f and g be defined as in (6.2.19) with $\alpha < 1$, $\beta < 1$, and slowly varying at zero functions $L_1(x)$ and $L_2(x)$ satisfying condition (6.2.20), and let the process $Z_T(\tau)$ be defined as in (6.2.21). Then, under the condition $\alpha + \beta > 1/2$, we have the following weak convergence in the space $\mathbb{C}[0, 1]$:*

$$Z_T(\tau) \Longrightarrow Z(\tau) \ \textit{as}\ T \to \infty,$$

where the limiting process $Z(\tau)$ is given by

$$Z(\tau) = \int_{\mathbb{R}^2}'' H_\tau(x_1, x_2) W(dx_1) W(dx_2), \tag{6.2.22}$$

with

$$H_\tau(x_1, x_2) = |x_1 x_2|^{-\alpha/2} \int_{\mathbb{R}} \left[\frac{e^{i\tau(x_1+u)} - 1}{i(x_1 + u)} \right] \cdot \left[\frac{e^{i\tau(x_2-u)} - 1}{i(x_2 - u)} \right] |u|^{-\beta} du \ , \tag{6.2.23}$$

where $W(\cdot)$ is a complex-valued Gaussian random measure with Lebesgue control measure, and the double prime in the integral (6.2.22) indicates that the integration excludes the diagonals $x_1 = \pm x_2$.

Remark 6.2.5 Let $\mathbb{P}_T$ and $\mathbb{P}$ denote the measures generated in the space $\mathbb{C}[0, 1]$ by the processes $Z_T(\tau)$ and $Z(\tau)$, as defined by Eqs. (6.2.21) and (6.2.22), respectively. Theorem 6.2.6 can be reformulated as follows: *under the conditions specified in Theorem 6.2.6, the measure $\mathbb{P}_T$ converges weakly to the measure $\mathbb{P}$ in the space $\mathbb{C}[0, 1]$ as $T \to \infty$*. Similar conclusions can also be drawn for Theorems 6.2.3 and 6.2.4.

6.3 Proofs of Theorems of Sect. 6.2

6.3.1 Preliminaries

In this section, we present several lemmas that will be used in the proofs of the theorems. The following result extends Lemma 5.4.2.

Lemma 6.3.1 *Let $Y(t)$ be a centered stationary Gaussian process with spectral density $f_Y(x) \in L^1(\mathbb{R}) \cap L^2(\mathbb{R})$. Consider the normalized process:*

$$L_T(t) := T^{-1/2} \left(\int_0^{Tt} Y^2(u) du - \mathbb{E}\left[\int_0^{Tt} Y^2(u) du \right] \right). \tag{6.3.1}$$

Then we have the following convergence of finite-dimensional distributions:

$$L_T(t) \xrightarrow{f.d.d.} \sigma_Y B(t) \ \text{as}\ T \to \infty, \quad \sigma_Y^2 = 4\pi \int_{-\infty}^{\infty} f_Y^2(x) dx, \tag{6.3.2}$$

where $B(t)$ is a standard Brownian motion.

Remark 6.3.1 Observe that the normalized processes $\widetilde{Q}_T(t)$ and $L_T(t)$, given by (6.1.4) and (6.3.1), respectively, can be expressed by double Wiener-Itô integrals (see Sect. 3.4.5). Additionally, for the definition and properties of Wiener-Itô stochastic integrals, see Sect. 3.4 and Major [222], Chapter 4).

In our proofs, we will utilize the following fact regarding the weak convergence of multiple Wiener-Itô integrals: If the covariance structure of a sequence of d-dimensional vectors of multiple Wiener-Itô integrals converges to that of a d-dimensional Gaussian vector, then multivariate convergence is implied by the univariate convergence of each component. In other words, componentwise convergence implies joint convergence (see Peccati and Tudor [241], Proposition 2).

Proof of Lemma 6.3.1 For a fixed t, the univariate convergence in distribution

$$L_T(t) \xrightarrow{d} N(0, t\sigma_Y^2) \quad \text{as} \quad T \to \infty$$

follows from Lemma 5.4.2. Therefore, in view of Remark 6.3.1, to prove (6.3.2), it remains to show that the covariance structure of the process $L_T(t)$ converges to that of $\sigma_Y B(t)$. Specifically, it suffices to show that for any $0 < s < t$,

$$\mathbb{E}\left[(L_T(t) - L_T(s))^2\right] \to \sigma_Y^2 \cdot (t-s) \quad \text{as} \quad T \to \infty. \tag{6.3.3}$$

Indeed, using the fact that for a Gaussian vector (G_1, G_2), we have

$$\mathrm{Cov}(G_1^2, G_2^2) = 2[\mathrm{Cov}(G_1, G_2)]^2,$$

and letting

$$r_Y(u) = \int_{\mathbb{R}} e^{ixu} f_Y(x)dx$$

be the covariance function of the process $Y(t)$, we can write

$$\mathbb{E}\left[(L_T(t) - L_T(s))^2\right] = 2(t-s)\int_{-T(t-s)}^{T(t-s)} \left(1 - \frac{|u|}{T(t-s)}\right) r_Y^2(u)du.$$

Since $f_Y(x) \in L^2(\mathbb{R})$, the Fourier transform $r_Y(u) = \widehat{f_Y}(u) \in L^2(\mathbb{R})$ as well. Therefore, by the dominated convergence theorem and Parseval-Plancherel's identity, we have as $T \to \infty$

$$\mathbb{E}\left[(L_T(t) - L_T(s))^2\right] \to 2(t-s)\int_{-\infty}^{\infty} r_Y^2(u)du$$

$$= 4\pi(t-s)\int_{-\infty}^{\infty} f_Y^2(x)dx = \sigma_Y^2(t-s). \tag{6.3.4}$$

Lemma 6.3.1 is proved. □

We will now describe an approximation approach that allows us to reduce the general quadratic functional $\widetilde{Q}_T(t)$ in Theorem 6.2.2 to the single integral functional $L_T(t)$ introduced in Lemma 6.3.1 (cf. the proof of Theorem 5.2.1 **(A)**).

By Theorem 3.2.2 (see also Theorem 16.7.2 from Ibragimov and Linnik [188]), the underlying process $X(t)$ admits a moving average representation:

$$X(t) = \int_{-\infty}^{\infty} \hat{a}(t-s)B(ds) \qquad \text{with} \qquad \int_{-\infty}^{\infty} |\hat{a}(t)|^2 dt < \infty, \tag{6.3.5}$$

where $B(t)$ is a standard Brownian motion and $\hat{a}(t)$ is such that its inverse Fourier transform $a(x)$ satisfies $f(x) = 2\pi |a(x)|^2$. Assuming the conditions (6.2.3) and (6.2.4), we define

$$b(x) = (2\pi)^{1/2} a(x)(g(x))^{1/2},$$

and observe that the function $b(x)$ is in $L^2(\mathbb{R})$ due to condition (6.2.3). Consider the stationary process

$$Y(t) = \int_{-\infty}^{\infty} \hat{b}(t-s)B(ds) \tag{6.3.6}$$

constructed using the Fourier transform $\hat{b}(t)$ of $b(x)$ and the *same* Brownian motion $B(t)$ as in (6.3.5). The process $Y(t)$ has the spectral density (see Eq. (5.4.25))

$$f_Y(x) = 2\pi f(x)g(x). \tag{6.3.7}$$

As an immediate consequence of Lemma 5.4.3, we obtain the following approximation result for the process $\widetilde{Q}_T(t)$.

Lemma 6.3.2 *Let $\widetilde{Q}_T(t)$ be defined as in (6.1.4) and $L_T(t)$ as in (6.3.1) with $Y(t)$ constructed as in (6.3.6). Then, under the conditions specified in (6.2.3) and (6.2.4), for any $t > 0$, we have*

$$\lim_{T\to\infty} \operatorname{Var}[\widetilde{Q}_T(t) - L_T(t)] = 0.$$

The following lemma will be utilized in the proof of Lemma 6.3.4.

Lemma 6.3.3 *Let $p_j \geq 1$, $j = 1, \ldots, k$, where $k \geq 2$ and $\sum_{j=1}^{k} \frac{1}{p_j} = k - 1$. If $f_j \in L^{p_j}(\mathbb{R})$ for $j = 1, \ldots, k$, then*

$$\int_{\mathbb{R}^{k-1}} |f_1(x_1)\ldots f_{k-1}(x_{k-1}) f_k(x_1 + \ldots + x_{k-1})| dx_1 \ldots dx_{k-1} \leq \prod_{j=1}^{k} \|f_j\|_{p_j},$$

where $\|\cdot\|_{p_j}$ is the norm in $L^{p_j}(\mathbb{R})$.

Proof Denote

$$F_m(y) := \int_{\mathbb{R}^{m-1}} |f_1(x_1)\dots f_{m-1}(x_{m-1})f_m(y-(x_1+\dots+x_{m-1}))|dx_1\dots dx_{m-1}.$$

Then, after changing variables and applying Hölder's inequality (Proposition 2.1.5), we obtain

$$\int_{\mathbb{R}^{k-1}} |f_1(x_1)\dots f_{k-1}(x_{k-1})f_k(x_1+\dots+x_{k-1})|dx_1\dots dx_{k-1}$$
$$= \int_{\mathbb{R}} f_k(x)F_{k-1}(x)dx \le \|f_k\|_{p_k}\|F_{k-1}\|_{r_{k-1}},$$

where $1/p_k + 1/r_{k-1} = 1$. Next, by applying Young's convolution inequality (Proposition 2.3.3) $k-1$ times, we obtain

$$\|F_{k-1}\|_{r_{k-1}} \le \|f_{k-1}\|_{p_{k-1}}\|F_{k-2}\|_{r_{k-2}} \le \dots \le \prod_{j=1}^{k-1}\|f_j\|_{p_j},$$

where $1+1/r_{k-1} = 1/p_{k-1}+1/r_{k-2}$, ..., $1+1/r_2 = 1/p_2+1/p_1$.

Summing these relations along with the relation $1/p_k+1/r_{k-1} = 1$ gives $1/p_1 + \dots + 1/p_k = k-1$. (Note that $r_j \ge 1$, $j = 1, \dots, k$). □

The following lemma will be used to establish tightness in the space $\mathbb{C}[0, 1]$ in Theorem 6.2.2.

Lemma 6.3.4 *Let the covariance function $r(t)$ and the generating kernel $a(t)$ satisfy condition (6.2.6), and let $\widetilde{Q}_T(t)$ be defined as in (6.1.4). Then, for all $0 \le s \le t \le 1$ and $T > 0$, there exists a constant $C > 0$ such that*

$$\mathbb{E}\left[|\widetilde{Q}_T(t) - \widetilde{Q}_T(s)|^2\right] \le C(t-s). \tag{6.3.8}$$

Proof For convenience we use the Wick product notation:

$$: X(u)X(v) := X(u)X(v) - \mathbb{E}\left[X(u)X(v)\right].$$

With this notation, for $0 \le s \le t \le 1$, we can write

$$\widetilde{Q}_T(t) - \widetilde{Q}_T(s) = \frac{1}{\sqrt{T}}\int_0^{Tt}\int_0^{Tt} a(u-v) : X(u)X(v) : dudv$$
$$- \frac{1}{\sqrt{T}}\int_0^{Ts}\int_0^{Ts} a(u-v) : X(u)X(v) : dudv$$

$$= \frac{1}{\sqrt{T}} \int_{Ts}^{Tt} \int_{Ts}^{Tt} a(u-v) : X(u)X(v) : dudv$$
$$+ \frac{2}{\sqrt{T}} \int_{0}^{Ts} \int_{Ts}^{Tt} a(u-v) : X(u)X(v) : dudv$$
$$:= A(s,t,T) + B(s,t,T).$$

Now we estimate $B(s, t, T)$ (the function $A(s, t, T)$ can be estimated in a similar manner). Using Theorem 3.9 from Janson [196], we have

$$\begin{aligned}\mathbb{E}\left[B^2(s,t,T)\right] &= \frac{4}{T}\int_0^{Ts} du_1 \int_{Ts}^{Tt} dv_1 \int_0^{Ts} du_2 \int_{Ts}^{Tt} dv_2 a(u_1-v_1)a(u_2-v_2)\\ &\times \mathbb{E}\left(: X(u_1)X(v_1) :: X(u_2)X(v_2) :\right)\\ &= \frac{4}{T}\int_0^{Ts} du_1 \int_{Ts}^{Tt} dv_1 \int_0^{Ts} du_2 \int_{Ts}^{Tt} dv_2 a(u_1-v_1)a(u_2-v_2)\\ &\times [r(u_1-u_2)r(v_1-v_2) + r(u_1-v_2)r(v_1-u_2)]\\ &:= B_1(s,t,T) + B_2(s,t,T).\end{aligned}$$

By the change of variables $x_1 = u_1 - v_1$, $x_2 = v_2 - u_2$, $x_3 = u_2 - u_1$, $x_4 = v_2$, and noting that $r(\cdot)$ and $a(\cdot)$ are even functions, we obtain

$$B_1(s,t,T) \le \frac{4}{T}\int_{Ts}^{Tt} dx_4 \int_{\mathbb{R}^3} |a(x_1)a(x_2)r(x_3)r(x_1+x_2+x_3)| dx_1 dx_2 dx_3.$$

Since $|r(t)| \le r(0)$, we have $r(\cdot) \in L^\infty(\mathbb{R})$. Additionally, $r(\cdot) \in L^p(\mathbb{R})$ by condition (6.2.6), where $1/p + 1/q \ge 3/2$. Therefore, by the L^p-interpolation theorem (see Proposition 2.1.1), one can choose $p' \ge p$ such that $r(\cdot) \in L^{p'}(\mathbb{R})$ and

$$\frac{1}{p'} + \frac{1}{p'} + \frac{1}{q} + \frac{1}{q} = 3, \quad \text{that is,} \quad \frac{1}{p'} + \frac{1}{q} = \frac{3}{2}.$$

Then, according to Lemma 6.3.3, we have

$$B_1(s,t,T) \le 4\|r\|_{p'}^2 \|a\|_q^2 (t-s).$$

Similarly, one can establish the bound $B_2(s, t, T) \le C(t - s)$, and hence $B(s, t, T) \le C(t - s)$. Thus, the inequality (6.3.8) is proved.

□

The lemmas that follow will be used in the proof of Theorem 6.2.6.

Lemma 6.3.5 *Define (cf. (4.6.35))*

$$\Delta_t(x) := \int_0^t e^{isx} ds = \frac{e^{itx} - 1}{ix}. \tag{6.3.9}$$

Then, for any $\delta \in (0, 1)$, there exists a constant $c > 0$ that depends only on δ, such that

$$|\Delta_t(x)| \le c|t|^\delta f_\delta(x), \quad t \in [0, 1], \ x \in \mathbb{R}, \tag{6.3.10}$$

where

$$f_\delta(x) = \begin{cases} |x|^{\delta-1} & \text{if } |x| > 1; \\ 1 & \text{if } |x| \le 1. \end{cases} \tag{6.3.11}$$

Proof In view of (6.3.9), we have

$$|\Delta_t(x)| \le \int_0^t |e^{isx}| ds = t.$$

So, under the constraint $t \in [0, 1]$, we have $|\Delta_t(x)| \le t \le t^\delta$. On the other hand, with some constant $C > 0$, we have $|e^{ix} - 1| \le C|x|^\delta$ for $\delta \in (0, 1)$. Therefore,

$$|\Delta_t(x)| \le \frac{|e^{itx} - 1|}{|x|} \le C|tx|^\delta |x|^{-1} = Ct^\delta |x|^{\delta-1}.$$

Combining this with (6.3.11), we obtain (6.3.10). □

Lemma 6.3.6 *Let $\gamma_i < 1$, $\gamma_i + \gamma_{i+1} > 1/2$, and let δ be such that*

$$0 \le \delta < \frac{\gamma_i + \gamma_{i+1}}{2},$$

where $i = 1, \dots, 4$ (with $\gamma_5 = \gamma_1$). Then

$$\int_{\mathbb{R}^4} f_\delta(y_1 - y_2) f_\delta(y_2 - y_3) f_\delta(y_3 - y_4) f_\delta(y_4 - y_1)$$
$$\times |y_1|^{-\gamma_1} |y_2|^{-\gamma_2} |y_3|^{-\gamma_3} |y_4|^{-\gamma_4} dy_1 dy_2 dy_3 dy_4 < \infty,$$

where $f_\delta(\cdot)$ is defined as in (6.3.11).

Proof The result follows from Proposition 2.6.1 and Remark 2.6.1 applied to the linear functionals:

$$L_1 = y_l - y_2, L_2 = y_2 - y_3, L_3 = y_3 - y_4, L_4 = y_4 - y_1,$$
$$L_5 = y_1, L_6 = y_2, L_7 = y_3, L_8 = y_4.$$

For details, see the proof of Lemma 1 in Terrin and Taqqu [301]. □

Lemma 6.3.6 can be used to establish the following result.

Lemma 6.3.7 *The function*

$$H_t^*(x_1, x_2) := |x_1|^{\alpha_1/2}|x_2|^{\alpha_2/2} \int_{\mathbb{R}} |\Delta_t(x_1+u)\Delta_t(x_2-u)||u|^{-\beta} du \tag{6.3.12}$$

belongs to $L^2(\mathbb{R}^2)$ *for all* $(\alpha_1, \alpha_2, \beta)$ *in the open region*

$$\{(\alpha_1, \alpha_2, \beta) : \alpha_1, \alpha_2, \beta < 1, \ \alpha_i + \beta > 1/2, \ i = 1, 2\}.$$

Proof It suffices to focus on the case where $t \in [0, 1]$. Otherwise, a change of variable can reduce it to this case. By a suitable change of variables and Lemma 6.3.5, we have

$$\begin{aligned}
\|H_t^*\|^2_{L^2(\mathbb{R}^2)} &= \int_{\mathbb{R}^4} \big|\Delta_t(y_1-y_2)\Delta_t(y_2-y_3)\Delta_t(y_3-y_4)\Delta_t(y_4-y_1)\big| \\
&\quad \times |y_1|^{-\alpha_1}|y_2|^{-\beta}|y_3|^{-\alpha_2}|y_4|^{-\beta} dy_1 dy_2 dy_3 dy_4 \\
&\le C \int_{\mathbb{R}^4} f_\delta(y_1-y_2) f_\delta(y_2-y_3) f_\delta(y_3-y_4) f_\delta(y_4-y_1) \\
&\quad \times |y_1|^{-\alpha_1}|y_2|^{-\beta}|y_3|^{-\alpha_2}|y_4|^{-\beta} dy_1 dy_2 dy_3 dy_4.
\end{aligned}$$

Applying Lemma 6.3.6 and observing that δ can be chosen to be arbitrarily small, the result follows. □

Lemma 6.3.8 *Define the function*

$$\begin{aligned}
H_{t,T}^*(x_1, x_2) &:= A_{1,T}(x_1, x_2)|x_1 x_2|^{-\alpha/2} \\
&\quad \times \int_{\mathbb{R}} |\Delta_t(x_1+u)\Delta_t(x_2-u)||u|^{-\beta} A_{2,T}(u)\, du,
\end{aligned} \tag{6.3.13}$$

where

$$A_{1,T}(x_1, x_2) = \sqrt{\frac{L_1(x_1/T)}{L_1(1/T)} \frac{L_1(x_2/T)}{L_1(1/T)}} \quad \text{and} \quad A_{2,T}(u) = \frac{L_2(u/T)}{L_2(1/T)}, \tag{6.3.14}$$

and $L_1(\cdot)$ *and* $L_2(\cdot)$ *are slowly varying at zero functions. Then, for large enough* T, *we have* $H_{t,T}^*(x_1, x_2) \in L^2(\mathbb{R}^2)$.

Proof By (6.2.20) and (6.3.14), for any $\epsilon > 0$, there exists a constant $C > 0$ such that for sufficiently large T,

$$|A_{1,T}(x_1, x_2)| \leq C(|x_1|^{\epsilon} + |x_1|^{-\epsilon})(|x_2|^{\epsilon} + |x_2|^{-\epsilon}) \tag{6.3.15}$$

and

$$|A_{2,T}(u)| \leq C(|u|^{\epsilon} + |u|^{-\epsilon}). \tag{6.3.16}$$

Hence, with some constant $C > 0$, we have

$$\begin{aligned} |H^*_{t,T}(x_1, x_2)| \leq & C \int_{\mathbb{R}} |\Delta_t(x_1 + u)\Delta_t(x_2 - u)||u|^{-\beta}(|u|^{\epsilon} + |u|^{-\epsilon})du \\ & \times |x_1 x_2|^{-\alpha/2}(|x_1|^{\epsilon} + |x_1|^{-\epsilon})(|x_2|^{\epsilon} + |x_2|^{-\epsilon}). \end{aligned} \tag{6.3.17}$$

By choosing ϵ small enough, in view of Lemma 6.3.7, we conclude that the right-hand side of (6.3.17) belongs to $L^2(\mathbb{R}^2)$, leading to the desired result. □

Lemma 6.3.9 *Let the process $Z_T(t)$ be defined as in (6.2.21), and let*

$$Z'_T(t) := \int''_{\mathbb{R}^2} H_{t,T}(x_1, x_2)\ W(dx_1)W(dx_2), \tag{6.3.18}$$

where

$$\begin{aligned} H_{t,T}(x_1, x_2) = & A_{1,T}(x_1, x_2)|x_1 x_2|^{-\alpha/2} \\ & \times \left[\int_{\mathbb{R}} \Delta_t(x_1 + u)\Delta_t(x_2 - u)|u|^{-\beta} A_{2,T}(u)\ du\right]. \end{aligned} \tag{6.3.19}$$

Then $Z_T(t) \overset{f.d.d.}{=} Z'_T(t)$, which means that the processes $Z_T(t)$ and $Z'_T(t)$ have the same finite-dimensional distributions.

Proof Using the spectral representation of $X(t)$ (see Sect. 3.1.4, formula (3.1.12)):

$$X(t) = \int_{\mathbb{R}} e^{itx}\sqrt{f(x)}W(dx),$$

where $W(\cdot)$ is a complex-valued Gaussian measure with a Lebesgue control measure, and the diagram formula (see, e.g., Major [222], Theorem 5.3), we obtain

$$\begin{aligned} X(u)X(v) - \mathbb{E}[X(u)X(v)] = & \int''_{\mathbb{R}^2} e^{i(ux_1+vx_2)}\sqrt{f(x_1)f(x_2)} \\ & \times W(dx_1)W(dx_2). \end{aligned} \tag{6.3.20}$$

By (6.1.1), (6.2.21), and (6.3.20), we have

$$Z_T(t) = \int_0^{Tt}\int_0^{Tt} \widehat{g}(u-v) \int_{\mathbb{R}^2}'' e^{i(ux_1+vx_2)} \\ \times \sqrt{f(x_1)f(x_2)}W(dx_1)W(dx_2)dudv. \tag{6.3.21}$$

Next, by the stochastic Fubini's lemma (see Proposition 3.4.4) and Lemma 6.3.8, we can change the order of integration in (6.3.21) to obtain the following:

$$\begin{aligned} Z_T(t) &= \frac{1}{T^{\alpha+\beta}L_1(1/T)L_2(1/T)} \int_{\mathbb{R}^2}'' \sqrt{f(x_1)f(x_2)} \\ &\times \int_0^{Tt}\int_0^{Tt}\int_{\mathbb{R}} e^{i(u-v)w} g(w)dw\, e^{i(ux_1+vx_2)} dudv\; W(dx_1)W(dx_2) \\ &= \frac{1}{T^{\alpha+\beta}L_1(1/T)L_2(1/T)} \int_{\mathbb{R}^2}'' \sqrt{f(x_1)f(x_2)} \\ &\times \int_{\mathbb{R}}\int_0^{Tt} e^{iu(x_1+w)}du \int_0^{Tt} e^{iv(x_2-w)}dv\; |w|^{-\beta}L(w)dw\; W(dx_1)W(dx_2) \\ &= \frac{1}{T^{\alpha+\beta}L_1(1/T)L_2(1/T)} \int_{\mathbb{R}^2}'' \sqrt{f(x_1)f(x_2)} \\ &\times \int_{\mathbb{R}} \Delta_{Tt}(x_1+w)\Delta_{Tt}(x_2-w)|w|^{-\beta}L_2(w)dw\; W(dx_1)W(dx_2), \end{aligned}$$

where $\Delta_t(\cdot)$ is defined as in (6.3.9).

Now we apply the change of variables $w \to u/T$, $x_1 \to x_1/T$, $x_2 \to x_2/T$, where the latter two changes of variables are subject to the rule $W(dx/T) \overset{d}{=} T^{-1/2}W(dx)$ (see Proposition 3.4.1), allowing us to obtain

$$Z_T(t) \overset{f.d.d.}{=} \frac{1}{T^{\alpha+\beta}L_1(1/T)L_2(1/T)} \int_{\mathbb{R}^2}'' \sqrt{f(x_1/T)f(x_2/T)} \\ \times \int_{\mathbb{R}} \Delta_t(x_1+u)\Delta_t(x_2-u)|w/T|^{-\beta}L_2(w/T)Tdw\; T^{-1}W(dx_1)W(dx_2). \tag{6.3.22}$$

Finally, taking into account the equality $f(x/T) = |x/T|^{-\alpha}L_1(x/T)$ and the equations in (6.3.14), we find that the right-hand side of (6.3.22) matches both (6.3.18) and (6.3.19). This completes the proof. □

The next lemma will be used to establish tightness in the space $\mathbb{C}[0,1]$ in Theorem 6.2.6.

Lemma 6.3.10 *Let δ be a fixed number within the range $(0, (\alpha + \beta)/2)$, and let the process $Z_T(t)$ be defined as in (6.2.21). Then, for all $0 \le s \le t \le 1$ and for sufficiently large T, there exists a constant $C > 0$ such that*

$$\mathbb{E}\left[|Z_T(t) - Z_T(s)|^2\right] \le C(t-s)^{2\delta}. \tag{6.3.23}$$

A similar inequality also applies to the corresponding limiting process $Z(t)$ defined by Eqs. (6.2.22) and (6.2.23).

Proof First, in view of Lemma 6.3.9, we have

$$\mathbb{E}\left[|Z_T(t) - Z_T(s)|^2\right] = \mathbb{E}\left[|Z'_T(t) - Z'_T(s)|^2\right]. \tag{6.3.24}$$

Next, using the linearity of the multiple stochastic integral, we can write

$$Z'_T(t) - Z'_T(s) = \int_{\mathbb{R}^2}'' H_{s,t,T}(x_1, x_2) W(dx_1) W(dx_2), \tag{6.3.25}$$

where

$$H_{s,t,T}(x_1, x_2) = A_{1,T}(x_1, x_2)|x_1 x_2|^{-\alpha/2}$$
$$\times \int_{\mathbb{R}} [\Delta_t(x_1+u)\Delta_t(x_2-u) - \Delta_s(x_1+u)\Delta_s(x_2-u)]\,|u|^{-\beta} A_{2,T}(u)du, \tag{6.3.26}$$

and the functions $A_{1,T}$ and $A_{2,T}$ are defined as in (6.3.14).

The term in the brackets of the integrand in (6.3.26) can be rewritten as follows:

$$\begin{aligned}
&\Delta_t(x_1+u)\Delta_t(x_2-u) - \Delta_s(x_1+u)\Delta_s(x_2-u)\\
&= \int_0^t \int_0^t e^{iw_1(x_1+u)} e^{iw_2(x_2-u)} dw_1 dw_2 - \int_0^s \int_0^s e^{iw_1(x_1+u)} e^{iw_2(x_2-u)} dw_1 dw_2\\
&= \int_0^s dw_1 \int_s^t dw_2 \ldots + \int_s^t dw_1 \int_0^s dw_2 \ldots + \int_s^t dw_1 \int_s^t dw_2 \ldots\\
&= \Delta_s(x_1+u)\Delta_{t-s}(x_2-u) + \Delta_{t-s}(x_1+u)\Delta_s(x_2-u)\\
&+ \Delta_{t-s}(x_1+u)\Delta_{t-s}(x_2-u).
\end{aligned}$$

Now we apply Lemma 6.3.5 to obtain:

$$\begin{aligned}
&|\Delta_t(x_1+u)\Delta_t(x_2-u) - \Delta_s(x_1+u)\Delta_s(x_2-u)|\\
&\le C[s^\delta(t-s)^\delta + (t-s)^\delta s^\delta + (t-s)^{2\delta}] f_\delta(x_1+u) f_\delta(x_2-u)\\
&\le C(t-s)^\delta f_\delta(x_1+u) f_\delta(x_2-u),
\end{aligned} \tag{6.3.27}$$

where the last inequality follows because $0 \leq s^{\delta} \leq 1$ and $0 \leq (t-s)^{\delta} \leq 1$.

Next, using formulas (3.4.18) and (6.3.24)–(6.3.27), we can write

$$\begin{aligned}
&\mathbb{E}\left[|Z_T(t) - Z_T(s)|^2\right] = \|H_{s,t,T}\|^2_{L^2(\mathbb{R}^2)} \\
&\leq C|t-s|^{2\delta} \int_{\mathbb{R}^2} dx_1 dx_2 A_{1,T}(x_1,x_2)^2 |x_1 x_2|^{-\alpha} \int_{\mathbb{R}^2} du_1 du_2 f_\delta(x_1+u_1) \\
&\times f_\delta(x_2-u_1) f_\delta(-x_1+u_2) f_\delta(-x_2-u_2) |u_1|^{-\beta} |u_2|^{-\beta} A_{2,T}(u_1) A_{2,T}(u_2) \\
&\leq C|t-s|^{2\delta} \int_{\mathbb{R}^4} A_{1,T}(y_1,y_3)^2 A_{2,T}(y_2) A_{2,T}(y_4) f_\delta(y_1-y_2) f_\delta(y_2-y_3) \\
&\times f_\delta(y_3-y_4) f_\delta(y_4-y_1) |y_1|^{-\alpha} |y_2|^{-\beta} |y_3|^{-\alpha} |y_4|^{-\beta} dy_1 dy_2 dy_3 dy_4, \qquad (6.3.28)
\end{aligned}$$

where we applied the change of variables: $y_1 = x_1$, $y_2 = -u_1$, $y_3 = -x_2$, and $y_4 = u_2$.

Since we assume that $\alpha < 1$, $\beta < 1$, and $\alpha + \beta > 1/2$, and the exponent ϵ in inequalities (6.3.15) and (6.3.16) can be chosen to be arbitrarily small, for a fixed δ such that $0 < \delta < (\alpha+\beta)/2$, we can apply Lemma 6.3.6 to conclude that the integral

$$\begin{aligned}
&\int_{\mathbb{R}^4} A_{1,T}(y_1,y_3)^2 A_{2,T}(y_2) A_{2,T}(y_4) f_\delta(y_1-y_2) f_\delta(y_2-y_3) \\
&\times f_\delta(y_3-y_4) f_\delta(y_4-y_1) |y_1|^{-\alpha} |y_2|^{-\beta} |y_3|^{-\alpha} |y_4|^{-\beta} dy_1 dy_2 dy_3 dy_4
\end{aligned}$$

is bounded for sufficiently large T. This, in view of (6.3.28), implies (6.3.23).

Thus, the proof for the process $Z_T(t)$ is complete. The proof for the process $Z(t)$ follows a similar approach; therefore, we will omit the details. □

6.3.2 *Proofs of Theorems*

Proof of Theorem 6.2.2 By Lemma 6.3.2, for any $0 \leq t_1 < \ldots < t_n$ and constants $c_1, \ldots, c_n$, we have

$$\lim_{T\to\infty} \operatorname{Var}\left[\sum_{j=1}^{n} c_j \left(\widetilde{Q}_T(t_j) - L_T(t_j)\right)\right] = 0.$$

Therefore, the convergence of the finite-dimensional distributions of $\widetilde{Q}_T(t)$ to those of Brownian motion $\sigma B(t)$ follows from Lemma 6.3.1, with $f_Y(\cdot)$ given in (6.3.7), and from the Cramér-Wold device (see Proposition 5.3.12). □

Proof of Theorem 6.2.3 In view of Proposition 6.1.5 (see also Proposition 6.1.3), to prove the theorem, we need to show the convergence of finite-dimensional distributions and tightness. The convergence of finite-dimensional distributions has already been established in Theorem 6.2.2. To prove tightness, observe that by Lemma 6.3.4 and the hypercontractivity inequality of multiple Wiener-Itô integrals (see Proposition 3.4.3), for any $T > 0$ and $0 \le s \le t \le 1$, there exists a constant $C > 0$ such that

$$\mathbb{E}\left[|\widetilde{Q}_T(t) - \widetilde{Q}_T(s)|^4\right] \le C_2\left(\mathbb{E}\left[|\widetilde{Q}_T(t) - \widetilde{Q}_T(s)|^2\right]\right)^2 \le C(t-s)^2. \quad (6.3.29)$$

Therefore, the tightness of the family of measures generated by the processes $\{\widetilde{Q}_T(t) : T > 0\}$ in the space $\mathbb{C}[0, 1]$ follows from Prohorov's second theorem (see Proposition 6.1.4 and Remark 6.1.1). □

Proof of Theorem 6.2.4 The convergence of finite-dimensional distributions follows from Theorem 6.2.2. Specifically, the assumptions on f and g in Theorem 6.2.4 imply the conditions (6.2.3) and (6.2.4) presented in Theorem 6.2.2 (see the proof of Theorem 5.2.2). Tightness can be verified in a manner similar to the proof of Theorem 6.2.3. □

Proof of Theorem 6.2.6 As in the proof of Theorem 6.2.3, we need to show the convergence of finite-dimensional distributions and establish tightness. We start by proving the convergence of finite-dimensional distributions, specifically that

$$Z_T(t) \xrightarrow{f.d.d.} Z(t) \quad \text{as} \quad T \to \infty, \quad (6.3.30)$$

where $Z_T(t)$ and $Z(t)$ are defined by Eqs. (6.2.21) and (6.2.22), respectively.

By Lemma 6.3.9, the process $Z_T(t)$ defined in (6.2.21) has the same finite-dimensional distributions as the process $Z'_T(t)$ defined in (6.3.18). Therefore, taking into account the linearity of the multiple Wiener-Itô integral and applying Lemma 6.3.9 along with Crámer-Wold device, to prove (6.3.30), it suffices to show that as $T \to \infty$,

$$H_{t,T}(x_1, x_2) \to H_t(x_1, x_2) \quad \text{in} \quad L^2(\mathbb{R}^2), \quad (6.3.31)$$

where $H_t(x_1, x_2)$ and $H_{t,T}(x_1, x_2)$ are defined as in Eqs. (6.2.23) and (6.3.19), respectively.

First, we show pointwise convergence for a.e. $(x_1, x_2) \in \mathbb{R}^2$, that is,

$$\begin{aligned} H_{t,T}(x_1, x_2) &= A_{1,T}(x_1, x_2)|x_1x_2|^{-\alpha/2} \\ &\times \int_{\mathbb{R}} \Delta_t(x_1+u)\Delta_t(x_2-u)|u|^{-\beta} A_{2,T}(u)du \to H_t(x_1, x_2) \quad (6.3.32) \\ &= |x_1x_2|^{-\alpha/2} \int_{\mathbb{R}} \Delta_t(x_1+u)\Delta_t(x_2-u)|u|^{-\beta}du \quad \text{as} \quad T \to \infty, \quad (6.3.33) \end{aligned}$$

where the functions $A_{1,T}$ and $A_{2,T}$ are defined as in (6.3.14).

Because $L_1(x)$ is a slowly varying at zero function, we have $A_{1,T}(x_1, x_2) \to 1$ as $T \to \infty$. To show that the integral in (6.3.32) converges to the integral in (6.3.33), we first note that, according to (6.3.14), $A_{2,T}(u) \to 1$ as $T \to \infty$, since $L_2(x)$ is also a slowly varying at zero function.

Therefore, it is sufficient to properly bound the integrand and apply the dominated convergence theorem. To this end, we observe that, by (6.3.16), for sufficiently large T, we have

$$\begin{aligned} g_T(u; x_1, x_2) &:= |\Delta_t(x_1 + u)||\Delta_t(x_2 - u)||u|^{-\beta} A_{2,T}(u) \\ &\leq C|\Delta_t(x_1 + u)||\Delta_t(x_2 - u)||u|^{-\beta}(|u|^{\epsilon} + |u|^{-\epsilon}) \\ &:= g_\epsilon(u; x_1, x_2). \end{aligned} \tag{6.3.34}$$

By choosing ϵ small enough and applying Fubini's theorem along with Lemma 6.3.7, we conclude that $g_\epsilon(\cdot\,; x_1, x_2) \in L^1(\mathbb{R})$ for a.e. $(x_1, x_2) \in \mathbb{R}^2$. Now, relation (6.3.31) follows from (6.3.17) and the dominated convergence theorem.

To prove tightness, we first observe that by the hypercontractivity inequality of the multiple Wiener-Itô integrals (see Proposition 3.4.3) and Lemma 6.3.10, for sufficiently large T and for any $0 \leq s \leq t \leq 1$, there exists a constant $C > 0$ such that

$$\mathbb{E}\left[|Z_T(t) - Z_T(s)|^4\right] \leq C_2 \left(\mathbb{E}\left[|Z_T(t) - Z_T(s)|^2\right]\right)^2 \leq C|t - s|^{4\delta}, \tag{6.3.35}$$

where δ is a fixed number within the range $0 < 4\delta < 2(\alpha + \beta)$. Since we are assuming that $\alpha + \beta > 1/2$, we can choose δ to satisfy $4\delta > 1$.

Note that inequalities similar to (6.3.35) also hold for the limiting process $Z(t)$.

In view of (6.3.35) and a similar inequality for $Z(t)$, it follows from Kolmogorov's continuity criterion (see, e.g., Bass [23], Theorem 8.1) that the processes $Z_T(t)$ and $Z(t)$ admit continuous versions when T is sufficiently large.

Now, the tightness of the family of measures generated by the processes $\{Z_T(t) : T > 0\}$ in the space $\mathbb{C}[0, 1]$ follows from Prohorov's second theorem (see Proposition 6.1.4 and Remark 6.1.1). This concludes the proof of Theorem 6.2.6.

□

6.4 Functional Limit Theorems for Lévy-Driven Linear Models

In this section, we present results involving functional central and non-central limit theorems for a suitably normalized stochastic process:

$$Q_T(\tau) := \int_0^{T\tau} \int_0^{T\tau} \widehat{g}(t - s) X(t) X(s) dt ds, \quad \tau \in [0, 1] \tag{6.4.1}$$

in the case where the underlying model $\{X(t),\ t \in \mathbb{R}\}$ is a continuous time linear process driven from *Lévy noise* with time-invariant filter $a(\cdot)$. Specifically,

$$X(t) = \int_{\mathbb{R}} a(t-s)d\xi(s), \quad \int_{\mathbb{R}} |a(s)|^2 ds < \infty, \tag{6.4.2}$$

where $\xi(s)$ is a Lévy process satisfying the conditions (see Sect. 3.2):

$$\mathbb{E}[\xi(s)] = 0,\ \mathbb{E}[\xi^2(1)] = 1 \text{ and } \mathbb{E}[\xi^4(1)] < \infty.$$

The theorems of this section extend the results stated in Sect. 6.2 for Wiener-driven processes. They show that under certain (L^p, L^q)-type conditions imposed on the filter $a(\cdot)$ and the generating kernel $\widehat{g}(\cdot)$ of the quadratic functional, the process $Q_T(\tau)$ satisfies a central limit theorem. Specifically, the finite-dimensional distributions of the standard $\sqrt{T}$-normalized process $Q_T(\tau)$ in (6.4.1) converge to those of a normalized standard Brownian motion. In contrast, when the functions $a(\cdot)$ and $\widehat{g}(\cdot)$ exhibit slow power decay, then we obtain a non-central limit theorem for $Q_T(\tau)$; that is, the finite-dimensional distributions of the process $Q_T(\tau)$, normalized by T^γ for some $\gamma > 1/2$, converge to those of a non-Gaussian non-stationary increment self-similar process, which can be represented by a double stochastic Wiener-Itô integral on $\mathbb{R}^2$.

Recall the following notation that will be used in this section. The symbol $*$ will represent the convolution:

$$(\phi * \psi)(u) = \int_{\mathbb{R}} \phi(u-x)\psi(x)dx,$$

while the symbol $\bar{*}$ is used to denote the reversed convolution:

$$(\phi^{\bar{*}2})(u) = (\phi \bar{*} \phi)(u) = \int_{\mathbb{R}} \phi(u+x)\phi(x)dx.$$

By $\mathcal{F}$ and $\mathcal{F}^{-1}$, we denote the Fourier transform and its inverse:

$$(\mathcal{F}\phi)(u) = \widehat{\phi}(u) = \int_{\mathbb{R}} e^{ixu}\phi(x)dx, \quad (\mathcal{F}^{-1}\phi)(u) = \frac{1}{2\pi}\int_{\mathbb{R}} e^{-ixu}\phi(x)dx.$$

We will use the following well-known identities for Fourier transforms:

$$\mathcal{F}(\phi * \psi) = \mathcal{F}(\phi) \cdot \mathcal{F}(\psi) \quad \text{and} \quad \mathcal{F}(\phi \bar{*} \psi) = \mathcal{F}(\phi) \cdot \overline{\mathcal{F}(\psi)}. \tag{6.4.3}$$

6.4.1 Central Limit Theorems

Let $\{X(t),\ t \in \mathbb{R}\}$ be a centered real-valued linear process given by (6.4.2) with filter $a(\cdot) \in L^2(\mathbb{R})$. Recall that the covariance function $r(t)$ and the spectral density $f(\lambda)$ of $X(t)$ are given by formulas (3.2.11) and (3.2.12), respectively.

The following theorem presents (L^p, L^q)-type time domain sufficient conditions for the process $Q_T(\tau)$ to satisfy the functional central limit theorem.

Theorem 6.4.1 *Let $X(t)$ be defined as in (6.4.2), and let $Q_T(\tau)$ be defined as in (6.4.1). Assume that*

$$a(\cdot) \in L^p(\mathbb{R}) \cap L^2(\mathbb{R}), \quad b(\cdot) := \widehat{g}(\cdot) \in L^q(\mathbb{R}) \tag{6.4.4}$$

with

$$1 \le p, q \le 2, \quad \frac{2}{p} + \frac{1}{q} \ge \frac{5}{2}. \tag{6.4.5}$$

Then, as $T \to \infty$, we have

$$\widetilde{Q}_T(\tau) := T^{-1/2} \left(Q_T(\tau) - \mathbb{E}[Q_T(\tau)]\right) \xrightarrow{f.d.d.} \sigma B(\tau). \tag{6.4.6}$$

Here, the symbol $\xrightarrow{f.d.d.}$ stands for convergence of finite-dimensional distributions, $B(\tau)$ is a standard Brownian motion, and

$$\sigma^2 = \int_{\mathbb{R}} [2K_A(v) + \kappa_4 K_B(v)]\, dv, \tag{6.4.7}$$

where κ_4 is the fourth cumulant of the random variable $\xi(1)$ and

$$K_A(v) = \left((a * b)^{\bar{*}2} \cdot a^{\bar{*}2}\right)(v), \quad K_B(v) = \left((a{*}b) \cdot a\right)^{\bar{*}2}(v). \tag{6.4.8}$$

Remark 6.4.1 Applying Young's convolution inequality (see Proposition 2.3.3) to the convolution in (3.2.11), we obtain $\|r\|_{p_1} \le \|a\|_p^2 < \infty$, where $1 + 1/p_1 = 2/p$. Thus, the conditions (6.4.4) and (6.4.5) imply that:

$$r(\cdot) \in L^{p_1}(\mathbb{R}), \quad b(\cdot) \in L^q(\mathbb{R}), \quad \frac{1}{p_1} + \frac{1}{q} = \frac{2}{p} - 1 + \frac{1}{q} \ge \frac{5}{2} - 1 = \frac{3}{2}. \tag{6.4.9}$$

Note that the condition (6.4.9) is sufficient for the convergence stated in Theorem 6.4.1 to hold when $\xi(t)$ is a Brownian motion (see Theorem 6.2.2). In fact, in this case, the convergence in Theorem 6.4.1 holds under an even weaker condition imposed on the generating function $g(\lambda)$ and the spectral density $f(\lambda)$ of the process $X(t)$ (see Theorem 6.2.3).

Remark 6.4.2 In contrast to cases where the model is either a discrete time linear process (see Theorem 5.2.3) or a continuous time Gaussian process (see Theorems 5.2.1 and 5.2.2), for Lévy-driven linear models, it is more convenient to impose the time domain conditions (6.4.4) and (6.4.5) on the functions $a(\cdot)$ and $\widehat{g}(\cdot)$, rather than on the spectral density $f(\lambda)$ and the generating function $g(\lambda)$. This approach allows us to analyze the random diagonal term that arises from the double stochastic integral with respect to a non-Gaussian Lévy process.

In the discrete time case, the random diagonal term is estimated by the full double sum (see Giraitis et al. [147], Theorem 6.3.1, and Giraitis and Surgailis [148], relation (2.3)), while in the continuous time Gaussian case, there is no such random diagonal term (see the proof of Lemma 5.4.3). In contrast, in the continuous time non-Gaussian case, a random diagonal term arises in the form of a single stochastic integral that cannot be controlled by the double integral, requiring separate treatment (see formula (6.5.28) in the proof of Theorem 6.4.1).

Remark 6.4.3 Observe that the long-run variance σ^2, as given by (6.4.7), can be expressed in terms of the spectral density $f(\lambda)$ and the generating function $g(\lambda)$, provided these functions satisfy certain regularity conditions. Indeed, using the equalities in (6.4.3) and Parseval-Plancherel's theorem, under suitable integrability conditions on $a(\cdot)$ and $b(\cdot)$, we can write

$$
\begin{aligned}
&\int_{\mathbb{R}} K_A(v)dv = \int_{\mathbb{R}} (a*b)^{\bar{*}2}(v)a^{\bar{*}2}(v)dv \\
&= \frac{1}{2\pi}\int_{\mathbb{R}} \mathcal{F}\left((a*b)^{\bar{*}2}\right)(\lambda)\overline{\mathcal{F}\left(a^{\bar{*}2}\right)(\lambda)}d\lambda = \frac{1}{2\pi}\int_{\mathbb{R}} |\mathcal{F}(a*b)(\lambda)|^2|\mathcal{F}(a)(\lambda)|^2 d\lambda \\
&= \frac{1}{2\pi}\int_{\mathbb{R}} |\widehat{a}(\lambda)\widehat{b}(\lambda)|^2|\widehat{a}(\lambda)|^2 d\lambda = 8\pi^3\int_{\mathbb{R}} f(\lambda)^2 g(\lambda)^2 d\lambda,
\end{aligned}
$$

where in the last equality we used the equalities $|\widehat{a}|^2 = 2\pi f$ and $\widehat{b} = 2\pi g$ (because $b(\cdot)$ is an even function). Similarly, we have

$$
\begin{aligned}
&\int_{\mathbb{R}} K_B(v)dv = \int_{\mathbb{R}} dv \int_{\mathbb{R}} dx\Big((a*b)\cdot a\Big)(x)\Big((a*b)\cdot a\Big)(x+v) \\
&= \left(\int_{\mathbb{R}} (a*b)(x)a(x)dx\right)^2 = \frac{1}{4\pi^2}\left(\int_{\mathbb{R}} \widehat{a}(x)\widehat{b}(x)\overline{\widehat{a}(x)}d\lambda\right)^2 \\
&= 4\pi^2\left[\int_{\mathbb{R}} f(\lambda)g(\lambda)d\lambda\right]^2.
\end{aligned}
$$

Thus, an alternative expression for the limiting variance σ^2 in Eq. (6.4.7) can be presented as follows:

$$
\sigma^2 = 16\pi^3\int_{\mathbb{R}} f(\lambda)^2 g(\lambda)^2 d\lambda + \kappa_4\left[2\pi\int_{\mathbb{R}} f(\lambda)g(\lambda)d\lambda\right]^2, \tag{6.4.10}
$$

which should be compared with formula (5.2.14) for an analogous expression in the discrete time case.

The following corollary presents (α, β)-type sufficient conditions for the assumptions in Theorem 6.4.1 to hold. These conditions involve bounding the tails of the functions $a(\cdot)$ and $b(\cdot)$ using appropriate power functions.

Corollary 6.4.1 *The convergence in (6.4.6) holds if the functions $a(\cdot)$ and $b(\cdot)$ satisfy the following conditions:*

$$a(\cdot),\ b(\cdot) \in L^{\infty}(\mathbb{R}), \quad |a(x)| \le c|x|^{\alpha/2-1}, \quad |b(x)| \le c|x|^{\beta-1} \tag{6.4.11}$$

with $0 < \alpha, \beta < 1$ and $\alpha + \beta < 1/2$.

6.4.2 Non-central Limit Theorems

We now present the non-central limit theorems. First, we make the following assumptions regarding the functions $a(\cdot)$ and $b(\cdot)$, as well as their Fourier transforms $\widehat{a}(\cdot)$ and $\widehat{b}(\cdot)$.

Assumption 6.4.1 The Fourier transform $\widehat{a}(\cdot)$ of $a(\cdot) \in L^2(\mathbb{R})$ satisfies

$$\widehat{a}(x) = A(x)|x|^{-\alpha/2} L_1^{1/2}(x),$$

where the function $L_1(x)$ is even, non-negative, slowly varying at zero, and bounded on intervals $[c, \infty)$ for any $c > 0$; the function $A(x)$ is complex-valued and satisfies $|A(x)| = 1$ and $\lim_{x\to 0^+} A(x) = A_0$ for some A_0 on the complex unit circle (since $\widehat{a}(-x) = \overline{\widehat{a}(x)}$, we also have $\lim_{x\to 0^-} A(x) = \overline{A_0}$).

Assumption 6.4.2 The generating function $\widehat{b}(\cdot)$ belongs to $L^1(\mathbb{R})$ and satisfies

$$\widehat{b}(x) = |x|^{-\beta} L_2(x),$$

where the function $L_2(x)$ is even, non-negative, slowly varying at zero, and bounded on intervals $[c, \infty)$ for any $c > 0$.

Assumption 6.4.3 The parameters α and β above satisfy

$$-1/2 < \alpha < 1, \quad -1/2 < \beta < 1, \quad \alpha + \beta > 1/2. \tag{6.4.12}$$

Assumption 6.4.4 There exist numbers α^* and β^* satisfying

$$0 < \alpha^*, \beta^* < 1 \quad 1 < \alpha^* + \beta^* < \alpha + \beta + 1/2,$$

such that

$$|a(x)| \le C|x|^{\alpha^*/2-1}, \quad |b(x)| \le C|x|^{\beta^*-1}.$$

Theorem 6.4.2 *Suppose that Assumptions 6.4.1–6.4.4 hold. Then, as $T \to \infty$, we have the following convergence of finite-dimensional distributions:*

$$Z_T(\tau) := \frac{1}{T^{\alpha+\beta} L_1(1/T) L_2(1/T)} (Q_T(\tau) - \mathbb{E} Q_T(\tau)) \xrightarrow{f.d.d.} Z_{\alpha,\beta}(\tau), \tag{6.4.13}$$

where

$$Z_{\alpha,\beta}(\tau) = \frac{1}{2\pi} \int_{\mathbb{R}^2}'' |x_1 x_2|^{-\alpha/2} \int_{\mathbb{R}} \frac{e^{i\tau(x_1+u)} - 1}{i(x_1+u)} \frac{e^{i\tau(x_2-u)} - 1}{i(x_2-u)} |u|^{-\beta} du \times W(dx_1) W(dx_2). \tag{6.4.14}$$

Here the double prime $''$ indicates the exclusion of the hyper-diagonals: $u_p = \pm u_q$, $p \ne q$, and $W(\cdot)$ is a complex-valued Brownian motion (see Sect. 3.4.3).

Remark 6.4.4 The regular variation conditions on $\widehat{a}(\cdot)$ and $\widehat{b}(\cdot)$ in Assumptions 6.4.1–6.4.3 generally do not follow from the corresponding regular variation conditions imposed on the inverse Fourier transforms $a(\cdot)$ and $b(\cdot)$. This implication holds only under certain additional assumptions regarding the slowly varying factors $L_1(\cdot)$ and $L_2(\cdot)$ of $a(\cdot)$ and $b(\cdot)$. For instance, it will hold if we have (see Bingham et al. [43], formula (4.3.7))

$$a(x) = x^{\alpha/2-1} \ell_1(x) \mathbf{1}_{[0,\infty)}(x), \quad b(x) = |x|^{\beta-1} \ell_2(x), \tag{6.4.15}$$

where $0 < \alpha < 1$, $0 < \beta < 1$, $\alpha + \beta > 1/2$, and $\ell_1(x)$ and $\ell_2(x)$ are even, non-negative functions that are locally bounded, slowly varying at infinity, and *quasi-monotone*. Recall that a slowly varying function $l(\cdot)$ is quasi-monotone if it has locally bounded variation, and for all $\delta > 0$, one has (see Sect. 2.5):

$$\int_0^x t^\delta |d\ell(t)| = O(x^\delta l(x)) \quad \text{as} \quad x \to \infty.$$

A sufficient condition for a slowly varying $\ell(x)$ with locally bounded variation to be quasi-monotone is that it belongs to the Zygmund's class (see Sect. 2.5, and Theorem 1.5.5 and Corollary 2.7.4 in Bingham et al. [43]). Additionally, Assumption 6.4.4 will be satisfied if (6.4.15) holds (see Proposition 2.5.2).

The next theorem presents time domain representations for the limiting process $Z_{\alpha,\beta}(\tau)$ defined in (6.4.14) for the case $\alpha, \beta \ge 0$.

Theorem 6.4.3 *The limiting process $Z_{\alpha,\beta}(\tau)$ in (6.4.14) admits the following time domain representations:*

(a) For $\alpha > 0$ and $\beta > 0$:

$$Z_{\alpha,\beta}(\tau) \overset{f.d.d.}{=} c_{\alpha,\beta} \int'_{\mathbb{R}^2} \int_0^\tau \int_0^\tau |u-v|^{\beta-1}(u-x_1)_+^{\alpha/2-1}(v-x_2)_+^{\alpha/2-1} \times du dv B(dx_1)B(dx_2), \tag{6.4.16}$$

where

$$c_{\alpha,\beta} = \frac{\Gamma(1-\beta)\sin(\beta\pi/2)}{\pi\Gamma(\alpha/2)^2}.$$

(b) For $\alpha > 1/2$ and $\beta = 0$:

$$Z_{\alpha,\beta}(\tau) \overset{f.d.d.}{=} c_{\alpha} \int'_{\mathbb{R}^2} \int_0^\tau (u-x_1)_+^{\alpha/2-1}(u-x_2)_+^{\alpha/2-1} \times du B(dx_1)B(dx_2), \tag{6.4.17}$$

where

$$c_{\alpha} = \frac{\sin(\alpha\pi/2)\Gamma(1-\alpha/2)}{\pi\Gamma(\alpha/2)}.$$

(c) For $\alpha = 0$ and $\beta > 1/2$:

$$Z_{\alpha,\beta}(\tau) \overset{f.d.d.}{=} c_{\beta} \int'_{[0,\tau]^2} |x_1-x_2|^{\beta-1} B(dx_1)B(dx_2), \tag{6.4.18}$$

where

$$c_{\beta} = \frac{\Gamma(1-\beta)\sin(\beta\pi/2)}{\pi}.$$

Here, $B(\cdot)$ is the real Brownian random measure, and the prime $'$ in the integrals indicates the exclusion of the diagonals.

Remark 6.4.5 In view of (6.4.1) and (6.4.15), the representation (6.4.16) provides explicit insight into the convergence stated in Theorem 6.4.2 (see Theorem 6.4.4 below). The Eq. (6.4.17) offers a time domain representation of the Rosenblatt process (see Pipiras and Taqqu [246], Section 4.6). For the spectral domain representation, see formula (6.2.15). The convergence in Theorem 6.4.2 serves as the continuous time analog of the discrete time case considered in Fox and Taqqu [99]. The representation (6.4.18) is derived from the fact that when $\alpha = 0$, the underlying process $X(t)$ exhibits short memory. In this case, one can expect that in the limit, the term $X(t)dt$ in (6.4.1) can be replaced by the continuous white noise $B(dt)$.

By employing the time domain representation (6.4.16), we can establish a non-central limit theorem for the case where $\alpha > 0$ and $\beta > 0$, without relying on the spectral domain. This approach simplifies the assumptions imposed on the functions $a(\cdot)$ and $b(\cdot)$.

Theorem 6.4.4 *Let the functions $a(\cdot)$ and $b(\cdot)$ be defined by (6.4.15), where $0 < \alpha < 1$, $0 < \beta < 1$, and $\alpha + \beta > 1/2$. Additionally, let $\ell_1(x)$ and $\ell_2(x)$ be even functions that are slowly varying at infinity and bounded on bounded intervals. Then, as $T \to \infty$, we have the following convergence of finite-dimensional distributions:*

$$\frac{1}{T^{\alpha+\beta}\ell_1(T)\ell_2(T)}\left(Q_T(\tau) - \mathbb{E}[Q_T(\tau)]\right) \xrightarrow{f.d.d.}$$

$$\int'_{\mathbb{R}^2}\int_0^\tau\int_0^\tau |u-v|^{\beta-1}(u-x_1)_+^{\alpha/2-1}(v-x_2)_+^{\alpha/2-1}du dv\; B(dx_1)B(dx_2).$$

Here, as before, $B(\cdot)$ denotes the real Brownian random measure, and the prime $'$ in the integral indicates the exclusion of the diagonals.

6.5 Proofs of Theorems of Sect. 6.4

6.5.1 Proofs of the Functional Central Limit Theorems

In this section, we prove the functional central limit theorems presented in Sect. 6.4.1, specifically Theorem 6.4.1 and Corollary 6.4.1.

6.5.1.1 Lemmas

We will first derive some preliminary results. We define

$$R_T(x_1, x_2) := \frac{1}{\sqrt{T}}\int_0^T\int_0^T b(u-v)a(u-x_1)a(v-x_2)du dv, \tag{6.5.1}$$

and

$$S_T(x_1, x_2) := \frac{1}{\sqrt{T}}\int_0^T [(a * b)(v-x_1)]\,[a(v-x_2)]\,dv. \tag{6.5.2}$$

Lemma 6.5.1 *Let the functions $a(\cdot)$ and $b(\cdot)$ satisfy conditions (6.4.4) and (6.4.5), and let the functions $R_T(\cdot,\cdot)$ and $S_T(\cdot,\cdot)$ be defined as in (6.5.1) and (6.5.2), respectively, with $x_1 \neq x_2$. The following assertions hold.*

(a) We have

$$\lim_{T\to\infty} \|S_T\|^2_{L^2(\mathbb{R}^2)} = \int_{\mathbb{R}} K_A(u)du, \tag{6.5.3}$$

where $K_A(\cdot)$ is defined as in (6.4.8).

(b) We have

$$\lim_{T\to\infty} \|R_T - S_T\|_{L^2(\mathbb{R}^2)} = 0. \tag{6.5.4}$$

(c) For any $M > 0$, there exists a function $c_M(\cdot,\cdot)$ supported on $[-2M, 2M]^2$, so that the function

$$S_T^M(x_1, x_2) = \frac{1}{\sqrt{T}} \int_0^T c_M(v - x_1, v - x_2)dv$$

satisfies the relation:

$$\lim_{M\to\infty} \limsup_{T\to\infty} \|R_T - S_T^M\|_{L^2(\mathbb{R}^2)} = 0. \tag{6.5.5}$$

Proof We first prove assertion (a). We will use the following notation: $|a|(x) = |a(x)|$, $|b|(x) = |b(x)|$, and $|c|(x) = |c(x)|$.

By (6.4.4) and (6.4.5), we have $a(\cdot) \in L^p(\mathbb{R}) \cap L^2(\mathbb{R})$. Therefore, by the Riesz-Thorin theorem (see Proposition 2.3.2), it follows that $a(\cdot) \in L^{p'}(\mathbb{R})$ for any $p \le p' \le 2$. Setting $p' = 2$, we find that $1 + 1/q \le 2$, which is less than $5/2$. This implies that there exists a number p' such that $2/p' + 1/q = 5/2$. Thus, without loss of generality, we can assume that

$$a(\cdot) \in L^p(\mathbb{R}), \quad b(\cdot) \in L^q(\mathbb{R}), \quad \frac{2}{p} + \frac{1}{q} = \frac{5}{2}. \tag{6.5.6}$$

Let p and q be defined as in (6.5.6). Define the numbers q_1, q_1^*, q_2 to satisfy the following equations:

$$\frac{1}{q_1} + \frac{1}{q_1^*} = 1, \quad 1 + \frac{1}{q_1^*} = \frac{2}{p}, \quad 1 + \frac{1}{q_1} = \frac{2}{q_2}, \quad 1 + \frac{1}{q_2} = \frac{1}{p} + \frac{1}{q}. \tag{6.5.7}$$

Starting from the last equality in (6.5.7), one can successively solve for q_2, q_1^*, and q_1. Then, by using (6.5.6), one can verify that the first equality in (6.5.7) holds.

Taking into account (6.5.7), the relation

$$\begin{aligned} |R_T(x_1, x_2)| \le& \frac{1}{\sqrt{T}} \int_0^T \int_{-\infty}^{+\infty} |b(u - v)a(u - x_1)a(v - x_2)|dudv \\ =& \frac{1}{\sqrt{T}} \int_0^T (|a| * |b|)(v - x_1)|a|(v - x_2)dv, \end{aligned} \tag{6.5.8}$$

and by using Hölder's inequality (Proposition 2.1.5) and Young's inequality for convolution (Proposition 2.3.3), we can write

$$
\begin{aligned}
\|R_T\|^2_{L^2(\mathbb{R}^2)} &\le \frac{1}{T}\int_{[0,T]^2} dv_1 dv_2 \int_{\mathbb{R}^2} dx_1 dx_2 (|a| * |b|)(v_1 - x_1) \\
&\quad \times |a|(v_1 - x_2)(|a| * |b|)(v_2 - x_1)|a|(v_2 - x_2) \\
&= \frac{1}{T}\int_{[0,T]^2} dv_1 dv_2 \Big((|a| * |b|)^{\bar{*}2} \cdot |a|^{\bar{*}2} \Big)(v_1 - v_2) \qquad (6.5.9) \\
&= \int_{-T}^{T} \left(1 - \frac{|v|}{T}\right) \Big((|a| * |b|)^{\bar{*}2} \cdot |a|^{\bar{*}2} \Big)(v)\, dv \\
&\le \int_{\mathbb{R}} \Big((|a| * |b|)^{\bar{*}2} \cdot |a|^{\bar{*}2} \Big)(v)\, dv \\
&\underset{\text{Hölder}}{\le} \|(|a| * |b|)^{\bar{*}2}\|_{q_1} \||a|^{\bar{*}2}\|_{q_1^*} \underset{\text{Young}}{\le} \|(|a| * |b|)^{\bar{*}2}\|_{q_1} \|a\|_p^2 \\
&\underset{\text{Young}}{\le} \||a| * |b|\|_{q_2}^2 \|a\|_p^2 \underset{\text{Young}}{\le} \|a\|_p^4 \|b\|_q^2. \qquad (6.5.10)
\end{aligned}
$$

Similarly, we obtain:

$$
\|S_T\|^2_{L^2(\mathbb{R}^2)} \le \|a\|_p^4 \|b\|_q^2. \tag{6.5.11}
$$

In view of (6.5.2), (6.5.11), and Fubini's theorem, we have

$$
\|S_T\|^2_{L^2(\mathbb{R}^2)} = \int_{-T}^{T} \left(1 - \frac{|v|}{T}\right) \Big((a * b)^{\bar{*}2} \cdot a^{\bar{*}2} \Big)(v)\, dv,
$$

which converges to the limit claimed in (6.5.3) by the dominated convergence theorem (Proposition 2.1.2). This completes the proof of assertion (a) of the lemma.

Now, we proceed to prove assertions (b) and (c) of the lemma. To this end, for a number $M > 0$, we set

$$
\begin{aligned}
a_M(x) &= a(x)\mathbf{1}_{[-M,M]}(x), \quad a_M^-(x) = a(x) - a_M(x), \\
b_M(x) &= b(x)\mathbf{1}_{[-M,M]}(x), \quad b_M^-(x) = b(x) - b_M(x),
\end{aligned}
$$

and define the function:

$$
R_T^M(x_1, x_2) := \frac{1}{\sqrt{T}} \int_0^T \int_0^T b_M(u - v) a_M(u - x_1) a_M(v - x_2)\, du dv. \tag{6.5.12}
$$

In view of (6.5.1), (6.5.12), and the identity

$$
\begin{aligned}
baa - b_M a_M a_M &= (baa - b_M aa) + (b_M aa - b_M a_M a) + (b_M a_M a - b_M a_M a_M) \\
&= b_M^- aa + b_M a_M^- a + b_M a_M a_M^-,
\end{aligned}
$$

we have

$$
\begin{aligned}
& R_T(x_1, x_2) - R_T^M(x_1, x_2) = \\
& \frac{1}{\sqrt{T}} \int_0^T \int_0^T dudv \; \big[b_M^-(u-v) a(u-x_1) a(v-x_2) + \\
& b_M(u-v) a_M^-(u-x_1) a(v-x_2) + b_M(u-v) a_M(u-x_1) a_M^-(v-x_2) \big].
\end{aligned}
$$

Similar to (6.5.10), we obtain the following inequality:

$$
\begin{aligned}
\| R_T - R_T^M \|_{L^2(\mathbb{R}^2)}^2 \le C \big(& \| b_M^- \|_q^2 \| a \|_p^4 + \| b_M \|_q^2 \| a_M^- \|_p^2 \| a \|_p^2 \\
& + \| b_M \|_q^2 \| a_M \|_p^2 \| a_M^- \|_p^2 \big),
\end{aligned}
$$

where the right-hand side does not involve T.

Since $\| a_M^- \|_p \to 0$ and $\| b_M^- \|_q \to 0$ as $M \to \infty$, we conclude that

$$
\lim_{M\to\infty} \limsup_{T\to\infty} \| R_T - R_T^M \|_{L^2(\mathbb{R}^2)} = 0. \tag{6.5.13}
$$

Now we set

$$
c_M(x_1, x_2) := (a_M * b_M)(x_1) a_M(x_2), \tag{6.5.14}
$$

and define the function:

$$
\begin{aligned}
S_T^M(x_1, x_2) :=& \frac{1}{\sqrt{T}} \int_0^T c_M(v - x_1, v - x_2) du \\
=& \frac{1}{\sqrt{T}} \int_0^T (a_M * b_M)(v - x_1) a_M(v - x_2) du.
\end{aligned}
$$

In the same manner that we derived (6.5.13), we can conclude that

$$
\lim_{M\to\infty} \limsup_{T\to\infty} \| S_T - S_T^M \|_{L^2(\mathbb{R}^2)} = 0. \tag{6.5.15}
$$

Observe that

$$
S_T^M(x_1, x_2) = \frac{1}{\sqrt{T}} \int_0^T dv \left(\int_{\mathbb{R}} du b_M(v-u) a_M(u - x_1) \right) a_M(v - x_2). \tag{6.5.16}
$$

Suppose that $T > M$. Considering Eqs. (6.5.12) and (6.5.16), and using the fact that the function $b_M(\cdot)$ is supported on $[-M, M]$, we can write

$$
\begin{aligned}
&S_T^M(x_1, x_2) - R_T^M(x_1, x_2) \\
&= \frac{1}{\sqrt{T}} \int_0^T dv \int_{\mathbb{R}\setminus[0,T]} du b_M(u-v) a_M(u-x_1) a_M(v-x_2) \\
&= \frac{1}{\sqrt{T}} \int_0^T dv \int_T^{\infty} du b_M(u-v) a_M(u-x_1) a_M(v-x_2) \\
&\quad + \frac{1}{\sqrt{T}} \int_0^T dv \int_{-\infty}^{0} du b_M(u-v) a_M(u-x_1) a_M(v-x_2) \\
&= \frac{1}{\sqrt{T}} \int_{T-M}^T dv \int_T^{\infty} du b_M(u-v) a_M(u-x_1) a_M(v-x_2) \\
&\quad + \frac{1}{\sqrt{T}} \int_0^M dv \int_{-M}^{0} du b_M(u-v) a_M(u-x_1) a_M(v-x_2) \\
&=: A_{T,1}^M(x_1, x_2) + A_{T,2}^M(x_1, x_2).
\end{aligned}
$$

Therefore, by applying arguments similar to those used in (6.5.8) and (6.5.10), one can conclude that

$$
\begin{aligned}
\|A_{T,1}^M\|_{L^2(\mathbb{R}^2)}^2 &\leq \frac{1}{T} \int_{[T-M,T]^2} dv_1 dv_2 \Big((|a| * |b|)^{\bar{*}2} \cdot |a|^{\bar{*}2} \Big)(v_1 - v_2) \\
&= \frac{1}{T} \int_{[0,M]^2} dv_1 dv_2 \Big((|a| * |b|)^{\bar{*}2} \cdot |a|^{\bar{*}2} \Big)(v_1 - v_2) \\
&\leq \frac{M}{T} \int_{\mathbb{R}} dv \Big((|a| * |b|)^{\bar{*}2} \cdot |a|^{\bar{*}2} \Big)(v) \to 0 \text{ as } T \to \infty,
\end{aligned}
$$

where the integral

$$
\int_{\mathbb{R}} dv \Big((|a| * |b|)^{\bar{*}2} \cdot |a|^{\bar{*}2} \Big)(v)
$$

is finite due to (6.5.10). Similarly, one can show that

$$
\|A_{T,2}^M\|_{L^2(\mathbb{R}^2)}^2 \to 0 \text{ as } T \to \infty.
$$

Therefore,

$$
\lim_{M\to\infty} \limsup_{T\to\infty} \|S_T^M - R_T^M\|_{L^2(\mathbb{R}^2)}^2 = 0. \tag{6.5.17}
$$

Combining the relations in (6.5.13), (6.5.15), and (6.5.17), we obtain the desired relations (6.5.4) and (6.5.5) with $c_M(\cdot,\cdot)$ as defined in (6.5.14). This completes the proof of Lemma 6.5.1. □

The next result is similar to Lemma 6.5.1, where $\mathbb{R}^2$ is replaced by $\mathbb{R}$. We define

$$R_T(x) = R_T(x,x) := \frac{1}{\sqrt{T}} \int_0^T \int_0^T b(u-v)a(u-x)a(v-x)dudv$$

and

$$S_T(x) = S_T(x,x) := \frac{1}{\sqrt{T}} \int_0^T (a*b)(v-x)a(v-x)dv,$$

where the functions $R_T(\cdot,\cdot)$ and $S_T(\cdot,\cdot)$ are defined as in Eqs. (6.5.1) and (6.5.2), respectively.

Lemma 6.5.2 *Let the functions $a(\cdot)$ and $b(\cdot)$ be defined as in (6.4.4), with p and q satisfying*

$$1 \le p, q \le 2, \quad \frac{2}{p} + \frac{1}{q} \ge 2. \tag{6.5.18}$$

Then the following assertions hold.

(a) We have

$$\lim_{T\to\infty} \|S_T\|^2_{L^2(\mathbb{R})} = \int_{\mathbb{R}} K_B(u)du, \tag{6.5.19}$$

where $K_B(\cdot)$ is defined as in (6.4.8).

(b) We have

$$\lim_{T\to\infty} \|R_T - S_T\|_{L^2(\mathbb{R})} = 0. \tag{6.5.20}$$

(c) For any $M > 0$, there exists a function $d_M(\cdot)$ supported on $[-2M, 2M]$, so that the function

$$S_T^M(x) = \frac{1}{\sqrt{T}} \int_0^T d_M(v-x)dv$$

satisfies the relation:

$$\lim_{M\to\infty} \limsup_{T\to\infty} \|R_T - S_T^M\|_{L^2(\mathbb{R})} = 0. \tag{6.5.21}$$

Remark 6.5.1 It is clear that the condition in (6.5.18) is a consequence of the condition in (6.4.5).

Proof of Lemma 6.5.2 The proof is similar to that of Lemma 6.5.1. We thus outline the key steps of the proof, omitting the details.

As in the proof of Lemma 6.5.1, in view of the Riesz-Thorin theorem, one can assume that

$$a(\cdot) \in L^p(\mathbb{R}), \;\; b(\cdot) \in L^q(\mathbb{R}), \quad \frac{2}{p} + \frac{1}{q} = 2. \tag{6.5.22}$$

Let p and q be defined as in (6.5.22). Define the number p^* to satisfy the following equations:

$$\frac{1}{p} + \frac{1}{p^*} = 1, \;\; 1 + \frac{1}{p^*} = \frac{1}{p} + \frac{1}{q}. \tag{6.5.23}$$

Observe that by the equality in (6.5.22), one has

$$\frac{1}{p} + \frac{1}{p^*} = \frac{1}{p} + \frac{1}{p} + \frac{1}{q} - 1 = 2 - 1 = 1.$$

Then, using Hölder's inequality and Young's inequality for convolution, we can write (note the difference between (6.5.9) and (6.5.24)):

$$\begin{aligned} \|R_T\|^2_{L^2(\mathbb{R})} &\leq \frac{1}{T}\int_{[0,T]^2} dv_1 dv_2 \int_{\mathbb{R}} dx (|a| * |b|)(v_1 - x) \\ &\quad \times |a|(v_1 - x)(|a| * |b|)(v_2 - x)|a|(v_2 - x) \\ &= \frac{1}{T}\int_{[0,T]^2} dv_1 dv_2 \Big((|a| * |b|) \cdot |a|\Big)^{\bar{*}2}(v_1 - v_2) \end{aligned} \tag{6.5.24}$$

$$\begin{aligned} &= \int_{-T}^{T} \left(1 - \frac{|v|}{T}\right) \Big((|a| * |b|) \cdot |a|\Big)^{\bar{*}2}(v)\, dv \\ &\leq \int_{\mathbb{R}} \Big((|a| * |b|) \cdot |a|\Big)^{\bar{*}2}(v)\, dv \underset{\text{Young}}{\leq} \|(|a| * |b|) \cdot |a|\|_1^2 \\ &\underset{\text{Hölder}}{\leq} \||a| * |b|\|^2_{p^*} \|a\|^2_p \underset{\text{Young}}{\leq} \|a\|^4_p \|b\|^2_q. \end{aligned} \tag{6.5.25}$$

Similarly, we obtain

$$\|S_T\|^2_{L^2(\mathbb{R}^2)} \leq \|a\|^4_p \|b\|^2_q. \tag{6.5.26}$$

Then, the assertion (a) of the lemma follows from (6.5.26), Fubini's theorem, and the dominated convergence theorem.

To prove assertions (b) and (c), we set

$$a_M(x) := a(x)\mathbf{1}_{[-M,M]}(x),$$
$$b_M(x) := b(x)\mathbf{1}_{[-M,M]}(x),$$

and consider the functions

$$R_T^M(x) := \frac{1}{\sqrt{T}} \int_0^T \int_0^T b_M(u-v)a_M(u-x)a_M(v-x)dudv,$$
$$S_T^M(x) := \frac{1}{\sqrt{T}} \int_0^T d_M(v-x)dv,$$

where

$$d_M(x) := \Big((a_M * b_M) \cdot a_M\Big)(x). \tag{6.5.27}$$

Using the arguments from the proof of Lemma 6.5.1 but now with $x_1 = x_2 = x$, it can be shown that

$$\lim_{M\to\infty} \limsup_{T\to\infty} \| R_T^M - R_T \|_{L^2(\mathbb{R})}^2 = 0,$$
$$\lim_{M\to\infty} \limsup_{T\to\infty} \| S_T^M - R_T^M \|_{L^2(\mathbb{R})}^2 = 0,$$
$$\lim_{M\to\infty} \limsup_{T\to\infty} \| S_T^M - S_T \|_{L^2(\mathbb{R})}^2 = 0,$$

which implies the desired relations (6.5.20) and (6.5.21) with $d_M(\cdot)$ defined in (6.5.27). Lemma 6.5.2 is proved. □

6.5.1.2 Proofs of the Theorems

We are now ready to prove the central limit theorems stated in Section 6.4.1 (Theorem 6.4.1 and Corollary 6.4.1).

Proof of Theorem 6.4.1 By using formulas (6.1.1), (6.1.4), (3.4.27), and Proposition 3.4.4, we express the process $\widetilde{Q}_T(t)$ as follows:

$$\widetilde{Q}_T(t) = A_T(t) + B_T(t),$$

where

$$A_T(t) := \int_{\mathbb{R}^2}' \frac{1}{\sqrt{T}} \int_0^{Tt} \int_0^{Tt} b(u-v)a(u-x_1)a(v-x_2)dudv\, \xi(dx_1)\xi(dx_2),$$

and

$$B_T(t) := \int_{\mathbb{R}} \frac{1}{\sqrt{T}} \int_0^{Tt} \int_0^{Tt} b(u-v)a(u-x)a(v-x)dudv\, \xi_c^{(2)}(dx). \quad (6.5.28)$$

Choosing $c_M(x_1, x_2)$ as specified in Lemma 6.5.1 and defining

$$A_T^M(t) := \int_{\mathbb{R}^2}' \frac{1}{\sqrt{T}} \int_0^{Tt} c_M(u-x_1, u-x_2)du\, \xi(dx_1)\xi(dx_2), \quad (6.5.29)$$

we can apply (3.4.20) and the relation (6.5.5) from Lemma 6.5.1 to conclude that for all $t > 0$:

$$\lim_{M\to\infty} \limsup_{T\to\infty} \mathbb{E}|A_T(t) - A_T^M(t)|^2 = 0. \quad (6.5.30)$$

Choosing $d_M(x)$ as in Lemma 6.5.2 and defining

$$B_T^M(t) := \int_{\mathbb{R}} \frac{1}{\sqrt{T}} \int_0^{Tt} d_M(u-x)du\, \xi_c^{(2)}(dx), \quad (6.5.31)$$

by (3.4.20) and the relation (6.5.21) from Lemma 6.5.2, we have for all $t > 0$:

$$\lim_{M\to\infty} \limsup_{T\to\infty} \mathbb{E}|B_T(t) - B_T^M(t)|^2 = 0. \quad (6.5.32)$$

To complete the proof of the theorem, in view of (6.5.30) and (6.5.32), it suffices to show that as $T \to \infty$,

$$\widetilde{Q}_T^M(t) := A_T^M(t) + B_T^M(t) \overset{f.d.d.}{\longrightarrow} \sigma_M B(t) \quad (6.5.33)$$

with $\sigma_M \geq 0$ satisfying

$$\lim_{M\to\infty} \sigma_M^2 = \lim_{T\to\infty} \mathrm{Var}[A_T(1) + B_T(1)] = \sigma^2. \quad (6.5.34)$$

To this end, first observe that, according to the stochastic Fubini's lemma (Proposition 3.4.4), one has

$$\widetilde{Q}_T^M(t) = \frac{1}{\sqrt{T}} \int_0^{Tt} Y_M(u)du,$$

where

$$Y_M(u) = \int_{\mathbb{R}^2}' c_M(u-x_1, u-x_2)\xi(dx_1)\xi(dx_2) + \int_{\mathbb{R}} d_M(u-x)\, \xi_c^{(2)}(dx),$$

and $\xi_c^{(2)}(\cdot)$ is the Teugels martingale defined in (3.4.28).

Note that the process $Y_M(u)$ is independent of the σ-field generated by the random variables $\{\xi(s) : s < u - 2M, \ s > u + 2M\}$ since the function $c_M(\cdot, \cdot)$ vanishes outside $[-2M, 2M]^2$ and the function $d_M(\cdot)$ vanishes outside $[-2M, 2M]$. This implies that $Y_M(u)$ is a stationary $4M$-dependent process.

Therefore, the convergence in (6.5.33) can be deduced from the central limit theorem for m-dependent processes (see Propositions 5.3.3, 5.3.4, and 5.3.6) by applying the discretization argument from Proposition 5.3.11 (see also the proof of Theorem 18.7.1 from Ibragimov and Linnik [188] and Theorem 5.2 from Billingsley [39].

To show the relation (6.5.34), it suffices to note that, based on the arguments preceding formula (3.4.31), the random variables $A_T(1)$ and $B_T(1)$ are uncorrelated.

Therefore, by (3.4.20) and (6.5.3) with $k = 2$, we obtain

$$\mathrm{Var}[A_T(1)] \to 2 \int_{\mathbb{R}} K_A(u) du.$$

Similarly, by (3.4.20), (3.4.29), and (6.5.19), we can derive

$$\mathrm{Var}[B_T(1)] \to \kappa_4 \int_{\mathbb{R}} K_B(u) du.$$

This completes the proof of Theorem 6.4.1. □

Proof of Corollary 6.4.1 In view of Theorem 6.4.1, it is sufficient to verify that the conditions (6.4.4) and (6.4.5) are satisfied.

First, noting the assumptions $0 < \alpha, \beta < 1$ and $\alpha + \beta < 1/2$, we can choose $1 \le p, q \le 2$ to satisfy

$$p(\alpha/2 - 1) < -1, \ q(\beta - 1) < -1 \iff \frac{2}{p} < 2 - \alpha, \ \frac{1}{q} < 1 - \beta, \tag{6.5.35}$$

implying that

$$\frac{2}{p} + \frac{1}{q} < 3 - \alpha - \beta. \tag{6.5.36}$$

Next, since $\alpha + \beta < 1/2$, we have $3 - \alpha - \beta > 5/2$. Therefore, in view of (6.5.36), the numbers p and q can be chosen to satisfy $2/p + 1/q \ge 5/2$. Thus, the condition (6.4.5) is satisfied.

To verify the condition (6.4.4), note that with the p and q chosen above, in view of condition (6.4.11), we have

$$a(\cdot) \in L^p(\mathbb{R}) \cap L^2(\mathbb{R}), \quad b(\cdot) \in L^q(\mathbb{R}),$$

showing that (6.4.4) is also satisfied. Corollary 6.4.1 is proved. □

6.5.2 Proofs of the Functional Non-central Limit Theorems

In this section, we prove the functional non-central limit theorems stated in Section 6.4.2 (Theorems 6.4.2–6.4.4).

6.5.2.1 Lemmas

We first establish some preliminary lemmas. The following lemma, which serves as a continuous analog of Propositions 14.3.2 and 14.3.3 from Giraitis et al. [147], plays a key role in our proofs. It provides conditions under which Lévy-Itô multiple stochastic integrals converge in distribution to Wiener-Itô multiple stochastic integrals. For definitions and properties of these integrals, see Sect. 3.4.3.

Lemma 6.5.3 *For $T > 0$ and $f_{j,T}(\cdot) \in L^2(\mathbb{R}^{k_j})$, $j = 1, \ldots, J$, we define*

$$h_{j,T}(x_1, \ldots, x_{k_j}) := T^{k/2} f_{j,T}(Tx_1, \ldots, Tx_{k_j}), \tag{6.5.37}$$

and assume that there exist functions $f_j \in L^2(\mathbb{R}^{k_j})$ such that as $T \to \infty$

$$\|h_{j,T} - f_j\|_{L^2(\mathbb{R}^{k_j})} \to 0, \quad j = 1, \ldots, k_j. \tag{6.5.38}$$

Then, for any Lévy process $\xi(\cdot)$ with $\mathbb{E}[\xi(1)] = 0$ and $\mathbb{E}[\xi^2(1)] = 1$, we have the following joint convergence in distribution:

$$\left(I^{\xi}_{k_1}(f_{1,T}), \ldots, I^{\xi}_{k_J}(f_{J,T})\right) \xrightarrow{d} \left(I^{B}_{k_1}(f_1), \ldots, I^{B}_{k_J}(f_J)\right). \tag{6.5.39}$$

Proof For simplicity, we prove the result in the case where $J = 1$ and drop the index j. In this case, the proof is similar to that of Proposition 14.3.2 from Giraitis et al. [147]. The general case $J > 1$, which corresponds to Proposition 14.3.3 of Giraitis et al. [147], can be derived using similar arguments along with the Cramér-Wold device (see Proposition 5.3.12).

Let $S_M(\mathbb{R}^k)$, $M \in \mathbb{N}$, be the class of functions that are piecewise constant on the $1/M$-grid of $[-M, M]^k$ and vanish on the diagonals (each piece of the grid has a linear length of $1/M$). We define $\mathcal{S}_k := \cup_{M=1}^{\infty} S_M(\mathbb{R}^k)$ and observe that $\mathcal{S}_k$ is a dense subset of $L^2(\mathbb{R}^k)$. Then, in view of (3.4.19), for any $\epsilon > 0$, there exists a function $f_\epsilon \in \mathcal{S}_k$ such that

$$\mathbb{E}\left[|I^B_k(f) - I^B_k(f_\epsilon)|^2\right] \le k!\|f - f_\epsilon\|^2_{L^2(\mathbb{R}^k)} \le \epsilon. \tag{6.5.40}$$

Define

$$f_{\epsilon,T}(x_1, \ldots, x_k) := T^{-k/2} f_\epsilon(x_1/T, \ldots, x_k/T), \tag{6.5.41}$$

and note that

$$\|h_T - f_\epsilon\|^2_{L^2(\mathbb{R}^k)} \le 2\|h_T - f\|^2_{L^2(\mathbb{R}^k)} + 2\|f - f_\epsilon\|^2_{L^2(\mathbb{R}^k)}. \tag{6.5.42}$$

By (6.5.38), we have

$$\lim_{T\to\infty} \|h_T - f\|_{L^2(\mathbb{R}^k)} = 0.$$

Therefore, in view of (3.4.19) and (6.5.42), and a change of variable, we can write

$$\limsup_{T\to\infty} \mathbb{E}\left[|I_k^\xi(f_T) - I_k^\xi(f_{\epsilon,T})|^2\right] \le k! \limsup_{T\to\infty} \|f_T - f_{\epsilon,T}\|^2_{L^2(\mathbb{R}^k)} =$$

$$k! \limsup_{T\to\infty} \|h_T - f_\epsilon\|^2_{L^2(\mathbb{R}^k)} \le 2k! \limsup_{T\to\infty} \|f - f_\epsilon\|^2_{L^2(\mathbb{R}^k)} \le 2\epsilon. \tag{6.5.43}$$

To complete the proof of the lemma, in view of formulas (6.5.40) and (6.5.43), and Theorem 8.6.2 of Resnick [253], it remains to show that

$$I_k^\xi(f_{\epsilon,T}) \xrightarrow{d} I_k^B(f_\epsilon) \quad \text{as} \quad T \to \infty. \tag{6.5.44}$$

Since $f_\epsilon(\cdot) \in \mathcal{S}_k$, we have

$$f_\epsilon(x_1, \ldots, x_k) = \sideset{}{'}\sum_{1\le i_1,\ldots,i_k\le N} c(i_1, \ldots, i_k) \mathbf{1}_{\Delta_{i_1}\times\ldots\times\Delta_{i_k}}(x_1, \ldots, x_k),$$

where $N > 0$, $c(i_1, \ldots, i_k) \in \mathbb{R}$, and the intervals Δ_i's are disjoint such that $\cup_i \Delta_i = [-M, M]$. The prime $'$ indicates that the sum does not include the diagonals $i_p = i_q$ for $p \ne q$. Then, we have

$$\begin{aligned} I_k^\xi(f_{\epsilon,T}) &= T^{-k/2} \int_{\mathbb{R}^k}' \sideset{}{'}\sum_{1\le i_1,\ldots,i_k\le N} c(i_1, \ldots, i_k) \mathbf{1}_{\Delta_{i_1}\times\ldots\times\Delta_{i_k}}(x_1/T, \ldots, x_k/T) \\ &\quad \times \xi(dx_1)\ldots\xi(dx_k) \\ &= \sideset{}{'}\sum_{1\le i_1,\ldots,i_k\le N} c(i_1, \ldots, i_k) \xi_T(\Delta_{i_1}) \ldots \xi_T(\Delta_{i_k}), \end{aligned} \tag{6.5.45}$$

where

$$\xi_T(\Delta_i) = \frac{1}{\sqrt{T}} \int_{T\Delta_i} \xi(dx).$$

Combining the discretization argument from Proposition 5.3.11 and Theorem 5.2 of Billingsley [39], it can be shown that the random variables $\xi_T(\Delta_i)$ satisfy the central limit theorem. Therefore, for any $N \geq 1$, we have as $T \to \infty$

$$\Big(\xi_T(\Delta_1), \ldots, \xi_T(\Delta_N)\Big) \xrightarrow{d} \Big(B(\Delta_1), \ldots, B(\Delta_N)\Big),$$

where $B(\cdot)$ is the Gaussian random measure that appears in the multiple Wiener-Itô integral (3.4.22).

Finally, by applying the continuous mapping theorem, we obtain from (6.5.45) that

$$I_k^{\xi}(f_{\epsilon,T}) \xrightarrow{d} \sum_{1\leq i_1,\ldots,i_k\leq N}' c(i_1, \ldots, i_k)B(\Delta_{i_1}) \ldots B(\Delta_{i_k}) = I_k^B(f_\epsilon),$$

which implies (6.5.44). Thus, Lemma 6.5.3 is proved. □

From Proposition 3.4.2 and Lemma 6.5.3, we can easily derive the following result, which is the spectral version of Lemma 6.5.3.

Corollary 6.5.1 *Let $\widehat{f}_{j,T}$ be the L^2-Fourier transform of the function $f_{j,T}$. We define*

$$\widehat{h}_{j,T}(x_1, \ldots, x_{k_j}) := T^{-k_j/2}\widehat{f}_{j,T}(x_1/T, \ldots, x_{k_j}/T), \tag{6.5.46}$$

and assume that there exist $\widehat{f}_j \in L^2(\mathbb{R}^{k_j})$ such that as $T \to \infty$,

$$\|\widehat{h}_{j,T} - \widehat{f}_j\|_{L^2(\mathbb{R}^{k_j})} \to 0, \quad j = 1, \ldots, J. \tag{6.5.47}$$

Then, for any Lévy process $\xi(\cdot)$ with $\mathbb{E}[\xi(1)] = 0$ and $\mathbb{E}[\xi^2(1)] = 1$, we have the following joint convergence in distribution:

$$\begin{aligned}&\Big(I_{k_1}^{\xi}(f_{1,T}), \ldots, I_{k_J}^{\xi}(f_{J,T})\Big)\\ &\xrightarrow{d} \Big((2\pi)^{-k_1/2}I_{k_1}^W\left(\widehat{f}_1 A^{\otimes k_1}\right), \ldots, (2\pi)^{-k_J/2}I_{k_J}^W\left(\widehat{f}_J A^{\otimes k_J}\right)\Big),\end{aligned} \tag{6.5.48}$$

with any $A(\cdot)$ satisfying the conditions of Proposition 3.4.2.

The following lemma establishes a change of integration order in cases where Fubini's theorem cannot be directly applied.

Lemma 6.5.4 *Let $a(\cdot)$, $b(\cdot)$, $\widehat{a}(\cdot)$ and $\widehat{b}(\cdot)$ be as defined in Assumptions 6.4.1–6.4.4. Define*

$$g_T(x_1, x_2) := \int_0^T \int_0^T b(u - v)a(u - x_1)a(v - x_2)dudv.$$

Then $g_T(\cdot,\cdot) \in L^2(\mathbb{R}^2)$ *and for the* L^2*-Fourier transform* $\widehat{g}_T$ *of* g_T*, we have*

$$\widehat{g}_T(w_1, w_2) := \int_{\mathbb{R}^2} e^{i(w_1x_1+w_2x_2)} g(x_1, x_2) dx_1 dx_2$$
$$= \frac{1}{2\pi} \widehat{a}(-w_1)\widehat{a}(-w_2) \int_{\mathbb{R}} \frac{e^{iT(w_1+w)} - 1}{i(w_1+w)} \frac{e^{iT(w_1-w)} - 1}{i(w_1-w)} \widehat{b}(w) dw$$

for a.e. $(w_1, w_2) \in \mathbb{R}^2$.

Proof First, by the Cauchy-Schwartz inequality and Assumption 6.4.4, we have

$$\|g_T\|^2_{L^2(\mathbb{R}^2)} \leq \int_{\mathbb{R}^2} dx_1 dx_2 \left(\int_0^T \int_0^T |b(u-v)a(u-x_1)a(v-x_2)| du dv \right)^2$$
$$\leq \|a\|^4_{L^2(\mathbb{R})} \left(\int_0^T \int_0^T |b(u-v)| du dv \right)^2 < \infty. \tag{6.5.49}$$

Let $a_M(x) = a(x)\mathbf{1}_{[-M,M]}(x)$, and let $\widehat{g}_{T,M}(w_1, w_2)$ be the L^2-Fourier transform of $g_{T,M}(x_1, x_2)$ given by

$$g_{T,M}(x_1, x_2) = \int_0^T \int_0^T b(u-v)a_M(u-x_1)a_M(v-x_2) du dv.$$

Since, as $M \to \infty$, $a_M(x)$ and $\widehat{a}_M(w)$ converge in L^2 to $a(x)$ and $\widehat{a}(w)$, respectively, one can find a subsequence $M_n \uparrow \infty$, so that $a_{M_n}(x)$ and $\widehat{a}_{M_n}(w)$ converge a.e. to their limits. Hence, by (6.5.49) and the dominated convergence theorem, we have as $n \to \infty$

$$\|g_{T,M_n} - g_T\|_{L^2(\mathbb{R}^2)} = \int_{\mathbb{R}^2} dx_1 dx_2 \Big(\int_0^{Tt} \int_0^{Tt} b(u-v)$$
$$\times \left[a_{M_n}(u-x_1)a_{M_n}(v-x_2) - a(u-x_1)a(v-x_2) \right] du dv \Big)^2 \to 0.$$

Therefore, one can choose a subsequence of M_n, also denoted by M_n, such that $g_{T,M_n}(x_1, x_2)$ converges to $g_T(x_1, x_2)$ a.e. for $(x_1, x_2) \in \mathbb{R}^2$ as well as in L^2-norm, and $\widehat{g}_{T,M_n}(w_1, w_2)$ converges to $\widehat{g}_T(w_1, w_2)$ a.e. for $(w_1, w_2) \in \mathbb{R}^2$ and in L^2-norm.

Since $b(\cdot)$ is an even function and $\widehat{b}(\cdot) \in L^1(\mathbb{R})$, one has

$$b(x) = \frac{1}{2\pi} \int_{\mathbb{R}} e^{-ixw} \widehat{b}(w) dw = \frac{1}{2\pi} \int_{\mathbb{R}} e^{ixw} \widehat{b}(w) dw.$$

Next, taking into account that $a_{M_n}(\cdot)$ has a finite support and hence is in $L^1(\mathbb{R})$, we can write

$$\int_{\mathbb{R}} a_{M_n}(u-x)e^{iwx}dx = e^{iwu}\int_{\mathbb{R}} a_{M_n}(u-x)e^{i(-w)(u-x)}dx = e^{iwu}\widehat{a}_{M_n}(-w).$$

Then, by Fubini's theorem, we can change the order of integration to obtain

$$\begin{aligned}
&\widehat{g}_{T,M_n}(w_1,w_2)\\
&=\int_{\mathbb{R}^2} e^{i(w_1x_1+w_2x_2)}dx_1dx_2\int_0^{Tt}\int_0^{Tt} b(u-v)a_{M_n}(u-x_1)a_{M_n}(v-x_2)dudv\\
&=\frac{1}{2\pi}\int_{\mathbb{R}} dw\int_0^{Tt}\int_0^{Tt} dudv\, e^{i(u-v)w}\widehat{b}(w)\, e^{iw_1u}\widehat{a}_{M_n}(-w_1)\, e^{iw_2v}\widehat{a}_{M_n}(-w_2)\\
&=\frac{1}{2\pi}\widehat{a}_{M_n}(-w_1)\widehat{a}_{M_n}(-w_2)\int_{\mathbb{R}}\frac{e^{iTt(w_1+w)}-1}{i(w_1+w)}\frac{e^{iTt(w_2-w)}-1}{i(w_2-w)}\widehat{b}(w)dw.
\end{aligned}$$

Finally, as $n\to\infty$, we have a.e. convergence of $\widehat{g}_{T,M_n}$ to $\widehat{g}_T$ and the a.e. convergence of $\widehat{a}_{M_n}$ to $\widehat{a}$ with M_n chosen as described above. Therefore, the result follows, and the proof is complete. □

Lemma 6.5.5 *Let the numbers α^* and β^* be as defined in Assumption 6.4.4. Then,*

$$\begin{aligned}
I := &\int_{[0,1]^4} du_1du_2du_3du_4\int_{\mathbb{R}} dx\Big(|u_1-u_2|^{\beta^*-1}|u_3-u_4|^{\beta^*-1}|u_1-x|^{\alpha^*/2-1}\\
&\times|u_2-x|^{\alpha^*/2-1}|u_3-x|^{\alpha^*/2-1}|u_4-x|^{\alpha^*/2-1}\Big)<\infty. \qquad (6.5.50)
\end{aligned}$$

Proof Setting $\alpha_0=\alpha^*/2-1\in(-1,-1/2)$ and $\beta_0=\beta^*-1\in(-1,0)$, and noting that, by the assumption $\alpha^*+\beta^*>1$, we have

$$2\alpha_0+\beta_0>-2. \qquad (6.5.51)$$

By the change of variables $v_1=u_1-u_2$, $v_2=u_3-u_4$, $v_3=u_1$, $v_4=u_1-x$, $v_5=u_3-x$, and by enlarging the integration region if necessary, we can use the equality

$$\int_{\mathbb{R}}|v|^{\alpha_0}|x-v|^{\alpha_0}dv=c_{\alpha_0}|x|^{2\alpha_0+1}$$

with some constant $c_{\alpha_0} > 0$, to bound the integral in (6.5.50) as follows:

$$\begin{aligned} I \leq & c \int_{[-1,1]^3} dv_1 dv_2 dv_3 \, |v_1|^{\beta_0} |v_2|^{\beta_0} \\ & \times \int_{\mathbb{R}} |v_4|^{\alpha_0} |v_4 - v_1|^{\alpha_0} dv_4 \int_{\mathbb{R}} |v_5|^{\alpha_0} |v_5 - v_2|^{\alpha_0} dv_5 \\ = & c \int_{[-1,1]^3} dv_1 dv_2 dv_3 \, |v_1|^{2\alpha_0+\beta_0+1} |v_2|^{2\alpha_0+\beta_0+1}. \end{aligned} \tag{6.5.52}$$

The last integral in (6.5.52) is finite because, according to (6.5.51), we have $2\alpha_0 + \beta_0 + 1 > -1$. □

The following lemma, which follows from Proposition 2.6.2 and Remark 2.6.1 (see also Corollary 1.1 in Terrin and Taqqu [303]), will be used in the proof of Theorem 6.4.3.

Lemma 6.5.6 *Let $m \geq 2$ be a natural number, and let $\alpha_1, \ldots, \alpha_m$ be real numbers satisfying*

$$\alpha_1, \ldots, \alpha_n > -1, \quad \sum_{i=1}^{m} \alpha_i + m > 1, \tag{6.5.53}$$

then

$$\int_{[0,1]^m} |x_1 - x_2|^{\alpha_1} |x_2 - x_3|^{\alpha_2} \ldots |x_{m-1} - x_m|^{\alpha_{m-1}} |x_m - x_1|^{\alpha_m} dx_1 \ldots dx_m < \infty.$$

6.5.2.2 Proofs of the Theorems

We are now ready to prove the non-central limit theorems stated in Section 6.4.2 (Theorems 6.4.2–6.4.4).

Proof of Theorem 6.4.2 As in the proof of Theorem 6.4.1, we express the process $\widetilde{Q}_T(t)$ as follows:

$$\widetilde{Q}_T(t) = A_T(t) + B_T(t), \tag{6.5.54}$$

where now

$$\begin{aligned} A_T(t) := & \int_{\mathbb{R}^2}' \frac{1}{T^{\alpha+\beta} L_1(1/T) L_2(1/T)} \int_0^{Tt} \int_0^{Tt} b(u-v) a(u-x_1) \\ & \times a(v-x_2) du dv \, \xi(dx_1) \xi(dx_2), \end{aligned} \tag{6.5.55}$$

and

$$B_T(t) := \int_{\mathbb{R}} \frac{1}{T^{\alpha+\beta}L_1(1/T)L_2(1/T)} \int_0^{Tt}\int_0^{Tt} b(u-v)a(u-x) \times a(v-x)dudv\, \xi_c^{(2)}(dx). \tag{6.5.56}$$

In view of relations (6.5.54)–(6.5.56), to prove the theorem, it suffices to show that

$$A_T(t) \xrightarrow{f.d.d.} Z_{\alpha,\beta}(t) \quad \text{as} \quad T \to \infty, \tag{6.5.57}$$

where the limiting process $Z_{\alpha,\beta}(t)$ is defined as in (6.4.14), and

$$\lim_{T\to\infty} \mathbb{E}[B_T^2(t)] = 0. \tag{6.5.58}$$

We begin by proving the relation (6.5.57), which we derive from Corollary 6.5.1. First, we express $A_T(t)$ as a Wiener-Itô integral: $A_T(t) = I_2^{\xi}(f_{T,t})$, where

$$f_{T,t} = \frac{1}{T^{\alpha+\beta}L_1(1/T)L_2(1/T)} \int_0^{Tt}\int_0^{Tt} b(u-v)a(u-x_1)a(v-x_2)dudv.$$

Denoting by $\widehat{f}_{T,t}$ the L^2-Fourier transform of $f_{T,t}$, and using Lemma 6.5.4, we have

$$\widehat{f}_{T,t}(w_1,w_2) = \frac{1}{2\pi T^{\alpha+\beta}L_1(1/T)L_2(1/T)}\widehat{a}(-w_1)\widehat{a}(-w_2) \times \int_{\mathbb{R}} \frac{e^{iTt(w_1+w)}-1}{i(w_1+w)}\frac{e^{iTt(w_1-w)}-1}{i(w_1-w)}\widehat{b}(w)dw.$$

By changing the variables w_1, w_2 and w by x_1, x_2 and u/T, respectively, we obtain

$$\widehat{h}_{T,t}(x_1,x_2) := T^{-1}\widehat{f}_{T,t}(x_1/T, x_2/T) = \frac{1}{2\pi}\frac{\widehat{a}(-x_1/T)}{T^{\alpha/2}L_1(1/T)^{1/2}}\frac{\widehat{a}(-x_2/T)}{T^{\alpha/2}L_1(1/T)^{1/2}} \times \int_{\mathbb{R}} \frac{e^{it(x_1+u)}-1}{i(x_1+u)}\frac{e^{it(x_1-u)}-1}{i(x_1-u)}\frac{\widehat{b}(u/T)}{T^{\beta}L_2(1/T)}du.$$

Next, by Assumptions 6.4.1 and 6.4.2 and the property of slowly varying functions: $\lim_{T\to\infty} L_i(x/T)L_i(1/T) = 1$, $i = 1, 2$, we have

$$\widehat{h}_{T,t}(x_1,x_2) \to \widehat{f}_t(x_1,x_2) := \frac{1}{2\pi}H(x_1)H(x_2)|x_1x_2|^{-\alpha/2} \times \int_{\mathbb{R}} \frac{e^{it(x_1+u)}-1}{i(x_1+u)}\frac{e^{it(x_2-u)}-1}{i(x_2-u)}|u|^{-\beta}du \tag{6.5.59}$$

for a.e. $(x_1, x_2) \in \mathbb{R}^2$, where $H(x) = \overline{A_0}$ if $x \geq 0$ and $H(x) = A_0$ if $x < 0$, with A_0 as defined in Assumption 6.4.1.

By (6.2.20), for sufficiently large T and some constant $C > 0$, we have

$$|\widehat{h}_{T,t}(x_1, x_2)| \leq h_t^*(x_1, x_2) := C|x_1x_2|^{-\alpha/2}(|x_1|^{\epsilon} + |x_1|^{-\epsilon})(|x_2|^{\epsilon} + |x_2|^{-\epsilon})$$
$$\times \int_{\mathbb{R}} \left| \frac{e^{it(x_1+u)} - 1}{i(x_1+u)} \frac{e^{it(x_2-u)} - 1}{i(x_2-u)} \right| |u|^{-\beta}(|u|^{\epsilon} + |u|^{-\epsilon})du. \tag{6.5.60}$$

By Lemma 6.3.9, for sufficiently small ϵ, the bounding function $h_t^*(x_1, x_2)$ in (6.5.60) belongs to $L^2(\mathbb{R}^2)$.

Therefore, by the dominated convergence theorem, we have

$$\lim_{T\to\infty} \|\widehat{h}_{T,t} - \widehat{f}_t\|_{L^2(\mathbb{R}^2)} = 0.$$

Now we can apply Corollary 6.5.1 to obtain (6.5.57). Note that the function $H(x)$ in (6.5.59) can be omitted, as it serves the same purpose as the function $A(\cdot)$ in Proposition 3.4.2. Thus, the proof of (6.5.57) is complete.

Now we turn to the proof of relation (6.5.58). For simplicity, we consider the case $t = 1$ and define

$$D_T(x) := \int_0^T \int_0^T b(u-v)a(u-x)a(v-x)dudv. \tag{6.5.61}$$

Then, by Assumption 6.4.4 and the change of variables $u_i \to Tu_i$ and $v_i \to Tv_i$, we obtain

$$\|D_T\|^2_{L^2(\mathbb{R})} \leq CT^{2\alpha^*+2\beta^*-1}D(\alpha^*, \beta^*), \tag{6.5.62}$$

where

$$D(\alpha^*, \beta^*) := \int_{[0,1]^4} du_1du_2dv_1dv_2 \int_{\mathbb{R}} dx|u_1 - v_1|^{\beta^*-1}|u_1 - x|^{\alpha^*/2-1}$$
$$\times|v_1 - x|^{\alpha^*/2-1}|u_2 - x|^{\alpha^*/2-1}|v_2 - x|^{\alpha^*/2-1}. \tag{6.5.63}$$

Next, by Lemma 6.5.5, the last integral is finite. Since the functions L_1 and L_2 in (6.5.56) are slowly varying, for any $\epsilon > 0$ and sufficiently large T, we have (see Bingham et al. [43], Proposition 1.3.6):

$$L_1(1/T) \geq T^{-\epsilon/4}, \quad L_2(1/T) \geq T^{-\epsilon/4}. \tag{6.5.64}$$

Therefore, in view of (6.5.56) and (6.5.61)–(6.5.64), we obtain

$$\mathbb{E}B_T^2(t) \leq CT^{2(\alpha^*+\beta^*-\alpha-\beta)+\epsilon-1},$$

where $2(\alpha^* + \beta^* - \alpha - \beta) + \epsilon - 1 < 0$ by Assumption 6.4.4, provided that ϵ is chosen small enough. This completes the proof of relation (6.5.58).

Theorem 6.4.2 is proved. □

Proof of Theorem 6.4.3 To prove assertion (a), we begin with the corresponding "time domain" kernel in (6.4.16), specifically:

$$f_t(x_1, x_2) = \int_0^t \int_0^t |u - v|^{\beta-1} (u - x_1)_+^{\alpha/2-1} (v - x_2)_+^{\alpha/2-1} du dv,$$

and observe that $f_t \in L^2(\mathbb{R}^2)$. Indeed, using the equality

$$\int_{\mathbb{R}} (u - x)_+^{\alpha/2-1} (v - x)_+^{\alpha/2-1} dx = C_\alpha |u - v|^{\alpha-1}$$

with $0 < \alpha < 1$ and some constant $C_\alpha > 0$, and Lemma 6.5.6 with $\alpha + \beta > 1/2$, we have

$$\int_{\mathbb{R}^2} f_t(x_1, x_2)^2 dx_1 dx_2 = C_\alpha^2 \int_{[0,t]^4} |u_1 - v_1|^{\beta-1} |u_1 - u_2|^{\alpha-1}$$
$$\times |u_2 - v_2|^{\beta-1} |v_1 - v_2|^{\alpha-1} du_1 dv_1 du_2 dv_2 < \infty.$$

To determine the Fourier transform $\widehat{f_t}$ of the function f_t, we truncate f_t as follows:

$$f_t^A(x_1, x_2) = \int_0^t \int_0^t |u - v|^{\beta-1} (u - x_1)_+^{\alpha/2-1} (v - x_2)_+^{\alpha/2-1}$$
$$\times \mathbf{1}_{\{u-x_1<A, v-x_2<A\}} du dv.$$

Then, by the dominated convergence theorem, as $A \to \infty$, we have

$$f_t^A(x_1, x_2) \to f_t(x_1, x_2) \quad \text{in} \quad L^2(\mathbb{R}^2).$$

Thus, by Parseval-Plancherel's theorem, we have as $A \to \infty$,

$$\widehat{f_t^A}(x_1, x_2) \to \widehat{f_t}(x_1, x_2) \quad \text{in} \quad L^2(\mathbb{R}^2).$$

Therefore, we can let $A \to \infty$ through a suitable subsequence to obtain

$$\widehat{f_t^A}(w_1, w_2) \to \widehat{f_t}(w_1, w_2) \quad a.e. \ (w_1, w_2) \in \mathbb{R}^2. \tag{6.5.65}$$

Next, we determine $\widehat{f_t}$ explicitly. To this end, we apply Fubini's theorem to obtain

$$\begin{aligned}\widehat{f_t^A}(w_1, w_2) :=& \int e^{i(w_1x_1+w_2x_2)} f_t^A(x_1, x_2)dx_1dx_2\\ =& \int_0^t \int_0^t |u-v|^{\beta-1} e^{i(w_1u+w_2v)}dudv\\ &\times \int_0^A e^{-iw_1y_1} y_1^{\alpha/2-1}dy_1 \times \int_0^A e^{-iw_2y_2} y_2^{\alpha/2-1}dy_2. \qquad (6.5.66)\end{aligned}$$

We first deal with the first integral on the right-hand side of (6.5.66), rewriting it in a more convenient form. To do this, note that according to formula 3.761.9 of Jeffrey and Zwillinger [197], we have

$$\begin{aligned}&\lim_{B\to\infty} \int_{-B}^{B} |w|^{-\beta} e^{ixw}dw = \int_{\mathbb{R}} |w|^{-\beta} e^{ixw}dw\\ =&2\int_0^\infty w^{-\beta}\cos(xw)dw = 2|x|^{\beta-1}\Gamma(1-\beta)\sin(\beta\pi/2).\end{aligned}$$

We define

$$M_\beta := \sup_{B>0} \left| \int_{-B}^{B} |w|^{-\beta} e^{iw}dw \right| < \infty$$

and make the change of variable $w' = (u-v)w$ to obtain

$$\int_{-B}^{B} |w|^{-\beta} e^{iw(u-v)}dw \le M_\beta |u-v|^{\beta-1}.$$

Note also that since $\beta > 0$, we have

$$\int_0^t \int_0^t dudv|u-v|^{\beta-1} < \infty.$$

Therefore, by using the dominated convergence theorem, we can write

$$\begin{aligned}&\int_0^t \int_0^t dudv|u-v|^{\beta-1}e^{i(w_1u+w_2v)}\\ =&\frac{1}{2\Gamma(1-\beta)\sin(\beta\pi/2)} \int_0^t\int_0^t dudv \lim_{B\to\infty}\int_{-B}^{B} |w|^{-\beta}e^{iw(u-v)}dw\, e^{i(w_1u+w_2v)}\\ =&\frac{1}{2\Gamma(1-\beta)\sin(\beta\pi/2)} \lim_{B\to\infty}\int_{-B}^{B} \frac{e^{it(w_1+w)}-1}{i(w_1+w)}\frac{e^{it(w_2-w)}-1}{i(w_2-w)}|w|^{-\beta}dw\\ =&\frac{1}{2\Gamma(1-\beta)\sin(\beta\pi/2)} \int_{-\infty}^{\infty} \frac{e^{it(w_1+w)}-1}{i(w_1+w)}\frac{e^{it(w_2-w)}-1}{i(w_2-w)}|w|^{-\beta}dw, \qquad (6.5.67)\end{aligned}$$

because in view of (6.5.60) and the fact that $h_t^*(x_1, x_2) \in L^2(\mathbb{R}^2)$, the last integral converges absolutely.

Next, we focus on the last two integrals in (6.5.66). According to formulas 3.761.4 and 3.761.9 from Jeffrey and Zwillinger [197], we have

$$
\begin{aligned}
\lim_{A\to\infty} \int_0^A e^{-iwy} y_+^{\alpha/2-1} dy &= \int_0^\infty e^{-iwy} y^{\alpha/2-1} dy \\
&= \int_0^\infty [\cos(wy) - i\sin(wy)] y^{\alpha/2-1} dy \\
&= |w|^{-\alpha/2}\Gamma(\alpha/2)\left[\cos(\alpha\pi/4) - i\operatorname{sign}(w)\sin(\alpha\pi/4)\right] \\
&= |w|^{-\alpha/2}\Gamma(\alpha/2)\exp\left[-i\operatorname{sign}(w)\alpha\pi/4\right]. \qquad (6.5.68)
\end{aligned}
$$

Combining (6.5.65)–(6.5.68), we obtain

$$
\begin{aligned}
\widehat{f}_t(w_1, w_2) &= \frac{\Gamma(\alpha/2)^2}{2\Gamma(1-\beta)\sin(\beta\pi/2)} \int_{-\infty}^{\infty} \frac{e^{it(w_1+w)} - 1}{i(w_1+w)} \frac{e^{it(w_2-w)} - 1}{i(w_2-w)} |w|^{-\beta} dw \\
&\quad \times \exp\left[-i(\operatorname{sign}(w_1) + \operatorname{sign}(w_2))\alpha\pi/4\right] |w_1 w_2|^{-\alpha/2}.
\end{aligned}
$$

The proof of assertion (a) of the theorem can be concluded using Proposition 3.4.2 and noting that the factor

$$
\exp\left[-i(\operatorname{sign}(w_1) + \operatorname{sign}(w_2))\alpha\pi/4\right]
$$

in the last formula can be omitted, as it serves the same role as $A^{\otimes 2}(w_1, w_2)$ in Proposition 3.4.2.

To prove assertion (b), we define

$$
\widehat{\mathbf{1}}_{[0,t]}(x) := \int_{\mathbb{R}} \mathbf{1}_{[0,t]}(w) e^{iwx} dw = \frac{e^{itx-1}}{ix}
$$

and use the property of the Fourier transform for convolutions to obtain:

$$
\begin{aligned}
&\int_{\mathbb{R}} \frac{e^{it(x_1+u)} - 1}{i(x_1+u)} \frac{e^{it(x_2-u)} - 1}{i(x_2-u)} du \\
&= \int_{\mathbb{R}} \widehat{\mathbf{1}}_{[0,t]}(x_1+u)\widehat{\mathbf{1}}_{[0,t]}(x_2-u) du = \left(\widehat{\mathbf{1}}_{[0,t]} * \widehat{\mathbf{1}}_{[0,t]}\right)(x_1+x_2) \\
&= \left(\widehat{\mathbf{1}_{[0,t]} \cdot \mathbf{1}_{[0,t]}}\right)(x_1+x_2) = \widehat{\mathbf{1}}_{[0,t]}(x_1+x_2) = \frac{e^{it(x_1+x_2)} - 1}{i(x_1+x_2)}.
\end{aligned}
$$

Thus, based on (6.4.14), the process $Z_{\alpha,0}$ can be expressed as follows:

$$Z_{\alpha,0}(t) = \frac{1}{2\pi} \int_{\mathbb{R}^2}'' |x_1 x_2|^{-\alpha/2} \frac{e^{it(x_1+x_2)} - 1}{i(x_1 + x_2)} W(dx_1)W(dx_2),$$

which is the spectral domain representation of the Rosenblatt process (see formula (6.2.15)). Therefore, the time domain representation stated in (6.4.17) follows from Theorem 4.2.10 of Pipiras and Taqqu [246].

To prove assertion (c) of the theorem, we set

$$f_t(x_1, x_2) := \mathbf{1}_{[0,t]\times[0,t]}(x_1, x_2)|x_1 - x_2|^{\beta-1},$$

where $\beta > 0$, and observe that by (6.5.67),

$$\widehat{f_t}(w_1, w_2) = \frac{1}{2\Gamma(1-\beta)\sin(\beta\pi/2)} \int_{-\infty}^{\infty} \frac{e^{it(w_1+w)} - 1}{i(w_1 + w)} \frac{e^{it(w_2-w)} - 1}{i(w_2 - w)} |w|^{-\beta} dw,$$

which, in view of Proposition 3.4.2, implies (6.4.14).

This concludes the proof of Theorem 6.4.3. □

Proof of Theorem 6.4.4 As in the proof of Theorem 6.4.2, we can write

$$\frac{1}{T^{\alpha+\beta}\ell_1(T)\ell_2(T)} (Q_T - \mathbb{E}[Q_T]) = A_T(t) + B_T(t),$$

where $A_T(t)$ and $B_T(t)$ are defined in (6.5.55) and (6.5.56), respectively, with $L_j(1/T)$ replaced by $\ell_j(T)$, $j = 1, 2$.

Since the condition (6.4.15) implies Assumption 6.4.4 of Theorem 6.4.2, as in the proof of Theorem 6.4.2, we can conclude that

$$\lim_{T\to\infty} \mathbb{E}[B_T^2(t)] = 0,$$

which implies that the term $B_T(t)$ is negligible.

Next, setting

$$f_{t,T}(x_1, x_2) := \frac{1}{T^{\alpha+\beta}\ell_1(T)\ell_2(T)} \int_0^{Tt} \int_0^{Tt} b(u - v)a(u - x_1)a(v - x_2)dudv,$$

we have $A_T(t) = I_2^{\xi}(f_{t,T})$. Therefore, in view of Lemma 6.5.3, we can write

$$\begin{aligned} h_{t,T}(x_1, x_2) &:= T f_{t,T}(Tx_1, Tx_2) \\ &= \frac{1}{T^{\alpha+\beta-1}\ell_1(T)\ell_2(T)} \int_0^{Tt} \int_0^{Tt} b(u - v)a(u - Tx_1)a(v - Tx_2)dudv \end{aligned}$$

$$= \int_0^t \int_0^t |u-v|^{\beta-1}(u-x_1)_+^{\alpha/2-1}(v-x_2)_+^{\alpha/2-1}$$
$$\times \frac{\ell_1(T(u-x_1))}{\ell_1(T)} \frac{\ell_1(T(v-x_2))}{\ell_1(T)} \frac{\ell_2(T(u-v))}{\ell_2(T)} du dv, \qquad (6.5.69)$$

where we have applied the change of variables $u \to uT$ and $v \to vT$.

To complete the proof of the theorem, in view of Lemma 6.5.3, it is sufficient to show that for every $t > 0$, as $T \to \infty$

$$h_{t,T}(x_1, x_2) \to f_t(x_1, x_2) \quad \text{in } L^2(\mathbb{R}^2), \qquad (6.5.70)$$

where

$$f_t(x_1, x_2) := \int_0^t \int_0^t |u-v|^{\beta-1}(u-x_1)_+^{\alpha/2-1}(v-x_2)_+^{\alpha/2-1} du dv. \qquad (6.5.71)$$

By the property of slowly varying functions, we have as $T \to \infty$

$$\frac{\ell_1(T(u-x_1))}{\ell_1(T)}, \frac{\ell_1(T(v-x_2))}{\ell_1(T)}, \frac{\ell_2(T(u-v))}{\ell_2(T)} \to 1$$

for $u > x_1$, $v > x_2$ and $u \neq v$ (recall that $\ell_2(\cdot)$ is an even function).

Next, by Potter's bound (6.2.20) (see also Proposition 2.5.2), for any $\epsilon > 0$, we have

$$\mathbf{1}_{\{u>x_1, v>x_2\}} \left| \frac{\ell_1(T(u-x_1))}{\ell_1(T)} \frac{\ell_1(T(v-x_2))}{\ell_1(T)} \frac{\ell_2(T(u-v))}{\ell_2(T)} \right|$$
$$\leq C R_\epsilon(x_1, x_2, u, v),$$

where

$$R_\epsilon(x_1, x_2, u, v) = [(u-x_1)_+^\epsilon + (u-x_1)_+^{-\epsilon}][(v-x_2)_+^\epsilon + (v-x_2)_+^{-\epsilon}]$$
$$\times [|u-v|^\epsilon + |u-v|^{-\epsilon}].$$

Thus, the function $h_{t,T}(x_1, x_2)$ in (6.5.69) is bounded by the following function:

$$f_{t,\epsilon}(x_1, x_2) = C \int_0^t \int_0^t |u-v|^{\beta-1}(u-x_1)_+^{\alpha/2-1}(v-x_2)_+^{\alpha/2-1}$$
$$\times R_\epsilon(x_1, x_2, u, v) du dv.$$

By selecting a sufficiently small $\epsilon > 0$, as in the proof of Theorem 6.4.3 (a), we can use Lemma 6.5.5 to show that $f_{t,\epsilon}(x_1, x_2) \in L^2(\mathbb{R}^2)$. Subsequently, we can apply the dominated convergence theorem to obtain the relation in (6.5.70).

This concludes the proof of Theorem 6.4.4. □

6.6 Bibliographical Notes and Comments

Section 6.1 For the results presented in this section concerning weak convergence and tightness, we refer to the seminal paper by Prohorov [250] (see also Ibragimov [180]). A comprehensive discussion of weak convergence in various function spaces and tightness can be found in the monograph by Billingsley [41] (see also Bass [23]). Here, we provide an overview of the essential concepts needed for our purposes.

Section 6.2 Theorem 6.2.1 was obtained by Giraitis and Taqqu [153] (see also Fox and Taqqu [100] and Giraitis and Taqqu [149]). Theorems 6.2.2–6.2.4 and 6.2.6 were proved in the paper by Bai et al. [21]. The method used in the proofs of Theorems 6.2.2–6.2.4 was developed by Ginovyan [120], and Ginovyan and Sahakyan [136], which is based on approximations of the traces of products of truncated Toeplitz operators (see Sects. 4.5 and 5.3.1). Theorem 6.2.5 was proved by Terrin and Taqqu [301] using power counting theorems (see Sect. 2.6).

Comparing Theorems 6.2.5 and 6.2.6, we observe that the limiting process $Z(t)$ is the same for both continuous and discrete time models. It is important to note that while the statements of Theorems 6.2.5 and 6.2.6 are similar, the proofs differ. The proof of Theorem 6.2.6 presented here is simpler and does not utilize the complex analysis employed by Terrin and Taqqu [301] in their proof of Theorem 6.2.5.

Section 6.3 This section contains the proofs of the theorems presented in Sect. 6.2. For the preliminary results stated in Sect. 6.3.1, we refer to the following papers: Bai et al. [21], Ginovyan [120], Ginovyan and Sahakyan [136], Giraitis and Taqqu [150], Peccati and Tudor [241], and Terrin and Taqqu [301].

Section 6.4 The functional central limit theorems—Theorem 6.4.1 and Corollary 6.4.1—were proved in the paper by Bai et al. [22]. Theorem 6.4.1 for the Wiener-driven model ($\kappa_4 = 0$) was established by Bai et al. [21].

The discrete time analog of Theorem 6.4.1 with $\tau = 1$ and ξ being Gaussian was established by Giraitis and Surgailis [148]. A special case of Theorem 6.4.1 with $\tau = 1$ and ξ being Gaussian was established by Ginovyan [120] and by Ginovyan and Sahakyan [136].

For Lévy-driven model with $\tau = 1$ and σ^2 given by (6.4.10), a version of Theorem 6.4.1 was stated in Avram et al. [15] (Theorem 4.1). They impose (L^p, L^q)-type conditions on the spectral density f and the generating function g and assume the existence of all moments of the driving Lévy process $\xi(t)$. Unfortunately, the details of the proof of Theorem 4.1 in Avram et al. [15] are omitted, leaving it unclear how the missing details of the method-of-moments proof can be executed, given the complexity of computing the moments of multiple integrals with respect to non-Gaussian Lévy noise (see Peccati and Taqqu [240], Chapter 7).

For a related study of the sample covariances of Lévy-driven moving average processes, we refer to the papers by Cohen and Lindner [65] and Spangenberg [286].

The functional non-central limit theorems—Theorems 6.4.2–6.4.4—were proved in Bai et al. [22].

Section 6.5 This section contains the proofs of theorems from Sect. 6.4. Lemma 6.5.3, which provides conditions for Lévy-Itô multiple stochastic integrals to converge in distribution to Wiener-Itô multiple stochastic integrals, plays a key role in the proofs of the non-central limit theorems (Theorems 6.4.2–6.4.4). It serves as a continuous analog of Propositions 14.3.2 and 14.3.3 from Giraitis et al. [147].

Non-central limit theorems for non-linear functions of Gaussian fields were established in Dobrushin and Major [84].

Chapter 7
Central Limit Theorems for Tapered Toeplitz Quadratic Functionals

In this chapter, we study the asymptotic distributions of tapered random Toeplitz type quadratic functionals and processes. We prove central limit theorems for these functionals in both discrete and continuous time cases, as well as a functional central limit theorem for a suitably normalized stochastic process generated by a tapered Toeplitz type quadratic functional of a Lévy-driven linear process. Additionally, we establish the moderate deviation principle (MDP) for tapered quadratic functionals.

To study the asymptotic distribution (as $T \to \infty$) of the tapered functional Q_T^h, given by (7.1.1), we employ the method of cumulants, the frequency domain approach, and the technique of truncated tapered Toeplitz matrices and operators developed in Sect. 4.8.

In Sect. 7.1, we introduce the tapered random Toeplitz quadratic functionals Q_T^h and outline the problem addressed in this chapter. In Sect. 7.2, we present central limit theorems for the tapered functionals Q_T^h, covering both the Gaussian model (Theorems 7.2.1 and 7.2.2) and the Lévy-driven linear model (Theorem 7.2.3). Section 7.3 contains the proofs of Theorems 7.2.1 and 7.2.2. In Sect. 7.4, we prove a functional central limit theorem (Theorem 7.4.1) for the tapered Toeplitz process $Q_T^h(t)$, $t \in [0, 1]$. From this theorem, we derive the proof of Theorem 7.2.3. In Sect. 7.5, we establish the moderate deviation principle (MDP) for tapered quadratic functionals.

7.1 The Problem

In this section, we consider a question regarding the asymptotic distribution (as $T \to \infty$) of the following *tapered* Toeplitz type quadratic functionals of a centered stationary process $X(u)$, $u \in \mathbb{U}$, with the spectral density $f(\lambda)$, $\lambda \in \Lambda$:

M. S. Ginovyan, *Random Toeplitz Functionals and Their Applications*,
Frontiers in Probability and the Statistical Sciences,
https://doi.org/10.1007/978-3-032-00961-6_7

$$
Q_T^h := \begin{cases} \sum_{t=1}^{T}\sum_{s=1}^{T}\widehat{g}(t-s)h_T(t)h_T(s)X(t)X(s) \text{ in the d.t. case,} \\ \\ \int_0^T\int_0^T \widehat{g}(t-s)h_T(t)h_T(s)X(t)X(s)\,dt\,ds \text{ in the c.t. case,} \end{cases} \tag{7.1.1}
$$

where $\widehat{g}(t)$ is the Fourier transform of an integrable even function $g(\lambda)$ and $h_T(t) := h(t/T)$ with a taper function $h(\cdot)$ satisfying Assumption 4.8.1.

The limit distributions of the functionals in (7.1.1) are entirely determined by the functions f, g, and h. Depending on their properties, these distributions can be either Gaussian (meaning that Q_T^h, with the appropriate normalization, satisfies the central limit theorem) or non-Gaussian.

We discuss the case where the limiting distributions are Gaussian and present sufficient conditions in terms of the functions f, g, and h that ensure central limit theorems for the standard normalized tapered quadratic functional Q_T^h.

Quadratic functionals of the form (7.1.1) arise in statistical inference problems, including both nonparametric and parametric estimation, as well as hypothesis testing related to the spectrum of the process $X(t)$ based on tapered data (for details, see Chaps. 8–10). For example, when we are interested in the nonparametric estimation of a linear integral functional in $L^p(\Lambda)$, $p > 1$, of the form:

$$
J = J(f) := \int_\Lambda f(\lambda)g(\lambda)d\lambda,
$$

where $g(\lambda) \in L^q(\Lambda)$ and $1/p + 1/q = 1$. A natural statistical estimator for $J(f)$, based on tapered data, is the linear integral functional of the tapered *empirical spectral density (periodogram)* I_T^h of the process $X(t)$:

$$
J_T^h = J(I_T^h) := \int_\Lambda I_T^h(\lambda)g(\lambda)d\lambda,
$$

which can be expressed in the following form:

$$
J_T^h = C_T^{-1}Q_T^h, \tag{7.1.2}
$$

where Q_T^h is defined as in (7.1.1) and $C_T := 2\pi H_{2,T}(0)$ with $H_{2,T}(\cdot)$ as in (4.8.3).

Thus, in order to study the asymptotic properties of the estimator J_T^h, we need to examine the asymptotic distribution (as $T \to \infty$) of the tapered Toeplitz type quadratic functional Q_T^h, as defined in (7.1.1).

7.2 Central Limit Theorems for Tapered Quadratic Functional Q_T^h

Let Q_T^h be the tapered quadratic functional defined by (7.1.1). We will use the following notation. By $\widetilde{Q}_T^h$, we denote the standard normalized quadratic functional:

$$\widetilde{Q}_T^h = T^{-1/2}\left(Q_T^h - \mathbb{E}[Q_T^h]\right). \tag{7.2.1}$$

Also, we set

$$\sigma_h^2 := 16\pi^3 H_4 \int_\Lambda f^2(\lambda) g^2(\lambda)\, d\lambda, \tag{7.2.2}$$

where H_4 is defined in Eq. (4.8.5). As before, the notation $\widetilde{Q}_T^h \xrightarrow{d} \eta \sim N(0, \sigma_h^2)$ as $T \to \infty$ indicates that the distribution of the random variable $\widetilde{Q}_T^h$ converges to a centered normal distribution with variance σ_h^2 as $T \to \infty$.

7.2.1 *CLT for Gaussian Models*

Theorems that follow extend the results of Theorems 5.2.1 and 5.2.2 to the tapered case. We assume that the model process $X(t)$ is Gaussian and, without loss of generality, that the generating function g is non-negative.

Let $A_T^h(f)$ denote either the $T \times T$ tapered Toeplitz matrix $B_T^h(f)$ or the T-truncated tapered Toeplitz operator $W_T^h(f)$, generated by the spectral density f and the taper function h, as defined by Eqs. (4.8.7) and (4.8.8), respectively. Similarly, let $A_T^h(g)$ denote either the $T \times T$ tapered Toeplitz matrix or the T-truncated tapered Toeplitz operator generated by the functions g and h. We assume that the taper function h satisfies Assumption 4.8.1.

Theorem 7.2.1 *Let f, g, $A_T^h(f)$, $A_T^h(g)$, and $\widetilde{Q}_T^h$ be defined as above. Each of the following conditions is sufficient for the functional Q_T^h to satisfy the CLT:*

$$\widetilde{Q}_T^h \xrightarrow{d} \eta \sim N(0, \sigma_h^2) \quad \text{as} \quad T \to \infty, \tag{7.2.3}$$

where σ_h^2 is given by (7.2.2).

(A) $f \cdot g \in L^1(\Lambda) \cap L^2(\Lambda)$ *and*

$$\chi_2(\widetilde{Q}_T^h) = \frac{2}{T} tr\left[A_T^h(f) A_T^h(g)\right]^2 \longrightarrow \sigma_h^2 \quad \text{as} \quad T \to \infty. \tag{7.2.4}$$

(B) *The function*

$$\varphi(\mathbf{u}) := \varphi(u_1, u_2, u_3) = \int_\Lambda f(\lambda)g(\lambda - u_1)f(\lambda - u_2)g(\lambda - u_3)\, d\lambda \qquad (7.2.5)$$

belongs to $L^2(\Lambda^3)$ *and is continuous at the point* $\mathbf{0} = (0, 0, 0)$.

(C) $f \in L^1(\Lambda)\cap L^p(\Lambda)(p \geq 2)$ *and* $g \in L^1(\Lambda)\cap L^q(\Lambda)(q \geq 2)$ *with* $1/p+1/q \leq 1/2$.

(D) $f \in L^1(\Lambda) \cap L^2(\Lambda),\ g \in L^1(\Lambda) \cap L^2(\Lambda),\ fg \in L^2(\Lambda)$, *and*

$$\int_\Lambda f^2(\lambda)g^2(\lambda - \mu)\, d\lambda \longrightarrow \int_\Lambda f^2(\lambda)g^2(\lambda)\, d\lambda \quad \text{as} \quad \mu \to 0.$$

To present the next theorem, we recall the class $SV_0(\mathbb{R})$ of slowly varying at zero functions (see Definition 2.5.1).

Theorem 7.2.2 *Assume that the functions* f *and* g *are integrable on* $\mathbb{R}$, *bounded outside any neighborhood of the origin, and satisfy the following conditions for some* $a > 0$:

$$f(\lambda) \leq |\lambda|^{-\alpha}L_1(\lambda), \quad |g(\lambda)| \leq |\lambda|^{-\beta}L_2(\lambda), \quad \lambda \in [-a, a], \qquad (7.2.6)$$

for some $\alpha < 1$, $\beta < 1$ *with* $\alpha + \beta \leq 1/2$, *where* $L_1(x)$ *and* $L_2(x)$ *are slowly varying at zero functions satisfying*

$$L_i \in SV_0(\mathbb{R}),\ \ \lambda^{-(\alpha+\beta)}L_i(\lambda) \in L^2[-a, a],\ i = 1, 2. \qquad (7.2.7)$$

Then Q_T^h *satisfies the CLT, meaning that the relation (7.2.3) holds, with* σ_h^2 *given by Eq. (7.2.2).*

7.2.2 CLT for Lévy-Driven Stationary Linear Models

Now, we assume that the underlying model $X(t)$ is a Lévy-driven stationary linear process defined by (3.2.10), where $a(\cdot)$ is a filter from $L^2(\mathbb{R})$, and $\xi(t)$ is a Lévy process satisfying the following conditions: $\mathbb{E}[\xi(t)] = 0$, $\mathbb{E}[\xi^2(1)] = 1$, and $\mathbb{E}[\xi^4(1)] < \infty$.

Theorem 7.2.3 *Assume that the filter* $a(\cdot)$ *and the generating kernel* $b(\cdot) := \widehat{g}(\cdot)$ *are such that*

$$a(\cdot) \in L^p(\mathbb{R}) \cap L^2(\mathbb{R}), \quad b(\cdot) \in L^q(\mathbb{R}) \qquad (7.2.8)$$

with

$$1 \le p, q \le 2, \quad \frac{2}{p} + \frac{1}{q} \ge \frac{5}{2}. \tag{7.2.9}$$

Then Q_T^h *satisfies the CLT:*

$$\widetilde{Q}_T^h \xrightarrow{d} \eta \sim N(0, \sigma_{L,h}^2) \quad \text{as} \quad T \to \infty, \tag{7.2.10}$$

where

$$\sigma_{L,h}^2 = 16\pi^3 H_4 \int_{\mathbb{R}} f^2(\lambda) g^2(\lambda) d\lambda + \kappa_4 4\pi^2 H_4 \left[\int_{\mathbb{R}} f(\lambda) g(\lambda) d\lambda \right]^2. \tag{7.2.11}$$

Here, H_4 *is defined as in* (4.8.5)*, and* κ_4 *is the fourth cumulant of the random variable* $\xi(1)$.

Remark 7.2.1 Notice that if the underlying process $X(t)$ is Gaussian, then in formula (7.2.11), we only have the first term. Therefore, $\sigma_{L,h}^2 = \sigma_h^2$ (see 7.2.2), since in this case $\kappa_4 = 0$. On the other hand, the condition (7.2.8) (along with (7.2.9)) is more restrictive than the conditions in Theorems 7.2.1 and 7.2.2. Thus, for Gaussian processes, Theorems 7.2.1 and 7.2.2 improve the result presented in Theorem 7.2.3.

7.3 Proofs of Theorems 7.2.1 and 7.2.2

In this section, we prove Theorems 7.2.1 and 7.2.2. The proof of Theorem 7.2.3 is presented in Sect. 7.4.

Proof of Theorem 7.2.1 Observe first that assertion **(B)** of the theorem follows from assertion **(A)** and Theorem 4.8.1 **(C1)**, assertion **(C)** follows from assertion **(A)** and Theorem 4.8.1 **(C2)**, and assertion **(D)** follows from assertion **(A)** and Theorem 4.8.2 **(C3)**.

Thus, we only need to prove assertion **(A)** of the theorem. For simplicity, we provide the proof in the continuous time case. The proof is similar to that of the non-tapered Theorem 5.2.1 **(A)** and follows from Lemmas 7.3.1–7.3.3 stated below, which are the tapered versions of Lemmas 5.4.1–5.4.3, respectively. □

The following lemma, a tapered version of Lemma 5.4.1, establishes a connection between the distribution of the tapered quadratic functional Q_T^h, as defined in (7.1.1), and the trace problem for tapered Toeplitz matrices and operators, which is discussed in Sect. 4.8.

Lemma 7.3.1 *The following assertions hold:*

(a) The quadratic functional Q_T^h in (7.1.1) has the same distribution as the sum $\sum_{k=1}^{\infty} \lambda_{k,T}\xi_k^2$, where $\{\xi_k, k \ge 1\}$ are independent $N(0,1)$ Gaussian random variables and $\{\lambda_{k,T} := \lambda_{k,T,h}, k \ge 1\}$ are the eigenvalues of the operator $A_T^h(f)A_T^h(g)$.

(b) The characteristic function $\varphi_T^h(t)$ of Q_T^h is given by

$$\varphi_T^h(t) = \prod_{k=1}^{\infty} |1 - 2it\lambda_{k,T}|^{-1/2}. \tag{7.3.1}$$

(c) The k-th order cumulant $\chi_k(Q_T^h)$ of Q_T^h is given by

$$\begin{aligned}\chi_k(Q_T^h) &= 2^{k-1}(k-1)! \sum_{j=1}^{\infty} \lambda_{j,T}^k \\ &= 2^{k-1}(k-1)!\, tr\, [A_T^h(f)A_T^h(g)]^k,\end{aligned} \tag{7.3.2}$$

where tr[A] stands for the trace of an operator A.

The product in (7.3.1) *and the sum in* (7.3.2) *are taken over $j = 1, \ldots, T$ in the discrete time case.*

Proof Follows from Proposition 5.3.2 with $R = A_T^h(f)$ and $A = A_T^h(g)$. □

The next lemma is a tapered counterpart of Lemma 5.4.2.

Lemma 7.3.2 *Let $\{Y(t), t \in \mathbb{R}\}$ be a real-valued, centered, separable stationary Gaussian process with spectral density $f_Y(\lambda) \in L^1(\mathbb{R}) \cap L^2(\mathbb{R})$, and let $h_T(t) = h(t/T)$ with a taper function h that satisfies Assumption 4.8.1. Define*

$$L_T^h := \int_0^T [h_T(t)Y(t)]^2\, dt. \tag{7.3.3}$$

Then, the distribution of the normalized quadratic functional

$$\widetilde{L}_T^h := T^{-1/2}\left(L_T^h - \mathbb{E}[L_T^h]\right) \tag{7.3.4}$$

converges (as $T \to \infty$) to the normal distribution $N(0, \sigma_Y^2)$ with variance

$$\sigma_Y^2 = 4\pi H_4 \int_{\mathbb{R}} f_Y^2(\lambda)\, d\lambda, \tag{7.3.5}$$

where H_4 is defined in (4.8.5).

Proof Let $R(t)$ be the covariance function of the process $Y(t)$. Let $\lambda_j = \lambda_{j,h}(T)$, $j \in \mathbb{N}$, be the eigenvalues of the operator $W_T^h(f_Y)$ (see 4.8.8), and let $e_j(t) =$

$e_j(t, T) \in L_2[0.T]$, $j \in \mathbb{N}$, be the corresponding orthonormal eigenfunctions. Specifically,

$$\int_0^T K(t-s)e_j(s)\,ds = \lambda_j e_j(t), \quad j \in \mathbb{N}, \tag{7.3.6}$$

where $K(t-s) := R(t-s)h_T(t)h_T(s)$. Since by Mercer's theorem (Proposition 2.7.9)

$$K(t-s) = \sum_{j=1}^{\infty} \lambda_j e_j(t) e_j(s) \tag{7.3.7}$$

with positive and summable eigenvalues $\{\lambda_j\}$, we have the Karhunen-Loéve expansion (see Proposition 3.5.1):

$$h_T(t)Y(t) = \sum_{j=1}^{\infty} \sqrt{\lambda_j}\xi_j e_j(t), \tag{7.3.8}$$

where $\{\xi_j\}$ are independent $N(0, 1)$ random variables. Therefore, by (7.3.3), (7.3.8), and Lemma 7.3.1 (a), we have

$$\widetilde{L}_T^h = T^{-1/2} \sum_{j=1}^{\infty} \lambda_j (\xi_j^2 - 1). \tag{7.3.9}$$

Denote by $\chi_k(\widetilde{L}_T^h)$ the k-th order cumulant of $\widetilde{L}_T^h$. Then, by Eq. (7.3.9) and Lemma 7.3.1 (c), we have

$$\chi_k(\widetilde{L}_T^h) = \begin{cases} 0, & \text{for } k = 1, \\ (k-1)!\,2^{k-1}T^{-k/2}\text{tr}[W_T^h(f_Y)]^k, & \text{for } k \geq 2. \end{cases} \tag{7.3.10}$$

By (7.3.10) and Theorem 4.9.4 (c), we have

$$\chi_2(\widetilde{L}_T^h) = \frac{2}{T}||W_T^h(f_Y)||_2^2 \longrightarrow 4\pi H_4 \int_{\mathbb{R}} f_Y^2(\lambda)\,d\lambda \quad \text{as} \quad T \to \infty. \tag{7.3.11}$$

Next, by (7.3.10) for $k \geq 3$, we have

$$|\chi_k(\widetilde{L}_T^h)| \leq C\,\frac{1}{T}||W_T^h(f_Y)||_2^2\, T^{1-k/2}\lambda_T^{k-2}, \tag{7.3.12}$$

where $\lambda_T = ||W_T^h(f_Y)||_\infty$. By Theorem 4.9.4 (b) and (c), the right-hand side of (7.3.12) tends to zero as $T \to \infty$. Lemma 7.3.2 is proved. □

Lemma 7.3.3 *Under the assumptions of Theorem 7.2.1* **(A)**, *we have*

$$\operatorname{Var}(Q_T^h - L_T^h) = o(T) \quad \text{as} \quad T \to \infty, \tag{7.3.13}$$

where Q_T^h and L_T^h are defined as in Eqs. (7.1.1) and (7.3.3), respectively.

Proof Using Eqs. (5.4.19) and (7.1.1), we express the functional Q_T^h as follows:

$$Q_T^h = \int_{\mathbb{R}^2} \left[\int_0^T \int_0^T \widehat{g}(t-s)\widehat{a}(t-u_1)\widehat{a}(s-u_2)h_T(t)h_T(s)\,dt\,ds \right] d\xi(u_1)\,d\xi(u_2). \tag{7.3.14}$$

Similarly, by (5.4.23) and (7.3.3), we have

$$L_T^h = \int_{\mathbb{R}^2} \left[\int_0^T \widehat{a}_1(t-u_1)\widehat{a}_1(t-u_2)h_T^2(t)\,dt \right] d\xi(u_1)\,d\xi(u_2). \tag{7.3.15}$$

Define

$$d_{1Th}(u_1,u_2) := \int_0^T \int_0^T \widehat{g}(t-s)\widehat{a}(t-u_1)\widehat{a}(s-u_2)h_T(t)h_T(s)\,dt\,ds \tag{7.3.16}$$

and

$$\begin{aligned} d_{2Th}(u_1,u_2) &:= \int_0^T \int_0^T \widehat{a}_1(t-u_1)\widehat{a}_1(s-u_2)h_T^2(t)\,dt\,ds \\ &= \int_0^T \widehat{a}_1(t-u_1)\widehat{a}_1(t-u_2)h_T^2(t)\,dt. \end{aligned} \tag{7.3.17}$$

Then, from Eqs. (7.3.14)–(7.3.17), we obtain

$$Q_T^h - L_T^h = \int_{\mathbb{R}^2} \left[d_{1Th}(u_1,u_2) - d_{2Th}(u_1,u_2) \right] d\xi(u_1)\,d\xi(u_2). \tag{7.3.18}$$

Since the underlying process $X(t)$ is Gaussian, we have

$$\operatorname{Var}(Q_T^h - L_T^h) = 2\int_{\mathbb{R}^2} \left[d_{1Th}(u_1,u_2) - d_{2Th}(u_1,u_2) \right]^2 du_1\,du_2. \tag{7.3.19}$$

Let the functions $p_1 := p(\lambda_1, \lambda_2, s)$ and $p_2 := p(\lambda_1, \lambda_2, s)$ be defined as in Eqs. (5.4.33) and (5.4.34), respectively. By applying Parseval-Plancherel's identity, we obtain

$$\begin{aligned}
&\int_{\mathbb{R}^2} d_{iTh}^2(u_1, u_2)\, du_1\, du_2 \\
&= (2\pi)^2 \int_{\mathbb{R}^2} \left| \int_{\mathbb{R}} H_{1,T}(\lambda_1 - s) H_{1,T}(s - \lambda_2) p_i(\lambda_1, \lambda_2, s)\, ds \right|^2 d\lambda_1\, d\lambda_2 \\
&= (2\pi)^2 T ||p_i||_T^2, \quad i = 1, 2, \qquad (7.3.20)
\end{aligned}$$

where $H_{1,T}(u)$ is given by (4.8.3), $||p||_T^2 = (p, p)_T$,

$$(p, p')_T = \int_{\mathbb{R}^4} p(\lambda_1, \lambda_2, \lambda_3) \overline{p'(\lambda_1, \lambda_2, \lambda_4)}\, d\mu_T^h, \qquad (7.3.21)$$

and the measure μ_T^h is defined by Eq. (4.8.50).

In view of Eqs. (7.3.19) and (7.3.20), we have

$$\operatorname{Var}(Q_T - L_T) = 8\pi^2 T \|p_1 - p_2\|_T^2. \qquad (7.3.22)$$

Let the functions $p_i^K(\mathbf{u})(i = 1, 2)$, where $K > 0$ and $\mathbf{u} = (u_1, u_2, u_3) \in \mathbb{R}^3$, be defined as in Eq. (5.4.39). Then, we have (cf. formula (5.4.40))

$$\|p_1 - p_2\|_T^2 \leq 3 \left(\|p_1^K - p_2^K\|_T^2 + \|p_1 - p_1^K\|_T^2 + \|p_2 - p_2^K\|_T^2 \right). \qquad (7.3.23)$$

By reiterating the arguments from the proof of Lemma 5.4.3 and utilizing Lemma 4.8.9 (instead of Lemma 4.6.11), it can be shown that the following relations hold:

$$\lim_{T \to \infty} \|p_1^K - p_2^K\|_T = 0, \qquad (7.3.24)$$

$$\lim_{T \to \infty} \|p_1 - p_1^K\|_T = 0, \qquad (7.3.25)$$

$$\lim_{T \to \infty} \|p_2 - p_2^K\|_T = 0. \qquad (7.3.26)$$

A combination of Eqs. (7.3.22)–(7.3.26) yields (7.3.13). This completes the proof of Lemma 7.3.3. □

Proof of Theorem 7.2.2 The result follows from Theorem 7.2.1 **(A)** and Theorem 4.8.2 **(C4)**. □

7.4 A Functional Central Limit Theorem for Tapered Toeplitz Process $Q_T^h(t)$, $t \in [0, 1]$

In this section, we examine the asymptotic distribution of a suitably normalized stochastic process $\{Q_T^h(t),\ t \in [0,1]\}$, which is generated by a tapered Toeplitz type quadratic functional of a Lévy-driven linear process $X(t)$. We show that under the conditions of Theorem 7.2.3, the process $Q_T^h(t)$ satisfies the functional central limit theorem. Specifically, the finite-dimensional distributions of the standard $\sqrt{T}$—normalized process $Q_T^h(t)$ converge to those of a normalized standard Brownian motion.

Let Q_T^h be the tapered quadratic functional defined by (7.1.1). We are interested in the asymptotic distribution (as $T \to \infty$) of the stochastic process $\{Q_T^h(t),\ t \in [0,1]\}$, generated by the functional Q_T^h:

$$Q_T^h(t) := \int_0^{Tt} \int_0^{Tt} h_T(u)h_T(v)X(u)X(v)dudv, \quad t \in [0,1]. \tag{7.4.1}$$

The next theorem, a tapered version of Theorem 6.4.1, provides sufficient conditions for the process $\{Q_T^h(t),\ t \in [0,1]\}$ to satisfy the functional central limit theorem.

Theorem 7.4.1 *Assume that the filter $a(\cdot)$ and the generating kernel $b(\cdot) := \widehat{g}(\cdot)$ satisfy the assumptions of Theorem 7.2.3. Then, the process $Q_T^h(t)$ defined in (7.4.1) obeys the functional central limit theorem. More precisely, we have*

$$\widetilde{Q}_T^h(t) := T^{-1/2}\left(Q_T^h(t) - \mathbb{E}[Q_T^h(t)]\right) \overset{f.d.d.}{\longrightarrow} \sigma_h(Q)B(t) \quad \text{as} \quad T \to \infty, \tag{7.4.2}$$

where the symbol $\overset{f.d.d.}{\longrightarrow}$ stands for convergence of finite-dimensional distributions, $B(t)$ is a standard Brownian motion, and

$$\sigma_h^2(Q) = H_4 \int_{\mathbb{R}} [2K_A(v) + \kappa_4 K_B(v)]\, dv. \tag{7.4.3}$$

Here, H_4 is defined as in (4.8.5), κ_4 is the fourth cumulant of $\xi(1)$, and $K_A(v)$ and $K_B(v)$ are defined as in (6.4.8).

Similar to the non-tapered case (see the proof of Theorem 6.4.1), to prove Theorem 7.4.1, we first establish two lemmas: the tapered versions of Lemmas 6.5.1 and 6.5.2. We define

$$R_T^h(x_1, x_2) := \frac{1}{\sqrt{T}} \int_0^T \int_0^T h_T(u)h_T(v)b(u-v)a(u-x_1)a(v-x_2)dudv, \tag{7.4.4}$$

and

$$S_T^h(x_1, x_2) := \frac{1}{\sqrt{T}} \int_0^T |h_T(v)|^2 \, [(a * b)(v - x_1)] \, [a(v - x_2)] \, dv. \tag{7.4.5}$$

The next lemma is a tapered version of Lemma 6.5.1.

Lemma 7.4.1 *Let the functions $R_T^h(x_1, x_2)$ and $S_T^h(x_1, x_2)$ be defined as in (7.4.4) and (7.4.5), respectively, with $x_1 \neq x_2$. Then, under the conditions of Theorem 7.4.1, the following assertions hold:*

(a) We have

$$\lim_{T\to\infty} \|S_T^h\|^2_{L^2(\mathbb{R}^2)} = H_4 \int_{\mathbb{R}} K_A(u)du, \tag{7.4.6}$$

where $K_A(\cdot)$ is defined as in (6.4.8).

(b) We have

$$\lim_{T\to\infty} \|R_T^h - S_T^h\|_{L^2(\mathbb{R}^2)} = 0. \tag{7.4.7}$$

(c) For any $M > 0$, there exists a function $c_M^h(\cdot, \cdot)$ supported on $[-2M, 2M]^2$ such that the function

$$S_{T,M}^h(x_1, x_2) = \frac{1}{\sqrt{T}} \int_0^T c_M^h(v - x_1, v - x_2)dv$$

satisfies the relation:

$$\lim_{M\to\infty} \limsup_{T\to\infty} \|R_T^h - S_{T,M}^h\|_{L^2(\mathbb{R}^2)} = 0. \tag{7.4.8}$$

Proof In view of the Riesz-Thorin theorem (Proposition 2.3.2), we can assume, without loss of generality, that

$$a(\cdot) \in L^p(\mathbb{R}), \quad b(\cdot) \in L^q(\mathbb{R}), \quad \frac{2}{p} + \frac{1}{q} = \frac{5}{2}. \tag{7.4.9}$$

Let p and q be as defined in (7.4.9). Define the numbers q_1, q_1^*, q_2 to satisfy the equations in (6.5.7).

To prove the relation (7.4.6), we first observe that

$$\begin{aligned}
\|S_T^h\|^2_{L^2(\mathbb{R}^2)} &= \frac{1}{T}\int_0^T\int_0^T |h_T(v_1)|^2|h_T(v_2)|^2 \int_{\mathbb{R}} [(a*b)(v_1-x_1)] \\
&\quad \times [(a*b)(v_2-x_1)]\,dx_1 \int_{\mathbb{R}} [a(v_1-x_2)]\,.\,[a(v_2-x_2)]\,dx_2dv_1dv_2 \\
&= \frac{1}{T}\int_{\mathbb{R}}\int_{\mathbb{R}} |h_T(v_1)|^2|h_T(v_2)|^2\Big((a*b)^{\bar{*}2}\cdot a^{\bar{*}2}\Big)(v_1-v_2)dv_1dv_2 \\
&= \frac{1}{T}\int_{\mathbb{R}} |h_T(v_2)|^2\int_{\mathbb{R}} |h_T(v+v_2)|^2\Big((a*b)^{\bar{*}2}\cdot a^{\bar{*}2}\Big)(v)dvdv_2 \\
&= \frac{1}{T}\int_{\mathbb{R}} \Big((a*b)^{\bar{*}2}\cdot a^{\bar{*}2}\Big)(v)\int_{\mathbb{R}} |h_T(u)|^2|h_T(v+v_2)|^2dv_2dv \\
&= \int_{\mathbb{R}} \Big((a*b)^{\bar{*}2}\cdot a^{\bar{*}2}\Big)(v)\int_0^1 |h(u)|^2|h(\frac{v}{T}+u)|^2dudv. \qquad (7.4.10)
\end{aligned}$$

Since the taper h is a bounded and continuous function, we can apply the dominated convergence theorem (Proposition 2.1.2) to conclude that

$$\lim_{T\to\infty}\int_0^1 |h(u)|^2|h(\frac{v}{T}+u)|^2du = H_4, \quad v\in\mathbb{R}, \qquad (7.4.11)$$

where H_4 is defined as in (4.8.6).

On the other hand, by applying Hölder's inequality (Proposition 2.1.5) and Young's inequality for convolution (Proposition 2.3.3), we can write (with $M_h = \|h\|_\infty < \infty$)

$$\begin{aligned}
\|S_T^h\|^2_{L^2(\mathbb{R}^2)} &\le M_h^4 \int_{\mathbb{R}} \Big((|a|*|b|)^{\bar{*}2}\cdot|a|^{\bar{*}2}\Big)(v)\,dv \\
&\underset{\text{Hölder}}{\le} M_h^4\|(|a|*|b|)^{\bar{*}2}\|_{q_1}\||a|^{\bar{*}2}\|_{q_1^*} \\
&\underset{\text{Young}}{\le} M_h^4\|(|a|*|b|)^{\bar{*}2}\|_{q_1}\|a\|_p^2 \\
&\underset{\text{Young}}{\le} M_h^4\||a|*|b|\|_{q_2}^2\|a\|_p^2 \underset{\text{Young}}{\le} M_h^4\|a\|_p^4\|b\|_q^2 < \infty. \qquad (7.4.12)
\end{aligned}$$

By the dominated convergence theorem, from (7.4.10) and (7.4.11), we obtain (7.4.6). This concludes the proof of assertion (a) of the lemma.

To prove assertion (b) of the lemma, we first observe that, similar to (7.4.12), we can show that

$$\|R_T^h\|^2_{L^2(\mathbb{R}^2)} \le M_h^4\|a\|_p^4\|b\|_q^2. \qquad (7.4.13)$$

Next, we have

$$
\begin{aligned}
&R_T^h(x_1, x_2) - S_T^h(x_1, x_2) \\
&= \frac{1}{\sqrt{T}} \int_0^T \int_0^T h_T(u) h_t(v) b(u-v) a(u-x_1) a(v-x_2) du dv \\
&- \frac{1}{\sqrt{T}} \int_0^T \int_{\mathbb{R}} |h_t(v)|^2 b(u-v) a(u-x_1) a(v-x_2) du dv \\
&= \frac{1}{\sqrt{T}} \int_0^T \int_0^T [h_T(u) - h_T(v)] h_T(v) b(u-v) a(u-x_1) a(v-x_2) du dv \\
&- \frac{1}{\sqrt{T}} \int_0^T \int_{\mathbb{R}\setminus[0,T]} |h_T(v)|^2 b(u-v) a(u-x_1) a(v-x_2) du dv \\
&=: I_T^1(x_1, x_2) - I_T^2(x_1, x_2).
\end{aligned}
\tag{7.4.14}
$$

Consider first the special case when the functions $a(\cdot)$ and $b(\cdot)$ have compact supports. Assuming that $\operatorname{supp} b \subset [-M, M]$, we have

$$
\begin{aligned}
I_T^1(x_1, x_2) \leq & \omega\left(\frac{2M}{T}, h\right) \frac{M_h}{\sqrt{T}} \int_0^T \int_{\mathbb{R}} |b(u-v) a(u-x_1) a(v-x_2)| \, du dv \\
= & \omega\left(\frac{2M}{T}, h\right) \frac{M_h}{\sqrt{T}} \int_0^T (|a| * |b|)(v-x_1) |a(v-x_2)| dv,
\end{aligned}
$$

where $\omega(\delta, h)$ is the modulus of continuity of the function h. As in (7.4.12) we can show that

$$
\|I_T^1\|_{L^2(\mathbb{R}^2)} \leq \omega\left(\frac{2M}{T}, h\right) M_h \|a\|_p^4 \|b\|_q^2 \to 0 \quad \text{as} \quad T \to \infty. \tag{7.4.15}
$$

For the function $I_T^2(x_1, x_2)$, we have

$$
\begin{aligned}
I_T^2(x_1, x_2) \leq & \frac{M_h^2}{\sqrt{T}} \int_0^T \int_T^\infty |b(u-v) a(u-x_1) a(v-x_2)| \, du dv \\
& + \frac{M_h^2}{\sqrt{T}} \int_0^T \int_{-\infty}^0 |b(u-v) a(u-x_1) a(v-x_2)| \, du dv \\
= & \frac{M_h^2}{\sqrt{T}} \int_{T-M}^T \int_T^{T+M} |b(u-v) a(u-x_1) a(v-x_2)| \, du dv \\
& + \frac{M_h^2}{\sqrt{T}} \int_0^M dv \int_{-M}^0 |b(u-v) a(u-x_1) a(v-x_2)| \, du dv \\
=: & I_{T,1}^2(x_1, x_2) + I_{T,2}^2(x_1, x_2).
\end{aligned}
$$

Now, we will estimate the function $I^2_{T,1}(x_1, x_2)$. First, we have

$$\begin{aligned} I^2_{T,1}(x_1, x_2) &\le \frac{M_h^2}{\sqrt{T}} \int_{T-M}^{T} \int_{\mathbb{R}} |b(u-v)a(u-x_1)du|\, |a(v-x_2)|\, dv \\ &= \frac{M_h^2}{\sqrt{T}} \int_{T-M}^{T} (|a| * |b|)(v-x_1)\, |a(v-x_2)|\, dv. \end{aligned}$$

Next, similar to (7.4.10), we obtain

$$\begin{aligned} \|I^2_{T,1}\|^2_{L^2(\mathbb{R}^2)} &\le \frac{M_h^2}{T} \int_{T-M}^{T} \int_{T-M}^{T} \left((|a| * |b|)^{\bar{*}2} \cdot |a|^{\bar{*}2}\right)(v_1 - v_2) dv_1 dv_2 \\ &\le \frac{M_h^2 M}{T} \int_{\mathbb{R}} \left((|a| * |b|)^{\bar{*}2} \cdot |a|^{\bar{*}2}\right)(v) dv \le \frac{M_h^2 M}{T} \|a\|_p^4 \|b\|_q^2 \to 0 \end{aligned}$$

as $T \to \infty$. Similarly, we can conclude that:

$$\|I^2_{T,2}\|_{L^2(\mathbb{R}^2)} \to 0 \quad \text{as} \quad T \to \infty.$$

Therefore,

$$\|I^2_T\|_{L^2(\mathbb{R}^2)} \to 0 \quad \text{as} \quad T \to \infty.$$

This, along with (7.4.14) and (7.4.15), proves (7.4.7) in the special case where the functions $a(\cdot)$ and $b(\cdot)$ have compact supports.

To prove (7.4.7) in the general case, where the supports of the functions $a(\cdot)$ and $b(\cdot)$ are not bounded, we set for $M > 0$:

$$\begin{aligned} a_M(x) &: = a(x)\mathbf{1}_{[-M,M]}(x), \qquad \bar{a}_M(x) := a(x) - a_M(x), \\ b_M(x) &: = b(x)\mathbf{1}_{[-M,M]}(x), \qquad \bar{b}_M(x) := b(x) - b_M(x), \end{aligned}$$

and define

$$\begin{aligned} R^h_{M,T}(x_1, x_2) &: = \frac{1}{\sqrt{T}} \int_0^T \int_0^T h_T(u) h_T(v) b_M(u-v) a(u-x_1) a(v-x_2) du dv, \\ S^h_{M,T}(x_1, x_2) &: = \frac{1}{\sqrt{T}} \int_0^T |h_T(v)|^2 \left[(a_M * b_M)(v-x_1)\right] \left[a_M(v-x_2)\right] dv. \end{aligned} \tag{7.4.16}$$

In view of (7.4.5), (7.4.16), and the identity

$$\begin{aligned} baa - b_M a_M a_M &= (baa - b_M aa) + (b_M aa - b_M a_M a) + (b_M a_M a - b_M a_M a_M) \\ &= \bar{b}_M aa + b_M \bar{a}_M a + b_M a_M \bar{a}_M, \end{aligned}$$

we have

$$\begin{aligned}
&S_T^h(x_1, x_2) - S_{M,T}^h(x_1, x_2) \\
&\quad = \frac{1}{\sqrt{T}} \int_0^T \int_{\mathbb{R}} |h_T(u)|^2 \big[\bar{b}_M(u-v)a(u-x_1)a(v-x_2) \\
&\quad + b_M(u-v)\bar{a}_M(u-x_1)a(v-x_2) \\
&\quad + b_M(u-v)a_M(u-x_1)\bar{a}_M(v-x_2)\big] du dv.
\end{aligned}$$

Similar to (7.4.12), we obtain

$$\begin{aligned}
\|S_T^h - S_{M,T}^h\|^2_{L^2(\mathbb{R}^2)} \le M_h^4 \big(&\|b_M^-\|_q^2 \|a\|_p^4 + \|b_M\|_q^2 \|a_M^-\|_p^2 \|a\|_p^2 \\
&+ \|b_M\|_q^2 \|a_M\|_p^2 \|a_M^-\|_p^2\big),
\end{aligned}$$

where the right-hand side does not depend on T.

Since $\|\bar{a}_M\|_p \to 0$ and $\|\bar{b}_M\|_q \to 0$ as $M \to \infty$, one obtains

$$\lim_{M\to\infty} \sup_{T>0} \|S_T^h - S_{M,T}^h\|_{L^2(\mathbb{R}^2)} = 0. \tag{7.4.17}$$

Similarly, we can conclude that

$$\lim_{M\to\infty} \sup_{T>0} \|R_T^h - R_{M,T}^h\|_{L^2(\mathbb{R}^2)} = 0, \tag{7.4.18}$$

which proves (7.4.7) in the general case.

To prove assertion (c), it suffices to define

$$c_M(x_1, x_2) = (a_M * b_M)(x_1)a_M(x_2) \tag{7.4.19}$$

and use (7.4.7), (7.4.17), and the inequality

$$\|R_T^h - S_{T,M}^h\|_{L^2(\mathbb{R}^2)} \le \|R_T^h - S_T^h\|_{L^2(\mathbb{R}^2)} + \|S_T^h - S_{T,M}^h\|_{L^2(\mathbb{R}^2)}.$$

Lemma 7.4.1 is proved. □

The next result, a tapered version of Lemma 6.5.2, is analogous to Lemma 7.4.1, where $\mathbb{R}^2$ is replaced by $\mathbb{R}$. We define

$$R_T^h(x) = R_T^h(x, x) = \frac{1}{\sqrt{T}} \int_0^T \int_0^T h_T(u)h_T(v)b(u-v)a(u-x)a(v-x)dudv$$

and

$$S_T^h(x) = S_T^h(x, x) = \frac{1}{\sqrt{T}} \int_0^T |h_T(v)|^2 (a * b)(v - x) a(v - x) dv,$$

where $R_T^h(\cdot, \cdot)$ and $S_T^h(\cdot, \cdot)$ are defined as in (7.4.4) and (7.4.5), respectively, with $x_1 = x_2 = x$.

Lemma 7.4.2 *Let the functions $a(\cdot)$ and $b(\cdot)$ be defined as in (7.2.8), with p and q satisfying the condition* (6.5.18). *Then, the following assertions hold:*

(a) We have

$$\lim_{T\to\infty} \|S_T^h\|_{L^2(\mathbb{R})}^2 = H_4 \int_{\mathbb{R}} K_B(u) du, \tag{7.4.20}$$

where $K_B(\cdot)$ is defined as in (6.4.8).

(b) We have

$$\lim_{T\to\infty} \|R_T^h - S_T^h\|_{L^2(\mathbb{R})} = 0. \tag{7.4.21}$$

(c) For any $M > 0$, there exists a function $d_M^h(\cdot)$ supported on $[-2M, 2M]$ such that the function

$$S_{T,M}^h(x) = \frac{1}{\sqrt{T}} \int_0^T d_M^h(v - x) dv$$

satisfies the relation:

$$\lim_{M\to\infty} \limsup_{T\to\infty} \|R_T^h - S_{T,M}^h\|_{L^2(\mathbb{R})} = 0. \tag{7.4.22}$$

Proof The proof is similar to that of Lemma 7.4.1 (see also the proof of Lemma 6.5.2). We thus outline the key steps of the proof, omitting the details.

As in the proof of Lemma 6.5.2, in view of the Riesz-Thorin theorem, we can assume without loss of generality that $2/p + 1/q = 2$. Let the number p^* be defined to satisfy the equations in (6.5.23).

To prove the relation (7.4.20), we first observe that

$$\|S_T^h\|_{L^2(\mathbb{R}^2)}^2 = \frac{1}{T} \int_0^T \int_0^T |h_T(v_1)|^2 |h_T(v_2)|^2$$
$$\times \int_{\mathbb{R}} [(a * b)(v_1 - x)] [(a * b)(v_2 - x)] [a(v_1 - x)] . [a(v_2 - x)] dx dv_1 dv_2$$

$$=\frac{1}{T}\int_{\mathbb{R}}\int_{\mathbb{R}}|h_T(v_1)|^2|h_T(v_2)|^2\Big((a*b)\cdot a\Big)^{\bar{*}2}(v_1-v_2)dv_1dv_2$$
$$=\frac{1}{T}\int_{\mathbb{R}}|h_T(v_2)|^2\int_{\mathbb{R}}|h_T(v+v_2)|^2\Big((a*b)\cdot a\Big)^{\bar{*}2}(v)dvdv_2$$
$$=\frac{1}{T}\int_{\mathbb{R}}\Big((a*b)\cdot a\Big)^{\bar{*}2}(v)\int_{\mathbb{R}}|h_T(u)|^2|h_T(v+v_2)|^2dv_2dv$$
$$=\int_{\mathbb{R}}\Big((a*b)\cdot a\Big)^{\bar{*}2}(v)\int_0^1|h(u)|^2|h(\frac{v}{T}+u)|^2dudv. \tag{7.4.23}$$

Next, using Hölder's inequality, Young's inequality for convolution, and the equations in (6.5.23), we can write (with $M_h=\|h\|_\infty<\infty$)

$$\|S_T^h\|^2_{L^2(\mathbb{R}^2)}\le M_h^4\int_{\mathbb{R}}\Big((a*b)\cdot a\Big)^{\bar{*}2}(v)\,dv\underset{\text{Young}}{\le}M_h^4\|(|a|*|b|)\cdot a\|_1^2$$
$$\underset{\text{Hölder}}{\le}M_h^4\|(|a|*|b|)\|^2_{p^*}\|a\|_p^2\underset{\text{Young}}{\le}M_h^4\|a\|_p^4\|b\|_q^2<\infty.$$

Finally, by using the relation in (7.4.11) and the dominated convergence theorem, we derive the desired relation (7.4.20) from (7.4.23). This concludes the proof of assertion (a).

The proofs of assertions (b) and (c) of the lemma are similar to those of assertions (b) and (c) in Lemma 7.4.1 and are therefore omitted. Thus, Lemma 7.4.2 is proved. □

Proof of Theorem 7.4.1 By using (3.4.27), (7.4.1), (7.4.2), and Proposition 3.4.4, we express the process $\widetilde{Q}_T^h(t)$ as follows:

$$\widetilde{Q}_T^h(t)=A_T^h(t)+B_T^h(t),$$

where

$$A_T^h(t):=\int'_{\mathbb{R}^2}\frac{1}{\sqrt{T}}\int_0^{Tt}\int_0^{Tt}h_T(u)h_T(v)b(u-v)a(u-x_1)$$
$$\times\, a(v-x_2)dudv\;\xi(dx_1)\xi(dx_2),$$

and

$$B_T^h(t):=\int_{\mathbb{R}}\frac{1}{\sqrt{T}}\int_0^{Tt}\int_0^{Tt}h_T(u)h_T(v)b(u-v)$$
$$\times\, a(u-x)a(v-x)dudv\;\xi_c^{(2)}(dx). \tag{7.4.24}$$

The remainder of the proof repeats the arguments presented in the proof of Theorem 6.4.1, utilizing Lemmas 7.4.1 and 7.4.2 in place of Lemmas 6.5.1 and 6.5.2. It is important to note that, in this case, by (3.4.20) (with $k = 2$) and (7.4.6), we have

$$\mathrm{Var}[A_T^h(1)] \to 2H_4 \int_{\mathbb{R}} K_A(u)du,$$

and by (3.4.20) (with $k = 2$), (3.4.29), and (7.4.20), we have

$$\mathrm{Var}[B_T^h(1)] \to H_4\kappa_4 \int_{\mathbb{R}} K_B(u)du.$$

This completes the proof of Theorem 7.4.1. □

Proof of Theorem 7.2.3 The result we derive from Theorem 7.4.1. First, it follows from Theorem 7.4.1 that as $T \to \infty$

$$\widetilde{Q}_T^h(1) := T^{-1/2}\left(Q_T^h(1) - EQ_T^h(1)\right) \xrightarrow{d} \eta_1 \sim N\left(0, \sigma_h^2(Q)\right), \tag{7.4.25}$$

where $\sigma_h^2(Q)$ is given by

$$\sigma_h^2(Q) = H_4 \int_{\mathbb{R}} [2K_A(u) + \kappa_4 K_B(u)]\, du. \tag{7.4.26}$$

Next, taking into account that (see Remark 6.4.3):

$$\int_{\mathbb{R}} K_A(u)du = 8\pi^3 \int_{\mathbb{R}} f^2(\lambda)g^2(\lambda)d\lambda \tag{7.4.27}$$

and

$$\int_{\mathbb{R}} K_B(u)du = 4\pi^2 \left[\int_{\mathbb{R}} f(\lambda)g(\lambda)d\lambda\right]^2, \tag{7.4.28}$$

from (7.2.11) and (7.4.26), we obtain $\sigma_h^2(Q) = \sigma_{L,h}^2$. Theorem 7.2.3 is proved. □

7.5 Moderate Deviation Principle for Tapered Random Toeplitz Quadratic Functionals

In this section, we establish the moderate deviation principle for tapered quadratic functionals. For the non-tapered case, we refer to Sect. 5.5.3.

Recall that a collection $\{\xi_T,\ T \in \mathbb{U}\}$ of real-valued random variables is said to satisfy the moderate deviation principle (MDP) with speed $a_T \to \infty$ and rate function $I : \mathbb{R} \to [0, \infty]$ if the collection $\{m_T \xi_T,\ T \in \mathbb{U}\}$, where $m_T = o(T^{1/2})$ as $T \to \infty$, satisfies the large deviation principle with the same speed a_T and rate function I (see Sect. 5.5.3).

Let $a(t, s)$ be a symmetric kernel defined on $[0, T] \times [0, T]$, and let $h_T(t) = h(t/T)$ with a taper function $h(t)$, $t \in [0, 1]$, that satisfies Assumption 4.8.1.

Denote by A_T a linear integral operator defined on the weighted space $L^2([0, T]; h_T)]$, generated by the kernel $a(t, s)$:

$$[A_T^h u](t) = \int_0^T a(t, s)u(s)h_T(s)ds, \quad u(s) \in L^2([0, T]; h_T). \tag{7.5.1}$$

Consider the following general *tapered* random quadratic functional of a centered real-valued Gaussian stationary process $X(t)$, $t \in \mathbb{R}$, with the spectral density $f(\lambda)$, $\lambda \in \mathbb{R}$:

$$\mathcal{A}_T^h := \int_0^T \int_0^T a(t, s)h_T(t)h_T(s)X(t)X(s)\, dt\, ds. \tag{7.5.2}$$

By $\widetilde{\mathcal{A}}_T^h$, we denote the normalized quadratic functional:

$$\widetilde{\mathcal{A}}_T^h := b_T^{-1} \left(\mathcal{A}_T^h - \mathbb{E}[\mathcal{A}_T^h] \right), \tag{7.5.3}$$

where b_T is a positive numerical function of $T \in \mathbb{R}^+$ such that $b_T \to \infty$ as $T \to \infty$ (typically $b_T = T$).

We are interested in the moderate deviation principle for the functional $\widetilde{\mathcal{A}}_T^h$.

We denote the T-truncated tapered Toeplitz operator generated by the spectral density f and the taper function h as $W_T^h(f)$ (see formula (4.8.8)). We then set up the following assumptions:

(A1)

$$\lim_{T\to\infty} \frac{2}{b_T} \mathrm{tr}\big[A_T^h W_T^h(f)\big]^2 := V^h \in [0, \infty), \tag{7.5.4}$$

where $\mathrm{tr}[A]$ stands for the trace of the operator A.

(A2) With some $\delta \in [0, 1/2)$ and a constant $c \in [0, \infty)$

$$\sup_{T\in\mathbb{R}} b_T^{-\delta} ||A_T^h W_T^h(f)||_\infty \le c. \tag{7.5.5}$$

The next theorem presents sufficient conditions for the general quadratic functional $\widetilde{\mathcal{A}}_T^h$ defined by (7.5.3) to satisfy the MDP.

Theorem 7.5.1 *Let assumptions (A1) and (A2) be satisfied with some $\delta \in [0, 1/2)$, and let a_T be a positive numerical function of $T \in \mathbb{R}^+$ such that*

$$1/a_T + a_T b_T^{2\delta-1} \to 0 \quad \text{as} \quad T \to \infty. \tag{7.5.6}$$

Then, the random variable $(b_T/a_T)^{1/2}\widetilde{\mathcal{A}}_T^h$ satisfies the large deviation principle with speed a_T and a rate function $I_V^h(x)$:

$$I_V^h(x) = \sup_{y\in\mathbb{R}}\{xy - y^2 V^h/2\}, \quad x \in \mathbb{R}, \tag{7.5.7}$$

where V^h is defined as in (7.5.4). *Specifically, for any closed set $F \subset \mathbb{R}$,*

$$\limsup_{T\to\infty} a_T^{-1} \log \mathbb{P}\left\{(b_T/a_T)^{1/2}\widetilde{\mathcal{A}}_T \in F\right\} \le -\inf_{x\in F} I^h(x),$$

and for any open set $G \subset \mathbb{R}$,

$$\liminf_{T\to\infty} a_T^{-1} \log \mathbb{P}\left\{(b_T/a_T)^{1/2}\widetilde{\mathcal{A}}_T \in G\right\} \ge -\inf_{x\in G} I^h(x).$$

Proof By Proposition 5.3.2, with $R = W_T^h(f)$ and $A = A_T^h$, the quadratic functional $\mathcal{A}_T^h$ in (7.5.2) has the same distribution as the sum

$$\sum_{k=1}^{\infty} \lambda_{k,T}\xi_k^2,$$

where $\{\xi_k, \ k \ge 1\}$ are independent $N(0, 1)$ Gaussian random variables and $\{\lambda_{k,T} := \lambda_{k,T}^h, \ k \ge 1\}$ are the eigenvalues of the operator $W_T^h(f)A_T^h$.

Therefore, assumptions (A1) and (A2) imply that:

$$b_T \operatorname{Var}[\widetilde{\mathcal{A}}_T^h] = \frac{2}{b_T}\sum_{j=1}^{\infty}\lambda_{j,T}^2 = \frac{2}{b_T}\operatorname{tr}[A_T^h W_T^h(f)]^2 \to V^h \quad \text{as} \quad T \to \infty, \tag{7.5.8}$$

and

$$\sup_j |\lambda_{j,T}| = ||A_T^h W_T^h(f)||_\infty \le c\, b_T^\delta \quad \text{for all} \quad T \in \mathbb{R}. \tag{7.5.9}$$

To apply the Gärtner-Ellis theorem (Proposition 5.5.1) with $Y_T^h := (b_T/a_T)^{1/2}\widetilde{\mathcal{A}}_T^h$, we only need to prove that for all $s \in \mathbb{R}$, the following relation holds as $T \to \infty$:

$$M_T(\mathcal{A}^h, s) := a_T^{-1} \log \mathbb{E}[\exp(s a_T (b_T/a_T)^{1/2})\widetilde{\mathcal{A}}_T^h] \to \frac{s^2 V^h}{2}. \tag{7.5.10}$$

We have

$$\begin{aligned} M_T(\mathcal{A}^h, s) =& \frac{1}{a_T}\sum_{k=1}^{\infty} \log \mathbb{E}\left[\exp\{s(a_T/b_T)^{1/2}\lambda_{k,T}(\xi^2-1)\}\right] \qquad (7.5.11)\\ =& \frac{1}{a_T}\sum_{k=1}^{\infty}\left[-\frac{1}{2}\log(1-2s(a_T/b_T)^{1/2}\lambda_{k,T}) - s(a_T/b_T)^{1/2}\lambda_{k,T}\right]. \end{aligned}$$

Denote $\mu_{j,T}^{(s)} := s(a_T/b_T)^{1/2}\lambda_{j,T}$. Observe that, by (7.5.9) and the choice of a_T, we have

$$\sup_j |\mu_{j,T}^{(s)}| \to 0 \quad \text{as} \quad T \to \infty. \tag{7.5.12}$$

Indeed, using the relation in (7.5.9), we find that

$$|\mu_{j,T}^{(s)}| \le c\,(a_T/b_T)^{1/2} b_T^{\delta} = c\left(a_T b_T^{2\delta-1}\right)^{1/2}. \tag{7.5.13}$$

The last term in (7.5.13) approaches zero as $T \to \infty$ because, according to the assumption (7.5.6), we have $a_T b_T^{2\delta-1} \to 0$ as $T \to \infty$.

From (7.5.12), we obtain

$$\sup_j |1 - \theta(\mu_{j,T}^{(s)})\mu_{j,T}^{(s)}| \to 1 \quad \text{as} \quad T \to \infty, \tag{7.5.14}$$

where $\theta(\mu_{j,T}^{(s)}) \in [0, 1]$.

Next, by Taylor's theorem, for $|w| < 1$, we have

$$\log(1-w) = -w - (1/2)w^2(1-\theta(w)w)^{-2}, \tag{7.5.15}$$

where $\theta(w) \in [0, 1]$.

Applying the Eq. (7.5.15) with $w = \mu_{j,T}^{(s)}$, and considering the relations (7.5.8), (7.5.11), and (7.5.14), we obtain (7.5.10). Thus, Theorem 7.5.1 is proved. □

Now let $a(t,s) = \widehat{g}(t-s)$, where $\widehat{g}(t)$ is the Fourier transform of some integrable even function $g(\lambda)$. In this case, the operator A_T^h defined by Eq. (7.5.1) with $a(t,s) = \widehat{g}(t-s)$ becomes a T-truncated Toeplitz operator generated by the functions g and h:

$$[W_T^h(g)u](t) = \int_0^T \hat{g}(t-s)u(s)h_T(s)ds, \quad u(s) \in L^2([0,T]; h_T). \tag{7.5.16}$$

We are interested in the moderate deviation principle for the following *tapered* Toeplitz type quadratic functional:

$$Q_T^h := \int_0^T \int_0^T \widehat{g}(t-s)h_T(t)h_T(s)X(t)X(s)\,dt\,ds. \tag{7.5.17}$$

Denote

$$\widetilde{Q}_T^h := T^{-1}\left(Q_T^h - \mathbb{E}[Q_T^h]\right). \tag{7.5.18}$$

For this case, we set up the following assumptions:

(B1) $f \cdot g \in L^1(\mathbb{R}) \cap L^2(\mathbb{R})$ and as $T \to \infty$

$$\frac{2}{T}\mathrm{tr}\left[W_T^h(f)W_T^h(g)\right]^2 \longrightarrow V^h(f,g) := 16\pi^3 H_4 \int_{\mathbb{R}} f^2(\lambda)g^2(\lambda)d\lambda, \tag{7.5.19}$$

where H_4 is defined as in (4.8.6).

(B2) With some $\delta \in [0,1/2)$ and a constant $c \in [0,\infty)$

$$\sup_{T\in\mathbb{R}} T^{-\delta}||W_T^H(f)W_T^h(g)||_\infty \le c. \tag{7.5.20}$$

As a consequence of Theorem 7.5.1, we obtain the following moderate deviation principle for the tapered Toeplitz type quadratic functional $\widetilde{Q}_T^h$ as presented in Eq. (7.5.18).

Theorem 7.5.2 *Assume that the functions f and g satisfy assumptions (B1) and (B2) with some $\delta \in [0,1/2)$. Let a_T be a positive numerical function of $T \in \mathbb{R}^+$ such that $1/a_T + a_T T^{2\delta-1} \to 0$ as $T \to \infty$. Then, the random variable $(b_T/a_T)^{1/2}\widetilde{Q}_T^h$ satisfies the LDP with speed a_T and a rate function $I^h(x)$:*

$$I^h(x) = \sup_{y\in\mathbb{R}}\{xy - y^2V^h(f,g)/2\},$$

where $V^h(f,g)$ is defined as in (7.5.19).

The next theorem presents constructive sufficient conditions for the tapered Toeplitz quadratic functional $\widetilde{Q}_T^h$ to satisfy the moderate deviation principle. This theorem is a tapered version of Theorem 5.5.4.

Theorem 7.5.3 *Assume that* $f \in L^1(\mathbb{R}) \cap L^p(\mathbb{R})$ $(p > 2)$ *and* $g \in L^1(\mathbb{R}) \cap L^q(\mathbb{R})$ $(q > 2)$ *with* $1/p + 1/q < 1/2$. *Let* a_T *be a positive numerical function of* $T \in \mathbb{R}^+$ *such that*

$$1/a_T + a_T T^{2(1/p+1/q)-1} \to 0 \quad \text{as} \quad T \to \infty. \tag{7.5.21}$$

Then, the random variable $(T/a_T)^{1/2} \widetilde{Q}_T^h$ *satisfies the LDP with speed* a_T *and a rate function* $I^h(x)$*:*

$$I^h(x) = \sup_{y \in \mathbb{R}} \{xy - y^2 V^h(f, g)/2\},$$

where

$$V^h(f, g) = 16\pi^3 H_4 \int_{\mathbb{R}} f^2(\lambda) g^2(\lambda)\, d\lambda, \tag{7.5.22}$$

and H_4 *is defined as in* (4.8.5).

Proof The result we deduce from Theorem 7.5.2. First, assumption (B1) of Theorem 7.5.2 is satisfied by Theorem 4.8.1 **(C2)** with $k = 4$ and $\Lambda = \mathbb{R}$.

Second, by Theorem 4.9.4 (a), we have

$$||W_T(f) W_T(g)||_\infty \le ||W_T(f)||_\infty ||W_T(g)||_\infty \le c\, T^{1/p+1/q}, \tag{7.5.23}$$

which implies that assumption (B2) of Theorem 7.5.2 is also satisfied with $\delta = 1/p + 1/q$, and the result follows. □

Corollary 7.5.1 *Assume that* $f \in L^1(\mathbb{R}) \cap L^p(\mathbb{R})$ *with* $p > 2$, *and* g *is of bounded variation on* $\mathbb{R}$. *Let* a_T *be a positive numerical function of* $T \in \mathbb{R}^+$ *such that* $a_T \to \infty$ *and* $1/a_T + a_T T^{2/p-1} \to 0$ *as* $T \to \infty$. *Then, the random variable* $(T/a_T)^{1/2} \widetilde{Q}_T^h$ *satisfies the LDP with the speed* a_T *and the rate function* $I^h(x)$ *as defined in Theorem 7.5.3.*

The following theorem is a consequence of Theorem 7.5.3.

Theorem 7.5.4 *Let* $\{Y(t), t \in \mathbb{R}\}$, *be a real-valued, centered stationary Gaussian process with the spectral density* $f_Y(\lambda) \in L^1(\mathbb{R}) \cap L^r(\mathbb{R})$, $r > 2$, *and let* m_T *be a positive numerical function of* $T \in \mathbb{R}^+$ *such that* $m_T \to \infty$, $m_T T^{-1/r} \to \infty$, *and* $m_T T^{-1/2} \to 0$ *as* $T \to \infty$. *Then the normalized quadratic functional* $\widetilde{L}_T^h$ *defined by Eq.* (7.3.4) *satisfies the LDP with speed* T/m_T^2 *and the rate function:*

$$I^h(x) = x^2/\sigma_h^2,$$

where

$$\sigma_h^2 = 16\pi^3 H_4 \int_{\mathbb{R}} f_Y^2(\lambda)\, d\lambda. \tag{7.5.24}$$

Observe that if $m_T = T^\delta$, *then* $\delta \in (1/r, 1/2)$.

Remark 7.5.1 The results of Theorems 7.5.1–7.5.4 remain valid for quadratic forms in discrete time stationary Gaussian processes.

7.6 Bibliographical Notes

Section 7.2 Theorems 7.2.1 and 7.2.2 were established in the paper by Ginovyan and Sahakyan [141]. Theorem 7.2.3 was obtained by Ginovyan and Sahakyan [140].

Central and non-central limit theorems for tapered quadratic forms of discrete time short and long memory Gaussian stationary fields have been proven in the papers by Alomari et al. [3], Doukhan et al. [86], and Ludeña and Lavielle [221].

Section 7.3 This section is devoted to the proofs of Theorems 7.2.1 and 7.2.2.

Section 7.4 Theorem 7.4.1, a functional central limit theorem for the tapered Toeplitz process $Q_T^h(t)$, $t \in [0, 1]$, is a tapered version of Theorem 6.4.1 and was established by Ginovyan and Sahakyan [140]. This theorem is utilized to prove Theorem 7.2.3.

Section 7.5 The results presented in this section—Theorems 7.5.1–7.5.4—were obtained by Ginovyan and Sahakyan [144].

The non-tapered versions of Theorems 7.5.1 and 7.5.3 for quadratic forms in discrete time stationary Gaussian processes were proved by Kakizawa [199] in the context of estimating the spectral density function and spectral functionals. Kakizawa [199] also noted that these results can be extended to the tapered case.

A non-tapered version of Theorem 7.5.4 is presented in the paper by Bryc and Dembo [58].

See also the bibliographical notes at the end of Sect. 5.5.3.

Chapter 8
Nonparametric Estimation of Spectral Functionals

Linear and nonlinear functionals of the periodogram (the empirical spectral density) play a key role in the parametric estimation of the spectrum of stationary processes, particularly when employing the minimum contrast estimation method with various contrast functionals. In this chapter, we address the nonparametric estimation problem for linear and certain nonlinear smooth spectral functionals. We review the asymptotic properties, including asymptotic unbiasedness, bias rate convergence, consistency, a central limit theorem, and asymptotic normality of the empirical spectral functionals based on both tapered and non-tapered data. We discuss the estimation problem under the local asymptotic normality (LAN) condition and establish the efficiency of the proposed estimators. Finally, we provide asymptotic bounds for the minimax mean square risk of the estimators for linear functionals. The results presented in this chapter are utilized in Chap. 9 to demonstrate the consistency and asymptotic normality of the minimum contrast estimators based on the Whittle contrast functionals for stationary linear models with tapered data.

In Sect. 8.1, we present the nonparametric estimation problem for spectral functionals that will be discussed in this chapter. Section 8.2 focuses on estimating linear spectral functionals, where we establish the asymptotic properties of the estimators, including asymptotic unbiasedness, consistency, and asymptotic normality. The estimation problem for nonlinear functionals in Gaussian models is addressed in Sect. 8.3. Section 8.4 is devoted to estimation under the LAN condition. Here, we introduce the concept of LAN and define the H-efficiency and IK-efficiency of estimators. For Gaussian models, we provide sufficient conditions for both H-efficiency and IK-efficiency. In Sect. 8.5, we derive upper and lower asymptotic bounds for the minimax mean square risks of the estimators for linear functionals.

M. S. Ginovyan, *Random Toeplitz Functionals and Their Applications*, Frontiers in Probability and the Statistical Sciences,
https://doi.org/10.1007/978-3-032-00961-6_8

8.1 Introduction: The Nonparametric Estimation Problem

Let $\{X(t),\ t \in \mathbb{U}\}$ be a centered stationary process with an *unknown* spectral density $f(\lambda)$, $\lambda \in \Lambda$. We assume that $f(\lambda)$ belongs to a specified (infinite-dimensional) class $\mathcal{F} \subset L^p := L^p(\Lambda)$ $(p \geq 1)$ of spectral densities that possess certain smoothness properties. Let $J(\cdot)$ be a *known* functional, the domain of definition of which contains $\mathcal{F}$. The problem of interest is to estimate the value $J(f)$ of the functional $J(\cdot)$ at an *unknown* "point" $f \in \mathcal{F}$ based on the observed data $\mathbf{X}_T := \{X(t),\ t \in D_T\}$, where $D_T := [0, T]$ in the continuous time case and $D_T := \{1, \ldots, T\}$ in the discrete time case, and investigate the asymptotic properties (as $T \to \infty$) of the proposed estimators, depending on the memory structure of the model $X(t)$ and the smoothness structure of the "parametric" set $\mathcal{F}$.

Note that the solution to the estimation problem above essentially depends on the type of estimand functional, $J(f)$, whether it is linear or nonlinear.

Important examples of linear functionals include the covariance function $(r(u))$ and the spectral function $(F(\mu))$ of the process $X(u)$:

$$J_r(f) = r(u) := \int_\Lambda e^{iu\lambda} f(\lambda)\, d\lambda.$$

$$J_F(f) = F(\mu) := \int_a^\mu f(\lambda)\, d\lambda,$$

where $a = -\infty$ in the continuous time case and $a = -\pi$ in the discrete time case.

A brief catalog of important nonlinear functionals includes:

1. The L^p norms of the spectral density f and its derivatives $f^{(k)}$ $(f^{(0)} = f)$:

$$J_{p,k}(f) = ||f^{(k)}||_p := \begin{cases} \left(\int_\Lambda |f^{(k)}(\lambda)|^p d\lambda\right)^{1/p} & \text{for } 1 \leq p < \infty \\ \sup_{\lambda \in \Lambda} |f^{(k)}(\lambda)|, & \text{for } p = \infty. \end{cases}$$

2. The Shannon neg-entropy functional $J_S(f)$:

$$J_S(f) = \int_\Lambda f(\lambda) \ln f(\lambda)\, d\lambda.$$

3. The Burg (or spectral) entropy functional $J_B(f)$:

$$J_B(f) = \int_\Lambda \ln f(\lambda)\, d\lambda.$$

4. The spectral indeterminism functional $J_I(f)$:

$$J_I(f) = \int_\Lambda f^{-1}(\lambda)\, d\lambda.$$

The functionals $J_B(f)$ and $J_I(f)$ arise in the prediction and interpolation of stationary processes (see, e.g., Rozanov [263]). For instance, in the discrete time case, the one-step prediction error σ^2 when predicting the random variable $X(1)$ based on the entire past $\{X(u),\ u \le 0\}$ is given by the well-known Kolmogorov-Szegő formula (see Appendix B.4, Brockwell and Davis [57], Section 5.8, and Grenander and Szegő [162], Section 10.8):

$$\sigma^2 = 2\pi \exp\left\{\frac{1}{2\pi}\int_{-\pi}^{\pi} \ln f(\lambda)\, d\lambda\right\} = 2\pi \exp\left\{\frac{1}{2\pi} J_B(f)\right\}.$$

8.2 Estimation of Linear Spectral Functionals

In this section, we assume that the estimand functional $J(f)$ is linear and continuous in the space $L^p(\Lambda)$, $p > 1$. Consequently, $J(f)$ can be represented as follows (see Proposition 2.3.5):

$$J := J(f) = J(f, g) = \int_\Lambda f(\lambda) g(\lambda) d\lambda, \tag{8.2.1}$$

where $g(\lambda) \in L^q(\Lambda)$, $1/p + 1/q = 1$, which will be referred to as the generating function for the functional $J(f, g)$.

We are interested in the nonparametric estimation of the functional $J(f)$ based on the tapered data:

$$\mathbf{X}_T^h := \{h_T(t) X(t),\ t \in D_T\}, \tag{8.2.2}$$

where $D_T := [0, T]$ in the continuous time case and $D_T := \{1, \ldots, T\}$ in the discrete time case, and

$$h_T(t) := h(t/T) \tag{8.2.3}$$

with a taper function $h(t)$, $t \in \mathbb{R}$, that satisfies Assumption 4.8.1.

As before, the case $h(t) = \mathbf{1}_{[0,1]}(t)$, where $\mathbf{1}_{[0,1]}(\cdot)$ denotes the indicator function of the interval $[0, 1]$, will be referred to as the *non-tapered* case.

The Estimator As an estimator for the functional $J(f)$, defined in (8.2.1), we consider the averaged periodogram (or the simple "plug-in" statistic) based on the tapered data (8.2.2). To define the estimator, we first introduce some notation.

We define the finite Fourier transform of the tapered data (8.2.2) as follows:

$$d_T^h(\lambda) := \begin{cases} \sum_{t=1}^{T} h_T(t)X(t)e^{-i\lambda t} & \text{in the d.t. case,} \\ \int_0^T h_T(t)X(t)e^{-i\lambda t}dt & \text{in the c.t. case.} \end{cases} \tag{8.2.4}$$

Then, we define the tapered periodogram $I_T^h(\lambda)$ of the process $X(t)$ by the following equation:

$$\begin{aligned} I_T^h(\lambda) :&= \frac{1}{C_T} d_T^h(\lambda)d_T^h(-\lambda) \\ &= \begin{cases} \frac{1}{C_T}\left|\sum_{t=1}^{T} h_T(t)X(t)e^{-i\lambda t}\right|^2 & \text{in the d.t. case,} \\ \frac{1}{C_T}\left|\int_0^T h_T(t)X(t)e^{-i\lambda t}dt\right|^2 & \text{in the c.t. case.} \end{cases} \end{aligned} \tag{8.2.5}$$

where

$$C_T := 2\pi H_{2,T}(0) \neq 0, \tag{8.2.6}$$

and $H_{2,T}(\cdot)$ is defined as in (4.8.4) with $k = 2$.

Notice that in the non-tapered case, where $h(t) = \mathbf{1}_{[0,1]}(t)$, we have $C_T = 2\pi T$.

As an estimator J_T^h for the functional $J(f)$ in (8.2.1), based on the tapered data (8.2.2), we consider the simple "plug-in" statistic:

$$J_T^h = J(I_T^h, g) := \int_\Lambda I_T^h(\lambda)g(\lambda)d\lambda. \tag{8.2.7}$$

In view of Eqs. (7.1.1), (8.2.5), and (8.2.7), we have

$$\begin{aligned} J_T^h &= \frac{1}{C_T}\begin{cases} \sum_{t=1}^{T}\sum_{s=1}^{T} b(t-s)h_T(t)h_T(s)X(t)X(s) & \text{in the d.t. case,} \\ \int_0^T\int_0^T b(t-s)h_T(t)h_T(s)X(t)X(s)\,dt\,ds & \text{in the c.t. case,} \end{cases} \\ &= C_T^{-1}Q_T^h, \end{aligned} \tag{8.2.8}$$

where C_T is defined in (8.2.6), $b(t)$ is the Fourier transform of the function $g(\lambda)$:

$$b(t) := \widehat{g}(t) = \int_\Lambda e^{i\lambda t}g(\lambda)d\lambda, \tag{8.2.9}$$

and Q_T^h is the tapered Toeplitz type quadratic functional defined in Eq. (7.1.1).

We will refer to $g(\lambda)$ and its Fourier transform $b(t) := \widehat{g}(t)$ as the *generating function* and *generating kernel* for the functional J_T^h, respectively.

Thus, the study of the asymptotic properties of the estimator J_T^h is based on the asymptotic distribution (as $T \to \infty$) of the tapered random Toeplitz type quadratic functional Q_T^h as given in (7.1.1). This topic was discussed in Chaps. 5 and 7 for the non-tapered and tapered cases, respectively.

8.2.1 Asymptotic Properties of the Spectral Estimators

8.2.1.1 Asymptotic Unbiasedness

We begin with the following assumption.

Assumption 8.2.1 The function

$$\Psi(u) = \int_{\Lambda} f(v)g(u+v)\,dv \tag{8.2.10}$$

belongs to the space $L^1(\Lambda) \cap L^2(\Lambda)$ and is continuous at the point $u = 0$.

Theorem 8.2.1 *Let the functionals* $J := J(f, g)$ *and* $J_T^h := J(I_T^h, g)$ *be defined by (8.2.1) and (8.2.7), respectively. Then, the statistic* J_T^h *is an asymptotically unbiased estimator for* $J(f)$*; that is, the following relation holds:*

$$\lim_{T\to\infty} (\mathbb{E}[J_T^h] - J) = 0. \tag{8.2.11}$$

Here, and in what follows, $\mathbb{E}[\cdot]$ *denotes the expectation operator.*

Remark 8.2.1 Using Hölder's inequality, it can be easily shown that if $f \in L^1(\Lambda)\cap L^{p_1}(\Lambda)$ and $g \in L^1(\Lambda)\cap L^{p_2}(\Lambda)$ with $1 \le p_1, p_2 \le \infty$, $1/p_1+1/p_2 \le 1$, then the relation (8.2.11) is satisfied.

Under additional smoothness conditions on the functions f and g, we can estimate the rate of convergence to zero in (8.2.11). First, we introduce some assumptions. Recall the L^p-Hölder classes $H_p(\beta)$ (see Sect. 2.1).

Assumption 8.2.2 The spectral density f and the generating function g are such that the pair (f, g) satisfies the condition $(\mathcal{H})$; that is, $(f, g) \in (\mathcal{H})$ (see Definition 2.1.1).

Recall that if $(f, g) \in (\mathcal{H})$, then there exist numbers p_1 ($p_1 > p$) and q_1 ($q_1 > q$), such that $H_p(\beta_1) \subset L_{p_1}$, $H_q(\beta_2) \subset L_{q_1}$ and $1/p_1 + 1/q_1 \le 1/2$ (see Proposition 2.1.9).

Assumption 8.2.3 The spectral density f and the generating function g are such that $f, g \in L^1(\Lambda) \cap L^2(\Lambda)$ and g has bounded variation.

The following theorem provides sufficient conditions for the proper rate of convergence of the bias $E[J_T^h] - J$ to zero, which is necessary for the asymptotic normality of the estimator J_T^h.

Theorem 8.2.2 *Let the functionals* $J := J(f, g)$ *and* $J_T^h := J(I_T^h, g)$ *be defined by (8.2.1) and (8.2.7), respectively. Then, under Assumption 8.2.2 (or 8.2.3), the following asymptotic relation holds:*

$$T^{1/2}(\mathbb{E}[J_T^h] - J) \to 0 \quad \text{as} \quad T \to \infty. \tag{8.2.12}$$

Remark 8.2.2 We say that an estimator J_T^h of J is *asymptotically unbiased of the order of* T^β, $\beta > 0$, if

$$\lim_{T\to\infty} T^\beta(\mathbb{E}[J_T^h] - J) = 0.$$

Thus, Theorem 8.2.2 states that the statistic J_T^h is an asymptotically unbiased estimator for J of the order of $T^{1/2}$.

8.2.1.2 Consistency

Recall that an estimator J_T^h of J is said to be:

(a) consistent if $J_T^h \to J$ in probability as $T \to \infty$
(b) mean square consistent if $\mathbb{E}(J_T^h - J)^2 \to 0$ as $T \to \infty$
(c) $\sqrt{T}$-consistent in the mean square sense if $\mathbb{E}[\sqrt{T}(J_T^h - J)]^2] = O(1)$ as $T \to \infty$
(d) $\sqrt{T}$-consistent if $\sqrt{T}(J_T^h - J)$ is bounded in probability

To present the corresponding results, we first introduce the following assumption.

Assumption 8.2.4

(a) (Discrete time case). The spectral density f and the generating function g satisfy the conditions: $f \in L^p(\Lambda)$ $(p \geq 1)$ and $g \in L^q(\Lambda)$ $(q \geq 1)$ with $1/p + 1/q \leq 1/2$.
(b) (Continuous time OIP-driven case). The spectral density f and the generating function g satisfy the conditions: $f \in L^1(\Lambda) \cap L^p(\Lambda)$ $(p \geq 1)$ and $g \in L^1(\Lambda) \cap L^q(\Lambda)$ $(q \geq 1)$ with $1/p + 1/q \leq 1/2$.
(c) (Continuous time Lévy-driven case). The filter $a(\cdot)$ in the MA(∞)-representation of $X(t)$ (see (3.2.10)) and the generating kernel $\widehat{g}(\cdot)$ satisfy the conditions:

$$a(\cdot) \in L^2(\Lambda) \cap L^p(\Lambda), \quad \widehat{g}(\cdot) \in L^q(\Lambda) \quad \text{with } 1 \leq p, q \leq 2,\ 2/p + 1/q \geq 5/2.$$

The following theorem describes the asymptotic behavior of the variance of J_T^h:

$$\mathrm{Var}(J_T^h) = \mathbb{E}\left[(J_T^h - \mathbb{E}[J_T^h])^2\right].$$

Theorem 8.2.3 *Let the functionals $J := J(f, g)$ and $J_T^h := J(I_T^h, g)$ be defined by (8.2.1) and (8.2.7), respectively. Then, under Assumption 8.2.4, the following asymptotic relation holds:*

$$\lim_{T\to\infty} T\,\mathrm{Var}(J_T^h) = \sigma_h^2(J), \tag{8.2.13}$$

where

$$\sigma_h^2(J) := 4\pi e(h) \int_\Lambda f^2(\lambda) g^2(\lambda) d\lambda + \kappa_4 e(h) \left[\int_\Lambda f(\lambda) g(\lambda) d\lambda\right]^2. \tag{8.2.14}$$

Here, κ_4 is the fourth cumulant of $\xi(1)$, and

$$e(h) := \lim_{T\to\infty} \frac{T H_{4,T}}{H_{2,T}^2} \tag{8.2.15}$$

is the tapering factor, where $H_{k,T}(\cdot)$ is defined as in (4.8.4).

Observe that, in the continuous time case, we have

$$e(h) = \lim_{T\to\infty} \frac{T H_{4,T}}{H_{2,T}^2} = \frac{H_4}{H_2^2} = \int_0^1 h^4(t)dt \left(\int_0^1 h^2(t)dt\right)^{-2}. \tag{8.2.16}$$

From Theorems 8.2.1 to 8.2.3 we infer the following result.

Theorem 8.2.4 *The following assertions hold:*

(a) Under Assumptions 8.2.1 and 8.2.4, the statistic J_T^h is a mean square consistent estimator for J.

(b) Under Assumptions 8.2.2 (or 8.2.3) and 8.2.4, the statistic J_T^h is a $\sqrt{T}$-consistent estimator for J in the mean square sense.

8.2.1.3 Asymptotic Normality

The following theorem provides sufficient conditions for the estimator J_T^h to satisfy the central limit theorem.

Theorem 8.2.5 *Let $J := J(f, g)$ and $J_T^h := J(I_T^h, g)$ be defined by (8.2.1) and (8.2.7), respectively. Then, under Assumption 8.2.4, the functional J_T^h satisfies the central limit theorem. More precisely, we have*

$$T^{1/2}\left[J_T^h - \mathbb{E}(J_T^h)\right] \xrightarrow{d} \eta \quad \text{as} \quad T \to \infty, \tag{8.2.17}$$

where the symbol $\xrightarrow{d}$ denotes convergence in distribution and η is a normally distributed random variable with mean zero and variance $\sigma_h^2(J)$ as given by (8.2.14).

Taking into account the equality

$$T^{1/2}\left[J_T^h - J\right] = T^{1/2}\left[\mathbb{E}(J_T^h) - J\right] + T^{1/2}\left[J_T^h - \mathbb{E}(J_T^h)\right], \tag{8.2.18}$$

as an immediate consequence of Theorems 8.2.2 and 8.2.5, we obtain the following result, which provides sufficient conditions for the simple "plug-in" statistic $J(I_T^h)$ to be a $\sqrt{T}$-consistent and asymptotically normal estimator for the linear spectral functional J.

Theorem 8.2.6 *Let the functionals $J := J(f, g)$ and $J_T^h := J(I_T^h, g)$ be defined by (8.2.1) and (8.2.7), respectively. Then, under Assumptions 8.2.2 (or 8.2.3) and 8.2.4, the statistic J_T^h is a $\sqrt{T}$-consistent and asymptotically normal estimator for the functional J. More precisely, we have*

$$T^{1/2}\left[J_T^h - J\right] \xrightarrow{d} \eta \quad \text{as} \quad T \to \infty, \tag{8.2.19}$$

where η is defined in Theorem 8.2.5. Specifically, η is a normally distributed random variable with mean zero and variance $\sigma_h^2(J)$ as given by (8.2.14).

Notice that if the underlying process $X(u)$ is Gaussian (in this case, $\kappa_4 = 0$), then in formula (8.2.14), we only have the first term. Furthermore, for the asymptotic normality of the estimator J_T^h, it is sufficient to satisfy only Assumption 8.2.2. More precisely, we have the following result.

Theorem 8.2.7 *Assume that the underlying process $X(u)$ is Gaussian, and that the functionals J and J_T^h are defined by (8.2.1) and (8.2.7), respectively. Then, under Assumption 8.2.2, the statistic J_T^h is a $\sqrt{T}$-consistent and asymptotically normal estimator for the functional J. Specifically, the relation* (8.2.19) *holds, where now η is a normally distributed random variable with mean zero and variance $\sigma_{h,G}^2(J)$, defined as*

$$\sigma_{h,G}^2(J) := 4\pi e(h) \int_\Lambda f^2(\lambda) g^2(\lambda) d\lambda, \tag{8.2.20}$$

where $e(h)$ is defined in (8.2.15).

Example 8.2.1 (Estimation of the Covariance Function) Assume that $X(t)$ is a discrete time Gaussian stationary process with spectral density f satisfying $f \in L^2(\mathbb{T})$. Let $g(\lambda) = e^{iu\lambda}$. Then, for the functional $J(f, g)$ in (8.2.1), we have:

$$J(f, g) = \int_{-\pi}^{\pi} e^{iu\lambda} f(\lambda)\, d\lambda := r(u).$$

Thus, in this special case, our problem reduces to estimating the covariance function $r(u) = \mathbb{E}[X(t+u)X(t)]$ of the process $X(t)$. It is clear that the assumptions of Theorem 8.2.6 are satisfied, allowing us to conclude that the simple "plug-in" statistic

$$J_T^h = \widehat{r}_T(u) = \int_{-\pi}^{\pi} e^{iu\lambda} I_T^h(\lambda)\, d\lambda$$

is an asymptotically normal estimator for $r(u)$, with an asymptotic variance $\sigma^2(u)$, defined as

$$\sigma^2(u) = 4\pi e(h) \int_{-\pi}^{\pi} \cos^2(u\lambda) f^2(\lambda)\, d\lambda,$$

where $e(h)$ is defined by Eq. (8.2.15).

Example 8.2.2 (Estimation of the Spectral Function) Assume that $X(t)$ is as described in Example 8.2.1 with $f \in L^2(\mathbb{T})$. Let $g(\lambda) = \mathbf{1}_{[0,\mu]}(\lambda)$ be the indicator function of the interval $[0, \mu]$. Then, we have

$$J(f, g) = \int_{-\pi}^{\pi} \mathbf{1}_{[0,\mu]}(\lambda) f(\lambda)\, d\lambda = \int_0^{\mu} f(\lambda)\, d\lambda := F(\mu).$$

Thus, in this case, the estimand functional is the spectral function $F(\mu)$ of the process $X(u)$. It is clear that the assumptions of Theorem 8.2.6 are satisfied. Therefore, we can conclude that the simple "plug-in" statistic

$$J_T^h = \widehat{F}_T(\mu) = \int_0^{\mu} I_T^h(\lambda)\, d\lambda \tag{8.2.21}$$

is an asymptotically normal estimator for $F(\mu)$, with an asymptotic variance $\sigma^2(\mu)$ given by:

$$\sigma^2(\mu) = 4\pi e(h) \int_0^{\mu} f^2(s)\, ds,$$

where $e(h)$ is defined by (8.2.15).

8.2.2 Proofs of Theorems of Sect. 8.2.1

We provide proofs for the continuous time case, as the proofs for the discrete time case are similar.

8.2.2.1 Preliminaries

We will first prove three lemmas.

Lemma 8.2.1 *The expected value* $\mathbb{E}[J_T^h]$ *of the estimator* J_T^h *is given by the following formula:*

$$\mathbb{E}[J_T^h] = \int_{\mathbb{R}} \int_{\mathbb{R}} f(\lambda) g(\lambda + \mu) \Phi_{2,T}^h(\mu) d\lambda d\mu, \tag{8.2.22}$$

where the kernel $\Phi_{2,T}^h(\cdot)$ *is defined by Eq. (4.8.32).*

Proof Using formulas (3.1.5), (4.8.32), and (8.2.7), we can write

$$\begin{aligned}
\mathbb{E}[J_T^h] &= \frac{1}{C_T} \int_0^T \int_0^T h_T(t) h_T(s) \widehat{g}(t-s) r(t-s)\, dt\, ds \\
&= \frac{1}{C_T} \int_0^T \int_0^T h_T(t) h_T(s) \int_{\mathbb{R}} e^{i\lambda(s-t)} g(\lambda) d\lambda \int_{\mathbb{R}} e^{i\mu(t-s)} f(\mu) d\mu\, dt\, ds \\
&= \frac{1}{C_T} \int_{\mathbb{R}} \int_{\mathbb{R}} g(\lambda) f(\mu) \int_0^T \int_0^T h_T(t) h_T(s) e^{i(\lambda-\mu)(s-t)}\, dt\, ds d\lambda d\mu \\
&= \frac{1}{C_T} \int_{\mathbb{R}} \int_{\mathbb{R}} g(\lambda) f(\mu) \left| \int_0^T h_T(t) e^{-i(\lambda-\mu)t} dt \right|^2 d\lambda d\mu \\
&= \frac{1}{C_T} \int_{\mathbb{R}} \int_{\mathbb{R}} f(\lambda) g(\mu) \left| H_{1,T}(\lambda-\mu) \right|^2 d\lambda d\mu \\
&= \int_{\mathbb{R}} \int_{\mathbb{R}} f(\lambda) g(\mu) \Phi_{2,T}^h(\lambda-\mu) d\lambda d\mu \\
&= \int_{\mathbb{R}} \int_{\mathbb{R}} f(\lambda) g(\lambda+\mu) \Phi_{2,T}^h(\mu) d\lambda d\mu,
\end{aligned}$$

and (8.2.22) follows. □

Observe that, in the non-tapered case, the kernel $\Phi_{2,T}^h(\mu)$ coincides with the classical Fejér kernel $F_T(\mu)$ for the real line (see 2.2.2).

Lemma 8.2.2 *Assume that Assumption 8.2.3 is satisfied. Then the following asymptotic relation holds as $T \to \infty$:*

$$\int_{\mathbb{R}}\int_{\mathbb{R}} f(\lambda)g(\lambda+\mu)\Phi^h_{2,T}(\mu)d\lambda d\mu = \int_{\mathbb{R}} f(\lambda)g(\lambda)d\lambda + o\left(T^{-1/2}\right), \tag{8.2.23}$$

where $\Phi^h_{2,T}(\mu)$ is given by (4.8.32).

Proof We have

$$\begin{aligned} I_T &:= \int_{\mathbb{R}^2} f(\lambda)g(\lambda+\mu)\Phi^h_{2,T}(\mu)d\lambda d\mu \\ &= \int_{\mathbb{R}} \psi(\mu)\Phi^h_{2,T}(\mu)d\mu = \frac{1}{2\pi}\int_{\mathbb{R}} \hat{\psi}(\zeta)\hat{\Phi}^h_{2,T}(\zeta)d\zeta, \end{aligned} \tag{8.2.24}$$

where

$$\psi(\mu) = \int_{\mathbb{R}} f(\lambda)g(\lambda+\mu)d\lambda, \quad \hat{\psi}(\zeta) = \hat{f}(\zeta)\hat{g}(\zeta), \quad \lambda,\zeta \in \mathbb{R}. \tag{8.2.25}$$

According to (4.8.32) and (4.8.37), we can write

$$\Phi^h_{2,T}(\mu) = \frac{H_{1,T}(\mu)H_{1,T}(-\mu)}{2\pi T H_{2,T}(0)} = \frac{T}{2\pi H_2}\hat{h}(T\mu)\hat{h}(-T\mu) = \frac{T}{2\pi H_2}\varphi(T\mu), \tag{8.2.26}$$

where $H_2 = \int_0^1 h^2(t)dt$ and $\varphi(\zeta) := \hat{h}(\zeta)\hat{h}(-\zeta)$.

Observe that for the function $q(\lambda) := \int_{\mathbb{R}} h(t)h(t-\lambda)dt$ we have $\hat{q} = \varphi \in L^1(\mathbb{R})$. Therefore,

$$q(\lambda) = \frac{1}{2\pi}\int_{\mathbb{R}} \varphi(\zeta)e^{i\lambda\zeta}d\zeta = \frac{1}{2\pi}\hat{\varphi}(-\lambda)$$

and

$$\hat{\varphi}(\zeta) = 2\pi q(-\zeta) = 2\pi\int_{\mathbb{R}} h(t)h(t+\zeta)dt = 2\pi\int_0^1 h(t)h(t+\zeta)dt.$$

Now, by Eq. (8.2.26), we have

$$\hat{\Phi}^h_{2,T}(\zeta) = \frac{1}{H_2}\int_0^1 h(t)h\left(t+\frac{\zeta}{T}\right)dt,$$

and hence, by (8.2.24) and (8.2.25), we obtain

$$I_T = \frac{1}{2\pi H_2}\int_{\mathbb{R}} \hat{f}(\zeta)\hat{g}(\zeta)\int_0^1 h(t)h\left(t+\frac{\zeta}{T}\right)dtd\zeta.$$

Therefore, we can write

$$
\begin{aligned}
I_T - \int_{\mathbb{R}} f(\lambda)g(\lambda)d\lambda &= I_T - \frac{1}{2\pi}\int_{\mathbb{R}} \hat{f}(\zeta)\hat{g}(\zeta)d\zeta \\
&= \frac{1}{2\pi H_2}\int_{\mathbb{R}} \hat{f}(\zeta)\hat{g}(\zeta)\int_0^1 \left[h(t)h\left(t+\frac{\zeta}{T}\right) - h^2(t)\right]dtd\zeta \\
&= \frac{1}{2\pi H_2}\int_{|\zeta|\le T} \hat{f}(\zeta)\hat{g}(\zeta)\int_0^1 h(t)\left[h\left(t+\frac{\zeta}{T}\right) - h(t)\right]dtd\zeta \\
&\quad - \frac{1}{2\pi}\int_{|\zeta|>T} \hat{f}(\zeta)\hat{g}(\zeta)d\zeta =: R_1 - R_2.
\end{aligned}
\tag{8.2.27}
$$

Since we assume that the functions g and h are of bounded variation, we have

$$
|\hat{g}(\zeta)| \le K\frac{1}{|\zeta|}, \quad |h(\zeta)| < K, \quad \int_0^1 |h(t+\zeta) - h(t)|\,dt \le K|\zeta|, \quad \zeta \in \mathbb{R},
$$

where the constant K does not depend on ζ. Hence, as $T \to \infty$, we have

$$
\begin{aligned}
|R_1| &\le K^3 \int_{|\zeta|\le T} \left|\hat{f}(\zeta)\cdot\frac{1}{\zeta}\cdot\frac{\zeta}{T}\right| d\zeta \\
&\le \frac{K^3}{T}\left[\int_{|\zeta|\le\sqrt{T}} |\hat{f}(\zeta)|d\zeta + \int_{\sqrt{T}<|\zeta|\le T} |\hat{f}(\zeta)|d\zeta\right] \qquad (8.2.28)\\
&\le \frac{K^3}{T}\left[\|\hat{f}\|_2 (2\sqrt{T})^{1/2} + \left\{\int_{|\zeta|>\sqrt{T}} |\hat{f}(\zeta)|^2 d\zeta\right\}^{1/2} T^{1/2}\right] = o\left(T^{-1/2}\right),
\end{aligned}
$$

and

$$
\begin{aligned}
|R_2| &\le K\int_{|\zeta|>T} \left|\hat{f}(\zeta)\frac{1}{\zeta}\right| d\zeta \qquad (8.2.29)\\
&\le K\left\{\int_{|\zeta|>T} \left|\hat{f}(\zeta)\right|^2 d\zeta\right\}^{1/2} \left\{\int_{|\zeta|>T} \frac{1}{\zeta^2}d\zeta\right\}^{1/2} = o\left(T^{-1/2}\right).
\end{aligned}
$$

By combining the relations (8.2.27)–(8.2.29), we obtain the desired result in (8.2.23). Lemma 8.2.2 is proved. □

Denote

$$
\Delta_{2,T}^h := \left|\int_{\mathbb{R}^2} f(\lambda)g(\lambda+\mu)\Phi_{2,T}^h(\mu)d\lambda d\mu - \int_{\mathbb{R}} f(\lambda)g(\lambda)d\lambda\right|. \tag{8.2.30}
$$

Lemma 8.2.3 *Assume that Assumption 8.2.2 is satisfied. Then the following inequality holds:*

$$\Delta_{2,T}^h \le C_h \begin{cases} T^{-(\beta_1+\beta_2)}, & \text{if } \beta_1+\beta_2<1 \\ T^{-1}\ln T, & \text{if } \beta_1+\beta_2=1 \\ T^{-1}, & \text{if } \beta_1+\beta_2>1, \end{cases} \qquad T>0, \tag{8.2.31}$$

where C_h is a constant depending on h.

Proof According to Eqs. (4.8.32) and (4.8.33), we have

$$\Phi_{2,T}^h(\mu) = \frac{H_{1,T}(\mu)H_{1,T}(-\mu)}{2\pi T H_2}, \tag{8.2.32}$$

where $H_2 = \int_0^1 h^2(t)dt > 0$.

Observe first that, similar to the non-tapered case (see Proposition 2.2.3), the tapered kernel $\Phi_{2,T}^h$ has the following properties (see Lemma 4.8.5):

$$\int_{\mathbb{R}} \Phi_{2,T}^h(\mu)(\mu)\,d\mu = 1, \qquad \int_{|\mu|\ge 1} \left|\Phi_{2,T}^h(\mu)\right| d\mu \le C_h T^{-1},$$

$$\int_{|\mu|\le 1} \left|\Phi_{2,T}^h(\mu)\mu^\alpha\right| d\mu \le C_h \begin{cases} T^{-\alpha}, & \text{if } 0<\alpha<1 \\ T^{-1}\ln T, & \text{if } \alpha=1 \\ T^{-1}, & \text{if } \alpha>1. \end{cases} \tag{8.2.33}$$

Since the function $\Phi_{2,T}^h(\mu)$ is even, we have

$$\int_{\mathbb{R}^2} f(\lambda)g(\lambda+\mu)\Phi_{2,T}^h(\mu)d\lambda d\mu = \int_{\mathbb{R}^2} f(\lambda+\mu)g(\lambda)\Phi_{2,T}^h(\mu)d\lambda d\mu.$$

Therefore, using the equality

$$\int_{\mathbb{R}} f(\lambda)g(\lambda)dt = \int_{\mathbb{R}} f(\lambda+\mu)g(\lambda+\mu)dt, \qquad u \in \mathbb{R},$$

we obtain

$$2\Delta_{2,T}^h = \left|\int_{\mathbb{R}} \Phi_{2,T}^h(\mu) \int_{\mathbb{R}} [f(\lambda+\mu)-f(\lambda)]\,[g(\lambda+\mu)-g(\lambda)]\,d\lambda d\mu\right|. \tag{8.2.34}$$

Next, using Hölder's inequality from (8.2.34), we have

$$2\Delta_{2,T}^h \le \int_{\mathbb{R}} \left|\Phi_{2,T}^h(\mu)\right| \cdot \|f(\mu+\cdot)-f(\cdot)\|_{L^p} \|g(\mu+\cdot)-g(\cdot)\|_{L^q} d\mu. \tag{8.2.35}$$

Therefore, based on the conditions of the lemma, we obtain

$$\Delta_{2,T}^{h} \leq C \int_{|\mu|\leq 1} \left|\Phi_{2,T}^{h}(\mu)\mu^{\beta_1+\beta_2}\right| d\mu$$
$$+ C\|f\|_{L^p}\|g\|_{L^q} \int_{|\mu|>1} \left|\Phi_{2,T}^{h}(\mu)\right| d\mu. \tag{8.2.36}$$

Now, the desired inequality in (8.2.31) follows from (8.2.33) and (8.2.36). Lemma 8.2.3 is proved. □

8.2.2.2 Proofs of Theorems

Proof of Theorem 8.2.1 In view of formulas (8.2.1), (8.2.10), and (8.2.22), for the bias $\mathbb{E}[J_T^h] - J$ of the estimator J_T^h, we have

$$\mathbb{E}[J_T^h] - J = \int_{\mathbb{R}}\int_{\mathbb{R}} f(\lambda)g(\lambda+\mu)\Phi_{2,T}^{h}(\mu)d\lambda d\mu - \int_{\mathbb{R}} f(\lambda)g(\lambda)d\lambda \tag{8.2.37}$$

$$= \int_{\mathbb{R}} \Psi(\mu)\Phi_{2,T}^{h}(\mu)d\lambda d\mu - \Psi(0). \tag{8.2.38}$$

Therefore, the asymptotic relation (8.2.11) follows from (8.2.38) and Lemma 4.8.6 with $k = 2$. Theorem 8.2.1 is proved. □

Proof of Theorem 8.2.2 The asymptotic relation (8.2.12) under Assumption 8.2.3 follows from (8.2.37) and Lemma 8.2.2. Under Assumption 8.2.2, it follows from (8.2.37) and Lemma 8.2.3. Theorem 8.2.2 is proved. □

Proof of Theorem 8.2.3 The result under Assumption 8.2.4 (a) (the discrete time case) follows from Theorem 5.2.3. The proof for the continuous time case, that is, under Assumption 8.2.4 (b), is similar to that of the discrete time case. The result under Assumption 8.2.4 (c) (the continuous time Lévy-driven case) follows from Theorem 7.2.3. □

Proof of Theorem 8.2.5 We need to prove that

$$T^{1/2}\left[J_T^h - E(J_T^h)\right] \xrightarrow{d} \eta \sim N\left(0, \sigma_h^2(J)\right) \quad \text{as} \quad T\to\infty, \tag{8.2.39}$$

where $\sigma_h^2(J)$ is defined in (8.2.14).

We show that the relation (8.2.39) follows from Theorem 7.2.3. First, observe that, in light of Eqs. (4.8.6), (7.1.1), (7.1.2), (7.2.1), and (8.2.6), we can write

$$T^{1/2}\left[J_T^h - \mathbb{E}(J_T^h)\right] = T^{1/2}C_T^{-1}\left[Q_T^h - \mathbb{E}(Q_T^h)\right]$$
$$= \frac{T^{1/2}}{2\pi H_2 T}\left[Q_T^h - \mathbb{E}(Q_T^h)\right] = \frac{1}{2\pi H_2}\widetilde{Q}_T^h. \tag{8.2.40}$$

It follows from Theorem 7.2.3 that

$$\widetilde{Q}_T^h \xrightarrow{d} \eta \sim N(0, \sigma_{L,h}^2) \quad \text{as} \quad T \to \infty, \tag{8.2.41}$$

where $\sigma_{L,h}^2$ is given by (see 7.2.11)

$$\sigma_{L,h}^2 = 16\pi^3 H_4 \int_{\mathbb{R}} f^2(\lambda) g^2(\lambda) d\lambda + \kappa_4 4\pi^2 H_4 \left[\int_{\mathbb{R}} f(\lambda) g(\lambda) d\lambda \right]^2, \tag{8.2.42}$$

and H_4 is defined in (4.8.6).

Therefore, in view of (8.2.40) and (8.2.42), we have

$$\begin{aligned} \sigma_h^2(J) = \frac{1}{4\pi^2 H_2^2} \sigma_{L,h}^2 &= 4\pi e(h) \int_{\mathbb{R}} f^2(\lambda) g^2(\lambda) d\lambda \\ &+ \kappa_4 e(h) \left[\int_{\mathbb{R}} f(\lambda) g(\lambda) d\lambda \right]^2, \end{aligned} \tag{8.2.43}$$

where $e(h)$ is defined as in (8.2.16). The relation (8.2.39), with $\sigma_h^2(J)$ given by (8.2.14), follows from (8.2.40), (8.2.41), and (8.2.43). Theorem 8.2.5 is proved. □

Proof of Theorem 8.2.7 Taking into account the equality in (8.2.18), we need to establish the following two asymptotic relations to prove the theorem:

$$T^{1/2} \left[\mathbb{E}(J_T^h) - J \right] \to 0 \quad \text{as} \quad T \to \infty, \tag{8.2.44}$$

$$T^{1/2} \left[J_T^h - E(J_T^h) \right] \xrightarrow{d} \eta \sim N\left(0, \sigma_{h,G}^2(J)\right) \quad \text{as} \quad T \to \infty, \tag{8.2.45}$$

where $\sigma_{h,G}^2(J)$ is given by (8.2.20).

Observe first that the relation (8.2.44) is an immediate consequence of Lemma 8.2.3 since under each of the conditions (a)–(d) in condition ($\mathcal{H}$) (see Definition 2.1.1), we have $\beta_1 + \beta_2 > 1/2$.

Next, since by Proposition 2.1.9, under the condition ($\mathcal{H}$), there exist numbers p_1 ($p_1 > p$) and q_1 ($q_1 > q$) such that $H_p(\beta_1) \subset L_{p_1}$, $H_q(\beta_2) \subset L_{q_1}$ and $1/p_1 + 1/q_1 \le 1/2$, we can apply Theorem 7.2.1 (**C**), to obtain that

$$Q_T^h := T^{-1/2} \left(Q_T^h - \mathbb{E}[Q_T^h] \right) \xrightarrow{d} \eta \sim N(0, \sigma_h^2) \quad \text{as} \quad T \to \infty, \tag{8.2.46}$$

where Q_T^h and $\sigma_h^2 = \sigma_h^2(Q)$ are given by (7.1.1) and (7.2.2), respectively.

Also, in view of (7.1.2), (7.2.2), and (8.2.6), we have

$$\sigma_{h,G}^2(J) = \frac{1}{4\pi^2 H_2^2} \sigma_h^2(Q) = 4\pi e(h) \int_{\mathbb{R}} f^2(\lambda) g^2(\lambda) d\lambda, \tag{8.2.47}$$

where $e(h)$ is given by (8.2.16). By combining (8.2.46) and (8.2.47), we obtain the relation (8.2.45). Theorem 8.2.7 is proved. □

8.3 Estimation of Nonlinear Functionals for Gaussian Models

8.3.1 Estimation of Nonlinear Functionals

The estimation problem becomes somewhat more complicated for nonlinear spectral functionals. The issue is that if the estimand functional $J = J(f)$ is nonlinear, then the simple "plug-in" statistic $J(I_T)$ is not necessarily a consistent estimator for the functional $J(f)$ (see, e.g., Hasminskii and Ibragimov [171], and Taniguchi and Kakizawa [295], Section 6.1). Therefore, instead of using the periodogram $I_T(\lambda)$, we need to employ a suitable sequence of consistent estimators $\widehat{f}_T$ of f.

For simplicity, we will focus on the non-tapered case, where $h(t) = \mathbf{1}_{[0,1]}(t)$.

We assume that the spectral density f belongs to the Hölder class $H_p(\beta)$, where $\beta > 0$ and $p \geq 1$. As an estimator for the unknown spectral density f, we consider the following kernel-type statistic:

$$\widehat{f}_T(\lambda) = \int_{-\infty}^{\infty} W_T(\lambda - \mu) I_T(\mu)\, d\mu, \tag{8.3.1}$$

where $I_T(\lambda)$ is the periodogram of $X(t)$ defined by (8.2.5) with $h(t) = \mathbf{1}_{[0,1]}(t)$. For the kernel $W_T(\lambda)$ we set down the following assumptions.

Assumption 8.3.1 $W_T(\lambda) = M_T W(M_T \lambda)$, where $M_T = O(T^\gamma)$, and $b_T := M_T^{-1}$ is the bandwidth. The choice of the number γ $(0 < \gamma < 1)$ will depend on the a priori knowledge of the spectral density f and the estimand functional J.

Assumption 8.3.2 $W(\lambda)$ is a bounded, even, non-negative function such that $W(\lambda) \equiv 0$ for $|\lambda| > 1$, and

$$\int_{-1}^{1} W(\lambda)\, d\lambda = 1, \quad \int_{-1}^{1} \lambda^k W(\lambda)\, d\lambda = 0, \quad k = 1, 2, \ldots, r,$$

where $r = [\beta]$ is the integer part of β.

We assume that the functional $J(f)$ is Fréchet differentiable in L^2 with derivative $J'(f;\lambda)$. Specifically, there exists a continuous linear functional $J'_f : L^2 \to \mathbb{R}$:

$$J'_f(\psi) = \int_\Lambda J'(f;\lambda)\psi(\lambda) d\lambda, \quad \psi \in L^2 \tag{8.3.2}$$

such that for $f, g \in L^2$, we have

$$|J(g) - J(f) - J'_f(g - f)| = o\left(||g - f||_2\right) \quad \text{as} \quad ||g - f||_2 \to 0. \tag{8.3.3}$$

Additionally, we assume that the derivative $J'(f)$ satisfies a Hölder condition: there exist constants $C > 0$ and δ $(0 < \delta \leq 1)$ such that for any $f_1, f_2 \in L^2$,

$$||J'(f_1) - J'(f_2)|| \leq C||f_1 - f_2||_2^{\delta}. \tag{8.3.4}$$

Furthermore, we assume that the derivative $J'(f)$ satisfies the following condition:

$$0 < ||J'(f)f||_2 < \infty. \tag{8.3.5}$$

The next result provides sufficient conditions for the "plug-in" statistic $J(\widehat{f}_T)$ to be a $\sqrt{T}$-consistent and asymptotically normal estimator for a nonlinear functional $J(f)$.

Theorem 8.3.1 *Let the spectral density f and the functional $J(f)$ be such that:*
(i) The pair $(f, g := J'(f))$ satisfies the condition (8.3.5) *and Assumption 8.2.2.*
(ii) The functional $J(\cdot)$ satisfies the condition (8.3.4) with $\delta \geq (2\beta - 1)^{-1}$.
Let the estimator $\widehat{f}_T$ for f be defined by (8.3.1) with the kernel $W_T(\lambda)$ satisfying Assumptions 8.3.1 and 8.3.2 with $\frac{1}{2\beta} < \gamma < \frac{\delta}{\delta+1}$.

Then, the "plug-in" statistic $J(\widehat{f}_T)$ is a $\sqrt{T}$-consistent and asymptotically normal estimator for the functional $J(f)$ with asymptotic variance $\sigma^2_{h,G}(J)$ as given in (8.2.20) with $g = J'(f)$). Specifically, we have

$$T^{1/2}\left[J(\widehat{f}_T) - J(f)\right] \xrightarrow{d} \eta \quad \text{as} \quad T \to \infty, \tag{8.3.6}$$

where η is a centered normally distributed random variable with variance $\sigma^2_{h,G}(J)$, defined by (8.2.20) with $g = J'(f)$) and $e(h) = 1$.

8.3.2 *Proof of Theorem 8.3.1*

We will first prove two lemmas.

Lemma 8.3.1 *Let $\psi(\lambda)$ be an even integrable function such that the pair (f, ψ) satisfies condition ($\mathcal{H}$) (see Definition 2.1.1) and $0 < ||f\psi||_2 < \infty$. Let $\widehat{f}_T(\lambda)$ be defined as in (8.3.1) with the kernel $W_T(\lambda)$ satisfying Assumptions 8.3.1 and 8.3.2, where $\frac{1}{2\beta} < \gamma < 1$. Then, the distribution of the random variable*

$$\eta_T := T^{1/2} \int_{\Lambda} \psi(\lambda)\left[\widehat{f}_T(\lambda) - f(\lambda)\right] d\lambda \tag{8.3.7}$$

as $T \to \infty$ *converges to the normal distribution* $N(0, \sigma^2)$, *where*

$$\sigma^2 = 4\pi \int_\Lambda \psi^2(\lambda) f^2(\lambda)\, d\lambda. \tag{8.3.8}$$

Proof It follows from Theorem 8.2.7 that the distribution of the random variable

$$\xi_T := T^{1/2} \int_\Lambda \psi(\lambda)\, [I_T(\lambda) - f(\lambda)]\, d\lambda \tag{8.3.9}$$

converges to a normal distribution $N(0, \sigma^2)$ as $T \to \infty$, with σ^2 defined in (8.3.8).

Therefore, to complete the proof, it is sufficient to show that

$$|\xi_T - \eta_T| = o_P(1) \quad \text{as} \quad T \to \infty. \tag{8.3.10}$$

By Eqs. (8.3.1) and (8.3.7), we have

$$\begin{aligned}
\eta_T &= T^{1/2} \int_\Lambda \psi(\lambda) \left[\int_\Lambda W_T(\lambda - \mu)[I_T(\mu) - f(\mu)] d\mu \right] d\lambda \\
&+ T^{1/2} \int_\Lambda \psi(\lambda) \left[\int_\Lambda W_T(\lambda - \mu) f(\mu) d\mu - f(\lambda) \right] d\lambda \\
&:= \eta_T^{(1)} + S_T \quad \text{(say)}.
\end{aligned} \tag{8.3.11}$$

Using the change of variable $M_T(\lambda - \mu) = t$ and the dominated convergence theorem, it can be shown that (cf. Ginovyan [129], Taniguchi [294])

$$|\eta_T^{(1)} - \xi_T| = o_P(1) \quad \text{as } T \to \infty. \tag{8.3.12}$$

Next, by applying Hölder's and Minkowski's generalized inequalities, we obtain

$$|S_T| \le T^{1/2} ||\psi||_q (1 + ||W_T||_1) E_{M_T,p}(f), \tag{8.3.13}$$

where $E_{A,p}(f)$ represents the best approximation (in the metric of L^p) of the function f by entire analytic functions of exponential type A in the continuous time case and by polynomials of degree A in the discrete time case.

The assumption $f \in H_p(\beta)$ implies that $E_{A,p}(f) \le CA^{-\beta}$ (see Proposition 2.1.6). Hence, taking into account that $M_T = O(T^\gamma)$, in view of inequality (8.3.13), we can conclude that

$$S_T = O\left(T^{1/2} M_T^{-\beta}\right) = O\left(T^{1/2 - \gamma\beta}\right) \to 0 \tag{8.3.14}$$

as $T \to \infty$, because by assumption, $\gamma > \frac{1}{2\beta}$.

A combination of Eqs. (8.3.11), (8.3.12), and (8.3.14) yields (8.3.10). This completes the proof of Lemma 8.3.1. □

Lemma 8.3.2 *Let $f \in H_p(\beta)$, and let $\widehat{f}_T(\lambda)$ be defined as in (8.3.1) with the kernel $W_T(\lambda)$ satisfying Assumptions 8.3.1 and 8.3.2 with $\frac{1}{2\beta} < \gamma < \frac{\delta}{\delta+1}$ and $\delta > 0$. The following relation holds:*

$$T^{1/2}||\widehat{f}_T - f||_2^{1+\delta} = o_P(1) \quad \text{as} \quad T \to \infty.$$

Proof First, observe that (cf. Hannan [169], Section 6.4, and Theorem 4 in Ginovyan [120])

$$E|\widehat{f}_T(\lambda) - f(\lambda)|^2 = O\left(M_T T^{-1}\right) + O\left(M_T^{-2\beta}\right) \tag{8.3.15}$$

uniformly in λ. Using Fubini's theorem from (8.3.15), we have

$$||\widehat{f}_T - f||_2 = O_P\left(M_T^{1/2} T^{-1/2}\right) + O_P\left(M_T^{-\beta}\right). \tag{8.3.16}$$

Hence, taking into account that $M_T = O(T^\gamma)$ from (8.3.16), we obtain

$$T^{1/2}||\widehat{f}_T - f||_2^{1+\delta} = O_P\left(T^{\frac{1}{2}+\left(\frac{\gamma}{2}-\frac{1}{2}\right)(1+\delta)}\right) + O_P\left(T^{\frac{1}{2}-\gamma\beta(1+\delta)}\right). \tag{8.3.17}$$

The assumptions of the lemma imply that $\frac{1}{2} + \left(\frac{\gamma}{2} - \frac{1}{2}\right)(1+\delta) < 0$ and $\frac{1}{2} - \gamma\beta(1+\delta) < 0$. Therefore, both terms on the right-hand side of (8.3.17) are $o_P(1)$ as $T \to \infty$, and the result follows. □

Proof of Theorem 8.3.1 It follows from Eqs. (8.3.2)–(8.3.4) that

$$\left|J(\widehat{f}_T) - J(f) - \int_\Lambda J'(f;\lambda)(\widehat{f}_T(\lambda) - f(\lambda))\,d\lambda\right|$$
$$\le ||\widehat{f}_T - f|| \sup_{0\le\theta\le1} ||J'(f + \theta(\widehat{f}_T - f)) - J'(f)|| \le C||\widehat{f}_T - f||^{1+\delta}.$$

Therefore,

$$T^{1/2}[J(\widehat{f}_T) - J(f)] = T^{1/2}\int_\Lambda J'(f;\lambda)\left[\widehat{f}_T(\lambda) - f(\lambda)\right]d\lambda$$
$$+ O_P\left(T^{1/2}||\widehat{f}_T - f||^{1+\delta}\right). \tag{8.3.18}$$

By Lemma 8.3.2 and Eq. (8.3.18), we have

$$T^{1/2}[J(\widehat{f}_T) - J(f)] = T^{1/2} \int_{\Lambda} J'(f;\lambda)\left[\widehat{f}_T(\lambda) - f(\lambda)\right] d\lambda + o_P(1). \tag{8.3.19}$$

Applying Lemma 8.3.1 with $\psi(\lambda) = J'(f;\lambda)$ from (8.3.19), the result follows. Theorem 8.3.1 is proved. □

8.4 Estimation Under the LAN Condition: Efficient Estimation

8.4.1 *LAN and Efficiency Concepts*

In this section, we assume that the underlying model $X(t)$ is a discrete time Gaussian process, and the estimation procedure is based on the non-tapered observations $\mathbf{X}_T = \{X(t), t = 1, \ldots, T\}$. We establish the local asymptotic normality of families of distributions generated by $\mathbf{X}_T$ and present variants of the Hájek-Ibragimov-Khas'minskii convolution theorem and the Hájek-Le Cam local asymptotic minimax theorem. Based on these theorems, we define the concepts of H-efficiency and IK-efficiency for estimators. We derive sufficient conditions for the corresponding "plug-in" statistics to be asymptotically H-efficient and IK-efficient estimators for both linear and nonlinear estimand spectral functionals.

8.4.1.1 Local Asymptotic Normality

The concept of local asymptotic normality (LAN) for families of distributions, introduced by Le Cam in 1960 (see Le Cam [207]), plays a crucial role in asymptotic estimation theory.

L. Le Cam, J. Hájek, I. Ibragimov and R. Khas'minskii, and others (see, for instance, Le Cam [208], Hájek [166, 167], Ibragimov and Khas'minskii [185, 187], Kutoyants [204], and references therein) have shown that many important properties of statistical estimators—such as the characterization of limiting distributions, lower bounds on accuracy, and asymptotic efficiency—can be derived from the LAN condition. The significance of the LAN concept in nonparametric estimation problems has been emphasized by Levit [210, 211], Ibragimov and Khas'minskii [185, 187], Millar [224], Kutoyants [204], and others.

Following Ibragimov and Khas'minskii [187], who proposed a definition of the LAN concept for situations where the underlying parametric set is a subset of a normed space or a smooth infinite-dimensional manifold, we define the LAN concept for our model.

Definition 8.4.1 Let $\mathbb{P}_{T,f}$ be the probability distribution of the observation $\mathbf{X}_T = \{X(t),\ t = 1, \dots, T\}$ with spectral density $f(\lambda)$. A family of distributions $\{\mathbb{P}_{T,f},\ f \in \mathcal{F}\}$ is called *locally asymptotically normal (LAN)* at a point $f_0 \in \mathcal{F}$ in the direction of the space $L^2 := L^2(\Lambda)$ with norming factors $A_T := A_T(f_0)$ if there exist a linear manifold $H := H(f_0) \subset L^2$ with closure $\overline{H} = L^2$ and a family $\{A_T\}$ of linear operators $A_T : L^2 \to L^2$ that satisfy the following conditions:

1. For any $h \in H$, $||A_T h||_2 \to 0$ as $T \to \infty$, where $|| \cdot ||_2$ denotes the L^2-norm.
2. For any $h \in H$, there exists a natural number $T(h)$ such that $f_0 + A_T h \in \mathcal{F}$ for all $T > T(h)$.
3. For any $h \in H$ and $T > T(h)$, the following relation holds:

$$\ln \frac{d\mathbb{P}_{T,f_0+A_T h}}{d\mathbb{P}_{T,f_0}}(\mathbf{X}_T) = \Delta_T(h, f_0) - \frac{1}{2}||h||_2^2 + \phi(T, h, f_0), \tag{8.4.1}$$

where $\Delta_T(h) := \Delta_T(h, f_0)$ is a random linear function on H, and for any $h \in H$ the variable $\Delta_T(h)$ is asymptotically $N(0, \|h\|_2^2)$-normally distributed, while $\phi(T, h, f_0) \to 0$ in $\mathbb{P}_{T,f_0}$-probability as $T \to \infty$.

Note that the presence of the LAN property depends on the point f_0, the space L^2, and the family of operators $\{A_T\}$. The choice of the manifold $H = H(f_0)$ may be somewhat arbitrary; we only require that $\overline{H} = L^2$.

We define the parametric set $\mathcal{F}$ as a subset of the space L^p $(p \geq 1)$ consisting of spectral densities that satisfy the Muckenhoupt condition (A_2) (see Definition 3.3.2) and belong to the Hölder class $H_p(\beta)$. We define $H = H(f)$ to be the linear manifold of bounded functions $h(\lambda)$ such that the pair (f, hf^{-1}) satisfies the condition $(\mathcal{H})$ (see Definition 2.1.1). Additionally, we define the operator $A_T : L^2 \to L^2$ as the multiplication operator by the function $T^{-1/2} f$, that is, $A_T h = [T^{-1/2} f] \cdot h$.

The next theorem provides sufficient conditions for the local asymptotic normality of the family of distributions $\{\mathbb{P}_{T,f},\ f \in \mathcal{F}\}$.

Theorem 8.4.1 *Let $\mathcal{F}$, H, and A_T be defined as above. Then the family of distributions $\{\mathbb{P}_{T,f},\ f \in \mathcal{F}\}$ satisfies the LAN condition at any point $f \in \mathcal{F}$ in the direction L^2, with norming factors A_T and*

$$\Delta_T(h) := \frac{T^{1/2}}{4\pi} \int_{-\pi}^{\pi} \left[\frac{I_T(\lambda)}{f(\lambda)} - 1 \right] h(\lambda)\, d\lambda, \tag{8.4.2}$$

where $I_T(\lambda)$ is the non-tapered periodogram of the process $X(t)$ defined by Eq. (8.2.5) *with $h(t) = \mathbf{1}_{[0,1]}(t)$.*

8.4.1.2 Characterization of the Limiting Distribution: *H*-Efficiency

We consider the problem of estimating the value $J(f)$ of a known functional $J(\cdot)$ at an unknown point $f \in \mathcal{F}$ based on an observation $\mathbf{X}_T$, which follows the distribution $\mathbb{P}_{T,f}$. We assume that the family $\{\mathbb{P}_{T,f},\ f \in \mathcal{F}\}$ satisfies the LAN condition at a point $f \in \mathcal{F}$ in the direction L^2 with norming factors A_T. Additionally, we assume that the functional $J(f)$ is Fréchet differentiable at $f \in L^2$ with the derivative $J'(f)$ satisfying the following condition uniformly for $f \in \mathcal{F}$:

$$0 < ||J'(f)f||_2 < \infty. \tag{8.4.3}$$

We now present a version of the Hájek-Ibragimov-Khas'minskii convolution theorem for regular estimators.

First, recall that (see, e.g., Ibragimov and Khas'minskii ([185], Section 2.9) an estimator $\widehat{J}_T := \widehat{J}_T(\mathbf{X}_T)$ of $J(f)$ is called *H-regular* at a point $f_0 \in \mathcal{F}$ if, for any $h \in H$, there exists a proper limiting distribution function F, independent of h, for the normed difference $T^{1/2}\left(\widehat{J}_T - J(f_h)\right)$, where $f_h = f_0 + A_T h$, in the sense of weak convergence:

$$\mathcal{L}\left\{T^{1/2}\left(\widehat{J}_T - J(f_h)\right)\big|\,\mathbb{P}_{T,f_h}\right\} \Longrightarrow F \quad \text{as} \quad T \to \infty.$$

The following theorem is derived from Theorem 8.4.1 and Theorem 3.1 in Ibragimov and Khas'minskii [187].

Theorem 8.4.2 *Let $\widehat{J}_T$ be an H-regular estimator of the functional $J(\cdot)$ at $f \in \mathcal{F}$. Assume that the pair $(f, J'(f))$ satisfies condition (8.4.3). Then, under the assumptions of Theorem 8.4.1, the limiting distribution F of the random variable $T^{1/2}\left(\widehat{J}_T - J(f)\right)$ is the convolution of a probability distribution G and a centered normal distribution with variance $\sigma^2 := 4\pi||J'(f)f||_2^2$; specifically, we have*

$$F = N(0, \sigma^2) * G. \tag{8.4.4}$$

By a well-known lemma of Anderson (see, e.g., Ibragimov and Khas'minskii [185], Section 2.10), the distribution F in (8.4.4) is less concentrated across symmetric intervals compared to the normal distribution $N(0, \sigma^2)$. This observation justifies the following definition of H-efficiency (cf. Millar [224], Dahlhaus and Wefelmeyer [79], Kutoyants [203], Section 2.1).

Definition 8.4.2 Let the family $\{\mathbb{P}_{T,f},\ f \in \mathcal{F}\}$ be LAN at a point $f \in \mathcal{F}$. An estimator $\widehat{J}_T$ of the functional $J(\cdot)$ is called *H-efficient* at a point f (in the class of H-regular estimators) with asymptotic variance $\sigma^2 := 4\pi||J'(f)f||_2^2$ if

$$\mathcal{L}\left\{T^{1/2}\left(\widehat{J}_T - J(f_h)\right)\big|\,\mathbb{P}_{T,f_h}\right\} \Longrightarrow N(0, \sigma^2) \quad \text{as} \quad T \to \infty.$$

Remark 8.4.1 This efficiency concept is a nonparametric version of Hájek's efficiency and has the same intuitive interpretation: the asymptotic distribution of any regular estimator $\widehat{J_T}$ of $J(\cdot)$ is always "more spread out" than the centered normal distribution with variance $4\pi\,||J'(f)f||_2^2$ (cf. Hájek [166], Millar [224]).

Remark 8.4.2 We also have the following characterization of H-regular and H-efficient estimators (cf. Dahlhaus and Wefelmeyer [79], Taniguchi and Kakizawa [295], Chapter 6):

If the family $\{\mathbb{P}_{T,\theta},\ f \in \mathcal{F}\}$ is LAN at $f \in \mathcal{F}$, then an estimator $\widehat{J_T}$ of $J(f)$ is H-regular and H-asymptotically efficient at f with asymptotic variance $4\pi\,||fJ'(f)||_2^2$ if and only if it admits the following stochastic approximation:

$$T^{1/2}[\widehat{J_T} - J(f)] - \Delta_T(fJ'(f)) = o_P(1) \quad \text{as} \quad T \to \infty, \tag{8.4.5}$$

where $\Delta_T(fJ'(f))$ is defined by (8.4.2) with $h = fJ'(f)$.

8.4.1.3 A Lower Bound for the Asymptotic Minimax Risks: IK-Efficiency

Denote by $\mathcal{J}_T$ the set of all estimators for the functional $J(\cdot)$ that are constructed on the basis of an observation $\mathbf{X}_T$. Let $\mathbf{W}$ denote the set of all loss functions $w : \mathbb{R} \to \mathbb{R}$ that are symmetric and non-decreasing on $\mathbb{R}^+ := [0, \infty)$, and satisfy the conditions: $w(x) \geq 0$ and $w(0) = 0$.

The next theorem, which follows from Theorem 4.1 in Ibragimov and Khas'minskii [187] and Theorem 8.4.1, presents a minimax lower bound for the risks of all possible estimators $\widehat{J_T}$ of $J(\cdot)$ in the neighborhood of a point $f \in \mathcal{F}$.

Theorem 8.4.3 *Assume that the pair $(f, J'(f))$ satisfies condition (8.4.3). Then, under the assumptions of Theorem 8.4.1, for all $w \in \mathbf{W}$, we have*

$$\liminf_{\delta\to 0}\ \lim_{T\to\infty}\ \inf_{\widehat{J_T}\in\mathcal{J}_T}\ \sup_{||\theta - f||_2<\delta}\ \mathbb{E}_\theta\left[w\left(T^{1/2}(\widehat{J_T} - J(f))\right)\right] \geq \mathbb{E}[w(\xi)], \tag{8.4.6}$$

where ξ is a centered normal random variable with variance $4\pi\,||J'(f)f||_2^2$ and $\mathbb{E}_\theta[\cdot]$ denotes the expectation with respect to the measure corresponding to the spectral density $\theta(\lambda)$.

Based on Theorem 8.4.3, we define the concept of asymptotically efficient estimators in the spirit of Ibragimov and Khas'minskii, referred to as *IK-efficiency*. For more details, see the seminal papers by Hasminskii and Ibragimov [171] and by Ibragimov and Khas'minskii [187].

Definition 8.4.3 Let the family $\{\mathbb{P}_{T,f},\ f \in \mathcal{F}\}$ be LAN at a point $f \in \mathcal{F}$. An estimator $\widehat{J}_T$ of the functional $J(\cdot)$ is called *IK-asymptotically efficient* at a point $f \in \mathcal{F}$ for the loss function $w(x) \in \mathbf{W}$, with asymptotic variance $\sigma^2 = 4\pi\,||J'(f)f||_2^2$ if the following relation holds:

$$\liminf_{\delta\to 0}\lim_{T\to\infty}\sup_{||\theta-f||_2<\delta}\mathbb{E}_\theta\left[w\left(T^{1/2}(\widehat{J}_T - J(f))\right)\right] = \mathbb{E}[w(\xi)], \qquad (8.4.7)$$

where ξ is defined as in Theorem 8.4.3.

Remark 8.4.3 An estimator $\widehat{J}_T$ of the functional $J(\cdot)$ that satisfies (8.4.7) is referred to as a *locally asymptotically minimax (LAM)* estimator (see, e.g., Levit [210], Kutoyants [203], Section 2.1).

Remark 8.4.4 Both definitions of efficiency, H-efficiency and IK-efficiency, require an asymptotically efficient estimator $\widehat{J}_T$ to exhibit local uniformity in the convergence of the normed difference $T^{1/2}\big(\widehat{J}_T - J(f)\big)$ to the centered normal random variable ξ with variance $4\pi||J'(f)f||_2^2$. When considering bounded loss functions $w(\cdot)$, these definitions are quite similar.

One advantage of the IK-efficiency over H-efficiency is that it compares all estimators constructed based on the observations $\mathbf{X}_T$, rather than restricting itself to regular estimators. Conversely, a key benefit of H-efficiency over IK-efficiency is its focus on limiting distributions instead of the limits of expectations (see Beran [32], and Kutoyants [203], Section 2.1).

For a detailed discussion of the definitions and relationships among various efficiency concepts, we refer to Chapter 2 of the monograph by Ibragimov and Khas'minskii [185].

8.4.2 Asymptotic Efficiency of Estimators

Let $\mathbf{W}_e$ denote the subset of loss functions $w \in \mathbf{W}$ that, for some constants $C_1 > 0$ and $C_2 > 0$, satisfy the condition $w(x) \le C_1\exp\{C_2|x|\}$.

The following result presents sufficient conditions for the "plug-in" statistic $J(I_T)$ to serve as an asymptotically efficient estimator for the linear functional $J(\cdot)$.

Theorem 8.4.4 *Let the functionals $J := J(f, g)$ and $\widehat{J}_T := J(I_T, g)$ be defined by (8.2.1) and (8.2.7), respectively. Assume that the conditions of Theorems 8.2.6 and 8.4.1 are satisfied. Then the statistic $\widehat{J}_T$ is:*

(a) An H-regular and H-efficient estimator of J with asymptotic variance $4\pi||fg||_2^2$.
(b) An IK-efficient estimator of J for loss functions $w(x) \in \mathbf{W}_e$ with asymptotic variance $4\pi||fg||_2^2$.

The next result provides sufficient conditions for the statistic $J(\widehat{f_T})$ to be an asymptotically efficient estimator for a nonlinear functional $J(\cdot)$.

Theorem 8.4.5 *Let the spectral density f, the estimator $\widehat{f_T}$ of f, and the functional $J(\cdot)$ satisfy the assumptions of Theorem 8.3.1 and the condition* (8.4.3). *Then, under the assumptions of Theorem 8.4.1, the statistic $J(\widehat{f_T})$ is:*

(a) An H-regular and H-efficient estimator of J with asymptotic variance $4\pi||J'(f)f||_2^2$.
(b) An IK-efficient estimator of J for loss functions $w(x) \in \mathbf{W}_e$ with asymptotic variance $4\pi||J'(f)f||_2^2$.

8.4.3 Examples

Example 8.4.1 (Estimation of the Covariance Function) Consider the problem of estimating the covariance function $r(u)$:

$$r(u) = J(f) = \int_{-\pi}^{\pi} e^{iu\lambda} f(\lambda)\, d\lambda$$

based on non-tapered data. In Example 8.2.1, it was shown that the "plug-in" statistic:

$$\widehat{r_T}(u) = J_T = \int_{-\pi}^{\pi} e^{iu\lambda} I_T(\lambda)\, d\lambda$$

is an asymptotically normal estimator for $r(u)$ with asymptotic variance:

$$\sigma^2(u) = 4\pi \int_{-\pi}^{\pi} \cos^2(u\lambda) f^2(\lambda)\, d\lambda.$$

It follows from Theorem 8.4.4 that the statistic $\widehat{r_T}(u)$ is also an asymptotically H-efficient and IK-efficient estimator for $r(u)$, with the asymptotic variance $\sigma^2(u)$.

Example 8.4.2 Consider the problem of estimating the integrated squared spectral density functional $J(f)$:

$$J(f) := ||f||_2^2 = \int_{-\infty}^{\infty} f^2(\lambda)\, d\lambda. \tag{8.4.8}$$

We have $J'(f) = 2f$. Hence, we can apply Theorem 8.4.5, to conclude that the "plug-in" statistic:

$$\widehat{J_T} = J(\widehat{f_T}) := \int_{-\infty}^{\infty} [\widehat{f_T}(\lambda)]^2\, d\lambda,$$

where $\widehat{f}_T(\lambda)$ is defined as in (8.3.1), serves as an asymptotically H-efficient and IK-efficient estimator for the functional given in (8.4.8), with the asymptotic variance: $\sigma^2 = 16\pi||f^2||_2^2$.

Example 8.4.3 Estimation of the Shannon neg-entropy functional $J_S(f)$:

$$J_S(f) = \int_{-\infty}^{\infty} f(\lambda) \ln f(\lambda)\, d\lambda. \tag{8.4.9}$$

In this case, we have $J'(f) = \ln f + 1$, and it follows from Theorem 8.4.5 that the "plug-in" statistic:

$$\widehat{J}_T = J(\widehat{f}_T) := \int_{-\infty}^{\infty} \widehat{f}_T(\lambda) \ln \widehat{f}_T(\lambda)\, d\lambda, \tag{8.4.10}$$

where $\widehat{f}_T(\lambda)$ is defined as in (8.3.1), is an asymptotically H-efficient and IK-efficient estimator for the functional given in (8.4.9), with the asymptotic variance:

$$\sigma^2 = 4\pi \int_{-\infty}^{\infty} (\ln f(\lambda) + 1)^2 f^2(\lambda) d\lambda.$$

8.4.4 Proofs of Theorems of Sect. 8.4

Before proceeding to the proofs, note that when we refer to uniformity with respect to f and J, we specifically mean the uniformity within the set of pairs (f, g) that satisfy the conditions ($\mathcal{H}$) and (8.4.3).

We first establish the following lemma.

Lemma 8.4.1 *Let the functionals $J := J(f, g)$ and $\widehat{J}_T := J(I_T, g)$ be defined by (8.2.1) and (8.2.7), respectively. Assume that the pair of functions (f, g) satisfies the conditions ($\mathcal{H}$) and $0 < ||fg||_2 < \infty$. Then, for all $w(x) \in \mathbf{W}_e$, uniformly in f and J, we have*

$$\lim_{T\to\infty} \mathbb{E}_f[w(T^{1/2}(\widehat{J}_T - J)] = \mathbb{E}[w(\xi)], \tag{8.4.11}$$

where ξ is a centered normal random variable with variance $4\pi||fg||_2^2$.

In particular, the following relations hold uniformly in f and J:

(i) $\lim_{T\to\infty} \sqrt{T}|\mathbb{E}_f[\widehat{J}_T] - J)| = 0.$
(ii) $\lim_{T\to\infty} T\mathrm{Var}(\widehat{J}_T) = 4\pi||fg||_2^2.$
(iii) Uniformly in $x \in \mathbb{R}$

$$\lim_{T\to\infty} |\mathbb{P}\{\sqrt{T}(\widehat{J}_T - J) < x\} - \mathbb{P}\{\xi < x\}| = 0.$$

Proof For the loss function $w(x) = x$, the relation (8.4.11) follows from Theorems 8.2.5 and 8.2.7, with $h(t) = \mathbf{1}_{[0,1]}(t)$ and $e(h) = 1$.

Therefore, to complete the proof of the lemma, it is sufficient to show that for all $T > T(a)$, we have

$$\mathbb{E}_f\left[\exp\{\pm a\tilde{J}_T\}\right] < \infty, \tag{8.4.12}$$

where $\tilde{J}_T = \sqrt{T}(J_T - \mathbb{E}_f[J_T])$.

Using Lemma 5.4.1, we can write

$$\begin{aligned}\mathbb{E}_f\left[\exp\{\pm a\tilde{J}_T\}\right] &= \mathbb{E}\left[\exp\left\{\pm a\sum_{j=1}^{T}\lambda_{j,T}(\xi_j^2 - 1)\right\}\right] \\ &= \exp\left\{\mp a\sum_{j=1}^{T}\lambda_{j,T}\right\}\prod_{j=1}^{T}(1 \mp 2a\lambda_{j,T})^{-1/2}\end{aligned} \tag{8.4.13}$$

where $\lambda_{1,T} \geq \cdots \geq \lambda_{T,T}$ are the eigenvalues of the matrix $T^{-1/2}B_T(f)\,B_T(g)$.

By Proposition 4.9.1, we obtain

$$\lim_{T\to\infty}\lambda_{1,T} = \lim_{T\to\infty}T^{-1/2}\|B_T(f)\,B_T(g)\|_\infty = 0. \tag{8.4.14}$$

Hence, using the inequality $(1 - x)^{-1} \leq \exp\{x + 2x^2\}$ for $|x| < 1/2$, we find from (8.4.13) that, for all sufficiently large T,

$$\mathbb{E}_f\left[\exp\{\pm a\tilde{J}_T\}\right] \leq \exp\left\{\frac{a^2}{T}\operatorname{tr}(B_T(f)\,B_T(g))^2\right\}, \tag{8.4.15}$$

where $\operatorname{tr}[A]$ stands for the trace of an operator A.

By Theorem 4.4.1 (**A2**), the term

$$\frac{1}{T}\operatorname{tr}(B_T(f)\,B_T(g))^2 \tag{8.4.16}$$

is bounded uniformly in f and J.

Therefore, (8.4.12) follows from (8.4.15), concluding the proof of the lemma. □

Proof of Theorem 8.4.1 First, observe that for all $h \in H$ and sufficiently large T, we have

$$f_0 + A_T(h) = f_0 + [T^{-1/2}f]\cdot h \in \mathcal{F}.$$

By Hölder's inequality, the assumptions of the theorem imply that $f(\lambda)\,h(\lambda) \in L^2$. Therefore, for all $h \in H$, we have

$$\|A_T(h)\|_2 = T^{-1/2}\,\|f\,h\|_2 \to 0 \quad \text{as} \quad \mathrm{T} \to \infty.$$

Thus, to complete the proof, it suffices to prove the following two assertions (see Definition 8.4.1):

(a) For any $h \in H$, the random variable $\phi(T, h, f)$, defined by

$$\phi(T, f, h) := L_T(\mathbf{X_T}, f, f_T) - \frac{T^{1/2}}{4\pi} \int_{-\pi}^{\pi} \frac{I_T(\lambda) - f(\lambda)}{f(\lambda)} h(\lambda)\,d\lambda + \frac{1}{8\pi} \int_{-\pi}^{\pi} h^2(\lambda)\,d\lambda \tag{8.4.17}$$

converges to 0 as $T \to \infty$ in $\mathbb{P}_{T,f}$-probability, where

$$L_T(\mathbf{X}_T, f, f_T) = \log \frac{d\mathbb{P}_{T,f_T}}{d\mathbb{P}_{T,f}}\,(\mathbf{X}_T)$$

is the logarithm of the likelihood ratio.

(b) For any $h \in H$, the random variable $\Delta_T(h, f)$ defined by Eq. (8.4.2) has asymptotically (as $T \to \infty$) normal distribution with mean zero and variance $\sigma^2 = \|h\|_2^2$.

In view of Chebyshev's inequality, to prove assertion (a), it suffices to show that the expectation and variance of the random variable $\phi(T, h, f)$ converge to 0 as $T \to \infty$. Specifically,

$$\mathbb{E}_f\,[\phi(T, f, h)] \longrightarrow 0 \quad \text{as} \quad \mathrm{T} \longrightarrow \infty \tag{8.4.18}$$

$$\mathrm{Var}_f\,[\phi(T, f, h)] \longrightarrow 0 \quad \text{as} \quad \mathrm{T} \longrightarrow \infty. \tag{8.4.19}$$

The proofs of the asymptotic relations (8.4.18) and (8.4.19) for short memory models can be found in Dzhaparidze [91, 93] (see also Davies [80]). For models that satisfy the Muckenhoupt condition, see Ginovyan [124].

Assertion (b) follows from Theorem 8.2.6. □

Proof of Theorem 8.4.4 Since, in this case, $J'(f) = g$, it follows from Eqs. (8.2.1), (8.2.7), and (8.4.2) that

$$T^{1/2}[\widehat{J}_T - J] = T^{1/2} \int_{\Lambda} [I_T(\lambda) - f(\lambda)]\,g(\lambda)\,d\lambda = \Delta_T(fg). \tag{8.4.20}$$

Therefore, assertion (a) of the theorem follows from Eq. (8.4.20) and Remark 8.4.2, while assertion (b) follows from Lemma 8.4.1. □

Proof of Theorem 8.4.5 It follows from Eqs. (8.3.7), (8.3.9), (8.3.10), and (8.3.19) that

$$T^{1/2}[J(\widehat{f}_T) - J] = T^{1/2} \int_\Lambda J'(f;\lambda)\,[I_T(\lambda) - f(\lambda)]\,d\lambda + o_P(1). \tag{8.4.21}$$

Next, by Eqs. (8.4.2) and (8.4.21), we have

$$T^{1/2}[J(\widehat{f}_T) - J] - \Delta_T(fJ'(f)) = o_P(1) \quad \text{as} \quad T \to \infty. \tag{8.4.22}$$

Therefore, assertion (a) of the theorem follows from Eq. (8.4.22) and Remark 8.4.2, while assertion (b) of the theorem follows from Lemma 8.4.1 with $g = J'(f)$. □

8.5 Asymptotic Bounds for the Minimax Mean Square Risks

We revisit the problem of estimating a linear, L^p-continuous spectral functional $J(f)$ based on non-tapered data. When $1 \le p \le 2$, the functional $J(f)$ is continuous in L^2. Hence, we can apply Theorem 8.4.4 to construct an asymptotically efficient estimator for $J(f)$. However, when $p > 2$, it is no longer possible to obtain an efficient estimator, as the right-hand side of inequality (8.4.6) approaches infinity.

Therefore, it becomes important to estimate the rate of decrease (as $T \to \infty$) of the minimax risk:

$$\inf_{\widehat{J}_T \in \mathcal{J}_T} \sup_{f \in \mathcal{F}} \mathbb{E}_f\left[w(\widehat{J}_T - J(f))\right],$$

where $\mathcal{F}$ is a given class of spectral densities and $\mathcal{J}_T$ is the set of all estimators of the functional $J(f)$ constructed based on a non-tapered observation $\mathbf{X}_T$.

It is clear that the bounds will depend on the number p and the smoothness properties of the functions from the class $\mathcal{F}$.

Here, we obtain exact asymptotic bounds for the minimax mean square risk:

$$\Delta_T^2 := \sup_{||J||=1} \inf_{\widehat{J}_T \in \mathcal{J}_T} \sup_{f \in \mathcal{F}} \mathbb{E}_f\left[|\widehat{J}_T - J(f)|^2\right]. \tag{8.5.1}$$

The following theorem provides upper asymptotic bounds for the risk Δ_T^2.

Theorem 8.5.1 *Let $\mathcal{F}_p(\beta)$ be the class of spectral densities f that belong to the Hölder class $H_p(\beta)$ $(p \ge 1,\ \beta > 0)$, and let $J(f)$ be a linear L^p-continuous* *functional defined by* (8.2.1). *Let the risk* Δ_T^2 *be defined as in (8.5.1). The following assertions hold:*

(a) If $p \geq 2$ *and* $\beta > 1/p$, *then*

$$\Delta_T^2 \leq CT^{-\frac{2p\beta}{p+2p\beta-2}}. \tag{8.5.2}$$

(b) If $p \geq 2$ *and* $\beta \leq 1/p$, *then*

$$\Delta_T^2 \leq CT^{-2\beta}. \tag{8.5.3}$$

Imposing an additional condition on the class $\mathcal{F}_p(\beta)$ of spectral densities f that ensures the LAN for the family of distributions $\{\mathbb{P}_{T,f}, f \in \mathcal{F}_p(\beta)\}$ (see Theorem 8.4.1), we can derive the corresponding lower bounds for the risk Δ_T^2.

Specifically, we have the following result.

Theorem 8.5.2 *Under the conditions of Theorems 8.4.1 and 8.5.1, the following assertions hold:*

(a) If $p \geq 2$ *and* $\beta > 1/p$, *then* $\Delta_T^2 \geq cT^{-\frac{2p\beta}{p+2p\beta-2}}$.
(b) If $p \geq 2$ *and* $\beta \leq 1/p$, *then* $\Delta_T^2 \geq cT^{-2\beta}$.

By combining Theorems 8.5.1 and 8.5.2, we obtain the following corollary, which provides exact asymptotic bounds for the risk Δ_T^2.

Corollary 8.5.1 *Under the conditions of Theorem 8.5.2, the following assertions hold as* $T \to \infty$:

(a) If $p \geq 2$ *and* $\beta > 1/p$, *then* $\Delta_T^2 \asymp T^{-\frac{2p\beta}{p+2p\beta-2}}$.
(b) If $p \geq 2$ *and* $\beta \leq 1/p$, *then* $\Delta_T^2 \asymp T^{-2\beta}$.

We prove Theorems 8.5.1 and 8.5.2 for the continuous time case; the proof for the discrete time case follows similarly.

Proof of Theorem 8.5.1 As an estimator of the linear functional $J(f)$, we consider the statistics $\widehat{J}_{TA}$, defined by

$$\widehat{J}_{TA} := \int_{\mathbb{R}} I_T(\lambda) g_A(\lambda) d\lambda, \tag{8.5.4}$$

where $A = A(T) \leq T$, $A(T) \to \infty$ as $T \to \infty$, $I_T(\lambda)$ is the periodogram given by (8.2.5) (with $(h(t) = \mathbf{1}_{(0,1)}(t))$, and $g_A(\lambda)$ is the Dirichlet singular integral corresponding to the function $g(\lambda)$ defined by Eq. (2.2.8).

We define

$$R_{TA} := \mathbb{E}[\widehat{J}_{TA}] - J(f)$$

and show that

$$|R_{TA}| \le C\,A^{-\beta} + \begin{cases} CT^{-\beta}, & \text{for } \beta < 1 \\ CT^{-1}\ln T, & \text{for } \beta = 1 \\ CT^{-1}, & \text{for } \beta > 1. \end{cases} \tag{8.5.5}$$

Indeed, according to Lemma 8.2.1 (with $(h(t) = \mathbf{1}_{(0,1)}(t))$, we have

$$\begin{aligned} |R_{TA}| &= \left| \int_{\mathbb{R}} \int_{\mathbb{R}} F_T(\lambda - \mu) f(\lambda) g(\mu) d\lambda d\mu - \int_{\mathbb{R}} f(\lambda) g(\lambda) d\lambda \right| \\ &\le \left| \int_{\mathbb{R}} \int_{\mathbb{R}} F_T(\lambda - \mu) f(\lambda) g(\mu) d\lambda d\mu - \int_{\mathbb{R}} f(\lambda) g_A(\lambda) d\lambda \right| \\ &+ \left| \int_{\mathbb{R}} f(\lambda)[g(\lambda) - g_A(\lambda)] d\lambda \right| =: M_1 + M_2, \end{aligned} \tag{8.5.6}$$

where $F_T(u)$ is the Fejér kernel defined by Eq. (2.2.2). Taking into account that

$$\int_{\mathbb{R}} f_A(\lambda)[g(\lambda) - g_A(\lambda)] d\lambda = 0,$$

by Hölder's inequality and Proposition 2.2.4 (a), we have that

$$\begin{aligned} M_2 &= \left| \int_{\mathbb{R}} [f(\lambda) - f_A(\lambda)][g(\lambda) - g_A(\lambda)] d\lambda \right| \\ &\le \|f - f_A\|_p \|g - g_A\|_q \le C A^{-\beta}. \end{aligned} \tag{8.5.7}$$

From (8.5.6), (8.5.7), and Lemma 8.2.3 (with $(h(t) = \mathbf{1}_{(0,1)}(t))$, we obtain (8.5.5).

Next, according to Lemma 5.4.1 part (c), we have

$$\text{Var}(\widehat{J}_{TA}) = \chi_2(\widehat{J}_{TA}) = 2T^{-2}||W_T(f)W_T(g_A)||_2^2. \tag{8.5.8}$$

From (8.5.5) and (8.5.8), we infer that

$$\begin{aligned} \mathbb{E}_f\left[|\widehat{J}_{TA} - J(f)|^2\right] &\le C\,A^{-2\beta} + 2T^{-2}\|W_T(f)W_T(g_A)\|_2^2 \\ &+ \begin{cases} CT^{-2\beta}, & \text{for } \beta < 1 \\ C(T^{-1}\ln T)^2, & \text{for } \beta = 1 \\ CT^{-2}, & \text{for } \beta > 1. \end{cases} \end{aligned} \tag{8.5.9}$$

Now, taking into account the properties of the functions f and g, we choose the number A in an optimal way.

Proof of Part (A) ($p \geq 2$ and $\beta > 1/p$). By setting $p_1 = 2$; $p_2 = \infty$, and then successively applying Proposition 2.7.6 (c) and Theorem 4.9.1, we find that

$$\begin{aligned}\mathrm{Var}(\widehat{J}_{TA}) &= 2T^{-2}||W_T(f)W_T(g_A)||_2^2 \\ &\leq 2T^{-2}||W_T(f)||_\infty^2||W_T(g_A)||_2^2 \leq C\,T^{-1}||f||_\infty^2||g_A||_2^2. \end{aligned} \tag{8.5.10}$$

Since $q = p/(p-1) \leq 2$, Proposition 2.2.4 (b) yields

$$||g_A||_2 \leq C\,A^{1/q-1/2}.$$

On the other hand, by Proposition 2.1.8, we have $||f||_\infty < \infty$.
Therefore,

$$\mathrm{Var}(\widehat{J}_{TA}) \leq C\,T^{-1}A^{2/q-1}. \tag{8.5.11}$$

By setting $A = T^{-p/(p-2+2p\beta)}$, we find from (8.5.9) and (8.5.11) that

$$\begin{aligned}\mathbb{E}_f\left[|\widehat{J}_{TA} - J(f)|^2\right] &\leq C\,T^{-\frac{2p\beta}{p-2+2p\beta}} \\ &+ \begin{cases} CT^{-2\beta}, & \text{for } \beta < 1 \\ C(T^{-1}\ln T)^2, & \text{for } \beta = 1 \\ CT^{-2}, & \text{for } \beta > 1. \end{cases} \leq C\,T^{-\frac{2p\beta}{p-2+2p\beta}}, \end{aligned} \tag{8.5.12}$$

and the estimate (8.5.2) follows.

Proof of Part (B) ($p \geq 2$ and $\beta \leq 1/p$). By setting $p_1 = 2$; $p_2 = \infty$, and then successively applying Proposition 2.7.6 (c) and Theorem 4.9.1, we find that

$$\mathrm{Var}(\widehat{J}_{TA}) \leq C\,T^{-1}||f||_\infty^2||g_A||_2^2. \tag{8.5.13}$$

Therefore, according to Proposition 2.2.4 (b) and (c), we have

$$\mathrm{Var}(\widehat{J}_{TA}) \leq C\,T^{-1}A^{2/q-1}A^{2/p-2\beta} = C\,T^{-1}A^{1-2\beta}. \tag{8.5.14}$$

Hence, by setting $A = T$, we obtain from (8.5.9) and (8.5.14) that

$$\begin{aligned}\mathbb{E}_f\left[|\widehat{J}_{TA} - J(f)|^2\right] &\leq C\,T^{-2\beta} \\ &+ \begin{cases} CT^{-2\beta}, & \text{for } \beta < 1 \\ C(T^{-1}\ln T)^2, & \text{for } \beta = 1 \\ CT^{-2}, & \text{for } \beta > 1. \end{cases} \leq C\,T^{-2\beta}, \end{aligned} \tag{8.5.15}$$

and the estimate (8.5.3) follows. This completes the proof of Theorem 8.5.1.

□

Proof of Theorem 8.5.2 We employ the Stein-Levit method, which converts the underlying nonparametric estimation problem into a well-defined parametric problem. We then establish a lower bound for this parametric problem using a variant of Hájek's inequality. For a detailed presentation of this method, we refer to the monograph by Ibragimov and Has'minskii [185], Section 4.1.

Let $\widehat{J}_{T,A}$ be the estimator of the functional $J(f)$ defined by Eq. (8.5.4).

We define

$$\mathcal{F}'_p(\beta) := \{f \in \mathcal{F}_p(\beta);\ f(\lambda) \geq c > 0\},$$

and consider the function

$$h_A(\lambda) := \frac{g_A(\lambda)}{\sqrt{T}||fg_A||_2^2}. \tag{8.5.16}$$

Observe that if $f(\lambda)$ is some spectral density function from the class $\mathcal{F}'_p(\beta)$ such that $f(\lambda)h_A(\lambda) \in \mathcal{F}_p(\beta)$, then for sufficiently large values of T the function

$$f_{T,h}(\lambda) := f(\lambda)\,(1 + h_A(\lambda))\,, \quad A = A(T),$$

will be a spectral density from the class $\mathcal{F}'_p(\beta)$.

Let $\mathbb{P}_{T,\theta}$ be the distribution of the observation $\mathbf{X}_T = \{X(u),\, 0 \leq u \leq T\}$ with spectral density $\theta(\lambda)$. By Theorem 8.4.1 the family of Gaussian distributions $\{\mathbb{P}_{T,\theta},\ \theta \in \mathcal{F}'_p(\beta)\}$ is LAN at a point $f \in \mathcal{F}'_p(\beta)$ in the direction of the space L^2.

Therefore, by applying the Stein-Levit method and Hájek's inequality (see Ibragimov and Khas'minskii [185], Theorem 2.12.1), it can be shown that

$$\sup_{\theta \in \mathcal{F}'_p(\beta)} \inf_{\widehat{J}_T \in \mathcal{J}_T} \mathbb{E}_\theta \left[|\widehat{J}_T - \Phi(\theta)|^2\right] \geq \frac{c}{T} \int_{\mathbb{R}} f^2(\lambda)\, g_A^2(\lambda)\, d\lambda, \tag{8.5.17}$$

where $\mathcal{J}_T$ is the set of all possible estimators of the functional $J(\theta)$, constructed based on the observation $\mathbf{X}_T$, and c is a positive constant.

Thus, to complete the proof of the theorem, we must choose $A = A(T)$ to satisfy the following conditions:

1. The function fh_A belongs to the class $\mathcal{F}_p(\beta)$.
2. The right-hand side of inequality (8.5.17) takes the form T^{-a}, where the number a is specified by the theorem.

We will outline the proof of assertion (a) of the theorem, while assertion (b) can be proven in a similar manner (for details, see Ginovyan [129, 130]).

Thus, assume that $f(\lambda) \in \mathcal{F}_p(\beta)$, where $p \geq 2$ and $\beta > 1/p$. We first show that $fh_A \in \mathcal{F}_p(\beta)$, where $h_A(\cdot)$ is defined as in (8.5.16).

Let $\beta = \alpha + r$, where $r \in \mathbb{N}_0$ and $0 < \alpha \le 1$. Computing the derivative $(fh_A)^{(r)}$ by the Leibniz formula, we find

$$J_r := \left\|(fh_A)^{(r)}(\cdot+\delta) - (fh_A)^{(r)}(\cdot)\right\|_p$$
$$\le C \sum_{k=0}^{r} \left\| f^{(k)}(\cdot+\delta)h_A^{(r-k)}(\cdot+\delta) - f^{(k)}(\cdot)h_A^{(r-k)}(\cdot)\right\|_p. \quad (8.5.18)$$

By applying Hölder's and Minkowski's inequalities, we obtain

$$J_r \le C\,||f||_\infty \left\|h_A^{(r)}(\cdot+\delta) - h_A^{(r)}(\cdot)\right\|_p + ||h_A^{(r)}||_\infty \,\|f(\cdot+\delta) - f(\cdot)\|_p$$
$$+ C \sum_{k=1}^{r} ||f^{(k)}||_p \left\|h_A^{(r-k)}(\cdot+\delta) - h_A^{(r-k)}(\cdot)\right\|_\infty$$
$$+ C \sum_{k=1}^{r} ||h_A^{(r-k)}||_\infty \left\|f^{(k)}(\cdot+\delta) - f^{(k)}(\cdot)\right\|_p. \quad (8.5.19)$$

Since $f \in \mathcal{F}_p(\beta)$ with $\beta = \alpha + r$, by Proposition 2.2.4 (c), we have

$$\|f^{(k)}\|_p \le C < \infty, \qquad k = 1, 2, \ldots, r. \quad (8.5.20)$$

Because $\beta > 1/p$, by the Hardy-Littlewood embedding theorem for Hölder classes (see Proposition 2.1.8 (b)), we obtain

$$\|f\|_\infty \le C < \infty. \quad (8.5.21)$$

The function $h_A(\cdot)$ is an entire function of exponential type A. Therefore, by Bernstein's inequality (see Proposition 2.1.7), we have

$$\|h_A^{(k)}\|_s \le A^k \|h_A\|_s., \quad 1 \le s \le \infty. \quad (8.5.22)$$

From (8.5.22) and the inequality (see Proposition 2.2.4 (d))

$$\|h_A\|_s \le C\, A^{1/t-1/s} \|h_A\|_t, \quad t < s \le \infty, \quad (8.5.23)$$

we find that

$$\|h_A^{(k)}\|_\infty \le A^k \|h_A\|_\infty \le A^{k+1/q} \|h_A\|_q. \quad (8.5.24)$$

From the inequalities (8.5.22)–(8.5.24) for $q_1 > q$ and $k \le r$, we obtain

$$\begin{aligned}\left\| h_A^{(k)}(\cdot+\delta) - h_A^{(k)}(\cdot)\right\|_{q_1} &= \left\| \int_x^{x+\delta} h_A^{(k+1)}(y)\,dy \right\|_{q_1} \\ &\le \min\left(\delta \left\| h_A^{(k+1)} \right\|_{q_1}, 2\left\| h_A^{(k)} \right\|_{q_1}\right) \\ &\le \min\left(\delta A^{k+1+1/q-1/q_1} \|h_A\|_q, A^{k+1/q-1/q_1} \|h_A\|_q\right). \end{aligned} \tag{8.5.25}$$

Therefore, taking into account that $\beta = \alpha + r$, for $q_1 > q$ and $k \le r$, we have

$$\left\| h_A^{(k)}(\cdot+\delta) - h_A^{(k)}(\cdot)\right\|_{q_1} \le C \cdot \delta^\alpha A^{\beta+1/q-1/p} \|h_A\|_q. \tag{8.5.26}$$

In view of relations (8.5.19)–(8.5.21), (8.5.24), and (8.5.26), we find for $r \ge 1$ that

$$J_r = \left\| (fh_A)^{(r)}(\cdot+\delta) - (fh_A)^{(r)}(\cdot)\right\|_p \le C \cdot \delta^\alpha A^{\beta+1/q-1/p} \|h_A\|_q. \tag{8.5.27}$$

Similarly, using (8.5.24) and (8.5.26), we obtain for $r = 0$

$$\begin{aligned} J_0 :&= \|(fh_A)(\cdot+\delta) - (fh_A)(\cdot)\|_p \\ &\le C \cdot \left\{ \|h_A(\cdot+\delta) - h_A(\cdot)\|_p + \delta^\alpha \|h_A\|_\infty \right\} \\ &\le C \cdot \delta^\alpha A^{\beta+1/q-1/p} \|h_A\|_q. \end{aligned} \tag{8.5.28}$$

Next, it follows from (8.5.16), (8.5.27), and (8.5.28) that for all $r \in \mathbb{N}_0$, $p \ge 2$, and $\beta > 1/p$ that

$$J_r := \left\| (fh_A)^{(r)}(\cdot+\delta) - (fh_A)^{(r)}(\cdot)\right\|_p \le C \cdot M_{A,T} \delta^\alpha, \tag{8.5.29}$$

where

$$M_{A,T} := T^{-1/2} \cdot A^{\beta+1/q-1/p} \|fg_A\|_2^{-2}. \tag{8.5.30}$$

Thus, the statement $fh_A \in \mathcal{F}_p(\beta)$ will hold if we can choose $A = A(T)$ such that $M_{A,T}$, as defined in (8.5.30), can be made as small as desired. We define

$$g_A(\lambda) := A^{-1/p} \cdot \frac{\sin(A\lambda)}{\lambda}, \quad \lambda \in \mathbb{R}. \tag{8.5.31}$$

Since $\int_{\mathbb{R}} g_A^2(\lambda)\,d\lambda = \pi\, A^{1-2/p}$, and by assumption $f(\lambda) \geq c > 0$, we have

$$\int_{\mathbb{R}} g_A^2(\lambda) f^2(\lambda)\,d\lambda \geq \pi c \cdot A^{1-2/p}. \tag{8.5.32}$$

Choosing $A = T^{p/(p-2+2p\beta)}$, and taking into account that $\frac{1}{q} = 1 - \frac{1}{p}$, we find from (8.5.30) and (8.5.32) that

$$M_{A,T} \leq C \cdot \frac{T^{-1/2} \cdot A^{\beta+1/q-1/p}}{A^{1-2/p}} = T^{\frac{2-p}{2(2p\beta+p-2)}}. \tag{8.5.33}$$

Since $p \geq 2$, we conclude from (8.5.29) and (8.5.33) that $fh_A \in \mathcal{F}_p(\beta)$. Finally, for $A = T^{p/(p-2+2p\beta)}$, from (8.5.17) and (8.5.32), we obtain

$$\Delta_T^2 \geq \frac{c}{T}\int_{\mathbb{R}} g_A^2(\lambda) f^2(\lambda)\,d\lambda \geq c\,T^{-1}\,A^{1-2/p} \geq c\,T^{-\frac{2p\beta}{2p\beta+p-2}}.$$

This completes the proof of assertion (a) of the theorem, thereby proving Theorem 8.5.2. □

8.6 Bibliographical Notes

Section 8.2 The problem of nonparametric estimation of linear spectral functionals based on non-tapered data for discrete time stationary Gaussian processes has been addressed in several books and papers. We quote merely the following references: Bentkus [26], Bentkus and Rudzkis [28], Brillinger [55], Dahlhaus [70, 76], Dahlhaus and Wefelmeyer [79], Dzhaparidze [93], Ginovyan [117, 127], Grenander and Rosenblatt [161], Guyon [165], Hasminskii and Ibragimov [171], Ibragimov and Khas'minskii [187], Millar [224], Pristley [249], Rosenblatt [261], and Taniguchi and Kakizawa [295]. Additional references can be found within these works.

For continuous time stationary Gaussian processes, this problem was studied in the papers by Ginovyan [118, 120, 121, 129, 130]. The tapered case was addressed by Ginovyan and Sahakyan [140, 142, 143], where the proofs of Theorems 8.2.1–8.2.7 can also be found.

Note that in the case where the underlying process $X(t)$ is nonlinear, the second term on the right-hand side of formula (8.2.14), which determines the asymptotic variance of the estimator, is expressed in terms of the tri-spectrum (the fourth order spectrum) of the process $X(t)$ (see, e.g., Dahlhaus [70], and Guyon [165], p.142).

The nonparametric estimation problem for linear spectral functionals of homogeneous Gaussian fields was considered by Ginovyan [125] (see also the papers by Anh et al. [7], Doukhan et al. [86], and Lavancier and Philippe [205]). In particular, Ginovyan [125] established the consistency and asymptotic normality of spectral estimators for both discrete and continuous time cases.

Section 8.3 The problem of construction of consistent and asymptotically normal nonparametric estimators for nonlinear smooth spectral functionals has been studied in the seminal paper by Hasminskii and Ibragimov [171] and by Ginovyan [129, 130] (see also Taniguchi [294] and Taniguchi and Kakizawa [295]). Theorem 8.3.1 was proved by Ginovyan [129].

A result similar to Theorem 8.3.1 regarding the estimation of nonlinear functionals of probability density functions has been obtained by Goldstein and Khas'minskii [156].

In the paper by von Sachs [310] (see also Janas and von Sachs [195]), the authors considered the nonparametric estimation problem for certain nonlinear integral functionals of spectral density, utilizing data tapers. They established conditions that ensure the consistency and asymptotic normality of the corresponding nonlinear functionals of the tapered periodogram. The results obtained were applied to the use of data tapers in nonparametric, peak-insensitive spectrum estimation.

Section 8.4 The LAN property for families of distributions generated by discrete time Gaussian stationary processes has been studied by Davies [80], Dzhaparidze [93], and Ginovyan [115, 124]. For continuous time processes, sufficient conditions for LAN were obtained by Solev and Zerbet [284].

Definition 8.4.1 is attributed to Ibragimov and Khas'minskii [187]. Theorem 8.4.1 was proved by Ginovyan [124]. Under different conditions, the result of Theorem 8.4.1 was also proved by Davies [80] and Dzhaparidze [93]. The proofs of Theorems 8.4.2 and 8.4.3 can be found in the seminal paper by Ibragimov and Khas'minskii [187]. Theorems 8.4.4 and 8.4.5 were proved by Ginovyan [129].

Efficient estimators for linear and nonlinear smooth spectral functionals, specifically when the spectral density belongs to the Sobolev class $H^{1/2}$, have been constructed by Hasminskii and Ibragimov [171]. Additionally, efficient estimators for functionals of the spectral density of homogeneous Gaussian random fields were obtained in Ludeña [220].

Section 8.5 The problem of bounding the minimax mean square risks of estimators for linear spectral functionals of stationary Gaussian processes was considered in the papers by Ginovyan [120, 129, 130].

Theorem 8.5.1 was first proved in Ginovyan [120]. In Ginovyan [122] can be found an extension of this theorem to the case of loss functions $w(x)$ satisfying $0 \leq w(x) \leq C_1 \exp\{C_2|x|\}$ with positive constants C_1 and C_2.

The proof of Theorem 8.5.2 can be found in Ginovyan [126] (see also Ginovyan [129, 130]).

Results similar to Theorems 8.5.1 and 8.5.2 regarding the estimation of linear functionals of probability density functions have been established by Ibragimov and Khas'minskii [186].

The Stein-Levit method, used in the proof of Theorem 8.5.2, was first proposed by Stein [287] and later developed by Levit [210–212]. For a book presentation of this method, see Ibragimov and Has'minskii [185], Section 4.1.

Chapter 9
Parametric Estimation: The Whittle Estimation Procedure

This chapter examines the statistical parametric estimation of a vector spectral parameter for both discrete and continuous time linear models using tapered data.

We assume that the spectral density $f(\lambda)$ of the observed process belongs to a specified parametric family of spectral densities $\mathcal{F} := \{f(\lambda, \theta) : \theta \in \Theta\}$, where $\theta := (\theta_1, \ldots, \theta_p)$ is an unknown parameter and Θ is a subset of the Euclidean space $\mathbb{R}^p$. The primary objective is to estimate the unknown parameter θ based on the tapered data (8.2.2) and to investigate the asymptotic properties of the proposed estimators as the sample size approaches infinity.

There are several statistical methods available for estimating the unknown spectral parameter θ, including the method of moments, the maximum likelihood method, Whittle's method, and the minimum contrast method, among others.

In this text, we focus on the Whittle method, which utilizes smoothed periodogram analysis in the frequency domain. We demonstrate that, under broad regularity conditions on the model's spectral density, the Whittle estimator based on tapered data is consistent and asymptotically normal.

Additionally, we explore simplified Whittle estimators and robust estimation procedures for spectral parameters.

In Sect. 9.1, we describe the Whittle estimation procedure, which is based on approximating the likelihood function using Szegő's theorems on the approximation of the determinants of Toeplitz matrices (see Propositions 2.4.1 and 2.4.2) and on the approximation of the inverses of Toeplitz matrices (see Theorem 4.4.6). In Sect. 9.2, we present the asymptotic properties of the Whittle estimator, specifically its consistency (Theorem 9.2.1) and asymptotic normality (Theorem 9.2.2). Section 9.3 is devoted to the proofs of Theorems 9.2.1 and 9.2.2. In Sect. 9.4, we explore simplified Whittle estimators for spectral parameters, demonstrating that these estimators are asymptotically equivalent to the Whittle estimator in the sense that they share the same asymptotic properties. Robust estimation procedures for spectral parameters in linear models with memory are addressed in Sect. 9.5. Finally, in Sect. 9.6, we provide additional information on the Whittle approximation of the

M. S. Ginovyan, *Random Toeplitz Functionals and Their Applications*,
Frontiers in Probability and the Statistical Sciences,
https://doi.org/10.1007/978-3-032-00961-6_9

likelihood function. We also briefly discuss more minimum contrast estimators, particularly Ibragimov's estimator, and include bibliographical notes.

9.1 The Whittle Estimation Procedure

The Whittle estimation procedure, originally devised for discrete time short memory stationary processes, is based on smoothed periodogram analysis in the frequency domain. It involves approximating the likelihood function and examining the asymptotic properties of empirical spectral functionals (see Whittle [313]). Since its discovery, the Whittle estimation method has played a significant role in the asymptotic theory of parametric estimation in the frequency domain and has attracted considerable interest from statisticians. Their aims have included weakening the conditions necessary to ensure the validity of the Whittle approximation for discrete time short memory models, finding analogs for long and intermediate memory models, identifying conditions under which the Whittle estimator is asymptotically equivalent to the exact maximum likelihood estimator, and extending the estimation procedure to continuous time models and random fields.

To explain the concept behind the Whittle estimation procedure, let's assume, for simplicity, that the underlying process $X(t)$ is a discrete time Gaussian process. Our goal is to estimate the parameter θ based on the non-tapered sample $\mathbf{X}_T := \{X(t),\ t = 1, \ldots, T\}$. A natural approach is to find the maximum likelihood estimator (MLE) $\widehat{\theta}_{T,MLE}$ of θ. This involves maximizing the likelihood function or minimizing the $-1/T\times$ log-likelihood function, denoted by $L_T(\theta)$, which, in this case, takes the following form:

$$L_T(\theta) = L_T(\theta, \mathbf{X}_T) := \frac{1}{2}\ln 2\pi + \frac{1}{2T}\ln\det B_T(f_\theta) + \frac{1}{2T}\mathbf{X}_T'[B_T(f_\theta)]^{-1}\mathbf{X}_T, \tag{9.1.1}$$

where $B_T(f_\theta)$ is the Toeplitz matrix generated by the spectral density f_θ of the observed process $X(t)$ (specifically, $B_T(f_\theta)$ serves as the covariance matrix of the process $X(t)$).

Unfortunately, the function $L_T(\theta)$ is difficult to manage, and no explicit expression for the estimator $\widehat{\theta}_{T,MLE}$ is available, even for simpler models such as ARMA models (see, e.g., Dzhaparidze [93] and the references therein).

Taking into account the complexity of obtaining a suitable statistical inference about the unknown parameter θ based on the exact expression for the log-likelihood function $L_T(\theta)$, Mann and Wald [223] in 1943 (for autoregressive processes) and Whittle [313] in the early 1950s (for general discrete time short memory stationary processes) suggested using, instead of the exact expression for $L_T(\theta)$, its *principal part* $U_{T,W}(\theta)$, which satisfies the following condition:

$$T^{1/2}(L_T(\theta) - U_{T,W}(\theta)) \to 0 \quad \text{as} \quad T \to \infty \quad \text{in probability.} \tag{9.1.2}$$

An approach proposed by P. Whittle in [313], known as the *Whittle estimation procedure*, involves approximating the term $\ln \det B_T(f_\theta)$ by the integral of the logarithm of f_θ, and the inverse matrix $[B_T(f_\theta)]^{-1}$ by the Toeplitz matrix $B_T(1/f_\theta)$.

Specifically, by utilizing Szegő's weak theorem (see Propositions 2.4.1), we obtain the following approximation for the term $\ln \det B_T(f_\theta)$:

$$\frac{1}{2T}\log\det B_T(f_\theta) = \frac{1}{4\pi}\int_{\mathbb{T}} \log f_\theta(\lambda)\,d\lambda + o(1) \quad \text{as} \quad T \to \infty. \tag{9.1.3}$$

Regarding the approximation of the inverse matrix $B_T^{-1}(f_\theta)$, we expect that for large T, the Toeplitz matrix $B_T(1/f_\theta)$ should approximate $B_T^{-1}(f_\theta)$.

Indeed, denote the entries of the matrix $B_T(1/f_\theta)$ by $a_\theta(k-j)$, $k, j = 1, \dots, T$, which are the Fourier coefficients of the function $f_\theta^{-1}/(2\pi)$:

$$a_\theta(k-j) = \frac{1}{(2\pi)^2}\int_{\mathbb{T}} e^{i(k-j)\lambda}\frac{1}{f_\theta(\lambda)}\,d\lambda. \tag{9.1.4}$$

Then, noting that $B_T(f_\theta) = \|r_\theta(k-j)\|$, $k, j = 1, \dots, T$, where

$$r_\theta(k-j) = \int_{\mathbb{T}} e^{i(k-j)\lambda} f_\theta(\lambda)\,d\lambda, \tag{9.1.5}$$

for large T, by using Parseval-Plancherel's theorem, for the entries $c_{k,j}$ of the matrix $B_T(1/f_\theta)B_T(f_\theta)$, we can write

$$\begin{aligned} c_{k,j} &= \sum_{t=1}^{T} a_\theta(k-t)r_\theta(t-j) \approx \sum_{t=-\infty}^{\infty} a_\theta(k-t)r_\theta(t-j) \\ &= \frac{1}{2\pi}\int_{\mathbb{T}} e^{-i(k-j)\lambda}\widehat{a_\theta}(\lambda)\overline{\widehat{r_\theta}(\lambda)}\,d\lambda \\ &= \frac{1}{2\pi}\int_{\mathbb{T}} e^{-i(k-j)\lambda} f_\theta(\lambda)\frac{1}{f_\theta(\lambda)}\,d\lambda = \delta_{k,j}, \end{aligned}$$

where $\delta_{k,j}$ is Kronecker's delta and $\approx$ indicates an approximation for large T.

This observation leads to the following approximation of the quadratic form in Eq. (9.1.1):

$$\frac{1}{2T}\mathbf{X}_T'[B_T(f_\theta)]^{-1}\mathbf{X}_T \approx \frac{1}{2T}\mathbf{X}_T' B_T(1/f_\theta)\mathbf{X}_T = \frac{1}{4\pi}\int_{-\pi}^{\pi}\frac{I_T(\lambda)}{f_\theta(\lambda)}d\lambda, \tag{9.1.6}$$

where $I_T(\lambda)$ is the ordinary (non-tapered) periodogram of the process $X(t)$ defined by (8.2.5) with $h(t) = \mathbf{1}_{[0,1]}(t)$.

A combination of Eqs. (9.1.3) and (9.1.6) results in the following approximation of the log-likelihood function $L_T(\theta)$ in (9.1.1), introduced by Whittle [313] and referred to as the *Whittle functional*:

$$U_{T,W}(\theta) = U_{T,W}(\theta, \mathbf{X}_T) := \frac{1}{4\pi} \int_{-\pi}^{\pi} \left[\ln f_\theta(\lambda) + \frac{I_T(\lambda)}{f_\theta(\lambda)} \right] d\lambda. \tag{9.1.7}$$

Now, by minimizing the Whittle functional $U_{T,W}(\theta)$ with respect to θ, we obtain an estimator $\widehat{\theta}_{T,W}$ for the unknown parameter θ, known as the *Whittle estimator*:

$$\widehat{\theta}_{T,W} := \operatorname*{Arg\,min}_{\theta \in \Theta} U_{T,W}(\theta). \tag{9.1.8}$$

In the asymptotic setting (as the sample size $T \to \infty$), it turns out that the maximum likelihood estimator $\widehat{\theta}_{T,MLE}$, based on the exact log-likelihood function $L_T(\theta)$, and the Whittle estimator $\widehat{\theta}_{T,W}$, based on the "principal part" $U_{T,W}(\theta)$ of $L_T(\theta)$, are asymptotically equivalent. This means that, under suitable conditions on the observed model $X(t)$, the Whittle estimator $\widehat{\theta}_{T,W}$ is consistent, asymptotically normal, and asymptotically Fisher-efficient.

In the continuous time context, the Whittle procedure for estimating a spectral parameter θ based on the sample $\mathbf{X}_T := \{X(t),\ 0 \le t \le T\}$ involves selecting the estimator $\widehat{\theta}_T$ that minimizes the *"continuous" Whittle functional*:

$$U_T(\theta) := \frac{1}{4\pi} \int_{\mathbb{R}} \left[\ln f(\lambda, \theta) + \frac{I_T(\lambda)}{f(\lambda, \theta)} \right] \cdot w(\lambda)\, d\lambda, \tag{9.1.9}$$

where $I_T(\lambda)$ is the continuous periodogram of $X(t)$ and $w(\lambda)$ is a weight function ($w(-\lambda) = w(\lambda)$, $w(\lambda) \ge 0$, $w(\lambda) \in L^1(\mathbb{R})$) for which the integral in (9.1.9) is well-defined. An example of commonly used weight function is $w(\lambda) = 1/(1+\lambda^2)$.

The Whittle procedure for estimating a spectral parameter θ based on the tapered sample (8.2.2) involves selecting the estimator $\widehat{\theta}_{T,h}$ that minimizes the *tapered Whittle functional*:

$$U_{T,h}(\theta) := \frac{1}{4\pi} \int_{\Lambda} \left[\log f(\lambda, \theta) + \frac{I_T^h(\lambda)}{f(\lambda, \theta)} \right] \cdot w(\lambda)\, d\lambda, \tag{9.1.10}$$

where $I_T^h(\lambda)$ is the tapered periodogram of $X(t)$, as defined in (8.2.5), and $w(\lambda)$ is a weight function for which the integral in (9.1.10) is well-defined.

Note that, in the discrete time case, we take $w(\lambda) \equiv 1$ as the weight function.

Thus, the Whittle estimator $\widehat{\theta}_{T,h}$ of the unknown parameter θ, based on the tapered sample (8.2.2), is defined by

$$\widehat{\theta}_{T,h} := \operatorname*{Arg\,min}_{\theta \in \Theta} U_{T,h}(\theta), \tag{9.1.11}$$

where $U_{T,h}(\theta)$ is given by Eq. (9.1.10).

Therefore, the tapered Whittle estimator $\widehat{\theta}_{T,h}$ of the unknown parameter θ is the root of the following system of estimating equations:

$$\begin{aligned}F_{h,i}(\theta) =&F_{T,h,i}(\theta) := (\partial/\partial\theta_i)U_{T,h}(\theta)\\ =&\frac{1}{4\pi}\int_\Lambda \Big[(\partial/\partial\theta_i)\log f(\lambda,\theta) + I_T^h(\lambda)(\partial/\partial\theta_i)f^{-1}(\lambda,\theta)\Big]\cdot w(\lambda)\,d\lambda = 0,\\ &i = 1,\dots,p.\end{aligned} \tag{9.1.12}$$

For additional information about the Whittle approximation, see Sect. 9.6.1.

9.2 Asymptotic Properties of the Whittle Estimator

The tapered Whittle estimator $\widehat{\theta}_{T,h}$ of θ possesses good asymptotic properties. To state these properties, we first introduce the following set of assumptions.

Assumption 9.2.1 *The true value θ_0 of the parameter θ belongs to a bounded closed set Θ, which is contained within an open set S in the p-dimensional Euclidean space $\mathbb{R}^p$. Additionally, $f(\lambda,\theta_1) \neq f(\lambda,\theta_2)$ whenever $\theta_1 \neq \theta_2$, almost everywhere in Λ with respect to the Lebesgue measure.*

Assumption 9.2.2 *The functions $f(\lambda,\theta)$, $f^{-1}(\lambda,\theta)$ and $\frac{\partial}{\partial\theta_k}f^{-1}(\lambda,\theta)$, $k = 1,\dots,p$, are continuous in (λ,θ).*

Assumption 9.2.3 *The functions $f := f(\lambda,\theta)$ and $g := w(\lambda)\frac{\partial}{\partial\theta_k}f^{-1}(\lambda,\theta)$ satisfy Assumptions 8.2.2 or 8.2.3 for all $k = 1,\dots,p$ and $\theta \in \Theta$.*

Assumption 9.2.4 *The functions f, g, $a := a(\lambda,\theta)$ (the MA(∞) coefficients), and $b := \widehat{g}$, where g is as in Assumption 9.2.3, satisfy Assumption 8.2.4.*

Assumption 9.2.5 *The functions $\frac{\partial^2}{\partial\theta_k\partial\theta_j}f^{-1}(\lambda,\theta)$ and $\frac{\partial^3}{\partial\theta_k\partial\theta_j\partial\theta_j}f^{-1}(\lambda,\theta)$, $k,j,l = 1,\dots,p$, are continuous in (λ,θ) for $\lambda \in \Lambda$, $\theta \in N_\delta(\theta_0)$, where $N_\delta(\theta_0) := \{\theta : |\theta-\theta_0| < \delta\}$ is some neighborhood of θ_0.*

Assumption 9.2.6 *The matrices*

$$W(\theta) = \|w_{ij}(\theta)\|,\ A(\theta) = \|a_{ij}(\theta)\|,\ B(\theta) = \|b_{ij}(\theta)\|,\ i,j = 1,\dots,p \tag{9.2.1}$$

are positive definite, where

$$w_{ij}(\theta) = \frac{1}{4\pi}\int_\Lambda \frac{\partial}{\partial\theta_i}\ln f(\lambda,\theta)\frac{\partial}{\partial\theta_j}\ln f(\lambda,\theta)w(\lambda)d\lambda, \tag{9.2.2}$$

$$a_{ij}(\theta) = \frac{1}{4\pi}\int_\Lambda \frac{\partial}{\partial\theta_i}\ln f(\lambda,\theta)\frac{\partial}{\partial\theta_j}\ln f(\lambda,\theta)w^2(\lambda)d\lambda, \tag{9.2.3}$$

$$b_{ij}(\theta) = \frac{\kappa_4}{16\pi^2} \int_\Lambda \frac{\partial}{\partial\theta_i} \ln f(\lambda, \theta) w(\lambda) d\lambda \int_\Lambda \frac{\partial}{\partial\theta_j} \ln f(\lambda, \theta) w(\lambda) d\lambda, \qquad (9.2.4)$$

and κ_4 is the fourth cumulant of the driving process $\xi(t)$ at $t = 1$.

The next theorem provides sufficient conditions for the tapered Whittle estimator $\widehat{\theta}_{T,h}$ to be consistent.

Theorem 9.2.1 *Let $\widehat{\theta}_{T,h}$ be the tapered Whittle estimator defined by* (9.1.11), *and let θ_0 be the true value of the parameter θ. Then, under Assumptions 9.2.1–9.2.4, the statistic $\widehat{\theta}_{T,h}$ is a consistent estimator for θ; that is, $\widehat{\theta}_{T,h} \to \theta_0$ in probability as $T \to \infty$.*

Having established the consistency of the Whittle estimator $\widehat{\theta}_{T,h}$, we can now derive the limiting distribution of $T^{1/2}\left(\widehat{\theta}_{T,h} - \theta_0\right)$ using the standard approach. This involves applying Taylor's formula, the mean value theorem, and Slutsky's theorem. Specifically, we have the following result, which shows that under the aforementioned assumptions, the tapered Whittle estimator $\widehat{\theta}_{T,h}$ is $\sqrt{T}$-consistent and asymptotically normal.

Theorem 9.2.2 *Suppose that Assumptions 9.2.1–9.2.6 are satisfied. Then, the Whittle estimator $\widehat{\theta}_{T,h}$ of the unknown spectral parameter θ, based on the tapered data* (8.2.2), *is $\sqrt{T}$-consistent and asymptotically normal. More precisely, we have*

$$T^{1/2}\left(\widehat{\theta}_{T,h} - \theta_0\right) \xrightarrow{d} N_p\left(0, e(h)\Gamma(\theta_0)\right) \quad \text{as} \quad T \to \infty, \qquad (9.2.5)$$

where $N_p(\cdot, \cdot)$ denotes the p-dimensional normal distribution and $\xrightarrow{d}$ stands for convergence in distribution. The matrix $\Gamma(\theta_0)$ is defined as follows:

$$\Gamma(\theta_0) = W^{-1}(\theta_0)\left(A(\theta_0) + B(\theta_0)\right) W^{-1}(\theta_0), \qquad (9.2.6)$$

where the matrices W, A, and B are defined in (9.2.1)–(9.2.4), and the tapering factor $e(h)$ is given by formula (8.2.15).

Remark 9.2.1 (The Variance Effect) Since tapering the data reduces the effective length of the data set, it is not surprising that the corresponding tapered estimators generally have larger variances than their non-tapered counterparts. Specifically, by applying the Cauchy-Schwarz inequality to the tapering factor $e(h)$ (defined by formula (8.2.15)), we find that $e(h) \geq 1$, with equality attained in the non-tapered case, that is, for $h(t) = \mathbf{1}_{[0,1]}(t)$. Thus, the use of tapers typically results in a loss of efficiency.

However, as observed by Dahlhaus [76] (pages 156 and 161), it is not correct to conclude that tapering always increases the variance of the estimators. This is because a taper function h can be chosen to satisfy $e(h) = 1$. For more details regarding the variance effect, see Appendix A.

In the classical asymptotic setting for discrete time Gaussian processes, it is possible to select the taper function $h(t)$ such that the resulting tapered estimator is asymptotically Fisher-efficient. To clarify this, we consider an important subclass of tapers in the following proposition (see Dahlhaus [72], Corollary 2.3, and Ginovyan and Sahakyan [142]).

Proposition 9.2.1 *Suppose, in addition to the assumptions of Theorem 9.2.2, that the process $X(t)$ $(t \in \mathbb{Z})$ is Gaussian and that $h_T(t) := h_T(t/T)$, where $h_T : [0,1] \to [0,\infty)$ are continuous functions of uniformly bounded variation $BV(h_T)$ that converge almost everywhere to a function $h(x)$. Then, the tapered Whittle estimator $\widehat{\theta}_{T,h}$ is asymptotically Fisher-efficient if and only if $h(x) = 1$ almost everywhere.*

Proof The result follows from Theorem 9.2.2 with the relation

$$\left| \frac{1}{T} \sum_{t=1}^{T} h_T^k(t/T) - \int_0^1 h_T^k(x)dx \right| \leq \frac{BV(h_T^k)}{T},$$

the dominated convergence theorem, and the fact that in the Cauchy-Schwarz inequality we have equality if and only if both factors are the same almost everywhere. □

Remark 9.2.2 In the discrete time case, we take the weight function $w(\lambda) \equiv 1$, and the matrices $A(\theta_0)$ and $W(\theta_0)$ coincide (see (9.2.1)–(9.2.3)). Consequently, the Eq. (9.2.6) simplifies to

$$\Gamma(\theta_0) = W^{-1}(\theta_0) \left(W(\theta_0) + B(\theta_0)\right) W^{-1}(\theta_0).$$

If, in addition, $B(\theta_0) = 0$ and the taper is as described in Proposition 9.2.1, then we have $e(h) = 1$ and $\Gamma(\theta_0) = W^{-1}(\theta_0)$. This implies that the tapered Whittle estimator $\widehat{\theta}_{T,h}$ is asymptotically Fisher-efficient (cf. Theorem 6.3 of Dahlhaus [71]). For instance, for the polynomial taper presented in Example A.3.2, we have $h_{\rho T}(x) \to \mathbf{1}_{[0,1]}(x)$ and $e(h) = \lim_{T\to\infty}(TH_{4,T})/H_{2,T}^2 = 1$ (see Dahlhaus [76]). Therefore, the corresponding tapered Whittle estimator $\widehat{\theta}_{T,h}$ is asymptotically Fisher-efficient.

Remark 9.2.3 In view of the definition of the matrix $B(\theta)$) (see Eqs. (9.2.1) and (9.2.4)), for discrete time processes ($w(\lambda) \equiv 1$), the equality $B(\theta) = 0$ holds in the following two cases (cf. Dahlhaus and Janas [77]):

Case 1 $\kappa_4 = 0$. This condition is satisfied, for example, when the innovations are assumed to be Gaussian.

Case 2 $\int_\Lambda \frac{\partial}{\partial\theta} \ln f(\lambda, \theta) d\lambda = 0$. This condition is satisfied for various parameterizations, as can be inferred from the Kolmogorov-Szegő formula (see Appendix B.4, and Brockwell and Davis [57], Section 5.8).

Using Theorem 8.2.6 and the arguments from the proof of Theorem 9.2.2, we can state the following strengthened version of Theorem 9.2.2 for Gaussian models.

Theorem 9.2.3 *Let the model $X(t)$ be Gaussian. Suppose that Assumptions 9.2.1, 9.2.2, and 9.2.5 are satisfied. Let the matrices $W(\theta)$ and $A(\theta)$ in Assumption 9.2.6 be positive definite, and the functions $f := f(\lambda, \theta)$ and $g := w(\lambda)\frac{\partial}{\partial\theta_k} f^{-1}(\lambda, \theta)$ satisfy Assumption 8.2.2 for all $k = 1, \ldots, p$ and $\theta \in \Theta$. Then, the Whittle estimator $\widehat{\theta}_{T,h}$ of the unknown spectral parameter θ, based on the tapered data (8.2.2), is $\sqrt{T}$-consistent and asymptotically normal. More precisely, we have*

$$T^{1/2}\left(\widehat{\theta}_{T,h} - \theta_0\right) \xrightarrow{d} N_p\left(0, e(h)\Gamma(\theta_0)\right) \quad \text{as} \quad T \to \infty, \tag{9.2.7}$$

where now

$$\Gamma(\theta_0) = W^{-1}(\theta_0)A(\theta_0)W^{-1}(\theta_0), \tag{9.2.8}$$

and the tapering factor $e(h)$ is given by formula (8.2.15).

Example 9.2.1 (Estimation of Parameters for Fractional Riesz-Bessel Motion) The Lévy-driven fractional Riesz-Bessel motion (fRBm) $X(t)$ is a continuous time linear process defined by (3.2.10), with a time-invariant filter $a(t)$ that is the Fourier transform of the following function:

$$\widehat{a}(\lambda) = c^{1/2}(i\lambda)^{-\alpha}(1 + i\lambda)^{-\beta}, \quad \lambda \in \mathbb{R},\ c > 0,\ 0 < \alpha < 1,\ \beta > 0. \tag{9.2.9}$$

The explicit form of the filter $a(t)$ in terms of the confluent hypergeometric function is provided by Avram et al. [15]. According to (3.2.12), the spectral density of fRBm $X(t)$ is given by (see (3.3.20)):

$$f(\lambda) = \frac{1}{2\pi}|\widehat{a}(\lambda)|^2 = \frac{c}{2\pi}|\lambda|^{-2\alpha}(1 + \lambda^2)^{-\beta}, \quad \lambda \in \mathbb{R}. \tag{9.2.10}$$

The exponent α determines the long-range dependence, while the exponent β indicates the second-order intermittency of the fRBm $X(t)$. Note that for $\alpha = 0$ and $\beta = 1$, we have the Ornstein-Uhlenbeck process driven by Lévy noise (see (3.3.12)). Additionally, the fRBm $X(t)$ is stationary if $0 < \alpha < 1/2$ and is non-stationary with stationary increments if $1/2 \leq \alpha < 1$ (see Sect. 3.3.4). We consider the stationary case and assume that $0 < \alpha < 1/2$ and $\beta > 1/2$.

Thus, the fRBm $X(t)$ involves the unknown vector parameter $\theta := (\alpha, \beta, c)' \in \Theta$, where Θ is a compact subset of $(0, 1/2) \times (1/2, \infty) \times (0, \infty)$. It is easy to verify that the spectral density in (9.2.10) and the weight function $w(\lambda) = (1 + \lambda^2)^{-1}$ satisfy the conditions of Theorem 9.2.3.

Therefore, for this model, the Whittle estimator $\widehat{\theta}_{T,h}$ of the unknown parameter $\theta := (\alpha, \beta, c)'$ is $\sqrt{T}$-consistent and asymptotically normal.

Note that the entries of the matrices $W(\theta)$, $A(\theta)$, and $B(\theta)$, as defined by Eqs. (9.2.1)–(9.2.4), for this model, were calculated in the paper by Avram et al. [15].

9.3 Proofs of Theorems 9.2.1 and 9.2.2

9.3.1 Preliminaries

First, we will prove some auxiliary lemmas.

Lemma 9.3.1 *Let θ_0 be the true value of the parameter θ, and let θ_1 be any other point of Θ. Let $U_{T,h}(\theta)$ be the weighted tapered Whittle functional given by Eq.* (9.1.10). *Then, under Assumptions 9.2.1–9.2.4, there exists a positive constant $K(\theta_0, \theta_1)$ such that*

$$\lim_{T\to\infty} \mathbb{P}\left(U_{T,h}(\theta_0) - U_{T,h}(\theta_1) < -K(\theta_0, \theta_1)\right) = 1. \tag{9.3.1}$$

Proof We define

$$W_{T,h} := U_{T,h}(\theta_0) - U_{T,h}(\theta_1). \tag{9.3.2}$$

Then, in view of Eqs. (8.2.22) and (9.1.10) for the expected value of $W_{T,h}$, we have

$$\begin{aligned} \mathbb{E}[W_{T,h}] &= \frac{1}{4\pi}\int_{\mathbb{R}} \log\frac{f(\lambda,\theta_0)}{f(\lambda,\theta_1)} w(\lambda)d\lambda \\ &\quad + \frac{1}{4\pi}\int_{\mathbb{R}}\int_{\mathbb{R}} \Phi_{2,T}^h(\lambda-\mu) f(\lambda,\theta_0) g(\mu,\theta_0,\theta_1) w(\lambda) d\lambda d\mu, \end{aligned} \tag{9.3.3}$$

where

$$g(\mu,\theta_0,\theta_1) := \frac{1}{f(\mu,\theta_0)} - \frac{1}{f(\mu,\theta_1)} = \frac{1}{f(\mu,\theta_0)}\left[1 - \frac{f(\mu,\theta_0)}{f(\mu,\theta_1)}\right]. \tag{9.3.4}$$

Therefore, by applying Theorem 8.2.1 (see also Remark 8.2.1) and using Eq. (9.3.4), we can derive from (9.3.3) that

$$\begin{aligned} \lim_{T\to\infty} \mathbb{E}[W_{T,h}] &= \frac{1}{4\pi}\int_{\mathbb{R}} \log\frac{f(\lambda,\theta_0)}{f(\lambda,\theta_1)} w(\lambda)d\lambda \\ &\quad + \frac{1}{4\pi}\int_{\mathbb{R}} \left[1 - \frac{f(\lambda,\theta_0)}{f(\lambda,\theta_1)}\right] w(\lambda)d\lambda. \end{aligned} \tag{9.3.5}$$

Next, using the inequality

$$\log\frac{f(\lambda,\theta_0)}{f(\lambda,\theta_1)}\le\frac{f(\lambda,\theta_0)}{f(\lambda,\theta_1)}-1, \tag{9.3.6}$$

we obtain from (9.3.5) that

$$\lim_{T\to\infty}\mathbb{E}[W_{T,h}]=-l(\theta_0,\theta_1),\quad \text{where}\quad l(\theta_0,\theta_1)>0. \tag{9.3.7}$$

Now we consider the variance of W_T^h. Applying Theorem 8.2.3 with $g(\lambda)=g(\lambda,\theta_0,\theta_1)w(\lambda)$, where $g(\mu,\theta_0,\theta_1)$ is defined in Eq. (9.3.4), we obtain

$$\lim_{T\to\infty}T\,\mathrm{Var}(W_{T,h})=\sigma_h^2(W)<\infty, \tag{9.3.8}$$

where

$$\begin{aligned}\sigma_h^2(W)&=4\pi e(h)\int_{\mathbb{R}}\left(1-\frac{f(\lambda,\theta_0)}{f(\lambda,\theta_1)}\right)^2 w^2(\lambda)d\lambda\\&\quad+\kappa_4 e(h)\left[\int_{\mathbb{R}}\left(1-\frac{f(\lambda,\theta_0)}{f(\lambda,\theta_1)}\right)w(\lambda)d\lambda\right]^2.\end{aligned} \tag{9.3.9}$$

Here, κ_4 is the fourth cumulant of $\xi(1)$, and $e(h)$ is defined as in (8.2.16).

The asymptotic relation in (9.3.1) now follows from relations (9.3.7) and (9.3.8) by applying Chebyshev's inequality; as $K(\theta_0,\theta_1)$ can be taken any quantity less than $l(\theta_0,\theta_1)$. Lemma 9.3.1 is proved. □

It follows from Lemma 9.3.1 that $\widehat{\theta}_{T,h}$ is a consistent estimator for θ when Θ is a finite set. To achieve consistency when Θ is an arbitrary bounded closed set, the functional $U_{T,h}(\theta)$ must satisfy a suitable continuity condition.

Lemma 9.3.2 *Let $\theta_1\in\Theta$ and $\theta_2\in S$ be chosen such that $|\theta_2-\theta_1|<\delta$ for $\delta>0$ (possibly depending on θ_2). Then, under the conditions of Lemma 9.3.1, there exists a number $T_0>0$ such that for all $T>T_0$, the following inequality holds:*

$$\left|U_{T,h}(\theta_1)-U_{T,h}(\theta_2)\right|\le H_T^h(\delta,\theta_1), \tag{9.3.10}$$

where $H_T^h:=H_T^h(\delta,\theta_1)$ is a random variable satisfying:

$$\lim_{\delta\to 0}\mathbb{E}[H_T^h]=0\quad \textit{uniformly in}\quad T>T_0, \tag{9.3.11}$$

and

$$\lim_{T\to\infty}\mathrm{Var}[H_T^h]=0\quad \textit{for each}\quad \delta>0. \tag{9.3.12}$$

Proof For $\theta_1 \in \Theta$ and $\theta_2 \in S$ satisfying $|\theta_2 - \theta_1| < \delta$, one can find a positive number $A(\theta_1)$ such that the inequality

$$\left|1 - \frac{f(\lambda, \theta_1)}{f(\lambda, \theta_2)}\right| \le \lambda^{-2} C(\delta, \theta_1) \tag{9.3.13}$$

holds for $|\lambda| > A(\theta_1)$, where $C(\delta, \theta_1)$ is a positive constant that converges to zero as $\delta \to 0$.

Then, using inequalities (9.3.6) and (9.3.13), we can write

$$\begin{aligned} &\left|U_{T,h}(\theta_1) - U_{T,h}(\theta_2)\right| \\ &\le \frac{1}{4\pi} \int_{|\lambda| \le A(\theta_1)} \left[f^{-1}(\lambda, \theta_1) - f^{-1}(\lambda, \theta_2)\right]\left[f(\lambda, \theta_1) - I_T^h(\lambda)\right] w(\lambda) d\lambda \\ &+ \frac{C(\delta, \theta_1)}{4\pi} \int_{|\lambda| > A(\theta_1)} \lambda^{-2} \left[1 - \frac{I_T^h(\lambda)}{f(\lambda, \theta_1)}\right] w(\lambda) d\lambda. \end{aligned} \tag{9.3.14}$$

Applying the mean value theorem, with some $\theta' \in [\theta_1, \theta_2]$, we obtain

$$\left|f^{-1}(\lambda, \theta_1) - f^{-1}(\lambda, \theta_2)\right| \le \delta \sum_{k=1}^{p} \left| \left(\frac{\partial}{\partial \theta_k} f^{-1}(\lambda, \theta) \right)_{\theta = \theta'} \right|. \tag{9.3.15}$$

We define

$$V_T(\theta_1, \delta(\theta_1)) := \sum_{k=1}^{p} \sup_{|\lambda| \le A(\theta_1),\, |\theta - \theta_1| < \delta(\theta_1)} \left| \frac{\partial}{\partial \theta_k} f^{-1}(\lambda, \theta) \right| < \infty, \tag{9.3.16}$$

where $\delta(\theta_1) < \delta$ is chosen such that the set $\{\theta : |\theta - \theta_1| < \delta(\theta_1)\}$ is contained in S.

Then, from (9.3.14) and (9.3.16), we obtain (9.3.10), where

$$\begin{aligned} H_T^h :&= \frac{\delta}{4\pi} V_T(\theta_1, \delta(\theta_1)) \int_{\mathbb{R}} \left[f(\lambda, \theta_1) + I_T^h(\lambda)\right] w(\lambda) d\lambda \\ &+ \frac{C(\delta, \theta_1)}{4\pi} \int_{|\lambda| > A(\theta_1)} \frac{f(\lambda, \theta_1) + I_T^h(\lambda)}{\lambda^2 f(\lambda, \theta_1)} w(\lambda) d\lambda. \end{aligned} \tag{9.3.17}$$

Finally, by applying Theorems 8.2.1 and 8.2.3, it is straightforward to verify that the relations (9.3.11) and (9.3.12) hold for H_T^h, as defined by (9.3.17).

Lemma 9.3.2 is proved. □

Lemma 9.3.3 *Let $H_T^h := H_T^h(\delta, \theta_1)$ be defined as in (9.3.17). Then, under the conditions of Lemma 9.3.1, the statistic $\widehat{\theta}_{T,h}$ is a consistent estimator for θ; that is, $\widehat{\theta}_{T,h} \to \theta_0$ in probability as $T \to \infty$.*

Proof Let $\theta_1 \neq \theta_0$. Then, by Lemma 9.3.1, we have

$$\lim_{T\to\infty} \mathbb{P}\left(U_{T,h}(\theta_0) - U_{T,h}(\theta_1) < -K(\theta_0, \theta_1)\right) = 1. \tag{9.3.18}$$

Also, from (9.3.11) and (9.3.12), we obtain

$$\lim_{T\to\infty} \mathbb{P}\left(H_T^h(\delta, \theta_1) < K(\theta_0, \theta_1)\right) = 1, \tag{9.3.19}$$

since we can first choose δ so that $\mathbb{E}[H_T^h(\delta, \theta_1)] \leq (1/2)K(\theta_0, \theta_1)$, we can then let $T \to \infty$, using Chebyshev's inequality.

Next, from (9.3.18) and (9.3.19), we obtain

$$U_{T,h}(\theta_0) - U_{T,h}(\theta_2) < 0 \quad \text{for} \quad |\theta_2 - \theta_1| < \delta_1. \tag{9.3.20}$$

Therefore,

$$\lim_{T\to\infty} \mathbb{P}\left(\sup_{\theta_2 \in N(\theta_1)} [U_{T,h}(\theta_0) - U_{T,h}(\theta_2)] < 0\right) = 1, \tag{9.3.21}$$

where $N(\theta_1) := \{\theta : |\theta - \theta_1| < \delta_1\}$.

Now, consider the collection of sets $\{N(\theta_1) : \theta_1 \neq \theta_0\}$, obtained by letting θ_1 to run over all points in Θ except θ_0, along with the set $N(\theta_0) := \{\theta : |\theta - \theta_0| < \delta_0\}$, where δ_0 is arbitrary. Together, these sets constitute an open covering of Θ. Since Θ is a compact set, this collection contains a finite open covering, denoted as $\{N(\theta_j) : 0 \leq j \leq m\}$.

Thus, the relation (9.3.21) provides:

$$\lim_{T\to\infty} \mathbb{P}\left(\sup_{\theta_2 \in \cup_{j=1}^m N(\theta_j)} [U_{T,h}(\theta_0) - U_{T,h}(\theta_2)] < 0\right) = 1, \tag{9.3.22}$$

or

$$\lim_{T\to\infty} \mathbb{P}\left(\inf_{\theta\in\Theta} U_{T,h}(\theta_0) = \inf_{\theta\in N(\theta_0)} U_{T,h}(\theta_0)\right) = 1. \tag{9.3.23}$$

Therefore

$$\lim_{T\to\infty} \mathbb{P}\left(|\widehat{\theta}_{T,h} - \theta_0| < \delta_0\right) = 1. \tag{9.3.24}$$

Since δ_0 can be arbitrarily small, we conclude that $\widehat{\theta}_{T,h} \xrightarrow{P} \theta_0$ as $T \to \infty$. □

9.3.2 Proofs of Theorems 9.2.1 and 9.2.2

Proof of Theorem 9.2.1 The result follows directly from Lemmas 9.3.1–9.3.3. □

In what follows, we will adopt the following notation: for a function $\psi := \psi(\lambda, \theta)$ and numbers $k, j, l = 1, \ldots, p$, we denote $\psi^{(k)}(\lambda, \theta) := \frac{\partial}{\partial \theta_k} \psi(\lambda, \theta)$, $\psi^{(kj)}(\lambda, \theta) := \frac{\partial^2}{\partial \theta_k \partial \theta_j} \psi(\lambda, \theta)$, and $\psi^{(kjl)}(\lambda, \theta) := \frac{\partial^3}{\partial \theta_k \partial \theta_j \partial \theta_j} \psi(\lambda, \theta)$. By $\psi^{(k)}(\lambda, \theta')$, $\psi^{(kj)}(\lambda, \theta')$, and $\psi^{(kjl)}(\lambda, \theta')$, we denote the values of the functions $\psi^{(k)}(\lambda, \theta)$, $\psi^{(kj)}(\lambda, \theta)$ and $\psi^{(kjl)}(\lambda, \theta)$ at the point $\theta = \theta'$, respectively. Additionally, $\frac{\partial}{\partial \theta} := (\frac{\partial}{\partial \theta_1}, \ldots, \frac{\partial}{\partial \theta_p})$, and, as before, the true value of the parameter θ we denote by θ_0.

Proof of Theorem 9.2.2 Observe first that, since the Whittle estimator $\widehat{\theta}_{T,h}$ is consistent (Theorem 9.2.1), we can obtain the limiting distribution of the random variable $T^{1/2}\left(\widehat{\theta}_{T,h} - \theta_0\right)$ under the assumption that

$$\widehat{\theta}_{T,h} \in N_\delta(\theta_0) := \{\theta : |\theta - \theta_0| < \delta\}.$$

Next, according to the definition of the estimator $\widehat{\theta}_{T,h}$, we must have $U_{T,h}^{(k)}(\widehat{\theta}_{T,h}) = 0$, $k = 1, \ldots, p$, where $U_{T,h}(\cdot)$ is the Whittle functional defined by (9.1.10).

Therefore, using Taylor's formula and the mean value theorem for $j = 1, \ldots, p$, we can write

$$-\sum_{k=1}^{p} U_{T,h}^{(kj)}(\theta_{T,h}^*) T^{1/2}\left(\widehat{\theta}_{T,h,k} - \theta_{0,k}\right) = T^{1/2} U_{T,h}^{(j)}(\theta_0), \tag{9.3.25}$$

where $\theta_{T,h}^* := t\widehat{\theta}_{T,h} + (1-t)\theta_0 \in N_\delta(\theta_0)$, $t \in [0, 1]$, $\delta \in (0, 1)$.

Thus, to complete the proof of the theorem, in view of Slutsky's arguments, we need to establish the following two assertions:

(a) For $k, j = 1, \ldots, p$, the following limiting relations hold:

$$p - \lim_{T \to \infty} U_{T,h}^{(kj)}(\theta_{T,h}^*) = p - \lim_{T \to \infty} U_{T,h}^{(kj)}(\theta_0) = w_{kj}(\theta_0), \tag{9.3.26}$$

where $p -$ lim denotes the limit in $\mathbb{P}_0$-probability, corresponding to spectral density $f(\lambda, \theta_0)$, and w_{kj} is given by (9.2.2),

(b) The limiting (as $T \to \infty$) distribution of the random variable $T^{1/2} \frac{\partial}{\partial \theta} U_{T,h}(\theta_0)$ is multivariate normal $N_p\left(0, e(h)(A(\theta_0) + B(\theta_0))\right)$, where the matrices $A(\theta_0)$ and $B(\theta_0)$ are given by (9.2.3) and (9.2.4), respectively, and the tapering factor $e(h)$ is given by formula (8.2.16).

Proof of (a). We first show that for $k, j = 1, \dots, p$,

$$\lim_{T\to\infty} \mathbb{E}_0\left[U_{T,h}^{(kj)}(\theta_0)\right] = w_{kj}(\theta_0), \tag{9.3.27}$$

where $\mathbb{E}_0[\xi]$ stands for expectation with respect to $\mathbb{P}_0$.

Indeed, in view of Eq. (9.1.10), for $k, j = 1, \dots, p$, we have

$$\begin{aligned} U_{T,h}^{(kj)}(\theta) &= \frac{1}{4\pi}\int_{\mathbb{R}}\left[\frac{\partial^2}{\partial\theta_k\partial\theta_j}\log f(\lambda,\theta) + I_T^h(\lambda)\frac{\partial^2}{\partial\theta_k\partial\theta_j}f^{-1}(\lambda,\theta)\right]w(\lambda)d\lambda \\ &= \frac{1}{4\pi}\int_{\mathbb{R}}\frac{I_T^h(\lambda)}{f(\lambda,\theta)}\frac{\partial}{\partial\theta_k}\ln f(\lambda,\theta)\frac{\partial}{\partial\theta_j}\ln f(\lambda,\theta)w(\lambda)d\lambda \\ &\quad - \frac{1}{4\pi}\int_{\mathbb{R}}\left(\frac{I_T^h(\lambda)}{f(\lambda,\theta)} - 1\right)\frac{\partial^2}{\partial\theta_k\partial\theta_j}\log f(\lambda,\theta)w(\lambda)d\lambda. \end{aligned} \tag{9.3.28}$$

Therefore, we have

$$\begin{aligned} \mathbb{E}_0\left[U_{T,h}^{(kj)}(\theta_0)\right] &= \frac{1}{4\pi}\mathbb{E}_0\left[\int_{\mathbb{R}}\frac{I_T^h(\lambda)}{f(\lambda,\theta)}\frac{\partial}{\partial\theta_k}\ln f(\lambda,\theta)\frac{\partial}{\partial\theta_j}\ln f(\lambda,\theta)w(\lambda)d\lambda\right] \\ &\quad - \frac{1}{4\pi}\mathbb{E}_0\left[\int_{\mathbb{R}}\left(\frac{I_T^h(\lambda)}{f(\lambda,\theta)} - 1\right)\frac{\partial^2}{\partial\theta_k\partial\theta_j}\log f(\lambda,\theta)w(\lambda)d\lambda\right]. \end{aligned} \tag{9.3.29}$$

Now, applying Theorem 8.2.1, we conclude that the first term on the right-hand side of Eq. (9.3.29) converges to $w_{kj}(\theta_0)$, $k, j = 1, \dots, p$, while the second term converges to zero as $T \to \infty$. Thus, the relation (9.3.27) follows.

Next, we show that for $k, j = 1, \dots, p$,

$$p - \lim_{T\to\infty}\left[U_{T,h}^{(kj)}(\theta_{T,h}^*) - \mathbb{E}_0\left(U_{T,h}^{(kj)}(\theta_0)\right)\right] = 0. \tag{9.3.30}$$

Indeed, in view of Eq. (9.3.28) for the variance of $U_{T,h}^{(kj)}(\theta_0)$ $(k, j = 1, \dots, p)$, we have

$$\operatorname{Var}\left(U_{T,h}^{(kj)}(\theta_0)\right) = \operatorname{Var}\left(\frac{1}{4\pi}\int_{\mathbb{R}} I_T^h(\lambda)\frac{\partial^2}{\partial\theta_k\partial\theta_j}f^{-1}(\lambda,\theta_0)w(\lambda)d\lambda\right), \tag{9.3.31}$$

which, by Theorem 8.2.3, tends to zero as $T \to \infty$. Therefore, the relation (9.3.30) follows from Chebyshev's inequality.

Finally, we show that

$$p - \lim_{T\to\infty}\left[U_{T,h}^{(kj)}(\theta_{T,h}^*) - U_{T,h}^{(kj)}(\theta_0)\right] = 0, \quad k, j = 1, \dots, p. \tag{9.3.32}$$

In view of Eq. (9.3.28), we have

$$\begin{aligned}
&\left|U_{T,h}^{(kj)}(\theta_{T,h}^*) - U_{T,h}^{(kj)}(\theta_0)\right| \\
&\le \frac{1}{4\pi}\left|\int_{\mathbb{R}}\left[\frac{\partial^2}{\partial\theta_k\partial\theta_j}\log f(\lambda,\theta_{T,h}^*) - \frac{\partial^2}{\partial\theta_k\partial\theta_j}\log f(\lambda,\theta_0)\right]w(\lambda)d\lambda\right| \\
&+ \frac{1}{4\pi}\left|\int_{\mathbb{R}} I_T^h(\lambda)\left[\frac{\partial^2}{\partial\theta_k\partial\theta_j}f^{-1}(\lambda,\theta_{T,h}^*) - \frac{\partial^2}{\partial\theta_k\partial\theta_j}f^{-1}(\lambda,\theta_0)\right]w(\lambda)d\lambda\right|.
\end{aligned} \tag{9.3.33}$$

Since we assume that the function $\frac{\partial^2}{\partial\theta_k\partial\theta_j}\log f(\lambda,\theta)$ is continuous in θ, the first term on the right-hand side of (9.3.33) tends to zero as $T\to\infty$. To estimate the second term on the right-hand side of (9.3.33), we apply the mean value theorem, to obtain that it is equal to

$$\frac{1}{4\pi}\left|\sum_{k=1}^{p}\left(\theta_{T,h,k}^* - \theta_{0,k}\right)\int_{\mathbb{R}} I_T^h(\lambda)\frac{\partial^3}{\partial\theta_k\partial\theta_j\theta_l}f^{-1}(\lambda,\theta_{T,h}^{**})w(\lambda)d\lambda\right|, \tag{9.3.34}$$

where $\theta_{T,h}^{**}$ is some value of the parameter from the interval $[\theta_{T,h}^*,\theta_0]$.

Since the estimator $\widehat{\theta}_{T,h}$ is consistent, we can assume that

$$\widehat{\theta}_{T,h}\in N_\delta(\theta_0) := \{\theta : |\theta-\theta_0| < \delta\}.$$

Consequently, $\theta_{T,h}^*$ and $\theta_{T,h}^{**}$ also belong to $N_\delta(\theta_0)$.

Therefore, by choosing δ small enough and leveraging the continuity of the function $\frac{\partial^3}{\partial\theta_k\partial\theta_j\theta_l}f^{-1}(\lambda,\theta)$, we obtain

$$\left|\frac{\partial^3}{\partial\theta_k\partial\theta_j\theta_l}f^{-1}(\lambda,\theta_{T,h}^{**})\right| \le 2\left|\frac{\partial^3}{\partial\theta_k\partial\theta_j\theta_l}f^{-1}(\lambda,\theta_0)\right|.$$

Thus, the second term on the right-hand side of (9.3.33) is less than or equal to

$$\frac{\delta}{2\pi}\sum_{k=1}^{p}\int_{\mathbb{R}} I_T^h(\lambda)\left|\frac{\partial^3}{\partial\theta_k\partial\theta_j\theta_l}f^{-1}(\lambda,\theta_0)\right|w(\lambda)d\lambda. \tag{9.3.35}$$

Next, taking into account that

$$\int_{\mathbb{R}} f(\lambda,\theta_0)\left|\frac{\partial^3}{\partial\theta_k\partial\theta_j\theta_l}f^{-1}(\lambda,\theta_0)\right|w(\lambda)d\lambda < \infty,$$

we can apply Theorems 8.2.1 and 8.2.3 and Chebyshev's inequality to conclude that

$$p - \lim_{T\to\infty} \frac{\delta}{2\pi} \sum_{k=1}^{p} \int_{\mathbb{R}} I_T^h(\lambda) \left| \frac{\partial^3}{\partial\theta_k \partial\theta_j \theta_l} f^{-1}(\lambda, \theta_0) \right| w(\lambda) d\lambda = 0. \quad (9.3.36)$$

Thus, the relation in (9.3.32) is established.

By combining the relations (9.3.27), (9.3.30), and (9.3.32), we obtain (9.3.26), thereby proving assertion (a).

Proof of (b). The result we derive from Theorem 8.2.6. To this end, observe first that, by the Cramér-Wold device (Proposition 5.3.12), it suffices to show that for any arbitrary nonzero vector $c := (c_1, \ldots, c_p) \in \mathbb{R}^p$, the limiting (as $T \to \infty$) distribution of the random variable:

$$\begin{aligned}\Delta_T := T^{1/2} c' \frac{\partial}{\partial\theta} U_{T,h}(\theta_0) &= \sum_{k=1}^{p} c_k U_{T,h}^{(k)}(\theta_0) \\ &= \frac{T^{1/2}}{4\pi} \sum_{k=1}^{p} c_k \int_{\mathbb{R}} \left[\frac{I_T^h(\lambda)}{f(\lambda, \theta_0)} - 1 \right] \frac{\partial}{\partial\theta_k} \log f(\lambda, \theta_0) w(\lambda) d\lambda \qquad (9.3.37)\end{aligned}$$

is normal $N(0, e(h)[a(\theta_0) + b(\theta_0)])$ with $a(\theta_0) := c'A(\theta_0)c$ and $b(\theta_0) := c'B(\theta_0)c$, where the matrices $A(\theta_0)$ and $B(\theta_0)$ are defined by Eqs. (9.2.3) and (9.2.4), respectively, and the tapering factor $e(h)$ is given by formula (8.2.16).

Applying Theorem 8.2.6 with $f := f(\lambda, \theta_0)$ and

$$g := \frac{1}{4\pi} \sum_{k=1}^{p} c_k f^{-1}(\lambda, \theta_0) \frac{\partial}{\partial\theta_k} \log f(\lambda, \theta_0) w(\lambda),$$

we obtain the desired result. Theorem 9.2.2 is proved. □

9.4 Simplified Whittle Estimators for Spectral Parameters

In this section, we consider simplified tapered Whittle estimators and show that, under broad regularity conditions on the model's spectral density, these estimators are asymptotically statistically equivalent to the tapered Whittle estimator discussed in Sects. 9.1–9.3.

9.4.1 The Le Cam-Dzhaparidze Simplified Estimators

We first describe the Le Cam-Dzhaparidze approach to constructing simplified estimators in a general setting (see Le Cam [206] and Dzhaparidze [90, 93]).

Let $\theta := (\theta_1, \ldots, \theta_p) \in \Theta \subset \mathbb{R}^p$ be an unknown vector parameter that appears:

(a) In the probability density function of some random variable X, or
(b) In the finite-dimensional probability densities of a random process $\{X(t), t \in \mathbb{U}\}$, where $\mathbb{U} = \mathbb{R}$ in the continuous time case and $\mathbb{U} = \mathbb{Z}$ in the discrete time case.

The goal is to estimate the value of the parameter θ based on the sample $\mathbf{X}_T$.

In case (a), $\mathbf{X}_T := \{X_1, \ldots, X_T\}$, where $X_1, \ldots, X_T$ are T independent observations of the random variable X.

In case (b), $\mathbf{X}_T$ is an observed finite realization of the process $X(t)$: $\mathbf{X}_T := \{X(t),\ t \in D_T\}$, where $D_T := [0, T]$ in the continuous time case and $D_T := \{1, \ldots, T\}$ in the discrete time case.

The commonly used methods for constructing estimators of the unknown parameter θ (such as the method of moments, maximum likelihood method, least squares method, Whittle method, etc.) typically involve finding the roots of a system of estimating equations. These equations, which may be nonlinear with respect to the unknown parameter $\theta = (\theta_1, \ldots, \theta_p)$, take the following form:

$$F_i(\mathbf{X}_T, \theta) = 0, \quad i = 1, \ldots p, \tag{9.4.1}$$

where $F_i(\theta) := F_i(\mathbf{X}_T, \theta)$ are certain functionals of $\mathbf{X}_T$ that depend on θ.

For example, the tapered Whittle estimator $\widehat{\theta}_{T,h}$ of θ, discussed in Sect. 9.1, is the root of the system of estimating equations given by (9.1.12).

These classical estimation methods, however, have two disadvantages. First, the system of estimating equations (9.4.1) has an explicit solution only in relatively simple situations, and finding the roots $\hat{\theta}_T$ of this system often becomes a challenging problem. Second, for the roots to be consistent, the estimating equations in (9.4.1) must exhibit good behavior across the parameter set.

Therefore, it is of considerable interest to find more easily constructed (simplified) estimators $\check{\theta}_T$ that are asymptotically statistically equivalent to $\hat{\theta}_T$, meaning they possess the same asymptotic properties as $T \to \infty$.

The problem of constructing simplified estimators based on the non-tapered data $\mathbf{X}_T$ dates back to the classical work of Le Cam [206] and was further developed by Dzhaparidze [90, 93] (see also the survey paper by Dzhaparidze and Yaglom [94]).

To describe the Le Cam-Dzhaparidze approach for constructing simplified estimators in a general setting, we first introduce the following set of assumptions (see Dzhaparidze [90]). In what follows, $\tau_T = \tau(T)$ represents a non-random function that increases without bound as $T \to \infty$.

Assumption 9.4.1 *The system of estimating equations in* (9.4.1) *has a root* $\hat{\theta}_T$ *which is a consistent estimator of* θ*; that is,* $p - lim_{T\to\infty}\hat{\theta}_T = \theta_0$.

Assumption 9.4.2 *For* $\theta \in \Theta$*, the derivatives* $F_i^{(k)}(\theta) = (\partial/\partial\theta_k)F_i(\theta)$*,* $i, k = 1, \dots p$*, exist. Furthermore, for any arbitrarily small* $\varepsilon > 0$ *and* $\delta > 0$*, the following inequality holds:*

$$\mathbb{P}\left(|F_i^{(k)}(\theta_0) - w_{ik}(\theta_0)| < \varepsilon\right) \geq 1 - \delta, \tag{9.4.2}$$

where $F_i(\theta)$ *is defined as in* (9.4.1) *and* $W(\theta) := \|w_{ik}(\theta),\ i, k = 1, \dots, p\|$ *is a non-random matrix that is non-degenerate for* $\theta = \theta_0$.

Assumption 9.4.3 *The second derivatives* $F_i^{(k,j)}(\theta) = (\partial^2/\partial\theta_k\partial\theta_j)F_i(\theta)$ *exist and are continuous for* $\theta \in \Theta$ *and* $i, k, j = 1, \dots, p$*. Furthermore, for any arbitrarily small* $\delta > 0$*, there exists a constant* $M < \infty$ *such that*

$$\mathbb{P}\left(|F_i^{(k,j)}(\theta)| < M\right) \geq 1 - \delta. \tag{9.4.3}$$

Assumption 9.4.4 *Along with* (9.4.2)*, for sufficiently large* T*, the following stronger inequality holds:*

$$\mathbb{P}\left(\sqrt{\tau_T}|F_i^{(k)}(\theta_0) - w_{ik}(\theta_0)| < \varepsilon\right) \geq 1 - \delta. \tag{9.4.4}$$

Assumption 9.4.5 *There exists a random matrix* $D_* := \|d_{ik}^*\|$*,* $i, k = 1, \dots, p$*, such that for any arbitrarily small* $\varepsilon > 0$ *and* $\delta > 0$*, the following inequality*

$$\mathbb{P}\left(\sqrt{\tau_T}|d_{ik}^* - d_{ik}(\theta_0)| < \varepsilon\right) \geq 1 - \delta \tag{9.4.5}$$

holds for sufficiently large T *and for all* $i, k = 1, \dots, p$*. Here,* $d_{ik}(\theta_0)$ *are the elements of the matrix* $D(\theta_0) := W^{-1}(\theta_0)$*, and* $W(\theta)$ *is defined as in Assumption 9.4.2.*

The following theorem was proved by Dzhaparidze [90, 93].

Theorem 9.4.1 *Let* $\mathbf{F}(\theta)$ *be a* p*-dimensional vector with elements* $F_i(\theta)$*,* $i = 1, \dots p$*, and let* $\mathcal{F}(\theta)$ *be a matrix with elements* $F_i^{(k)}(\theta)i, k = 1, \dots p$*. Additionally, let* θ_T^* *be an arbitrary* τ_T^**-consistent estimator of* θ*, where* $\sqrt{\tau_T}/\tau_T^* \to 0$ *as* $T \to \infty$*. Then, the following assertions hold:*

(a) *Under Assumptions 9.4.1–9.4.3, the estimator* $\check{\theta}_{1,T} := \theta_T^* - \mathcal{F}^{-1}(\theta_T^*)F(\theta_T^*)$ *is asymptotically equivalent to* $\hat{\theta}_T$ *in the sense that*

$$p - \lim_{T\to\infty} \tau_T\left(\hat{\theta}_T - \check{\theta}_{1,T}\right) = 0.$$

(b) *Under Assumptions 9.4.1–9.4.5, the estimators of the form* $\check{\theta}_{2,T} := \theta_T^* - D_* F(\theta_T^*)$ *are asymptotically equivalent to* $\hat{\theta}_T$*; that is,*

$$p - \lim_{T\to\infty} \tau_T \left(\hat{\theta}_T - \check{\theta}_{2,T}\right) = 0.$$

9.4.2 Simplified Whittle Estimators for Spectral Parameters with Tapered Data

In Sect. 9.2 (see Theorems 9.2.1–9.2.3), it was shown that the tapered Whittle estimator $\widehat{\theta}_{T,h}$ of θ possesses good asymptotic properties. Specifically, the estimator $\widehat{\theta}_{T,h}$ is $\sqrt{T}$-consistent and asymptotically normal. Moreover, for discrete time Gaussian models, it is also asymptotically Fisher-efficient (see Remark 9.2.1).

However, the estimating equations in (9.1.12) that determine the tapered Whittle estimator $\widehat{\theta}_{T,h}$ are generally nonlinear, making it challenging to find the estimator $\widehat{\theta}_{T,h}$. Therefore, it is important to find simpler estimators of the parameter θ that retain the same asymptotic properties as $\widehat{\theta}_{T,h}$. The estimators proposed here are asymptotically equivalent to the tapered Whittle estimator $\widehat{\theta}_{T,h}$ under broad regularity conditions on the spectral density function $f(\lambda, \theta)$ of the observed model.

Theorem 9.4.2 *Let* $\mathbf{F}_h(\theta)$ *be a p-dimensional vector with elements* $F_{h,i}(\theta)$ $(i = 1, \dots p)$ *given by* (9.1.12). *Let* $\theta_{T,h}^*$ *be an arbitrary* τ_T^**-consistent estimator of* θ*, where* $\sqrt[4]{T}/\tau_T^* \to 0$ *as* $T \to \infty$*. Additionally, let* $D_* := \|d_{ik}^*\|$*,* $i, k = 1, \dots, p$*, be a random matrix whose elements* d_{ik}^* *satisfy the condition* (9.4.5). *Then, under the assumptions of Theorem 9.2.2, the estimators of the form*

$$\check{\theta}_{T,h} := \theta_{T,h}^* - D_* \mathbf{F}_h(\theta_{T,h}^*) \tag{9.4.6}$$

are asymptotically equivalent to the tapered Whittle estimator $\hat{\theta}_{T,h}$ *in the sense that*

$$p - \lim_{T\to\infty} \sqrt{T} \left(\hat{\theta}_{T,h} - \check{\theta}_{T,h}\right) = 0.$$

Proof The result we derive from Theorem 9.4.1 by using Theorems 8.2.1 and 8.2.3. Specifically, we show that Assumptions 9.4.2–9.4.4 are satisfied for the functions $F_{h,i}(\theta)$ $(i = 1, \dots p)$ as defined in (9.1.12).

First, by applying Theorem 8.2.1 we can easily conclude that for $k, j = 1, \dots, p$,

$$\begin{aligned} \lim_{T\to\infty} \mathbb{E}_0 \left[U_{T,h}^{(kj)}(\theta_0)\right] &= w_{kj}(\theta_0) \\ &= \frac{1}{4\pi} \int_\Lambda \frac{\partial}{\partial\theta_k} \ln f(\lambda, \theta_0) \frac{\partial}{\partial\theta_j} \ln f(\lambda, \theta_0) w(\lambda) d\lambda, \end{aligned} \tag{9.4.7}$$

where $\mathbb{E}_0[\xi]$ denotes the expectation with respect to the probability $\mathbb{P}_0$, which corresponds to the spectral density $f(\lambda, \theta_0)$, and $U_{T,h}^{(kj)}(\theta) = (\partial^2/\partial\theta_k\partial\theta_j)U_{T,h}(\theta)$ with $U_{T,h}(\theta)$ defined as in (9.1.10) (see the proof of Theorem 9.2.1 for details).

Next, by applying Theorem 8.2.3, for the variance of $U_{T,h}^{(kj)}(\theta_0)$ $(k, j = 1, \ldots, p)$, we have

$$\lim_{T\to\infty} \sqrt{T}\mathrm{Var}\left(U_{T,h}^{(kj)}(\theta_0)\right) = 0. \tag{9.4.8}$$

Therefore, by Chebyshev's inequality, it follows that, for sufficiently large T,

$$\mathbb{P}\left(\sqrt[4]{T}|F_{h,i}^{(k)}(\theta_0) - w_{ik}(\theta_0)| < \varepsilon\right) \geq 1 - \delta, \tag{9.4.9}$$

where $\varepsilon > 0$ and $\delta > 0$ are any arbitrary small numbers. Hence, Assumption 9.4.4 is satisfied with $\tau_T = \sqrt{T}$.

Next, since the matrix $W(\theta) := \|w_{ik}(\theta),\ i, k = 1, \ldots, p\|$ is assumed to be non-degenerate for $\theta = \theta_0$, Assumption 9.4.2 also holds.

Finally, using Theorems 8.2.1 and 8.2.3, we can easily infer that the function

$$F_{T,h,i}^{(kj)}(\theta) = \frac{1}{4\pi}\int_\Lambda I_T^h(\lambda)\frac{\partial^3}{\partial\theta_i\partial\theta_k\partial\theta_j} f^{-1}(\lambda, \theta)w(\lambda)d\lambda$$

satisfies Assumption 9.4.3. Thus, the result follows from Theorem 9.4.1(b). Theorem 9.4.2 is proved. □

Corollary 9.4.1 *Let $\mathcal{F}(\theta)$ be a matrix with elements $F_{h,i}^{(k)}(\theta)$ $(i, k = 1, \ldots p)$. Then, under the conditions of Theorem 9.4.2, the estimator*

$$\check{\theta}_{1,T,h} := \theta_{T,h}^* - \mathcal{F}^{-1}(\theta_{T,h}^*)\mathbf{F}_h(\theta_{T,h}^*) \tag{9.4.10}$$

is also asymptotically equivalent to the estimator $\hat{\theta}_{T,h}$.

Corollary 9.4.2 *Assume that there exist continuous derivatives $(\partial/\partial\theta_i)w_{k,j}(\theta)$ $(i, k, j = 1, \ldots, p)$ satisfying $|(\partial/\partial\theta_i)w_{i,k}(\theta)| < C$, where the constant C does not depend on θ. Then, under the conditions of Theorem 9.4.2, the estimator*

$$\check{\theta}_{2,T,h} := \theta_{T,h}^* - W^{-1}(\theta_{T,h}^*)\mathbf{F}_h(\theta_{T,h}^*) \tag{9.4.11}$$

is also asymptotically equivalent to the estimator $\hat{\theta}_{T,h}$.

Remark 9.4.1 It is easy to see that, similar to the non-tapered case (see Dzhaparidze [90, 93]), the estimators $\check{\theta}_{1,T,h}$ and $\check{\theta}_{2,T,h}$ can be constructed relatively easily.

In fact, to find them, it is necessary to have some τ_T^*-consistent estimator with $\sqrt[4]{T}/\tau_T^* \to 0$ as $T \to \infty$, and to determine the matrices $\mathcal{F}^{-1}(\theta)$ and $W^{-1}(\theta)$, respectively.

Note that the estimators $\check{\theta}_{T,h}$, $\check{\theta}_{1,T,h}$ and $\check{\theta}_{2,T,h}$ are relevant only when solving the system of estimating equations given in (9.1.12) is too difficult. In cases where the equations in (9.1.12) are linear (and thus easily solved), the estimator $\check{\theta}_{1,T,h}$ coincides with $\hat{\theta}_{T,h}$.

It is important to note that using simplified estimators can present certain challenges, particularly when calculating inverse matrices and second-order derivatives. These difficulties become more pronounced as the dimension of the parameter θ increases. In such cases, it is advisable to employ numerical iteration methods, such as the Newton-Raphson method or the Davidon-Fletcher-Powell method.

For more information on iteration methods applied in statistics, we refer to the papers by Beinicke and Dzhaparidze [24] and by Dzhaparidze [92].

9.5 Robust Estimation for Linear Models with Memory

In this section, we examine the robustness of inferences drawn from linear stationary models that may exhibit long memory and be influenced by a small trend. We demonstrate that a smoothed periodogram approach, applicable to both parametric and nonparametric estimation procedures, remains robust even in the presence of a small trend in the model. To streamline our discussion, we focus on the continuous time, non-tapered case.

9.5.1 Robustness to Small Trends of Estimation

Let $\{Y(t),\ t \in \mathbb{R}\}$ be a centered stationary process that possesses a spectral density $f(\lambda)$, $\lambda \in \mathbb{R}$. We assume that f is known except for an unknown vector parameter $\theta \in \Theta \subset \mathbb{R}^p$, or that f is completely unknown and belongs to a specified class $\mathcal{F}$ of spectral densities. Our goal is to make inferences about θ or the value $J(f)$ of a given functional $J(\cdot)$ at an unknown point $f \in \mathcal{F}$ when the actual observed data is in a contaminated form:

$$X(t) = Y(t) + M(t), \quad 0 \le t \le T, \tag{9.5.1}$$

where $M(t)$ is a deterministic trend.

We are particularly focused on determining whether the conclusions drawn in Chaps. 8 and 9 regarding the stationary model $Y(t)$ remain valid when the actual observed data is in a contaminated form (9.5.1). Specifically, we aim to assess the robustness of these conclusions in light of this departure from stationarity.

We show that if the trend $M(t)$ is considered "small," the asymptotic properties of the estimators for θ and $J(f)$, which are derived from the stationary model $Y(t)$, remain valid for the contaminated model $X(t)$. In particular, both parametric and nonparametric estimation procedures are robust when substituting the stationary model $Y(t)$ with the non-stationary model $X(t)$.

To achieve this, we first establish an asymptotic relationship between the stationary and contaminated periodograms.

9.5.2 An Asymptotic Relationship Between Stationary and Contaminated Periodograms

The next result shows that a small trend $M(t)$ of the form

$$|M(t)| \leq C|t|^{-\beta}, \quad \beta > 1/4, \quad t \in \mathbb{R}, \tag{9.5.2}$$

where C is a positive constant, does not affect the asymptotic properties of the empirical spectral linear functionals of the periodogram.

Theorem 9.5.1 *Let $\{Y(t),\ t \in \mathbb{R}\}$ be a centered, stationary process, $\{M(t),\ t \in \mathbb{R}\}$ be a deterministic trend, and $X(t) := Y(t) + M(t)$. Let $I_{TY}(\lambda)$ and $I_{TX}(\lambda)$ be the periodograms of $Y(t)$ and $X(t)$, respectively. Additionally, let $g(\lambda),\ \lambda \in \mathbb{R}$, be an even integrable function. Assume that the trend $M(t)$ and the Fourier transform $a(t) := \widehat{g}(t)$ of g are such that $M(t)$ is locally integrable on $\mathbb{R}$ and*

$$|M(t)| \leq C|t|^{-\beta}, \qquad |a(t)| \leq C|t|^{-\gamma}, \quad t \in \mathbb{R}, \quad 2\beta + \gamma > 3/2, \tag{9.5.3}$$

for some constants $C > 0$, $\gamma > 0$, and $\beta > 1/4$. Then the following asymptotic relation holds:

$$T^{1/2} \int_{\mathbb{R}} g(\lambda)\, [I_{TX}(\lambda) - I_{TY}(\lambda)]\, d\lambda \xrightarrow{P} 0 \quad \text{as} \quad T \to \infty, \tag{9.5.4}$$

provided that one of the following conditions is satisfied:

(i) The process $Y(t)$ has short or intermediate memory, that is, the covariance function $r(t)$ of $Y(t)$ satisfies $r \in L^1(\mathbb{R})$, and $\beta + \gamma > 1$.

(ii) The process $Y(t)$ has long memory with covariance function $r(t)$ satisfying

$$|r(t)| \leq C|t|^{-\alpha}, \quad t \in \mathbb{R}, \quad \alpha + \gamma \geq 3/2 \tag{9.5.5}$$

for some constants $C > 0$, $0 < \alpha \leq 1$, and $\alpha + 2\beta > 1$ if $\beta < 1 < \gamma$.

Remark 9.5.1 It is easy to verify that the statement of Theorem 9.5.1 holds, particularly when the parameters α, β, and γ satisfy the following conditions:

In case *(i)*: $\beta > 1/2,\ \gamma \geq 1/2$.
In case *(ii)*: $\alpha \geq 3/4,\ \beta > 3/8,\ \gamma \geq 3/4$.

9.5.3 Robustness to Small Trends in Nonparametric Estimation

The following theorem demonstrates that a small trend of the form (9.5.2) does not impact the asymptotic properties of the estimator for a linear spectral functional $J(f)$. In other words, the nonparametric estimation procedure is robust against the presence of a small trend in the model.

Theorem 9.5.2 *Suppose that the assumptions of Theorems 8.2.6 and 9.5.1 are satisfied. Then, the statistic $J(I_{TX})$ is a $\sqrt{T}$-consistent and asymptotically normal estimator for the functional $J(f)$, with asymptotic variance σ^2 as defined in (8.2.14) with $h(t) = \mathbf{1}_{[0,1]}(t)$. Specifically, the asymptotic relation (8.2.19) holds with $I_{TY}(\lambda)$ replaced by the contaminated periodogram $I_{TX}(\lambda)$:*

$$T^{1/2}\left[J(I_{TX}) - J(f)\right] \xrightarrow{d} \eta \quad as \quad T \to \infty, \tag{9.5.6}$$

where η follows a normal distribution, $N(0, \sigma^2)$, with σ^2 given by (8.2.14) with $h(t) = \mathbf{1}_{[0,1]}(t)$.

The following theorem shows that the nonparametric estimation procedure is robust to the presence of a small trend, even when the estimand spectral functional $J(f)$ is nonlinear.

Theorem 9.5.3 *Suppose that the assumptions of Theorems 8.3.1 and 9.5.1 are satisfied. Then, the statistic $J(\widehat{f}_{TX})$, where $\widehat{f}_{TX}$ is the kernel estimator of f constructed from the contaminated periodogram $I_{TX}(\lambda)$, is a $\sqrt{T}$-consistent and asymptotically normal estimator for the nonlinear functional $J(f)$ with an asymptotic variance σ^2 given by (8.2.14) with $g = J'(f)$) and $h(t) = \mathbf{1}_{[0,1]}(t)$. Specifically, the following asymptotic relation*

$$T^{1/2}\left[J(\widehat{f}_{TX}) - J(f)\right] \xrightarrow{d} \eta \quad \text{as} \quad T \to \infty \tag{9.5.7}$$

holds, where η is $N(0, \sigma^2)$ with σ^2 given by (8.2.14) with $g = J'(f)$.

9.5.4 Robustness to Small Trends in Parametric Estimation

The next result demonstrates that a small trend of the form (9.5.2) does not affect the asymptotic properties of the Whittle estimator for an unknown spectral parameter θ. In other words, the Whittle parametric estimation procedure is robust to the presence of a small trend in the model.

Theorem 9.5.4 *Suppose the assumptions of Theorem 9.5.1 hold with $g = f^{-1}(\lambda, \theta) \cdot w(\lambda)$. Then, under the conditions of Theorem 9.2.2, the Whittle estimator $\widehat{\theta}_{TX}$, constructed based on the contaminated periodogram $I_{T,X}(\lambda)$, is a $\sqrt{T}$-consistent and asymptotically normal estimator for the unknown spectral parameter*

θ. Specifically, the asymptotic relation (9.2.5) holds with $I_{TY}(\lambda)$ replaced by the contaminated periodogram $I_{TX}(\lambda)$:

$$T^{1/2}\left(\widehat{\theta}_{TX}-\theta_0\right)\xrightarrow{d} N_p\left(0, R(\theta_0)\right) \quad \text{as} \quad T\to\infty, \tag{9.5.8}$$

where the matrix $R(\theta_0)$ is defined in Eq. (9.2.6).

9.5.5 Proofs of Theorems 9.5.1–9.5.4

Proof of Theorem 9.5.1 In view of Eqs. (8.2.5) and (9.5.1), we can write

$$\begin{aligned} I_{T,X}(\lambda)-I_{T,Y}(\lambda) &= \frac{1}{2\pi T}\left(\left|\int_0^T e^{i\lambda t}X(t)\,dt\right|^2-\left|\int_0^T e^{i\lambda t}Y(t)\,dt\right|^2\right)\\ &= \frac{1}{2\pi T}\left(\left|\int_0^T e^{i\lambda t}[Y(t)+M(t)]\,dt\right|^2-\left|\int_0^T e^{i\lambda t}Y(t)\,dt\right|^2\right)\\ &= \frac{1}{2\pi T}\int_0^T\int_0^T e^{i\lambda(t-s)}\left[Y(t)M(s)+Y(s)M(t)+M(t)M(s)\right]dtds \end{aligned}$$

and

$$\begin{aligned} &\int_{-\infty}^{+\infty} g(\lambda,\theta)\left[I_{T,X}(\lambda)-I_{T,Y}(\lambda)\right]d\lambda\\ &= \frac{1}{T}\int_0^T\int_0^T \left[Y(t)M(s)+Y(s)M(t)+M(t)M(s)\right]a(t-s)\,dtds, \end{aligned} \tag{9.5.9}$$

where $a(t) := \widehat{g}(t)$ is the Fourier transform of the function $g(t)$.

Thus, to complete the proof, it suffices to show that under the conditions of the theorem, the following two relations hold:

$$T^{-1/2}\int_0^T\int_0^T M(t)M(s)a(t-s)\,dtds\to 0 \quad \text{as} \quad T\to\infty \tag{9.5.10}$$

and

$$T^{-1/2}\int_0^T\int_0^T Y(t)M(s)a(t-s)\,dtds\xrightarrow{P} 0 \quad \text{as} \quad T\to\infty. \tag{9.5.11}$$

Proof of relation (9.5.10). For $T > 2$, we write

$$
\begin{aligned}
I(T) &:= \int_0^T \int_0^T |M(t)M(s)a(t-s)| \, dtds \\
&= \int_0^1 \int_0^2 + \int_0^1 \int_2^T + \int_1^T \int_0^{1/2} + \int_1^T \int_{1/2}^T \\
&=: I_1(T) + I_2(T) + I_3(T) + I_4(T),
\end{aligned}
\tag{9.5.12}
$$

and estimate the integrals $I_i(T)$, $i = 1, 2, 3, 4$, separately.

Observe first that the Fourier transform $a(t) := \widehat{g}(t)$ is a bounded function on $\mathbb{R}$, since g is integrable on $\mathbb{R}$. Taking into account that, by assumption, the trend $M(t)$ is locally integrable on $\mathbb{R}$, we obtain the following estimate for the integral $I_1(T)$:

$$
I_1(T) \le C\|a\|_\infty \int_0^1 |M(s)| \, ds \int_0^2 |M(t)| \, dt \le C < \infty, \quad T > 2. \tag{9.5.13}
$$

Next, in view of conditions in (9.5.3), for $0 < s < 1$ and $t > 2$, we have

$$
|a(t-s)| \le C(t-s)^{-\gamma} \le Ct^{-\gamma}.
$$

Thus, taking into account that $\beta + \gamma > 1$, we can estimate the integral $I_2(T)$ as follows:

$$
I_2(T) \le C \int_0^1 |M(s)| \, ds \int_2^T \frac{1}{t^{\beta+\gamma}} \, dt \le C < \infty, \quad T > 2. \tag{9.5.14}
$$

Similarly, for the integral $I_3(T)$, we have

$$
I_3(T) \le C < \infty, \qquad T > 2. \tag{9.5.15}
$$

To estimate the integral $I_4(T)$, we first observe that, in view of conditions in (9.5.3), for $1 < s < T$, we can write

$$
\begin{aligned}
\psi(s) = \psi(T, s) &:= \int_{1/2}^T |M(t)a(t-s)| \, dt \\
&\le C \Bigg[\int_{s-1}^{s+1} \frac{|a(t-s)|}{t^\beta} dt + \int_{s+1}^{2s} \frac{1}{t^\beta (t-s)^\gamma} dt \\
&\quad + \int_{2s}^T \frac{1}{t^\beta (t-s)^\gamma} dt + \int_{1/2}^{s/2} \frac{1}{t^\beta (s-t)^\gamma} dt + \int_{s/2}^{s-1} \frac{1}{t^\beta (s-t)^\gamma} dt \Bigg] \\
&\le C \Bigg[\|a\|_\infty \cdot s^{-\beta} + s^{-\beta} \int_1^s \frac{1}{\tau^\gamma} d\tau + \int_{2s}^T \frac{1}{t^{\beta+\gamma}} dt
\end{aligned}
$$

$$
\begin{aligned}
&+s^{-\gamma}\int_{1/2}^{s/2}\frac{1}{t^{\beta}}dt+s^{-\beta}\int_{1}^{s/2}\frac{1}{\tau^{\gamma}}d\tau\Big] \\
&\le C\Big[s^{-\beta}+L(\gamma,T)s^{1-\beta-\gamma}+L(\beta+\gamma,T)\left(T^{1-\beta-\gamma}+s^{1-\beta-\gamma}\right) \\
&\quad +L(\beta,T)\left(s^{1-\beta-\gamma}+s^{-\gamma}\right)+L(\gamma,T)s^{1-\beta-\gamma}\Big] \\
&\le C\log T\cdot\left(T^{1-\beta-\gamma}+s^{1-\beta-\gamma}+s^{-\beta}+s^{-\gamma}\right), \qquad (9.5.16)
\end{aligned}
$$

where the function $L(u,T)$ is defined by

$$
L(u,T)=\begin{cases}\log T & \text{if} \quad u=1,\\ 1 & \text{otherwise.}\end{cases}
$$

Taking into account that $\beta+\gamma>1$, from (9.5.16) we obtain

$$
\psi(s)\le C\log T\cdot\left(s^{1-\beta-\gamma}+s^{-\beta}+s^{-\gamma}\right),\quad 1<s<T, \qquad (9.5.17)
$$

and consequently, for $T>2$, the integral $I_4(T)$ can be estimated as follows:

$$
\begin{aligned}
I_4(T)&=\int_1^T|M(s)\psi(s)|\,ds \\
&\le C\log T\left[\int_1^T\frac{1}{s^{2\beta+\gamma-1}}\,ds+\int_1^T\frac{1}{s^{2\beta}}\,ds+\int_1^T\frac{1}{s^{\beta+\gamma}}\,ds\right] \\
&\le C\log T\left[1+L(2\beta+\gamma-1,T)\,T^{2-2\beta-\gamma}+L(2\beta,T)\,T^{1-2\beta}+T^{1-\beta-\gamma}\right] \\
&\le C\log^2 T\left(1+T^{2-2\beta-\gamma}+T^{1-2\beta}\right). \qquad (9.5.18)
\end{aligned}
$$

Finally, taking into account the assumptions that $2\beta+\gamma>3/2$ and $\beta>1/4$, from (9.5.12)–(9.5.15) and (9.5.18), we obtain

$$
T^{-1/2}\cdot I(T)\le C\log^2 T\left(T^{-1/2}+T^{3/2-2\beta-\gamma}+T^{1/2-2\beta}\right)\to 0\quad \text{as}\quad T\to\infty,
$$

which implies (9.5.10).

Proof of relation (9.5.11). Observe first that the inequality

$$
\beta+\gamma>1 \qquad (9.5.19)
$$

also holds in case *(ii)*, since by (9.5.3) and (9.5.5), we have $2\beta + 2\gamma \geq 2\beta + \gamma + 3/2 - \alpha > 3/2 + 1/2 = 2$. Denote

$$\nu(s) = \nu(T, s) := \int_0^T M(t)a(t-s)\,dt, \quad 0 < s < T,$$

and observe that

$$\int_0^{1/2} |M(t)a(t-s)|\,dt \leq C \int_0^{1/2} |M(t)| \frac{1}{(s-1/2)^\gamma}\,dt \leq C \cdot s^{-\gamma}, \quad 1 < s < T.$$

Similar to (9.5.17), we have

$$|\nu(s)| \leq C \log T \cdot \left(s^{1-\beta-\gamma} + s^{-\beta} + s^{-\gamma}\right), \quad 1 < s < T. \tag{9.5.20}$$

On the other hand, by conditions in (9.5.3), for $T > 2$ and $0 < s < 1$, we have

$$|\nu(s)| \leq C \left[\|a\|_\infty \int_0^2 |M(t)|\,dt + \int_2^T \frac{1}{t^\beta (t-1)^\gamma} dt \right] \leq C \log T. \tag{9.5.21}$$

From (9.5.3) and (9.5.20), it follows that (9.5.21) holds for $0 < s < T$.

Now, we denote

$$\xi(T) := T^{-1/2} \int_0^T \int_0^T Y(s)M(t)a(t-s)\,dtds = T^{-1/2} \int_0^T Y(s)\nu(s)ds,$$

and observe that

$$\begin{aligned}\mathbb{E}[\xi^2(T)] &= T^{-1} \int_0^T \int_0^T E\{Y(s)Y(\tau)\}\nu(s)\nu(\tau)dsd\tau \\ &= T^{-1} \int_0^T \int_0^T \nu(s)\nu(\tau)r(s-\tau)dsd\tau.\end{aligned}$$

Therefore, based on Chebyshev's inequality, to prove (9.5.11), it is sufficient to show that

$$S(T) := \int_0^T \int_0^T |\nu(s)\nu(\tau)r(s-\tau)|\,dsd\tau = o(T) \quad \text{as} \quad T \to \infty. \tag{9.5.22}$$

In case *(i)*, when the process $Y(t)$ has short or intermediate memory, and hence $r \in L^1(\mathbb{R})$, from (9.5.21) for $T > 2$, we obtain

$$\begin{aligned}|S(T)| &\le C \log T \int_0^T |\nu(s)| \int_0^T |r(s-\tau)|\, d\tau ds \\ &\le C \log T \int_0^T |\nu(s)|\, ds. \end{aligned} \tag{9.5.23}$$

In view of inequality (9.5.20), the last integral in (9.5.23) can be estimated as follows:

$$\begin{aligned}&\int_0^T |\nu(s)|\, ds \le C \log^2 T \left[\int_0^1 ds + \int_1^T \left(s^{1-\beta-\gamma} + s^{-\beta} + s^{-\gamma}\right) ds\right] \\ &\le C \log^2 T \left[1 + L(\beta+\gamma-1, T)\, T^{2-\beta-\gamma} + L(\beta, T)\, T^{1-\beta} + L(\gamma, T)\, T^{1-\gamma}\right] \\ &\le C \log^3 T \left(1 + T^{1-\beta} + T^{1-\gamma} + T^{2-\beta-\gamma}\right). \end{aligned} \tag{9.5.24}$$

Therefore, taking into account that $\beta + \gamma > 1$, from (9.5.23) and (9.5.24), we obtain

$$S(T) = o(T) \quad \text{as} \quad T \to \infty.$$

In case *(ii)*, when the process $Y(t)$ exhibits long memory, we can use Eqs. (9.5.5), (9.5.19), (9.5.20), and (9.5.21), to obtain that for $1 < \tau < T$,

$$\begin{aligned}q(\tau) &:= \int_0^T |\nu(s) r(s-\tau)| ds \le C \log T \int_0^{1/2} \left(\tau - \frac{1}{2}\right)^{-\alpha} ds \\ &+ C \log T \left[\int_{1/2}^T \frac{|r(s-\tau)|}{s^{\beta+\gamma-1}} ds + \int_{1/2}^T \frac{|r(s-\tau)|}{s^{\beta}} ds + \int_{1/2}^T \frac{|r(s-\tau)|}{s^{\gamma}} ds\right]. \end{aligned} \tag{9.5.25}$$

Taking into account that r is bounded ($|r(t)| \le r(0) = E|Y(t)|^2 < \infty,\ t \in \mathbb{R}$), and using similar arguments as in (9.5.16), we can conclude from the condition in (9.5.5) that for any $\eta > 0$,

$$\begin{aligned}\int_{1/2}^T \frac{|r(t-\tau)|}{t^\eta} dt &\le C \log T \left(T^{1-\alpha-\eta} + \tau^{1-\alpha-\eta} + \tau^{-\alpha} + \tau^{-\eta}\right) \\ &\le C \log T \left(1 + T^{1-\alpha-\eta}\right), \quad 1 < \tau < T. \end{aligned}$$

Applying this inequality for $\eta = \beta + \gamma - 1$, $\eta = \beta$, and $\eta = \gamma$, and taking into account that $\alpha + \gamma > 1$, we obtain from (9.5.25) that

$$q(\tau) \le C \log^2 T \left(1 + T^{2-\alpha-\beta-\gamma} + T^{1-\alpha-\beta}\right), \quad 1 < \tau < T. \tag{9.5.26}$$

On the other hand, from (9.5.20) and (9.5.21), for $T > 2$ and $0 < \tau < 1$, we have

$$\begin{aligned} q(\tau) &\le C\left[\log T \int_0^2 |r(s-\tau)|\,ds + \int_2^T \frac{|\nu(s)|}{(s-1)^\alpha} ds\right] \\ &\le C \log T \left(1 + T^{2-\alpha-\beta-\gamma} + T^{1-\alpha-\beta}\right) \le C \log T \left(1 + T^{1-\beta}\right), \end{aligned} \tag{9.5.27}$$

since $\alpha + \gamma > 1$ and $\alpha > 0$.

Next, we express the integral $S(T)$ in (9.5.22) as follows:

$$S(T) = \int_0^T |\nu(\tau)|\, q(\tau)\, d\tau = \int_0^1 + \int_1^T =: S_1(T) + S_2(T), \tag{9.5.28}$$

and estimate the integrals $S_1(T)$ and $S_2(T)$.

First, according to (9.5.21), (9.5.27), and the condition $\beta > 0$, for the integral $S_1(T)$, we have

$$S_1(T) \le C \log^2 T \left(1 + T^{1-\beta}\right) = o(T) \quad \text{as} \quad T \to \infty. \tag{9.5.29}$$

To estimate the integral $S_2(T)$, we consider three cases and use conditions (9.5.3), (9.5.5), and (9.5.19), along with inequalities (9.5.20) and (9.5.26).

Case 1. If $\beta \ge 1$, then we have

$$|\nu(\tau)| \le C \log T \left(\tau^{-\beta} + \tau^{-\gamma}\right), \quad q(\tau) \le C \log^2 T, \quad 1 < \tau < T,$$

and hence

$$S_2(T) \le C \log^3 T \left(1 + T^{1-\beta} + T^{1-\gamma}\right) = o(T) \quad \text{as} \quad T \to \infty. \tag{9.5.30}$$

Case 2. If $\beta < 1 < \gamma$, then we have

$$|\nu(\tau)| \le C \log T \cdot \tau^{-\beta}, \quad q(\tau) \le C \log^2 T \left(1 + T^{1-\alpha-\beta}\right) \quad 1 < \tau < T,$$

and hence

$$\begin{aligned} S_2(T) &\le C \log^3 T \left(1 + T^{1-\beta}\right)\left(1 + T^{1-\alpha-\beta}\right) \\ &\le C \log^3 T \left(1 + T^{1-\alpha-\beta} + T^{1-\beta} + T^{2-\alpha-2\beta}\right) = o(T) \end{aligned} \tag{9.5.31}$$

as $T \to \infty$, since, in this case, by assumption $\alpha + 2\beta > 1$.

Case 3. If $\beta < 1$ and $\gamma \le 1$, then we have

$$|\nu(\tau)| \le C \log T \cdot \tau^{1-\beta-\gamma}, \quad q(\tau) \le C \log^2 T \left(1 + T^{2-\alpha-\beta-\gamma}\right), \quad 1 < \tau < T,$$

and hence

$$\begin{aligned} S_2(T) &\le C \log^3 T \left(1 + T^{2-\beta-\gamma}\right) \left(1 + T^{2-\alpha-\beta-\gamma}\right) \\ &\le C \log^3 T \left(1 + T^{2-\alpha-\beta-\gamma} + T^{2-\beta-\gamma} + T^{4-\alpha-2\beta-2\gamma}\right) = o(T) \end{aligned} \tag{9.5.32}$$

as $T \to \infty$, since $\beta + \gamma > 1$ and $\alpha + 2\beta + 2\gamma = (2\beta + \gamma) + (\alpha + \gamma) > 3$.

Finally, from (9.5.28) to (9.5.32), we obtain

$$S(T) = o(T) \quad \text{as} \quad T \to \infty.$$

Thus, the relation (9.5.22), and consequently (9.5.11), has been established.

Theorem 9.5.1 is proved. □

Proof of Theorem 9.5.2 In view of Eqs. (8.2.1) and (8.2.7), we can write

$$\begin{aligned} T^{1/2} [J(I_{TX}) - J(f)] &= T^{1/2} \left[\int_{\mathbb{R}} I_{TX}(\lambda) g(\lambda) d\lambda - \int_{\mathbb{R}} f(\lambda) g(\lambda) d\lambda\right] \\ &= T^{1/2} \left[\int_{\mathbb{R}} I_{TX}(\lambda) g(\lambda) d\lambda - \int_{\mathbb{R}} I_{TY}(\lambda) g(\lambda) d\lambda\right] \\ &+ T^{1/2} \left[\int_{\mathbb{R}} I_{TY}(\lambda) g(\lambda) d\lambda - \int_{\mathbb{R}} f(\lambda) g(\lambda) d\lambda\right] \\ &= T^{1/2} \int_{\mathbb{R}} g(\lambda) \left[I_{TX}(\lambda) - I_{TY}(\lambda)\right] d\lambda + T^{1/2} [J(I_{TY}) - J(f)]. \end{aligned} \tag{9.5.33}$$

Next, according to Theorem 9.5.1, the first term on the right-hand side of (9.5.33) converges to zero in probability as $T \to \infty$. Meanwhile, by Theorem 8.2.6, the second term on the right-hand side of (9.5.33) converges in distribution to η. Therefore, the result follows. □

Proof of Theorem 9.5.3 We write

$$\begin{aligned} T^{1/2} \left[J(\widehat{f}_{TX}) - J(f)\right] &= T^{1/2} \left[J(\widehat{f}_{TY}) - J(f)\right] \\ &+ T^{1/2} \left[J(\widehat{f}_{TX}) - J(\widehat{f}_{TY})\right], \end{aligned} \tag{9.5.34}$$

and observe that, by Theorem 8.3.1, the first term on the right-hand side of (9.5.34) converges in distribution to η as $T \to \infty$.

Next, it follows from the proof of Theorem 8.3.1 (see formula (8.3.19)) that

$$T^{1/2}[J(\widehat{f}_{TY}) - J(f)] = T^{1/2} \int_{\mathbb{R}} J'(f;\lambda)\,[I_{TY}(\lambda) - f(\lambda)]\,d\lambda + o_P(1),$$

$$T^{1/2}[J(\widehat{f}_{TX}) - J(f)] = T^{1/2} \int_{\mathbb{R}} J'(f;\lambda)\,[I_{TX}(\lambda) - f(\lambda)]\,d\lambda + o_P(1).$$

Therefore, we have

$$T^{1/2}\left[J(\widehat{f}_{TX}) - J(\widehat{f}_{TY})\right] = T^{1/2} \int_{\mathbb{R}} J'(f;\lambda)\,[I_{TX}(\lambda) - I_{TY}(\lambda)]\,d\lambda + o_P(1).$$

Now we can apply Theorem 9.5.1 with $g = J'(f))$ to conclude that the second term on the right-hand side of (9.5.34) converges to zero in probability as $T \to \infty$. Therefore, the result follows. □

Proof of Theorem 9.5.4 Let $U_{TY}(\cdot)$ be the Whittle functional, and $\widehat{\theta}_{TY}$ be the Whittle estimator constructed based on the observation $\mathbf{Y}_T = \{Y(t),\, 0 \le t \le T\}$ (see formulas (9.1.10) and (9.1.11)). By Taylor's formula, for $|\widehat{\theta}_T^* - \theta_0| < |\widehat{\theta}_{TY} - \theta_0|$ and for sufficiently large T, we can write

$$T^{1/2}\left[\widehat{\theta}_{TY} - \theta_0\right] = -T^{1/2}\left[\frac{\partial^2}{\partial\theta\partial\theta'}U_{TY}(\theta_T^*)\right]^{-1}\left[\frac{\partial}{\partial\theta}U_{TY}(\theta_0)\right] + o_P(1). \tag{9.5.35}$$

Next, by Theorem 9.5.1, we have

$$U_{TX}(\theta_T) = U_{TY}(\theta_T) + o_P(1). \tag{9.5.36}$$

Now let $U_{TX}(\cdot)$ and $\widehat{\theta}_{TX}$ denote the Whittle functional and the Whittle estimator, respectively, constructed based on the contaminated observation $\mathbf{X}_T = \{X(t),\, 0 \le t \le T\}$. By applying Taylor's formula and considering the relations (9.5.35) and (9.5.36), we can infer that

$$T^{1/2}\left[\widehat{\theta}_{TX} - \theta_0\right] = T^{1/2}\left[\widehat{\theta}_{TY} - \theta_0\right] + o_P(1),$$

which shows that the estimator $\widehat{\theta}_{TX}$ has the same asymptotic properties as $\widehat{\theta}_{TY}$.

Therefore, the result follows from Theorems 9.2.1 and 9.2.2. □

9.6 Complements and Bibliographical Notes

9.6.1 More About the Whittle Approximation

In Sect. 9.1, it was observed that if the Whittle approximation (9.1.2) holds, with $U_{T,W}(\theta)$ as defined in (9.1.7), then the maximum likelihood estimator $\widehat{\theta}_{T,MLE}$ based on the exact log-likelihood function $L_T(\theta)$ and the Whittle estimator $\widehat{\theta}_T$ based on the "principle part" $U_{T,W}(\theta)$ are asymptotically equivalent. Therefore, it is crucial to establish sufficient conditions for the validity of the Whittle approximation (9.1.2).

The Whittle approximation was first derived by Whittle [313, 314] for discrete time Gaussian short memory models. He obtained this approximation using purely heuristic arguments and did not specify any exact conditions for the validity of relation (9.1.2). Since then, many authors have established sufficient conditions for (9.1.2) in various discrete time models (see, e.g., Azencott and Dacunha-Castelle [17], Clevenson [64], Coursol and Dacunha-Castelle [66, 67], Dzhaparidze [89, 93], Dzhaparidze and Yaglom [94]), Ginovyan [111], and references therein).

We present several results related to this topic. Denote

$$\eta_T := T(L_T - U_{T,W}) \tag{9.6.1}$$

where $L_T := L_T(\theta)$ and $U_T := U_{T,W}(\theta)$ are defined as in Eqs. (9.1.1) and (9.1.7), respectively.

For moving average models of order r (MA(r), $r < \infty$), Whittle's approximation was rigorously proved by Clevenson [64]. He showed, in particular, that

$$\eta_T = o_P(T) \quad \text{as} \quad T \to \infty.$$

Recall the Sobolev class of functions $H^{1/2}$ as defined in Eq. (2.4.3). Dzhaparidze [93] (Section 1.2, Theorem 1) proved that if f is the spectral density of a short memory process and satisfies the condition $f \in H^{1/2}$, then the asymptotic relation (9.1.2) holds.

A stronger result for short memory nondeterministic models with spectral density $f \in H^{1/2}$ was obtained by Coursol and Dacunha-Castelle [66]. By utilizing Szegő's strong theorem (see Proposition 2.4.2), the expected value and variance of the random variable η_T were explicitly calculated in terms of the Fourier coefficients of the functions f, $1/f$, and $\log f$, and the following relation was then proved:

$$\eta_T = O_P(1) \quad \text{as} \quad T \to \infty. \tag{9.6.2}$$

Sufficient conditions for the Whittle approximation (9.1.2) for a class of discrete time models exhibiting long or intermediate memory were established by Ginovyan [111, 114]. For models with spectral densities that have fixed polynomial zeros, sufficient conditions for (9.1.2) were derived by Dzhaparidze [93] (Section 1.2, Theorem 2) and by Ginovyan [112].

In their survey paper, Dzhaparidze and Yaglom [94] propose a conjecture on page 45, stating that the Whittle approximation (9.1.2) and the asymptotic relation (9.6.2) may be valid under much broader conditions. They suggest that these results may be valid whenever the underlying model is a nondeterministic process, specifically when the logarithm of the spectral density is integrable. However, to the best of our knowledge, this conjecture has not yet been proven or disproven.

In the continuous time case, approximating the likelihood function, similar to the Whittle approximation in (9.1.7), becomes a complicated problem. This challenge arises because, unlike in the discrete time case, there is no measure analogous to the Lebesgue measure in continuous time with respect to which the probability distributions of the observed process are absolutely continuous. Consequently, we must compute the density of the probability distribution of the sample with respect to a fixed dominating measure.

Specifically, let $\{X(t),\ t \in \mathbb{R}\}$ be a continuous time Gaussian process with spectral density $f(\lambda)$, and let $\mathbf{X}_T := \{X(t), 0 \leq t \leq T\}$. Let $\mathbb{P}_{f,T}$ be the associated Gaussian measure defined on the space of functions of t $(0 \leq t \leq T)$. Then, the likelihood function $p_{f,T}(\mathbf{X}_T)$ is defined as the Radon-Nikodym derivative $d\mathbb{P}_{f,T}/d\mathbb{P}_{g,T}$ of the measure $\mathbb{P}_{f,T}$ with respect to some standard dominating measure $\mathbb{P}_{g,T}$ on the same space. Thus, both the likelihood function $p_{f,T}(\mathbf{X}_T)$ and the log-likelihood function $L_T(\mathbf{X}_T) := \ln p_{f,T}(\mathbf{X}_T)$ are functionals depending on $\mathbf{X}_T$.

To the best of our knowledge, the equation (similar to the Whittle Eq. (9.1.7)) for the *main part* $U_{T,W}$ of the log-likelihood L_T, has, so far, been rigorously derived only for a specific class of continuous time Gaussian stationary processes. This class includes processes that have everywhere positive rational spectral densities of the form (3.3.11), specifically the class of Gaussian CARMA processes. For this class of processes, Pisarenko [247] (see also Dzhaparidze and Yaglom [94], page 50) found an appropriate analog of the Whittle approximation $U_{T,W}$.

Specifically, let $\mathbb{P}_{f,T}$ and $\mathbb{P}_{g,T}$ be two Gaussian measures corresponding to stationary Gaussian processes with positive rational spectral density functions f and g of the form (3.3.11), such that the Radon-Nikodym derivative $d\mathbb{P}_{f,T}/d\mathbb{P}_{g,T}$ exists. As an approximation to the $-1/T\times$log-likelihood function $L_T(\mathbf{X}_T)$, Pisarenko [247] proposed the following functional:

$$\widetilde{U}_T(\mathbf{X}_T) := \frac{1}{4\pi}\int_{\mathbb{R}}\left[\ln\frac{f(\lambda)}{g(\lambda)} + \left(\frac{1}{f(\lambda)} - \frac{1}{g(\lambda)}\right) I_T(\lambda)\right] d\lambda, \tag{9.6.3}$$

where $I_T(\lambda)$ is the continuous periodogram of $X(t)$, and the integral

$$\int_{\mathbb{R}}\left(f^{-1}(\lambda) - g^{-1}(\lambda)\right) d\lambda$$

is understood in a special, yet quite natural sense (for details, see Pisarenko [247]).

In Pisarenko [247], it was stated (without a detailed proof) that the following relation holds (cf. (9.6.2)):

$$\widetilde{\eta}_T := T(L_T - \widetilde{U}_T) = O_P(1) \quad \text{as} \quad T \to \infty,$$

where $L_T := L_T(\mathbf{X}_T)$ is the exact log-likelihood and $\widetilde{U}_T := \widetilde{U}_T(\mathbf{X}_T)$ is defined in Eq. (9.6.3).

Another expression for the asymptotic principal part $U_{T,W}$ of the exact log-likelihood L_T for a pair of stationary Gaussian processes with positive rational spectral density functions f and g was proposed by Pham-Dinh [244]. His equation for $U_{T,W}$ is based on a representation of the derivative $d\mathbb{P}_{f,T}/d\mathbb{P}_{g,T}$ as a functional of a specific Wiener process with independent increments.

An approximation to the likelihood function for generalized Gaussian stationary processes was obtained by Solev [281, 282]. He proposed an approximation $\widehat{U}_T(\mathbf{X}_T)$ for $L_T(\mathbf{X}_T)$ in terms of reproducing kernels and derived sufficient conditions to ensure the asymptotic relation:

$$T(L_T - \widehat{U}_T) = O_P(1) \quad \text{as} \quad T \to \infty.$$

In this text, for continuous time processes, as a Whittle approximation we consider the functional (see (9.1.9)):

$$U_{T,w} := \frac{1}{4\pi} \int_{\mathbb{R}} \left[\ln f(\lambda, \theta) + \frac{I_T(\lambda)}{f(\lambda)} \right] \cdot w(\lambda)\, d\lambda, \tag{9.6.4}$$

where $w(\lambda)$ is a weight function ($w(-\lambda) = w(\lambda)$, $w(\lambda) \geq 0$, $w(\lambda) \in L^1(\mathbb{R})$) for which the integral in (9.6.4) is well-defined.

Note that in the discrete time case, where the spectral density $f(\lambda)$ is defined on $\mathbb{T}$, the weight function $w(\lambda)$ is not necessary (i.e., $w(\lambda) = 1$). In contrast, for the continuous time case, the slow decay of the spectral density $f(\lambda)$ at infinity necessitates the use of the weight function $w(\lambda)$ in Eq. (9.6.4) to ensure that the equation is well-defined.

A commonly used weight function is $w(\lambda) = 1/(1 + \lambda^2)$. The Eq. (9.6.4) with this weight function can be justified by applying the transformation $\mu = 2 \arctan \lambda$ under which integrals of the form $\int_{\mathbb{T}} \psi(\mu) d\mu$ transform into $\int_{\mathbb{R}} \widetilde{\psi}(\lambda)(1+\lambda^2)^{-1} d\lambda$ (see, e.g., Gao [103], Rozanov [263], Section 3.1, and Dym and McKean [88]).

In the aforementioned papers, as well as in this text, the question of how well the functional $U_{T,w}$, defined in (9.6.4), approximates the $-\frac{1}{T}\times$ log-likelihood function has not been addressed.

Specifically, the extent to which $U_{T,w}$ approximates the $-\frac{1}{T}\times$ log-likelihood function remains unexplored. Consequently, it is important to establish sufficient conditions that ensure asymptotic relations similar to those in (9.1.2) and (9.6.2) for the approximation $U_{T,w}$ defined in (9.6.4).

9.6.2 *Minimum Contrast Estimators: Ibragimov's Estimator*

We revisit the parametric estimation problem, focusing on the non-tapered continuous time case for simplicity. Specifically, we assume that the spectral density $f(\lambda)$ $\lambda \in \mathbb{R}$ of the observed model $\{X(t),\ t \in \mathbb{R}\}$ belongs to a given parametric family of spectral densities $\mathcal{F} := \{f(\lambda, \theta) : \theta \in \Theta\}$, where θ is an unknown parameter and

Θ is a subset of $\mathbb{R}$. The problem of interest is to estimate the unknown parameter θ based on the observed data $\mathbf{X}_T := \{X(t), \ t \in [0, T]\}$, using the minimum contrast estimation method.

Minimum contrast estimators are generally defined as follows:

$$\widehat{\theta}_T := \operatorname*{Arg\,min}_{\theta \in \Theta} D(f(\cdot, \theta), \widehat{f}_T(\cdot)), \tag{9.6.5}$$

where

$$D(f, g) := \int_{\mathbb{R}} K(f(\lambda), g(\lambda)) \cdot w(\lambda) d\lambda \tag{9.6.6}$$

is a criterion for measuring the distance (or discrepancy) between the functions f and g, with the corresponding function $K(\cdot, \cdot)$. Here, $\widehat{f}_T(\cdot)$ is a nonparametric estimator of the spectral density f and $w(\lambda)$ is a weight function (see, e.g., Taniguchi [294], and Taniguchi and Kakizawa [295]).

Examples of functions $K(f, g)$, which are of the form $K(f/g)$ can be found in Taniguchi [294]. In particular, the function $K(x) = \log x + 1/x$ yields the following criterion:

$$D(f, \widehat{f}_T) := \int_{\mathbb{R}} \left[\log f(\lambda, \theta) - \log \widehat{f}_T(\lambda) + \frac{\widehat{f}_T(\lambda)}{f(\lambda, \theta)}) \right] \cdot w(\lambda) d\lambda. \tag{9.6.7}$$

This criterion, with $\widehat{f}_T(\lambda) = I_T(\lambda)$, where $I_T(\lambda)$ is the periodogram of the observed process $X(t)$, is equivalent to the Whittle functional (see Eq. (9.1.9)):

$$D(f, I_T) := \int_{\mathbb{R}} \left[\log f(\lambda, \theta) + \frac{I_T(\lambda)}{f(\lambda, \theta)}) \right] \cdot w(\lambda) d\lambda. \tag{9.6.8}$$

Following Guyon [165], we will adopt a more general approach to define minimum contrast estimators. To achieve this, we will first introduce the concepts of the contrast function and the contrast process.

A non-random, non-negative function $K(\theta, \theta_0)$, $\theta_0 \in S$, $\theta \in \Theta$, is called a *contrast function* if it has a unique minimum at the point $\theta = \theta_0$, where θ_0 is the true value of the parameter θ.

A *contrast process* $U_T(\theta) := U(\mathbf{X}_T, \theta)$, for a contrast function $K(\theta, \theta_0)$ and observations $\mathbf{X}_T := \{X(t), \ t \in [0, T]\}$, is defined as a random process that satisfies the following relation:

$$\liminf_{T \to \infty} [U_T(\theta) - U_T(\theta_0)] \geq K(\theta, \theta_0) \tag{9.6.9}$$

in $\mathbb{P}_0$ probability as $T \to \infty$, where $\mathbb{P}_0$ is the probability measure corresponding to the spectral density $f(\lambda, \theta_0)$.

The minimum contrast estimator $\widehat{\theta}_T$ is defined as the point that minimizes the contrast process $U_T(\theta)$:

$$\widehat{\theta}_T := \operatorname*{Arg\,min}_{\theta\in\Theta} U_T(\theta). \tag{9.6.10}$$

The Whittle estimator is a minimum contrast estimator. The contrast function $K(\theta, \theta_0)$ and the contrast process $U_T(\theta)$ are defined by the following equations (see Sect. 9.1):

$$K(\theta, \theta_0) := \int_{\mathbb{R}} \left[\frac{f(\lambda, \theta_0)}{f(\lambda, \theta)} - 1 - \log \frac{f(\lambda, \theta_0)}{f(\lambda, \theta)} \right] \cdot w(\lambda)\, d\lambda.$$

$$U_T(\theta) := \int_{\mathbb{R}} \left[\ln f(\lambda, \theta) + \frac{I_T(\lambda)}{f(\lambda, \theta)} \right] \cdot w(\lambda)\, d\lambda.$$

Note that the Whittle functional and the Whittle estimator were originally introduced in the context of the maximum likelihood approach (see Sect. 9.1).

Another important minimum contrast estimator is based on *Ibragimov's contrast function* and the *contrast process*.

We assume the existence of a weight function $w(\lambda)$, $\lambda \in \mathbb{R}$, such that $f(\lambda, \theta)w(\lambda) \in L^1(\mathbb{R})$ for all $\theta \in \Theta$. We set

$$\sigma^2(\theta) := \int_{\mathbb{R}} f(\lambda, \theta) w(\lambda) d\lambda, \tag{9.6.11}$$

and consider the following factorization of the spectral density:

$$f(\lambda, \theta) = \sigma^2(\theta)\psi(\lambda, \theta), \quad \lambda \in \mathbb{R},\ \theta \in \Theta. \tag{9.6.12}$$

For the function $\psi(\lambda, \theta)$ in (9.6.12), we have

$$\int_{\mathbb{R}} \psi(\lambda, \theta) w(\lambda) d\lambda = 1. \tag{9.6.13}$$

The Ibragimov contrast function $K_I(\theta, \theta_0)$ and the contrast process $U_{I,T}(\theta)$ are defined by the following equations:

$$K_I(\theta, \theta_0) := \int_{\mathbb{R}} f(\lambda, \theta_0) \log \frac{\psi(\lambda, \theta_0)}{\psi(\lambda, \theta)} w(\lambda)\, d\lambda, \quad \theta_0 \in S,\ \theta \in \Theta. \tag{9.6.14}$$

$$U_{I,T}(\theta) := - \int_{\mathbb{R}} I_T(\lambda) \psi(\lambda, \theta) w(\lambda)\, d\lambda, \quad \theta \in \Theta. \tag{9.6.15}$$

The minimum contrast estimator $\widehat{\theta}_{I,T}$, based on Ibragimov's contrast process $U_{I,T}(\theta)$, is defined as follows:

$$\widehat{\theta}_{I,T} := \operatorname*{Arg\,min}_{\theta\in\Theta} U_{I,T}(\theta). \tag{9.6.16}$$

A contrast process of the type (9.6.15), with the weight function $w(\lambda) = 1/(1 + \lambda^2)$, was first proposed by Ibragimov [183] as an alternative to the Whittle contrast process. For this weight function, Ibragimov [183] established sufficient conditions for the consistency of the minimum contrast estimator based on the contrast process (9.6.15).

Since then, Ibragimov's minimum contrast estimation procedure for general weight functions $w(\lambda)$ has been examined in a series of papers by V. Anh, F. Avram, N. Leonenko, and L. Sakhno (see, e.g., Anh et al. [7–9], Avram et al. [15], Sakhno [266], and references therein). These papers established sufficient conditions for Ibragimov's minimum contrast estimators, based on both tapered and non-tapered data, to be consistent and asymptotically normal for stationary random processes and homogeneous random fields.

9.6.3 Bibliographical Notes

Section 9.1 The Whittle estimation method, developed by Peter Whittle in his 1951 PhD thesis (see Whittle [313]), has made significant contributions to the asymptotic theory of parametric estimation in the frequency domain. This method is widely used in time series analysis and signal processing for estimating spectral parameters and detecting signals.

For non-tapered discrete time Gaussian and linear stationary models, the Whittle estimation procedure has been examined by many authors (see, e.g., Dahlhaus [74], Dzhaparidze [89, 91, 93], Fox and Taqqu [101], Giraitis and Surgailis [148], Giraitis and Taqqu [152], Guyon [165], Hannan [170], Heyde and Gay [174], Taniguchi and Kakizawa [295], Walker [312], and references therein). It has been demonstrated that for these models, the Whittle approach yields consistent and asymptotically normal estimators under short, intermediate, and long memory assumptions. Furthermore, it has been shown that in the Gaussian case, the Whittle estimator is also asymptotically Fisher-efficient.

For continuous time models, the Whittle estimation procedure has been considered in several studies, including those by Anh et al. [9], Avram et al. [15], Casas and Gao [61], Dzhaparidze [89, 91, 93], Dzhaparidze and Yaglom [94], Gao [103, 104], Gao et al. [105, 106], Leonenko and Sakhno [213], and Tsai and Chan [304], where additional references can also be found. In this context, it has been established that the Whittle estimator is consistent and asymptotically normal.

The Whittle estimation procedure based on the discrete time tapered data has been studied by Alomari et al. [3], Dahlhaus [70, 72, 73], Dahlhaus and Künsch

[78], Guyon [165], and Ludeña and Lavielle [221]. For a Lévy-driven continuous time linear model with a possibly unbounded or vanishing spectral density function, the consistency and asymptotic normality of the Whittle estimator were established by Ginovyan [132] and Ginovyan and Sahakyan [142] (see also Avram et al. [15]).

A closely related question to the Whittle estimation procedure discussed in this chapter is the estimation of spectral parameters in linear or nonlinear regression models with short or long memory errors. For discrete time linear regression models with non-Gaussian long memory moving average errors, this question was studied by Koul and Surgailis [202], who established the asymptotic normality of the Whittle estimator for unknown spectral parameters. In continuous time setting, the problem was addressed by Ivanov and Prikhod'ko [194], where they proved the consistency and asymptotic normality of the Whittle and Ibragimov's minimum contrast estimators for unknown spectral parameters in nonlinear regression models with stationary Gaussian noise.

Section 9.2 Theorems 9.2.1–9.2.3 were proved by Ginovyan [132]. Proposition 9.2.1 was established by Dahlhaus [72]. Example 9.2.1 is attributed to Avram et al. [15]. Results similar to Theorems 9.2.1–9.2.3, for discrete time models based on non-tapered data, have been obtained by Davies [80], Dzhaparidze [91, 93], Dzhaparidze and Yaglom [94], and Walker [312].

In the survey paper by Dzhaparidze and Yaglom [94] (page 68), it was shown that, in the non-tapered discrete time case, the covariance structure (9.2.6) in Theorem 9.2.2 is the smallest obtainable for a large class of estimators, even in the non-Gaussian case. Using this result, Proposition 9.2.1 can also be extended to non-Gaussian processes (see also Dahlhaus [72]).

Section 9.4 The problem of constructing simplified estimators with good asymptotic properties based on non-tapered data dates back to the classical work of Le Cam [206]. This area was further developed by Dzhaparidze [90, 93] (see also Dzhaparidze and Yaglom [94]). Theorem 9.4.1 was established by Dzhaparidze [90] (see also Dzhaparidze [93]). Theorem 9.4.2 is due to Ginovyan [134]. For iterative methods, such as the Newton-Raphson method and the Davidon-Fletcher-Powell method, we refer to the papers by Beinicke and Dzhaparidze [24] and by Dzhaparidze [92] (see also Dzhaparidze and Yaglom [94]).

Section 9.5 Theorems 9.5.1–9.5.4 were proved by Ginovyan and Sahakyan [139]. The discrete time version of Theorem 9.5.1 (with additional conditions $\gamma = 1$ in case *(i)*, and $\gamma > 1,\ \alpha < 1/2$ in case *(ii)*), was proved by Heyde and Dai [173] (see also Taniguchi and Kakizawa [295], Theorems 6.4.1 and 6.4.2). Using the same arguments from the proof of Theorem 9.5.1, it can be shown that the complete discrete time analog of Theorem 9.5.1 also holds true.

Chapter 10
Goodness-of-Fit Tests Based on the Tapered Data

This chapter focuses on constructing goodness-of-fit tests for testing the hypothesis H_0 that the hypothetical spectral density of a stationary Gaussian process $X(t)$ has a specified form, using tapered data. We show that when the hypothetical spectral density of $X(t)$ does not depend on unknown parameters—indicating that the null hypothesis H_0 is simple—the proposed test statistic follows a limiting chi-square distribution. In contrast, when the null hypothesis H_0 is composite, meaning that the hypothetical spectral density of $X(t)$ depends on an unknown parameter, we select an appropriate estimator for this parameter and describe the limiting distribution of the test statistic. This distribution is similar to those obtained by Chernoff and Lehmann [62] for independent observations and by Dzhaparidze [93] for discrete time, short memory stationary observations. The testing procedure applies to both short and long memory models.

In Sect. 10.1, we outline the hypothesis testing problem discussed in this chapter. Section 10.2 constructs a goodness-of-fit test for simple hypotheses, while Sect. 10.3 presents a goodness-of-fit test for composite hypotheses. Finally, Sect. 10.4 provides the proofs of the theorems stated in Sects. 10.2 and 10.3.

10.1 Introduction

Let $\{X(u),\ u \in \mathbb{U}\}$ be a centered real-valued stationary Gaussian process with spectral density $f(\lambda)$, $\lambda \in \Lambda$, and covariance function $r(t)$, $t \in \mathbb{U}$. We consider both the continuous time case, where $\mathbb{U} = \mathbb{R}$ and $\Lambda = \mathbb{R}$, and the discrete time case, where $\mathbb{U} = \mathbb{Z}$ and $\Lambda = [-\pi.\pi)$.

We address the following problem of hypothesis testing. Based on the tapered data $\mathbf{X}_T^h := \{h_T(t)X(t),\ t \in D_T\}$, where $D_T := [0, T]$ in the continuous time case and $D_T := \{1, \dots, T\}$ in the discrete time case, we aim to construct goodness-of-fit tests for the hypothesis H_0, which asserts that the spectral density of the process

M. S. Ginovyan, *Random Toeplitz Functionals and Their Applications*,
Frontiers in Probability and the Statistical Sciences,
https://doi.org/10.1007/978-3-032-00961-6_10

$X(t)$ has the specified form $f(\lambda)$. We will distinguish between the following two cases:

(a) The hypothesis H_0 is *simple*, meaning that the hypothetical spectral density $f(\lambda)$ of $X(t)$ does not depend on any unknown parameters.
(b) The hypothesis H_0 is *composite*, indicating that the hypothetical spectral density $f(\lambda)$ of $X(t)$ depends on an unknown p-dimensional vector parameter $\theta = (\theta_1, \ldots, \theta_p)'$. Specifically, we have $f(\lambda) = f(\lambda, \theta)$, $\lambda \in \Lambda$, $\theta \in \Theta \subset \mathbb{R}^p$.

To test the hypothesis H_0, it is logical to introduce a measure of divergence (or disparity) between the hypothetical and empirical spectral densities. This measure can then be used to construct a goodness-of-fit test based on its distribution.

10.2 A Goodness-of-Fit Test for Simple Hypothesis

We first consider case (a) of a simple hypothesis H_0. As a measure of divergence between the hypothetical spectral density $f(\lambda)$ and the tapered empirical spectral density $I_T^h(\lambda)$, we consider the m-dimensional random vector:

$$\Phi_T^h := (\Phi_{1T}^h, \ldots, \Phi_{mT}^h) \tag{10.2.1}$$

with elements

$$\Phi_{jT}^h := \Phi_{jT}(\mathbf{X}_T^h) = \frac{\sqrt{T}}{\sqrt{4\pi e(h)}} \int_\Lambda \left[\frac{I_T^h(\lambda)}{f(\lambda)} - 1\right] \varphi_j(\lambda) d\lambda, \quad j = 1, 2, \ldots, m, \tag{10.2.2}$$

where $e(h)$ is the tapering factor, defined in (8.2.16), and $\{\varphi_j(\lambda),\ j = 1, 2, \ldots, m\}$ is an orthonormal system on Λ:

$$\int_\Lambda \varphi_k(\lambda)\varphi_j(\lambda)\, d\lambda = \delta_{kj} = \begin{cases} 1 \text{ for } k = j, \\ 0 \text{ for } k \neq j. \end{cases} \tag{10.2.3}$$

In Lemma 10.4.1 below, we demonstrate that under broad conditions on $f(\lambda)$ and $\varphi_j(\lambda)$, the random vector Φ_T^h defined in Eqs. (10.2.1) and (10.2.2) has an asymptotically $N(0, I_m)$-normal distribution as $T \to \infty$, where I_m denotes the $m \times m$ identity matrix.

Thus, in the case of a simple hypothesis H_0, we can use the following statistic:

$$S_T^h = S_T(\mathbf{X}_T^h) := \Phi_T'(\mathbf{X}_T^h)\Phi_T(\mathbf{X}_T^h) = \sum_{j=1}^m \Phi_{jT}^2(\mathbf{X}_T^h), \tag{10.2.4}$$

which, as $T \to \infty$, will follow a χ^2-distribution with m degrees of freedom.

The following theorem provides sufficient conditions for the statistic S_T^h, as defined in (10.2.4), to have a limiting χ^2-distribution with m degrees of freedom as $T \to \infty$.

Theorem 10.2.1 *Let the spectral density $f(\lambda)$ of the model and the orthonormal functions $\{\varphi_j(\lambda),\ j = 1, 2, \dots, m\}$ be such that $(f, g_j) \in (\mathcal{H})$ for all $j = 1, \dots, m$, where $g_j = \varphi_j/f$ (see Definition 2.1.1 for condition $(\mathcal{H})$). Then, the limiting distribution of the statistic $S_T^h = S_T(\mathbf{X}_T^h)$ is a χ^2-distribution with m degrees of freedom as $T \to \infty$.*

Thus, by fixing a significance level α, we can consider a class of goodness-of-fit tests for the simple hypothesis H_0 regarding the form of the spectral density f with an asymptotic level of significance α. These tests are determined by critical regions of the form:

$$\{\mathbf{X}_T^h : S_T(\mathbf{X}_T^h) > d_\alpha\},$$

where $S_T(\mathbf{X}_T^h)$ is defined by (10.2.4), and d_α is the α-quantile of χ^2-distribution with m degrees of freedom. Specifically, d_α is determined from the condition:

$$P(\chi^2 > d_\alpha) = \int_{d_\alpha}^{\infty} k_m(x)\, dx = \alpha,$$

where $k_m(x)$ is the probability density function of the χ^2-distribution with m degrees of freedom.

10.3 A Goodness-of-Fit Test for Composite Hypothesis

In the case of a composite hypothesis H_0, where the hypothetical spectral density $f(\lambda, \theta)$ of the observed process $X(t)$ depends on an unknown p-dimensional parameter $\theta = (\theta_1, \dots, \theta_p)'$, constructing goodness-of-fit tests becomes more complex. This complexity arises because the unknown parameter θ must first be estimated. It is important to note that in this scenario, the limiting distribution of the test statistic will vary depending on the properties of the estimator of θ and, in general, will not follow a χ^2-distribution.

For testing a composite hypothesis H_0, we can again use a statistic of the type (10.2.4), but with a statistical estimator $\widehat{\theta}_T$ for the unknown θ. The corresponding statistic can be expressed as follows:

$$S_T^h(\widehat{\theta}_T) = S_T(\mathbf{X}_T^h, \widehat{\theta}_T) := \Phi_T'(\mathbf{X}_T^h, \widehat{\theta}_T)\Phi_T(\mathbf{X}_T^h, \widehat{\theta}_T) = \sum_{j=1}^{m} \Phi_{jT}^2(\mathbf{X}_T^h, \widehat{\theta}_T), \tag{10.3.1}$$

where

$$\Phi_T^h = \Phi_T^h(\mathbf{X}_T^h, \widehat{\theta}_T) := \left(\Phi_{1T}(\mathbf{X}_T^h, \widehat{\theta}_T), \ldots, \Phi_{mT}(\mathbf{X}_T^h, \widehat{\theta}_T)\right) \tag{10.3.2}$$

with elements

$$\Phi_{jT}(\mathbf{X}_T^h, \widehat{\theta}_T) := \frac{\sqrt{T}}{\sqrt{4\pi e(h)}} \int_\Lambda \left[\frac{I_T^h(\lambda)}{f(\lambda, \widehat{\theta}_T)} - 1 \right] \varphi_j(\lambda) d\lambda, \quad j = 1, 2, \ldots, m. \tag{10.3.3}$$

Thus, we need to select an appropriate statistical estimator $\widehat{\theta}_T$ for the unknown θ and determine the limiting distribution of the statistic (10.3.1). Once we establish the limiting distribution of this statistic, we can consider a class of goodness-of-fit tests for the composite hypothesis H_0 regarding the form of the spectral density f at given significance level α. These tests are characterized by critical regions of the form:

$$\{\mathbf{X}_T^h : \ S_T(\mathbf{X}_T^h, \widehat{\theta}_T) > d_\alpha\},$$

where d_α is the α-quantile of the limiting distribution of the statistic (10.3.1). Specifically, d_α is determined from the following equation:

$$\int_{d_\alpha}^{\infty} \widehat{k}_m(x)\, dx = \alpha,$$

where $\widehat{k}_m(x)$ is the probability density function of the limiting distribution of the statistic $S_T^h(\widehat{\theta}_T)$ defined by (10.3.1).

To describe the limiting distribution of the statistic (10.3.1) we introduce the following assumptions.

Assumption (A1) The true value θ_0 of the parameter θ belongs to a bounded closed set Θ, which is contained in an open set S in the p-dimensional Euclidean space $\mathbb{R}^p$.

Assumption (A2) If θ_1 and θ_2 are two distinct points of Θ, then $f(\lambda, \theta_1) \neq f(\lambda, \theta_2)$ almost everywhere in Λ with respect to the Lebesgue measure.

Assumption (A3) For $\theta \in \Theta$, $(f, g_j) \in (\mathcal{H})$ for all $j = 1, 2, \ldots, m$, where $f := f(\lambda, \theta)$ and $g_j := \varphi_j(\lambda)/f(\lambda, \theta)$.

Assumption (A4) For $\theta \in \Theta$, $(f, h_{kj}) \in (\mathcal{H})$ for all $k = 1, 2, \ldots, p$ and $j = 1, 2, \ldots, m$, where $f := f(\lambda, \theta)$ and $h_{kj} := \frac{\varphi_j(\lambda)}{f(\lambda,\theta)} \frac{\partial}{\partial\theta_k} \ln f(\lambda, \theta)$.

Assumption (A5) The $(p \times p)$ matrix $\Gamma(\theta_0) = ||\gamma_{kj}(\theta_0)||$, $k, j = 1, \ldots, p$, with elements:

$$\gamma_{kj}(\theta_0) := \frac{1}{4\pi} \int_\Lambda \left[\frac{\partial}{\partial\theta_k} \ln f(\lambda, \theta) \right]_{\theta=\theta_0} \left[\frac{\partial}{\partial\theta_j} \ln f(\lambda, \theta) \right]_{\theta=\theta_0} d\lambda \tag{10.3.4}$$

is non-singular.

Assumption (A6) There exists a $\sqrt{T}$-consistent estimator $\widehat{\theta}_T$ for the parameter θ such that the following asymptotic relation holds:

$$\sqrt{T}\,(\widehat{\theta}_T - \theta_0) - \Gamma^{-1}(\theta_0)\Delta_T(\theta_0) = o_P(1), \tag{10.3.5}$$

where $\Gamma^{-1}(\theta_0)$ is the inverse of the matrix $\Gamma(\theta_0)$ defined in Assumption (A5), and

$$\Delta_T(\theta) := \big(\Delta_{1T}(\theta), \ldots, \Delta_{pT}(\theta)\big)' \tag{10.3.6}$$

is a p-dimensional random vector with components

$$\Delta_{kT}(\theta) := \frac{\sqrt{T}}{\sqrt{4\pi e(h)}} \int\limits_{\Lambda} \left[\frac{I_T^h(\lambda)}{f(\lambda, \theta)} - 1\right] \frac{\partial}{\partial \theta_k} \ln f(\lambda, \theta)\, d\lambda, \quad k = 1, \ldots, p. \tag{10.3.7}$$

The term $o_P(1)$ in relation (10.3.5) tends to zero in probability as $T \to \infty$. (Recall that an estimator $\widehat{\theta}_T$ for θ is called $\sqrt{T}$—consistent if $\sqrt{T}(\widehat{\theta}_T - \theta)$ is bounded in probability.)

As an estimator $\widehat{\theta}_T$ for θ that satisfies the relation (10.3.5), we can consider minimum contrast estimators, particularly the tapered Whittle estimator.

Let $B(\theta) = ||b_{jk}(\theta)||$, $j = 1, \ldots, m$, $k = 1 \ldots, p$, be a $(m \times p)$ matrix of rank p $(m > p)$ with elements:

$$b_{jk}(\theta) := \frac{1}{\sqrt{4\pi e(h)}} \int_{\Lambda} \varphi_j(\lambda) \frac{\partial}{\partial \theta_k} \ln f(\lambda, \theta)\, d\lambda, \tag{10.3.8}$$

where $\varphi_j(\lambda)$ $(j = 1, \ldots, m)$ are the functions from Eq. (10.2.2), and $e(h)$ is defined as in (8.2.16).

Theorem 10.3.1 *Under Assumptions (A1)–(A6), the limiting distribution (as $T \to \infty$) of the statistic $S_T(\mathbf{X}_T^h, \widehat{\theta}_T)$, given by (10.3.1), coincides with the distribution of the random variable*

$$\sum_{j=1}^{m-p} \xi_j^2 + \sum_{j=1}^{p} \nu_j\, \xi_{m-p+j}^2,$$

where ξ_j, $j = 1, 2, \ldots, m$, are i.i.d. $N(0, 1)$ random variables, and the numbers ν_k $(0 \le \nu_k < 1)$, $k = 1, 2, \ldots, p$, are the roots relative to ν of the following equation:

$$\det\big[(1 - \nu)\Gamma(\theta_0) - B'(\theta_0)\, B(\theta_0)\big] = 0. \tag{10.3.9}$$

Example 10.3.1 Let $X(t)$ be a discrete time autoregressive process of order p. Specifically, $X(t)$ is a stationary process with the spectral density:

$$f(\lambda) = \frac{1}{2\pi} \cdot \left|\alpha_p(e^{-i\lambda})\right|^{-2},$$

where $\alpha_p(z) = 1 - \theta_1 z - \cdots - \theta_p z^p$.

Consider the functions $\varphi_j(\lambda)$ ($j = 1, \ldots, m$) defined by

$$\varphi_j(\lambda) := \begin{cases} ce^{-ij\lambda}\alpha_p(e^{-i\lambda})\left|\alpha_p(e^{-i\lambda})\right|^{-1} & \text{for } j > p, \\ 0 & \text{for } j \le p, \end{cases}$$

where c is a normalizing constant.

As a taper, we use the Tukey-Hanning taper function $h(t)$ (see Remark 4.8.1), and as an estimator for the unknown parameter $\theta := (\theta_1, \ldots, \theta_p)$, we consider the tapered Whittle estimator. It is straightforward to verify that the conditions of Theorem 10.3.1 are satisfied and that $B(\theta) = 0$.

Thus, in this special case, the limiting distribution of the statistic in (10.3.1) is a χ^2-distribution with $m - p$ degrees of freedom as $T \to \infty$.

10.4 Proofs of Theorems of Sects. 10.2 and 10.3

We first prove several lemmas.

Lemma 10.4.1 *Under the assumptions of Theorem 10.2.1, the random vector Φ_T^h, defined by Eqs.* (10.2.1) *and* (10.2.2) *has asymptotically $N(0, I_m)$-normal distribution as $T \to \infty$, where I_m is the $m \times m$ identity matrix.*

Proof The result follows from Theorem 8.2.7 and the Cramér-Wold device (see Proposition 5.3.12). □

Consider the $(m + p)$—dimensional random vector

$$\Psi_T^h(\theta) := (\Phi_T^h(\theta), \Delta(\theta))', \tag{10.4.1}$$

where $\Phi_T^h(\theta)$ is defined by Eqs. (10.2.1) and (10.2.2) with $f(\lambda) = f(\lambda, \theta)$, and $\Delta(\theta)$ is defined by Eqs. (10.3.6) and (10.3.7).

Lemma 10.4.2 *Under the assumptions of Theorem 10.3.1, the random vector $\Psi_T^h(\theta)$ defined by Eq.* (10.4.1) *has asymptotically $N(0, G(\theta_0))$-normal distribution as $T \to \infty$, where*

$$G(\theta_0) = \begin{pmatrix} I_m & B(\theta_0) \\ B'(\theta_0) & \Gamma(\theta_0) \end{pmatrix}.$$

Here, I_m is the $m \times m$ identity matrix, $\Gamma(\theta_0)$ is the $(p \times p)$ matrix with elements given by (10.3.4), and $B(\theta)$ is the $(m \times p)$ matrix with elements given by (10.3.8).

Proof The result follows from Theorem 8.2.7 and the Cramér-Wold device. □

The result presented in the next lemma follows from Proposition 5.3.2 (see also Chernoff and Lehmann [62] and Chibisov [63]).

Lemma 10.4.3 *Assume that a random vector $\eta_T = (\eta_{1T}, \dots, \eta_{nT})'$ has a limiting $N(0, A)$-normal distribution (as $T \to \infty$). Then, the limiting distribution of the random variable $\eta_T'\eta_T = \sum_{j=1}^n \eta_{jT}^2$ coincides with the distribution of the sum $\sum_{j=1}^n \lambda_j \xi_j^2$, where ξ_j, $j = 1, \dots, n$, are i.i.d. $N(0, 1)$ random variables, and the numbers λ_j $(j = 1, \dots, n)$ are the eigenvalues of the matrix A. In particular, if the matrix A is idempotent, that is, $A^2 = A$, then $\eta_T'\eta_T$ has a limiting χ^2-distribution with $k = tr[A]$ degrees of freedom, where $tr[A]$ denotes the trace of the matrix A.*

Lemma 10.4.4 *Let $\widehat{\theta}_T$ be a $\sqrt{T}$—consistent estimator for the unknown parameter θ. Then, under the assumptions of Theorem 10.3.1, the following asymptotic relation holds:*

$$\Phi_T^h(\widehat{\theta}_T) = \Phi_T^h(\theta_0) - \sqrt{T}B(\theta_0)(\widehat{\theta}_T - \theta_0) + o_P(1), \tag{10.4.2}$$

where $B(\theta)$ is the $(m \times p)$ matrix with elements given by (10.3.8) and the term $o_P(1)$ tends to zero in probability as $T \to \infty$.

Proof Using the mean value theorem, in view of (10.3.3), we can write

$$\Phi_{jT}^h(\widehat{\theta}_T) - \Phi_{jT}^h(\theta_0) = b_{jk,T}(\theta_*) \sum_{k=1}^p \sqrt{T}(\widehat{\theta}_{kT} - \theta_{k0}), \tag{10.4.3}$$

where

$$b_{jk,T}(\theta_*) := \frac{1}{\sqrt{4\pi e(h)}} \int_\Lambda I_T^h(t) \left[\frac{1}{f(t,\theta)} \frac{\partial}{\partial \theta_k} \ln f(t,\theta)\right]_{\theta=\theta_*} \varphi_j(t)\, dt, \tag{10.4.4}$$

and $\theta_* \in (\theta_0, \widehat{\theta}_T)$. Since $\widehat{\theta}_T$ is a $\sqrt{T}$—consistent estimator for θ, in view of (10.3.8), to prove (10.4.2) it is sufficient to show that

$$b_{jk,T}(\theta_*) = b_{jk}(\theta_0) + o_P(1) \quad \text{as} \quad T \to \infty. \tag{10.4.5}$$

We denote $g_* = g_*(t) = \left[\frac{1}{f(t,\theta)}\frac{\partial}{\partial\theta_k}\ln f(t,\theta)\right]_{\theta=\theta_*}\varphi_j(t)$ and $f_0 = f(t,\theta_0)$. Then, for the expectation of the random variable $b_{jk,T}(\theta_*)$ in Eq. (10.4.4), we have as $T \to \infty$

$$\mathbb{E}\big[b_{jk,T}(\theta_*)\big] = \frac{1}{\sqrt{4\pi e(h)}}\frac{1}{2\pi T H_2}\mathrm{tr}\big[\mathbb{A}_T^h(f_0)\mathbb{A}_T^h(g_*)\big]$$
$$\longrightarrow \frac{1}{\sqrt{4\pi e(h)}}\int_\Lambda \varphi_j(\lambda)\frac{\partial}{\partial\theta_k}\ln f(\lambda,\theta)\,d\lambda = b_{jk}(\theta_0). \qquad (10.4.6)$$

In (10.4.6), the first relation follows from Lemma 4.8.4 with $k = 2$, $\psi_1 = f_0$ and $\psi_2 = g_*$. The second relation follows from Proposition 2.1.9 and Theorem 4.8.1 **(C2)** with $k = 2$, $\psi_1 = f_0$ and $\psi_2 = g_*$.

Next, for the variance of the random variable $b_{jk,T}(\theta_*)$, we have as $T \to \infty$

$$\mathrm{Var}\big[b_{jk,T}(\theta_*)\big] = \frac{1}{32\pi^4 T^2 H_4}\mathrm{tr}\Big[\Big(\mathbb{A}_T^h(f_0)\mathbb{A}_T^h(g_*)\Big)^2\Big] \longrightarrow 0. \qquad (10.4.7)$$

In (10.4.7), the first relation follows from Lemma 4.8.4 with $k = 4$, $\psi_1 = \psi_2 = f_0$ and $\psi_3 = \psi_4 = g_*$. The second relation follows from Proposition 2.1.9 and Theorem 4.8.1 **(C2)** with $k = 4$, $\psi_1 = \psi_2 = f_0$ and $\psi_3 = \psi_4 = g_*$.

Now, the relation (10.4.5) follows from (10.4.6), (10.4.7), and the Chebyshev inequality.

Finally, by combining the relations (10.4.3) and (10.4.5), we obtain (10.4.2). Lemma 10.4.4 is proved. □

Proof of Theorem 10.2.1 The result follows directly from Lemma 10.4.1 and the definition of the χ^2-distribution. □

Proof of Theorem 10.3.1 By Eq. (10.3.5) and Lemma 10.4.4, with $\Phi_T(\cdot) := \Phi_T^h(\cdot)$, we can write

$$\Phi_T(\widehat{\theta_T}) = \Phi_T(\theta_0) - B(\theta_0)\,\Gamma^{-1}(\theta_0)\Delta_T(\theta_0) + o_P(1) \quad \text{as} \quad T \to \infty.$$

The last relation can be expressed as follows:

$$\Phi_T(\widehat{\theta_T}) = U_T(\theta_0) + [V_T(\theta_0) - W_T(\theta_0)] + o_P(1) \quad \text{as} \quad T \to \infty, \qquad (10.4.8)$$

where

$$U_T(\theta_0) = A(\theta_0)\Phi_T(\theta_0) \quad \text{with}$$
$$A(\theta_0) := I_m - B(\theta_0)\big(B'(\theta_0)B(\theta_0)\big)^{-1}B'(\theta_0), \qquad (10.4.9)$$
$$V_T(\theta_0) = B(\theta_0)\big(B'(\theta_0)B(\theta_0)\big)^{-1}B'(\theta_0)\Phi_T(\theta_0),$$
$$W_T(\theta_0) = B(\theta_0)\,\Gamma^{-1}(\theta_0)\Delta_T(\theta_0).$$

It is easy to see that

$$A(\theta_0)B(\theta_0) = 0, \tag{10.4.10}$$

which implies that $U_T'(\theta_0)V_T(\theta_0) = U_T'(\theta_0)W_T(\theta_0) = 0$.

Therefore, from relation (10.4.8), we find that

$$\begin{aligned}\Phi_T'(\widehat{\theta}_T)\Phi_T(\widehat{\theta}_T) &= U_T'(\theta_0)U_T(\theta_0)\\ &+[V_T(\theta_0) - W_T(\theta_0)]'[V_T(\theta_0) - W_T(\theta_0)] + o_P(1). \end{aligned}\tag{10.4.11}$$

It follows from Lemma 10.4.2 and Eq. (10.4.10) that

$$\mathbb{E}[U_T(\theta_0)(V_T(\theta_0) - W_T(\theta_0))'] \longrightarrow 0 \quad \text{as} \quad T \to \infty.$$

Therefore, the terms on the right-hand side of (10.4.11) are asymptotically independent random variables. Noting that the matrix $A(\theta_0)$ in Eq. (10.4.9) is idempotent ($A^2(\theta_0) = A(\theta_0)$), and that $\mathrm{tr}[A(\theta_0)] = m - p$, in view of Lemmas 10.4.2 and 10.4.3, we conclude that the first term on the right-hand side of (10.4.11) follows a limiting χ^2-distribution with $m - p$ degrees of freedom.

To describe the limiting distribution of the second term on the right-hand side of (10.4.11), we first note that, by Lemma 10.4.2, we have

$$\begin{aligned}&\mathbb{E}[(V_T(\theta_0) - W_T(\theta_0))(V_T(\theta_0) - W_T(\theta_0))']\\ &\longrightarrow B(\theta_0)\big[(B'(\theta_0)B(\theta_0))^{-1} - \Gamma^{-1}(\theta_0)\big]B'(\theta_0) \quad \text{as} \quad T \to \infty. \end{aligned}\tag{10.4.12}$$

Next, using Lemma 10.4.3, we can easily infer that the limiting distribution of the second term on the right-hand side of (10.4.11) coincides with the distribution of the sum

$$\sum_{j=1}^{p} \nu_j \, \xi_{m-p+j}^2,$$

where ξ_j, $j = 1, 2, \ldots, m$, are i.i.d. $N(0, 1)$ random variables, and the numbers ν_k ($k = 1, 2, \ldots, p$) are the nonzero eigenvalues of the matrix on the right-hand side of (10.4.12).

By Proposition 2.7.3, the numbers ν_k ($k = 1, 2, \ldots, p$) coincide with the nonzero eigenvalues of the matrix

$$B'(\theta_0)B(\theta_0)\big[(B'(\theta_0)B(\theta_0))^{-1} - \Gamma^{-1}(\theta_0)\big],$$

which means that ν_k are the roots with respect to ν of the following equation:

$$\det\Big[B'(\theta_0)B(\theta_0)\big[(B'(\theta_0)B(\theta_0))^{-1} - \Gamma^{-1}(\theta_0)\big] - \nu I_p\Big] = 0. \tag{10.4.13}$$

Since the matrix $\Gamma(\theta_0)$ is non-singular, Eq. (10.4.13) is equivalent to (10.3.9).

To show that the numbers ν_k, $k = 1, 2, \ldots, p$, satisfy the condition $0 \le \nu_k < 1$, note that, by Lemma 10.4.2, we have

$$\mathbb{E}\big[(\Delta_T(\theta_0) - B'(\theta_0)\Phi_T(\theta_0)\big]\big[\Delta_T(\theta_0) - B'(\theta_0)\Phi_T(\theta_0)\big]'$$
$$\longrightarrow \Gamma(\theta_0) - B'(\theta_0)B(\theta_0) \quad \text{as} \quad T \to \infty.$$

This implies that $\Gamma(\theta_0) - B'(\theta_0)B(\theta_0) \ge 0$. (Recall that for a matrix A, the notation $A \ge 0 (> 0)$ means that A is non-negative (positive) definite.)

Therefore, $(1 - \nu)\Gamma(\theta_0) - B'(\theta_0)B(\theta_0) > 0$ for $\nu < 0$.

On the other hand, since $\Gamma(\theta_0) > 0$ and $B'(\theta_0)B(\theta_0) > 0$, we have $(1-\nu)\Gamma(\theta_0) - B'(\theta_0)B(\theta_0) < 0$ for $\nu \ge 1$.

Thus, $0 \le \nu_k < 1$ for $k = 1, 2, \ldots, p$. This completes the proof of Theorem 10.3.1. □

10.5 Bibliographical Notes

The problem of constructing goodness-of-fit tests for various models has been examined by numerous authors.

For independent observations, this problem was addressed by Chernoff and Lehmann [62], Chibisov [63], and Cramér [68].

In the case of observations generated by discrete time Gaussian stationary processes, it was investigated by Hannan [168], Dzhaparidze [91, 93], Osidze [232, 233], and Ginovyan [128].

For continuous time Gaussian stationary observations, the problem was discussed by Osidze [232, 233] and by Ginovyan [131].

For discrete time short memory Gaussian stationary processes, the non-tapered version of Theorem 10.2.1 was first proven by Hannan [168] (p. 94) under different conditions (see also Dzhaparidze [91, 93] and Osidze [232, 233].

In cases where the spectral density exhibits singularities (zeros and/or poles), the non-tapered counterpart of Theorem 10.2.1 was established by Ginovyan [128] for the discrete time case and by Ginovyan [131] for the continuous time case.

For independent observations, the non-tapered version of Theorem 10.3.1 was first established by Chernoff and Lehmann [62] (see also Chibisov [63]).

For observations generated by discrete time short memory Gaussian stationary processes, the result was stated by Osidze [232] (see also Dzhaparidze [91, 93]).

In cases where the spectral density exhibits singularities, indicating that the model displays long or intermediate memory, the non-tapered counterpart of Theorem 10.3.1 was established by Ginovyan [128] for the discrete time case and by Ginovyan [131] for the continuous time case.

Theorems 10.2.1 and 10.3.1 were first established by Ginovyan [133].

Appendix A
Tapering: Motivations and Benefits

A.1 Motivations

Tapering is a technique designed to improve the accuracy of spectral estimators by applying specific preliminary transformations to the original data prior to the estimation procedure. The primary objective of tapering is to minimize bias, leakage, and trough effects in spectral estimators, which are issues that arise from the periodogram.

It is well known (see, e.g., Brillinger [55], Priestley [249], and Dahlhaus [76]) that the periodogram is an unreliable statistical estimator of unknown spectral density due to two main drawbacks.

First, the periodogram is not a consistent estimator of the unknown spectral density, and the periodogram values at different frequencies are nearly uncorrelated. Additionally, the periodogram is influenced by leakage and trough effects, which arise from strong peaks and troughs in the spectrum.

To address these disadvantages, we will assume, for simplicity, that the underlying process $X(t)$ is a discrete time stationary process with an unknown spectral density f. Our goal is to estimate f based on the sample $\mathbf{X}_T := \{X(t), t = 1, \dots, T\}$. Similar arguments also apply in the continuous time case.

As an estimator of the unknown $f(\lambda)$, consider the periodogram $I_T(\lambda)$ of the process $X(t)$ (see Sect. 8.2):

$$I_T(\lambda) := \frac{1}{2\pi T} \left| \sum_{t=1}^{T} X(u) e^{-it\lambda} \right|^2 . \tag{A.1.1}$$

M. S. Ginovyan, *Random Toeplitz Functionals and Their Applications*, Frontiers in Probability and the Statistical Sciences,
https://doi.org/10.1007/978-3-032-00961-6

Then, for the expected value of $I_T(\lambda)$, we have the convolution:

$$\mathbb{E}[I_T(\lambda)] = \int_{-\pi}^{\pi} f(\lambda - \mu) F_T(\mu) d\mu, \qquad \text{(A.1.2)}$$

where $F_T(\mu)$ is the classical Fejér kernel given by Eq. (2.2.2).

To improve the estimator for $f(\lambda)$, we could begin by replacing the periodogram $I_T(\lambda)$ with a more efficient nonparametric estimator $\widehat{f_T}(\lambda)$. In particular, we can utilize the smoothed periodogram estimator that employs a suitably chosen smoothing spectral window $W(\cdot)$:

$$\widehat{f_T}(\lambda) := \int_{-\pi}^{\pi} I_T(\mu + \lambda) W_N(\mu) d\mu, \qquad \text{(A.1.3)}$$

where, e.g., $W_N(\mu) = NW(N\mu)$ with $N/T \to 0$ as $T \to \infty$.

Although the standard asymptotic theory provides satisfactory results for the estimator $\widehat{f_T}(\lambda)$, it can exhibit poor performance in certain situations. For instance, it is well known (see, e.g., Pristley [249], page 558) that when the underlying model exhibits short memory with a continuous spectral density, the periodogram bias is typically negligible compared to the bias introduced by the smoothing window.

However, this conclusion does not hold when the spectral window has discontinuities. This is one situation where tapering should be applied.

The second disadvantage of the periodogram—the leakage effect—is attributed to the properties of the Fejér kernel $F_T(\mu)$. It is known (see, e.g., Dahlhaus [73]) that the function $F_T(\mu)$ exhibits side peaks of magnitude $O(T^{-1})$ at frequencies $\mu = 2\pi(s + 1/2)/T$ with $\mu \in (0, \pi]$.

In the convolution (A.1.2), if such a side peak is multiplied by a strong peak of the spectral density f, it may lead to an expectation that is too large, resulting in an overestimation of the spectrum. As a consequence, lower peaks may be masked and potentially overlooked. Therefore, without the use of a data taper, the estimator cannot resolve the lower peaks of the spectrum.

This phenomenon is known as the *nonparametric leakage effect.*

A notable example of the leakage effect can be found in Bloomfield [44], Sections 5.2 and 5.3, where variable star data were analyzed.

Take into account that the Whittle approximation of the log-likelihood function can be interpreted as the information divergence between the periodogram $I_T(\lambda)$ and the spectral density $f(\lambda)$. As a result, the leakage effect is also carried over to the parametric estimation procedure.

This phenomenon is known as the *parametric leakage effect* (see, e.g., Dahlhaus [73]).

To prevent leakage and other undesirable effects discussed below, we apply a tapering technique to the original data prior to the estimation procedure.

Tapering involves replacing the Fejér kernel $F_T(\mu)$ with a more general *tapered kernel.* The aim is to choose a taper function that results in a tapered kernel with significantly smaller side lobes compared to the classical Fejér kernel $F_T(\mu)$.

A.2 Benefits of Tapering

The use of data tapers in nonparametric time series analysis was first suggested by Tukey [305]. Since then, this concept has been explored by numerous authors for both parametric and nonparametric estimation procedures.

The benefits of tapering data have been widely reported in the literature (see, e.g., Alomari et al. [3], Anh et al. [10], Bloomfield [44], Brillinger [55], Dahlhaus [70]–[73], [76], Dahlhaus and Künsch [78], Ginovyan [133], Ginovyan and Sahakyan [140], Guyon [165], Hannan [169], Janas and von Sachs [195], Ludeña and Lavielle [221], Pristley [249], von Sachs [310], and references therein).

For instance, data tapers are introduced to mitigate the aforementioned "leakage effects," resulting in a more accurate estimation of the model's spectrum, especially when it features strong peaks.

Additionally, tapering can minimize the so-called trough effects, enhancing the accuracy of spectrum estimation in cases where strong troughs are present. The trough effect, characterized by values close to zero—often due to moving average (MA) roots near the unit circle—was first observed and discussed by Dahlhaus [73].

Another application of data tapers is in situations where some data values are missing. Additionally, the use of tapers leads to bias reduction, which is especially important when dealing with spatial data. In this context, tapers can be employed to reduce the so-called edge effects (for details, see Dahlhaus and Künsch [78], and Guyon [165]).

A.3 The Leakage, Trough, and Variance Effects: Dahlhaus's Workshop

The arguments presented in Sect. 9.1 (see also Dahlhaus [72], Dzhaparidze and Yaglom [94]) show that, under broad conditions on the observed model, the exact maximum likelihood estimator and the Whittle estimator are asymptotically equivalent. However, in cases with small sample sizes, the Whittle estimator may perform poorly. For instance, in the context of autoregressive processes, it is known that the Whittle estimator aligns with the Yule-Walker estimator (see Dzhaparidze and Yaglom, [94], Example 5.1). However, the Yule-Walker estimators tend to perform poorly when a root of the corresponding characteristic equation is close to the unit circle (see Priestley, [249], page 351).

Thus, asymptotic results do not adequately capture the effects observed in small sample scenarios. On the other hand, due to the extreme technical complexity involved, exact calculations for finite sample sizes T are only practical in very specific situations, such as AR(1) processes.

To address these challenges, Dahlhaus developed specific asymptotic models in papers [73, 76] that effectively cover the small sample situations for both parametric and nonparametric cases. Within this asymptotic framework, Dahlhaus rigorously demonstrated the advantages of tapering. In particular, using the proposed models, Dahlhaus theoretically explained the leakage and trough effects associated with both the parametric Whittle estimator and the nonparametric spectral estimator. He showed that these effects can be eliminated by applying suitable data tapers.

It is a widely held belief (see, e.g., Brillinger [55], page 151; Hannan [169], page 272; and Pristley [249], page 562) that tapering may reduce the bias of estimators while increasing their variance. This phenomenon is known as the *variance effect.*

However, Dahlhaus [73, 76] challenges this common view by demonstrating that data tapers can not only reduce bias but also decrease the variance and cumulants of all orders of the corresponding estimators (see Dahlhaus [73], page 829, and [76], page 161). For the variance effect see also Remark 9.2.1.

Dahlhaus's approach is based on the requirement that estimators perform well as the number of processes increases with the sample size. Specifically, by analyzing equation (A.1.2) and noting that the side peaks of the Fejér kernel $F_T(\mu)$ for $\mu \in (0, \pi]$ are of order $O(T^{-1})$, it was observed that the leakage effect disappears asymptotically. To maintain this small sample effect, Dahlhaus recommended allowing the main peak of the spectrum to increase with the sample size T and examined the properties of the estimators related to the lower peaks of the spectrum.

For example, when considering the spectral density of an AR(p) process:

$$f(\lambda) = \frac{\sigma^2}{2\pi} \cdot \prod_{k=1}^{p} |1 - z_k e^{-i\lambda})|^{-2}, \tag{A.3.1}$$

where z_k are the roots of the characteristic equation, based on simulation results, Dahlhaus observed that it is relatively straightforward to estimate the strongest peak, specifically, the roots that are closest to the unit circle. From this observation, he concluded that it is reasonable to assume these roots are known, while the remaining roots need to be estimated.

Based on these arguments, Dahlhaus [73, 76] proposed the following parametric estimation model: the spectral density $f_\theta(\lambda)$ is expressed as a product:

$$f_\theta(\lambda) = g_T(\lambda) f_\theta^*(\lambda), \tag{A.3.2}$$

where $g_T(\lambda)$ is a known function characterized by strong peaks or troughs, and its magnitude may increase with the sample size T. In contrast, the function $f_\theta^*(\lambda)$ is bounded, independent of T, and its parameter θ needs to be estimated.

To investigate the properties of the tapered Whittle estimators, Dahlhaus [73] introduced an increasing class of stochastic processes, denoted by $\mathfrak{X}_T$ (see Definition 2.2 in Dahlhaus [73]), and imposed additional smoothness conditions on the taper function $h(\cdot)$.

Specifically, Dahlhaus [73] defined the class of taper functions $h_T(t) := h_T(t/T)$ of degree (k, κ), where $k \in \mathbb{N}_0 := \{0, 1, 2, \ldots\}$ and $\kappa \in [0, 1/2)$ (see Definition 5.1 in Dahlhaus [73]). We denote this class by $D(k, \kappa)$. Note that in the non-tapered case, $h_T(t) := \mathbf{1}_{[0,1]}(t/T)$, which corresponds to a degree of (0, 0).

The following two commonly used taper functions belong to the class $D(k, \kappa)$.

Example A.3.1 (Tukey-Hanning Cosine Bell Taper) Consider the function:

$$h_\rho(x) := \begin{cases} (1/2)[1 - \cos(2\pi x/\rho)] & x \in [0, \rho/2), \\ 1 & x \in [\rho/2, 1/2], \\ h_\rho(1-x) & x \in (1/2, 1]. \end{cases} \tag{A.3.3}$$

In Dahlhaus [73] it was shown that if ρ is fixed, then the taper $h_T(t) := h_\rho(t/T)$ is of degree $(2, 0)$. Furthermore, if $\rho = \rho_T = T^{-\kappa/3}$, then the taper $h_T(t) := h_{\rho_T}(t/T)$ is of degree $(2, \kappa)$.

Example A.3.2 (Polynomial Taper) Consider the function:

$$h_\rho(x) := \begin{cases} 4^k(x/\rho)^k(1 - x/\rho)^k & x \in [0, \rho/2), \\ 1 & x \in [\rho/2, 1/2], \\ h_\rho(1-x) & x \in (1/2, 1]. \end{cases} \tag{A.3.4}$$

In Dahlhaus [73] (see also Dahlhaus [76]) it was shown that if ρ is fixed, then the taper $h_T(t) := h_\rho(t/T)$ is of degree $(k, 0)$. If $\rho = \rho_T = T^{-\kappa/(k+1)}$, then the taper $h_T(t) := h_{\rho_T}(t/T)$ is of degree (k, κ).

Observe that in both examples we have $h_{\rho_T}(x) \to \mathbf{1}_{[0,1]}(x)$ as $T \to \infty$, and

$$\lim_{T\to\infty} (TH_{4,T})/H_{2,T}^2 = 1, \tag{A.3.5}$$

where $H_{k,T}$ is defined in Eq. (4.8.4).

In Section 6 of Dahlhaus [73], the behavior of the tapered Whittle estimators was examined for double-indexed sequences of processes $X(t, T) \in \mathfrak{X}_T$ with a spectral density of the form (A.3.2), and taper functions $h_T(t)$ from the class $D(k, \kappa)$.

Specifically, it was proved that if $X(t, T)$ is a sequence of stationary Gaussian processes and $h_T(t)$ is a taper function from the class $D(k, \kappa)$ satisfying (A.3.5) (e.g., the tapers in Examples A.3.1 and A.3.2), then the tapered Whittle estimator is Fisher-efficient. Consequently, it is equivalent to the exact maximum likelihood estimator (for details, see Theorem 6.3 in Dahlhaus [73]).

Furthermore, it was shown that the classical non-tapered Whittle estimator may be inconsistent if sharp peaks are present in the spectrum (Dahlhaus [73], Theorem 7.1). The same conclusion applies if strong troughs are present in the spectrum (Dahlhaus [73], Theorem 7.2).

Observe that the Dahlhaus model is also important for studying time series that exhibit long-range dependence. Recall that long-range dependence (or long memory) is typically associated with the spectral density $f(\lambda)$ having a singularity of the form $|\lambda|^{-1+\varepsilon}$ in the neighborhood of zero (see Sect. 3.3). The Dahlhaus model includes spectral densities of the following form:

$$f(\lambda) = |1 - (1 - 1/T)e^{-i\lambda}|^{-2s} f_\theta^*(\lambda), \tag{A.3.6}$$

which, for large T, is approximately of the form $|\lambda|^{-2s} f_\theta^*(\lambda)$ for small λ (see Dahlhaus [73], page 816).

A.4 The Bias and Edge Effect

Edge effects are a significant issue in spatial statistics, particularly because the number of boundary points increases as the number of dimensions grows.

Assume that a discrete time d-dimensional centered random field $\{X(t),\ t \in \mathbb{Z}^d\}$ is observed on the rectangle $P_T := \{1, \dots, T_1\} \times \cdots \times \{1, \dots, T_d\}$ with a sample size $T := \prod_{k=1}^{d} T_k$.

In his analysis of the standard parametric Whittle estimator, Guyon [164] observed that when the estimation of the underlying unknown parameter is based on the sample covariances $\hat{r}_T(k)$, a bias arises due to boundary effects (edge effect).

Specifically, if $T_k \to \infty$ at the same rate in all directions, this bias is of order $O(T^{-1/d})$. For $d \geq 2$, this bias is of the same order or greater than the standard deviation of $\hat{r}_T(k)$, which is typically $O(T^{-1/2})$.

To address this edge effect, Guyon [164] proposed the use of an unbiased modification of the sample covariances, denoted as $\hat{r}_T^*(k)$, and provided a $\sqrt{T}$-consistent and asymptotically normal estimator of the underlying parameter.

Dahlhaus and Künsch [78] observed that the Guyon estimator $\hat{r}_T^*(k)$ has several undesirable properties. First, it is not always positive definite, which means that the spectral estimators based on $\hat{r}_T^*(k)$ may be negative. The second disadvantage of $\hat{r}_T^*(k)$ is its large variance, particularly for larger lags. They showed that these disadvantages of $\hat{r}_T^*(k)$ can be avoided while still achieving an asymptotically negligible bias by using data tapers (for details, see Dahlhaus and Künsch [78], Theorem 1).

In the paper by Anh et al. [10], the authors discuss the bias and edge effects of estimators for spectral functionals in both discrete and continuous time scenarios. They demonstrated that, under certain smoothness conditions imposed on the taper function h, the bias is of order $O(T^{-2})$ (for further details, see Anh et al. [10], Theorem 2.1).

Additionally, the paper by Ludeña and Lavielle [221] discusses the edge effect for the Whittle estimator in a class of long-memory stationary Gaussian fields.

Appendix B
A Few Problems Not Studied Here

The book intentionally omits several fascinating and important problems closely related to its topics, some of which have significant practical and theoretical implications. We will briefly introduce a few of these topics and provide key bibliographical references for further exploration. Note that some of these topics could constitute the content of a separate book.

B.1 The Szegő and Avram-Parter Type Theorems

In the theory of Toeplitz matrices, the Szegő and Avram-Parter type theorems establish the limits of the arithmetic means of specific test functions evaluated at the eigenvalues of Hermitian Toeplitz matrices and at the singular values of arbitrary Toeplitz matrices, respectively. These limits are examined as the matrix dimension approaches infinity.

Let $B_T := B_T(f)$ be the $(T \times T)$ truncated Toeplitz matrix generated by a real-valued function $f \in L^1(\mathbb{T})$. Denote the singular values of $B_T(f)$ by $s_1(B_T) \leq \cdots \leq s_T(B_T)$ and the eigenvalues by $\lambda_1(B_T) \leq \cdots \leq \lambda_T(B_T)$, counted with their algebraic multiplicities (see Sect. 2.7). Let $F(\lambda)$ be a real-valued function defined on $\mathbb{R}$.

The Szegő type theorems state that, under certain conditions on the functions f and F, including the requirement that f is real-valued, the following asymptotic relation holds:

$$\lim_{T\to\infty} \frac{1}{T} \sum_{k=1}^{T} F(\lambda_k(B_T(f))) = \frac{1}{2\pi} \int_{-\pi}^{\pi} F(f(\lambda))\, d\lambda. \tag{B.1.1}$$

M. S. Ginovyan, *Random Toeplitz Functionals and Their Applications*,
Frontiers in Probability and the Statistical Sciences,
https://doi.org/10.1007/978-3-032-00961-6

Observe that in the special case where $F(\lambda) = \ln \lambda$, the relation (B.1.1) is equivalent to Szegő's weak theorem (see Proposition 2.4.1).

Theorems of the Avram-Parter type pertain to complex-valued functions f and state that, under appropriate assumptions on the functions f and F, the following relation holds:

$$\lim_{T\to\infty} \frac{1}{T} \sum_{k=1}^{T} F(s_k(B_T(f))) = \frac{1}{2\pi} \int_{-\pi}^{\pi} F(|f(\lambda)|)\, d\lambda. \tag{B.1.2}$$

The function F in (B.1.1) and (B.1.2), referred to as a test function, is assumed to be real-valued and continuous on $\mathbb{R} := (-\infty, \infty)$ ($F \in \mathbb{C}(\mathbb{R})$) when considering Szegő's approximation (B.1.1), and continuous on $\mathbb{R}^+ := [0, \infty)$ ($F \in \mathbb{C}(\mathbb{R}^+)$) when dealing with the Avram-Parter approximation (B.1.2).

The asymptotic relation (B.1.1) originates from G. Szegő (1920), who proved it for real-valued functions $f \in L^\infty(\mathbb{T})$ and for compactly supported continuous functions F on $\mathbb{R}$ (see, e.g., Grenander and Szegő [162], Section 5.2, and Szegő [290]).

The asymptotic relation (B.1.2) was first established by Parter [235] for all $F \in C(\mathbb{R}^+)$ under the assumptions that $f \in L^\infty(\mathbb{T})$ and that f is locally self-adjoint. This means that f can be expressed as $f = f_1 f_2$, where f_1 is a continuous 2π-periodic function and f_2 is a real-valued function.

Avram [13] extended Parter's result by proving that the relation (B.1.2) holds for all $f \in L^\infty(\mathbb{T})$ and all $F \in C(\mathbb{R}^+)$.

Subsequently, the asymptotic relations (B.1.1) and (B.1.2) became the focus of many researchers aiming to prove these relations for broader classes of generating and test functions. For a detailed analysis of the topic, as well as up-to-date results and references, we refer to the book by Böttcher and Silbermann [52] (Section 5.6) and the papers by Böttcher et al. [49, 50].

Here, we highlight several significant results on this topic.

Tyrtyshnikov [306] (see also Zamarashkin and Tyrtyshnikov [319]) showed that the relations (B.1.1) and (B.1.2) hold for all continuous functions F with compact support, provided that $f \in L^2(\mathbb{T})$ and f is real-valued in the context of (B.1.1).

Serra Capizzano [275] derived the relation (B.1.2) under the assumption that $f \in L^p(\mathbb{T})$ ($1 \le p < \infty$), and that $F \in C(\mathbb{R}^+)$ satisfies the condition:

$$F(t) = O(t^p) \quad \text{as} \quad t \to \infty.$$

Serra Capizzano's result implies, in particular, that the relation (B.1.2) is valid for all $f \in L^1(\mathbb{T})$ under the sole assumption that $F(t) = O(t)$. This encompasses all the results concerning (B.1.2) mentioned previously.

In all the quoted results regarding the Avram-Parter approximation (B.1.2), it is assumed that the integral in (B.1.2) is finite. However, the paper by Böttcher et al. [50] demonstrates that the relation (B.1.2) holds true even when the integral in (B.1.2) is infinite.

The continuous analogues of relations (B.1.1) and (B.1.2), specifically the Szegő and Avram-Parter type theorems for Toeplitz (Wiener-Hopf) operators, were presented in the paper by Böttcher et al. [48].

B.2 Limit Theorems for Quadratic Forms in Bivariate Appell Polynomials

The bivariate Appell polynomials serve as a useful tool for establishing both central and non-central limit theorems for quadratic functionals in random variables of the form $G(X(t))$, where $X(t)$ is a linear or Gaussian process and G is a polynomial. While Hermite polynomials suffice when $X(t)$ is Gaussian, Appell polynomials become relevant when $X(t)$ is a linear process. Notably, the Hermite polynomials are Appell polynomials associated with the Gaussian distribution.

The objective is to describe the limiting distributions of quadratic functionals of the form:

$$Q_{T,m,n} := \sum_{t=1}^{T}\sum_{s=1}^{T} \widehat{g}(t-s)P_{m,n}(X(t), X(s)), \quad T \in \mathbb{N}, \tag{B.2.1}$$

where $\widehat{g}(t)$ is the Fourier transform of a real, even, integrable function g, and $X(t)$ is a discrete time linear process of the form (3.2.7). The term $P_{m,n}(X(t), X(s))$ refers to a bivariate Appell polynomial (Wick power) of the linear variables $X(t)$ and $X(s)$, with $m, n \geq 0$ and $m+n \geq 1$. For the definition and properties of Appell polynomials, see, for example, Giraitis and Taqqu [149, 151].

Observe that $Q_{T,1,1} = Q_T$, where Q_T is the Toeplitz quadratic functional defined by Eq. (5.1.1) in the discrete time context. Consequently, the limit theorems for functionals $Q_{T,m,n}$ extend the limit theorems for Q_T established in Chap. 5.

Central and non-central limit theorems for the quadratic functional $Q_{T,m,n}$, which involves Hermite or Appell polynomials, have been studied in several papers. Here, we cite some of these works, where additional references can be found.

Central limit theorems for the quadratic functional $Q_{T,m,n}$ have been established in the paper by Avram and Fox [14], as well as in a series of papers by Giraitis and Taqqu [149–151, 153]. In particular, the following results, among others, were obtained in Giraitis and Taqqu [151], which involve time domain sufficient conditions for the functional $Q_{T,m,n}$ in (B.2.1) to satisfy the central limit theorem.

Theorem B.2.1 *Assume that*

$$\sigma_{m,n}^2 := \sum_{k_1,k_2\in\mathbb{Z}} \left|\widehat{g}(k_1)\widehat{g}(k_2)\mathrm{Cov}\left(P_{m,n}(X(t), X(t+k_1), P_{m,n}(X(0), X(k_2))\right)\right| < \infty. \tag{B.2.2}$$

If $\widehat{g}(0) = 0$, *assume in addition that*

$$\sum_{t \in \mathbb{Z}} |r(t)|^{m+n} < \infty,$$

where $r(t)$ *is the covariance function of the process* $X(t)$. *Then the CLT for the functional* $Q_{T,m,n}$ *holds, specifically:*

$$Q_T \xrightarrow{d} \eta \sim N(0, \sigma^2_{m,n}) \quad \text{as} \quad T \to \infty \tag{B.2.3}$$

with the limiting variance $\sigma^2_{m,n}$ *as defined in* (B.2.2).

The next result provides constructive sufficient conditions in terms of the covariance function $r(t)$ and the generating kernel $\widehat{g}(t)$.

Theorem B.2.2 *Assume that* $r \in L^p(\mathbb{T})$ *and* $\widehat{g} \in L^q(\mathbb{T})$, *where* $p, q \geq 1$ *and satisfy*

$$\min(mp^{-1}, 1) + \min(np^{-1}, 1) + 2q^{-1} \geq 3. \tag{B.2.4}$$

Then, the CLT for the functional $Q_{T,m,n}$ *holds, with the limiting variance defined in Eq.* (B.2.2).

Observe that Theorem B.2.2 extends Theorem 5.2.5.

Frequency domain sufficient conditions for the functional $Q_{T,m,n}$ to satisfy the central limit theorem were obtained by Giraitis and Taqqu [149].

Non-central limit theorems for $Q_{T,m,n}$ have been established in the works of Avram and Taqqu [16], Giraitis and Taqqu [149, 153], and Giraitis et al. [154].

B.3 Estimation of Spectral Density and Spectral Functions

Estimating spectral density and spectral functions is a classical inference problem in stationary models. Many foundational texts have examined the properties of statistical estimators for spectral density and spectral functions under various conditions related to the observed models. Notable works in this area include those by Anderson [4], Brillinger [55], Brockwell and Davis [57], Grenander and Rosenblatt [161], Hannan [168, 169], Priestley [249], and Rosenblatt [261], among others.

Estimation of Spectral Density Function The well-known estimator for the spectral density function $f(\lambda)$ is given by the following formula (see Sects. 8.3 and A.1):

$$\widehat{f_T}(\lambda) = \int_{\Lambda} g_M(\lambda - \mu) I_T(\mu)\, d\mu, \tag{B.3.1}$$

where $M = M(T)$ is a positive real number that increases to infinity at a suitable rate depending on T. Note that $\widehat{f_T}(\lambda)$ in Eq. (B.3.1) is a Toeplitz quadratic functional of the type given in (5.1.1), but now with a time-dependent generating function $g_M(\mu)$, commonly referred to as a spectral window.

Asymptotic normality of the random variable

$$\eta_T := (T/M)^{1/2}[\widehat{f_T}(\lambda)) - f(\lambda)] \tag{B.3.2}$$

has been well-established under various regularity conditions (see, e.g., Bentkus [25, 26], Bentkus and Rudzkis [29], Brillinger [55], Ibragimov [180], Parzen [236, 237], and Saulis and Statulevicius [271], Section 5.3.

Exponential inequalities for the tail probabilities:

$$\mathbb{P}[\sup_{\lambda\in[0,\pi]} |\widehat{f_T}(\lambda)) - f(\lambda)| \geq x], \quad x > 0. \tag{B.3.3}$$

were obtained in the papers by Bentkus [27], Bentkus and Rudzkis [28], and Bentkus et al. [30] (see also Saulis and Statulevicius [271], Section 5.3).

Results on large and moderate deviations for the random variable η_T in (B.3.2) with a speed $a_T \to \infty$ such that $a_T = o(T)$ or $a_T = o(T/M)$ were proved in the papers by Bentkus [27], Bentkus and Rudzkis [28], and Kakizawa [199].

Edgeworth expansions for the distribution of the random variable η_T, as defined in (B.3.2), have been obtained by Bentkus and Rudzkis [29] and by Velasco and Robinson [308].

A minimax lower bound on the risk for spectral density estimators for discrete time Gaussian stationary processes was obtained by Samarov [267].

In the referenced works, spectral density estimators were mainly examined for discrete time, short memory Gaussian, and linear processes.

In the papers by Liu and Wu [218] and Wu and Shao [316], the authors established the asymptotic properties of spectral density estimators under very mild conditions, including for nonlinear and non-strong mixing processes.

In the paper by Rousseau et al. [262], a Bayesian approach was applied to estimate the spectral density $f(\lambda)$. In particular, for discrete time long and intermediate memory Gaussian processes, the consistency of the Bayesian nonparametric estimators of $f(\lambda)$ was established under certain general conditions imposed on both the prior and the true spectral density.

The asymptotic properties of spectral density estimators based on tapered data have been studied by Brillinger [55] and by Dahlhaus [70, 72, 76].

Estimation of Spectral Function The estimation of the spectral function $F(\lambda)$ is typically focused on the interval $\lambda \in [0, \pi]$ in the discrete time case and $\lambda \in \mathbb{R}^+$ in the continuous time case. We assume that $F(\lambda)$ is absolutely continuous, with a spectral density $f(\lambda)$, and is normalized such that $F(0) = 0$. Several statistical estimators for the spectral function $F(\lambda)$ exist, the most well-known of which is the following estimator:

$$\widehat{F}_T(\lambda) = \int_0^\lambda I_T(\mu)\,d\mu, \tag{B.3.4}$$

where $I_T(\mu)$ is the non-tapered periodogram.

Note that the estimator $\widehat{F}_T(\lambda)$ is a Toeplitz quadratic functional of the type given in (5.1.1) with the indicator generating function $g(\mu) = \mathbf{1}_{[0,\lambda]}(\mu)$ (see Example 8.2.2).

In the seminal paper by Ibragimov [180], several properties of the estimator $\widehat{F}_T(\lambda)$ were established, including the following functional limit theorem.

Theorem B.3.1 *Let $X(t)$ be a Gaussian stationary process with a spectral function $F(\lambda)$. Assume that $F(\lambda)$ has no intervals of constancy and that $f \in L^{2+\delta}(\mathbb{T})$. Then, the measures $\mathbb{P}_T$, generated in the space $\mathbb{C}[0,\pi]$ by the random processes:*

$$\xi_T(\lambda) := T^{1/2}[\widehat{F}_T(\lambda) - F(\lambda)] \tag{B.3.5}$$

converge weakly (as $T \to \infty$) to the measure $\mathbb{P}$, generated by the centered Gaussian random process $\xi(\lambda)$, with a covariance function given by:

$$\mathbb{E}[\xi_T(\lambda)\xi_T(\mu)] = 2\pi \int_0^{\min(\lambda,\mu)} f^2(v)\,dv. \tag{B.3.6}$$

In Ibragimov [180] (see also Ibragimov [182]), it was shown that the result in Theorem B.3.1 cannot be significantly strengthened.

Specifically, for a Gaussian stationary process $X(t)$ with spectral density $f(\lambda) = |\lambda|^{-a}$, $0 < a < 1$, it can be shown that $\sup_{0\le\lambda\le\pi} |\widehat{F}_T(\lambda) - F(\lambda)|$ converges to zero with probability 1. However, for $a > 1/2$, the nature of the convergence depends significantly on a.

More precisely, let $\mathbb{P}_T^{(\alpha)}$ be the measure in the space $\mathbb{C}[0,\pi]$ generated by the random process

$$\eta_T(\lambda) := T^{1-\alpha}|\widehat{F}_T(\lambda) - F(\lambda)|.$$

Then, the following assertions hold (for details, see Ibragimov [180]):

(a) If $\alpha < a$, the sequence $\mathbb{P}_T^{(\alpha)}$ is not compact and $\max_{0\le\lambda\le\pi} \eta_T(\lambda) \xrightarrow{P} \infty$ as $T \to \infty$.
(b) If $\alpha > a$, the sequence $\mathbb{P}_T^{(\alpha)}$ converges to the measure concentrated on the zero element of $\mathbb{C}[0,\pi]$.
(c) If $\alpha = a$, the limiting measure of $\mathbb{P}_T^{(\alpha)}$ is not generated by a Gaussian process.

In the paper by Bentkus [25], the asymptotic properties of the estimator $\widehat{F}_T(\lambda)$ were established for non-Gaussian stationary processes.

Lower bounds for the estimators of the spectral function of discrete time stationary Gaussian processes were obtained in the paper by Levit and Samarov [213]. As a consequence, they derived conditions for the asymptotic unimprovability of the estimator $\widehat{F}_T(\lambda)$, as defined in Eq. (B.3.4), for a broad class of loss functions.

B.4 Asymptotic Behavior of the Prediction Error Variance for Stationary Sequences

Let $X(t)$ be a centered discrete time stationary process with spectral measure $\mu(\lambda)$ and spectral density $f(\lambda)$, $\lambda \in [-\pi, \pi)$. The finite linear prediction problem is defined as follows.

Suppose we observe a finite realization of the process $X(t)$:

$$\{X(t),\ -T \le t \le -1\}, \quad T \in \mathbb{N} := \{1, 2, \ldots\}.$$

We aim to make a one-step ahead prediction, specifically to predict the unobserved random variable $X(0)$, using the *linear predictor* $Y = \sum_{k=1}^{T} c_k X(-k)$.

The coefficients c_k, $k = 1, 2, \ldots, T$, are chosen to minimize the mean-squared error: $\mathbb{E}\,|X(0) - Y|^2$, where $\mathbb{E}[\cdot]$ denotes the expectation operator.

If such minimizing constants $\widehat{c}_k := \widehat{c}_{k,T}$ can be found, then the random variable

$$\widehat{X}_T(0) := \sum_{k=1}^{T} \widehat{c}_k X(-k)$$

is referred to as *the best linear one-step ahead predictor* of $X(0)$ based on the observed finite past: $X(-T), \ldots, X(-1)$. The minimum mean-squared error:

$$\sigma_T^2(\mu) := \mathbb{E}\left|X(0) - \widehat{X}_T(0)\right|^2 \ge 0$$

is called *the best linear one-step ahead prediction error* of $X(t)$ based on the past of length T.

One of the main problems in the prediction theory of stationary processes is describing the asymptotic behavior of the prediction error $\sigma_T^2(\mu)$ as $T \to \infty$. This behavior depends on the regularity nature of the observed process $X(t)$, specifically whether it is deterministic or nondeterministic.

Observe that $\sigma_{T+1}^2(\mu) \le \sigma_T^2(\mu)$, $T \in \mathbb{N}$, which implies that the limit of $\sigma_T^2(\mu)$ as $T \to \infty$ exists. We denote by $\sigma^2(\mu) := \sigma_\infty^2(\mu)$ the prediction error of $X(0)$ based on the entire infinite past of the process: $\{X(t), t \le -1\}$.

From a predictive perspective, it is crucial to distinguish between two types of processes. The first type includes processes that allow for *error-free prediction* by using the entire infinite past, characterized by $\sigma^2(\mu) = 0$. Such processes are termed *deterministic* or *singular*. In contrast, the second type of processes, characterized by $\sigma^2(\mu) > 0$, is classified as *nondeterministic* or *regular*.

The well-known Kolmogorov-Szegő theorem offers a spectral characterization of both deterministic and nondeterministic processes (see, e.g., Grenander and Szegő [162], page 44, and Ibragimov and Rozanov [189], page 35).

Theorem B.4.1 (Kolmogorov-Szegő Theorem) *Let $X(t)$ be a non-degenerate stationary process with spectral measure μ of the form* (3.1.6). *The following assertion holds:*

$$\lim_{T\to\infty} \sigma_T^2(\mu) = \lim_{T\to\infty} \sigma_T^2(f) = \sigma^2(f) = 2\pi G(f), \tag{B.4.1}$$

where $G(f)$ is the geometric mean of f, namely,

$$G(f) := \begin{cases} \exp\left\{\frac{1}{2\pi}\int_{-\pi}^{\pi} \ln f(\lambda)\, d\lambda\right\} & \text{if } \ln f \in L^1(\mathbb{T}) \\ 0, & \text{otherwise.} \end{cases} \tag{B.4.2}$$

It follows from formulas (B.4.1) and (B.4.2) that a stationary process $X(t)$ is deterministic if $\ln f \notin L^1(\mathbb{T})$ and is nondeterministic if $\ln f \in L^1(\mathbb{T})$.

The condition $\ln f \in L^1(\mathbb{T})$ is known as *Szegő's condition*, which is also referred to as the *non-determinism condition*.

Define the *"relative" prediction error*

$$\delta_T(f) := \sigma_T^2(f) - \sigma^2(f),$$

and observe that $\delta_T(f)$ is non-negative and approaches zero as $T \to \infty$. But what about the rate of convergence of $\delta_T(f)$ to zero as $T \to \infty$?

Specifically, the prediction problem we are interested in is *to describe the rate at which $\delta_T(f)$ decreases to zero* as $T \to \infty$, depending on the regularity of the observed process $X(t)$.

It turns out that for nondeterministic processes, the asymptotic behavior of the prediction error is influenced by the dependence (memory) structure of the observed process $X(t)$ and the differential properties of its spectral density f.

In contrast, for deterministic processes, this behavior is determined by the geometric properties of the spectrum of the process $X(t)$ and the singularities of its spectral density f.

The prediction problem mentioned above originates from the classical works of A. N. Kolmogorov, G. Szegő, and N. Wiener. Since then, it has been explored by numerous authors across various classes of nondeterministic processes. A substantial number of both early and recent references on this subject can be found in the survey papers by Babayan et al. [20], Babayan and Ginovyan [19], Bingham [42], and Ginovyan [123].

For nondeterministic processes, it has been shown that for exponential decay ($\delta_T(f) = O(e^{-cT})$, $c > 0$), or hyperbolic decay ($\delta_T(f) = O(T^{-\gamma})$, $\gamma > 1$), the underlying model $X(t)$ must be a short memory process with a sufficiently smooth spectral density $f(\lambda)$ (see Ibragimov [181]).

Furthermore, if the model is not a short memory process—that is, if the spectral density f has zeros or is unbounded—then, in general, we cannot expect $\delta_T(f)$ to decrease to zero faster than $1/T$ as $T \to \infty$ (see Grenander and Rosenblatt [160]).

For instance, for the ARFIMA(p, d, q) model, which exhibits long memory for $0 < d < 1/2$, we have $\delta_T(f) \sim d^2/T$ as $T \to \infty$ (see Inoue [191]).

Only a few works focus on the speed of convergence of $\delta_T(f) = \sigma_T^2(f)$ to zero as $T \to \infty$, specifically the asymptotic behavior of the prediction error for deterministic processes (see Babayan and Ginovyan [19] and references therein). One needs to go back to the classical work of M. Rosenblatt [257]. In this paper, using the techniques of orthogonal polynomials on the unit circle and Toeplitz determinants, M. Rosenblatt investigated the asymptotic behavior of the prediction error $\sigma_T^2(f)$ for deterministic processes in the following two specific cases:

(a) The spectral density $f(\lambda)$ is continuous and positive on an interval of $[-\pi, \pi)$ and is zero elsewhere.
(b) The spectral density $f(\lambda)$ has a very high order of contact with zero at the points $\lambda = 0, \pm\pi$, and is strictly positive otherwise.

In case (a), M. Rosenblatt proved in [257] that the prediction error $\sigma_T^2(f)$ decreases to zero exponentially as $T \to \infty$.

In case (b), for a specific deterministic process $X(t)$, Rosenblatt proved in [257] that the prediction error $\sigma_T^2(f)$ decreases to zero hyperbolically (like a power) as $T \to \infty$.

Later, problems (a) and (b) were examined in a series of papers by Babayan and Ginovyan (see references in Babayan et al. [20] and Babayan and Ginovyan [19]). They extended Rosenblatt's results to broader classes of spectral densities and derived upper bounds for the prediction error $\sigma_T^2(f)$.

It is important to note that there is a relationship between the prediction error $\sigma_T^2(f)$ and the minimal eigenvalue $\lambda_{1,T}(f)$ of a truncated Toeplitz matrix $B_T(f)$ generated by the spectral density f.

Specifically, for any $T \in \mathbb{N}$ and a constant $M_f > 0$, the following inequalities hold (see Pourahmadi [248] and Serra [274]):

$$\lambda_{1,T}(f) \le \sigma_T^2(f) \le M_f \frac{\lambda_{1,T}(f)}{\lambda_{1,T-1}(f)}. \tag{B.4.3}$$

These inequalities illustrate how we can derive information about the minimal eigenvalue $\lambda_{1,T}(f)$ from the prediction error $\sigma_T^2(f)$ and vice versa.

For more details on this topic we refer readers to the survey papers by Babayan and Ginovyan [19] and Bingham [42].

B.5 Large-Sample Estimation of the Mean for a Stationary Sequence

Consider the stochastic model

$$Y(t) = m + X(t), \quad t \in \mathbb{Z},$$

where m is the constant *unknown* mean of $Y(t)$, and the noise $X(t)$ is assumed to be a centered stationary process with a spectral density function $f(\lambda)$, $\lambda \in [-\pi, \pi)$, and a covariance function $r(t)$, $t \in \mathbb{Z}$.

The problem of interest is the estimation of the unknown mean m by using unbiased linear estimators $\widehat{m}_n$, based on a random sample $\{Y(t), t = 0, 1, \dots, T\}$:

$$\widehat{m}_T = \sum_{k=0}^{T} c_k Y(k), \quad \sum_{k=0}^{T} c_k = 1. \tag{B.5.1}$$

Of particular interest is the *best linear unbiased estimator (BLUE)*, denoted as $\widehat{m}_{T,BLUE}$. This estimator takes the form given in Eq. (B.5.1), where the weights c_k, $k = 0, 1, \dots, T$, are chosen to minimize the variance:

$$\text{Var}\,(\widehat{m}_T) = E|\widehat{m}_T - m|^2 = \sum_{j,k=0}^{T} c_j \bar{c}_k r(j-k) \tag{B.5.2}$$

subject to the constraint that $\sum_{k=0}^{T} c_k = 1$.

We assume that the process $X(t)$ is non-degenerate, meaning that $E|X(t)|^2 > 0$, which implies that the BLUE exists and is unique (see, e.g., Grenander and Szegő [162], Sections 11.1).

Our focus here is on the asymptotic behavior of the variance:

$$\sigma_T^2(f) := \text{Var}_f\left(\widehat{m}_{T,BLUE}\right) \quad \text{as} \quad T \to \infty.$$

The first significant result in this area can be traced back to the classical works of U. Grenander (see Grenander [159] and Grenander and Szegő [162], Sections 11.1–11.3), who showed that $\widehat{m}_{T,BLUE}$ has an asymptotic variance $2\pi f(0)/T$, provided that the spectral density $f(\lambda)$ is positive and continuous.

Since then, many authors have examined the properties of the BLUE and compared it with other estimators across various time series models, including those with long and intermediate memory. Notable contributions in this area include works by Adenstedt [1], Adenstedt and Eisenberg [2], Grenander [159], Grenander and Rosenblatt [160, 161], Grenander and Szegő [162], Samarov and Taqqu [269], and Vitale [309], among others.

Summarizing the results from the references above, we conclude that the variance $\sigma_T^2(f)$ depends asymptotically only on the behavior of the spectral density f near the origin. For certain classes of spectral densities that satisfy the condition:

$$f(\lambda) \sim \lambda^{\nu} \quad (\nu > -1) \quad \text{as} \quad T \to \infty,$$

the variance $\sigma_T^2(f)$ decreases hyperbolically. Specifically, we have

$$\sigma_T^2(f) \sim T^{-\nu-1} \quad (\nu > -1) \quad \text{as} \quad T \to \infty.$$

This raises a natural question:

Is it possible to increase the order of zero of the spectral density f at the origin to achieve exponential decay of the variance $\sigma_T^2(f)$ as $T \to \infty$?

In the paper by Babayan and Ginovyan [18], the authors negatively addressed this question by demonstrating that if the spectral density f vanishes only at the origin, then achieving exponential decay of $\sigma_T^2(f)$ is impossible, regardless of how high the order of the zero of f at the origin is.

Furthermore, it has been proven that for the variance $\sigma_T^2(f)$ to exhibit exponential decay as $T \to \infty$, the underlying process $X(t)$ must be purely deterministic. Specifically, a necessary condition for the variance $\sigma_T^2(f)$ to decrease to zero exponentially as $T \to \infty$ is that the spectral density $f(\lambda)$ vanishes on a set of positive Lebesgue measure in any vicinity of zero.

It is important to highlight a similarity between the problem discussed here and the prediction problem outlined in Sect. B.4.

The main distinction is that, in the prediction problem, the asymptotic behavior of the prediction error variance is influenced by the spectral density f across the entire interval $[-\pi, \pi)$, rather than just at the origin.

References

1. Adenstedt, R. K. (1974). On large-sample estimation for the mean of a stationary random sequence. *The Annals of Mathematical Statistics, 2*, 1095–1107.
2. Adenstedt, R. K., & Eisenberg B. (1974). Linear estimation of regression coefficients. *Quarterly of Applied Mathematics, 32*(3), 317–327.
3. Alomari, H. M., Frías, M. P., Leonenko, N. N., Ruiz-Medina, M. D. Sakhno, L., & Torres, A. (2017). Asymptotic properties of parameter estimates for random fields with tapered data. *Electronic Journal of Statistics, 11*, 3332–3367.
4. Anderson, T. (1971). *The statistical analysis of time series*. Wiley.
5. Anh, V. V., Angulo, J. M., & Ruiz-Medina, M. D. (1999). Possible long-range dependence in fractional random fields. *Journal of Statistical Planning and Inference, 80*, 95–110.
6. Anh, V. V., Leonenko, N. N., & McVinish, R. (2001). Models for fractional Riesz-Bessel motion and related processes. *Fractals, 9*, 329–346.
7. Anh, V. V., Leonenko, N. N., & Sakhno, L. M. (2004). On a class of minimum contrast estimators for fractional stochastic processes and fields. *Journal of Statistical Planning and Inference, 123*, 161–185.
8. Anh, V. V., Leonenko, N. N., & Sakhno, L. M. (2004). Quasi-likelihood-based higher-order spectral estimation of random fields with possible long-range dependence. Stochastic methods and their applications. *Journal of Applied Probability, 41*, 35–53.
9. Anh, V. V., Leonenko, N. N., & Sakhno, L. M. (2007). Minimum contrast estimation of random processes based on information of second and third orders. *Journal of Statistical Planning and Inference, 137*, 1302–1331.
10. Anh, V. V., Leonenko, N. N., & Sakhno, L. M. (2009). Evaluation of bias in high-order spectral estimation. *Theory of Probability and Mathematical Statistics, 80*, 1–14.
11. Aronszajn, N. (1950). Theory of reproducing kernels. *Transactions of the American Mathematical Society, 68*, 307–404.
12. Ash, R. B., & Gardner, M. F. (1975). *Topics in Stochastic Processes*. Academic Press.
13. Avram, F. (1988). On bilinear forms in Gaussian random variables and Toeplitz matrices. *Probability Theory and Related Fields, 79*, 37–45.
14. Avram, F., & Fox, R. (1992). Central limit theorems for sums of Wick products of stationary sequences. *Transactions of the American Mathematical Society, 330*, 651–663.
15. Avram, F., Leonenko, N. N., & Sakhno, L. (2010). On a Szegö type limit theorem, the Hölder-Young-Brascamp-Lieb inequality, and the asymptotic theory of integrals and quadratic forms of stationary fields. *ESAIM: Probability and Statistics, 14*, 210–255.

M. S. Ginovyan, *Random Toeplitz Functionals and Their Applications*,
Frontiers in Probability and the Statistical Sciences,
https://doi.org/10.1007/978-3-032-00961-6

16. Avram, F., & Taqqu, M. S. (1987). Noncentral limit theorems and Appell polynomials. *Annals of Probability, 15*, 767–775.
17. Azencott R., & Dacunha–Castelle D. (1986). *Series of irregular observations: Forecasting and model building*. Springer.
18. Babayan, N. M., & Ginovyan, M. S. (2019). On exponential decay of the variance of BLUE of mean for stationary sequences. *Studia Scientiarum Mathematicarum Hungarica, 56*(4), 482–491.
19. Babayan, N. M., & Ginovyan, M. S. (2023). Asymptotic behavior of the prediction error for stationary sequences. *Probability Surveys, 20*, 664–721.
20. Babayan, N. M., Ginovyan, M. S., & Taqqu, M. S. (2021). Extensions of Rosenblatt's results on the asymptotic behavior of the prediction error for deterministic stationary sequences. *Journal of Time Series Analysis, 42*, 622–652.
21. Bai, S., Ginovyan, M. S., & Taqqu, M. S. (2015). Functional limit theorems for Toeplitz quadratic functionals of continuous time Gaussian stationary processes. *Statistics and Probability Letters, 104*, 58–67.
22. Bai, S., Ginovyan, M. S., & Taqqu, M. S. (2016). Limit theorems for quadratic forms of Lévy-driven continuous time linear processes. *Stochastic Processes and their Applications, 126*, 1036–1065.
23. Bass, R. F. (2011). *Stochastic processes*. Cambridge University Press.
24. Beinicke, G., & Dzhaparidze, K. O. (1982). On parameter estimation by the Davidon-Fletcher-Powell method. *Theory of Probability and Its Applications, 27*, 396–402.
25. Bentkus, R. (1971). On asymptotic behavior of the estimate of a spectral function of a Gaussian multi-dimensional stationary sequence. *Litovskii Matematicheskii Sbornik, 11*, 745–760.
26. Bentkus, R. (1972). On the error of the estimate of the spectral function of a stationary process. *Lithuanian Mathematical Journal, 12*, 55–71.
27. Bentkus, R. (1978). Large deviations for estimates of spectrum of stationary time series. *Lithuanian Mathematical Journal, 18*, 214–226.
28. Bentkus, R., & Rudzkis, R. (1976). Large deviations for the spectral estimate a stationary Gaussian sequence. *Lithuanian Mathematical Journal, 16*, 519–529.
29. Bentkus, R., & Rudzkis, R. (1982). On the distribution of some statistical estimates of spectral density. *Theory of Probability and Its Applications, 27*, 795–814.
30. Bentkus, R., Rudzkis, R., & Statulevichus, V. (1974). Exponential inequalities for estimates of the spectrum of a stationary Gaussian time series. *Lithuanian Mathematical Journal, 15*, 25–39.
31. Beran, J. (1994). *Statistics for long-memory processes*. Chapman & Hall.
32. Beran, J. (1995). The role of Hájek's convolution theorem in statistical theory, *Kybernetika, 31*, 221–237.
33. Beran, J., Feng, Y., Ghosh, S., & Kulik, R. (2013). *Long-memory processes: Probabilistic properties and statistical methods*. Springer.
34. Bercu, B., Bony, J.-F., & Bruneau, V. (1997). Large deviations for Gaussian stationary processes and semi-classical analysis. *Séminaire de Probabilités, XLIV*, (Vol. 2046, pp. 409–428). Springer.
35. Bercu, B., Gamboa, F., & Rouault, A. (1997). Large deviations for quadratic forms of Gaussian stationary processes. *Stochastic Processes and their Applications, 71*, 75–90.
36. Bergh, J., & Löfström, J. (1976). *Interpolation spaces. An introduction* (2nd ed.). Springer.
37. Berlient, A., & Thomas-Agnan, C. (2004). *Reproducing Kernel Hilbert Spaces in probability and statistics*. Kluwer Academic Publishers.
38. Bhatia, R. (1997). *Matrix analysis*. Springer.
39. Billingsley, P. (1956). The invariance principle for dependent random variables. *Transactions of the American Mathematical Society, 83*(1), 250–268.
40. Billingsley, P. (1986). *Probability and measures* (2nd ed.). Wiley.
41. Billingsley, P. (1999). *Convergence of probability measures*. Wiley.
42. Bingham, N. H. (2012). Szegő's theorem and its probabilistic descendants. *Probability Surveys, 9*, 287–324.

43. Bingham, N. H., Goldie, C. M., & Teugels, J. L. (1989). *Regular variation*. Cambridge University Press.
44. Bloomfield, P. (1976). *Fourier analysis of time series*. Wiley.
45. Bogachev, V. I. (2007) *Measure theory* (Vol. 1). Springer.
46. Bondon, P., & Palma, W. (2006). A class of antipersistent processes. *Journal of Time Series Analysis, 28*, 261–273.
47. Böttcher, A. (2006). Schatten norms of Toeplitz matrices with Fisher-Hartwig singularities *The Electronic Journal of Linear Algebra, 15*, 251–259.
48. Böttcher, A., Brunner, H., Iserles, A., & Nørsett, S. (2010). On the singular values and eigenvalues of the Fox–Li and related operators. *New York Journal of Mathematics, 16*, 539–561.
49. Böttcher, A., Grudsky, S., & Maksimenko, E. (2008). Pushing the envelope of the test functions in the Szegő and Avram–Parter theorems. *Linear Algebra and Its Applications, 429*, 346–366.
50. Böttcher, A., Grudsky, S., & Schwartz, M. (2008). Some problems concerning the test functions in the Szegő and Avram-Parter theorems. *Operator Theory: Advances and Applications, 187*, 81–93.
51. Böttcher, A., & Karlovich, Yu. I. (1997). *Carleson curves, muckenhoupt weights, and toeplitz operators*, Birkhäuser.
52. Böttcher, A., & Silbermann, B. (1999). *Introduction to large truncated Toeplitz matrices*. Springer.
53. Böttcher, A., & Silbermann, B. (2005). *Analysis of toeplitz operators* (2nd ed.). Springer.
54. Breuer, P., & Major, P. (1983). Central limit theorems for non-linear functionals of Gaussian fields. *Journal of Multivariate Analysis, 13*, 425–441.
55. Brillinger, D. R. (1981). *Time series: Data analysis and theory*. Holden Day.
56. Brockwell, P. J. (2014). Recent results in the theory and applications of CARMA processes. *Annals of the Institute of Statistical Mathematics, 66*(4), 647–685.
57. Brockwell, P. J., & Davis, R. A. (1991). *Time series: Theory and methods* (2nd ed.). Springer.
58. Bryc, W., & Dembo, A. (1997). Large deviations for quadratic functionals of Gaussian processes. *Journal of Theoretical Probability, 10*, 307–332.
59. Bucklew, J. A. (1990). *Large deviations techniques in decision, simulation, and estimation*. Wiley.
60. Butzer, P. L., & Nessel, R. J. (1971). *Fourier analysis and approximation I*. Academic Press.
61. Casas, I., & Gao, J. (2008). Econometric estimation in long-range dependent volatility models: Theory and practice. *Journal of Econometrics, 147*, 72–83.
62. Chernoff, H., & Lehmann, E. L. (1954). The use of maximum likelihood estimates in χ^2–tests for goodness of fit. *Annals of Mathematical Statistics, 25*(3), 579–586.
63. Chibisov, D. M. (1971). Some tests of chi-square type for continuous distributions. *Theory of Probability and Its Applications, 16*(1), 1–20.
64. Clevenson, M. L. (1970). *Asymptotically efficient estimate of the parameters of a moving average time series*. Technical Report 15, Department of Statistics, Stanford University, Stanford, California.
65. Cohen, S., & Lindner, A. (2013). A central limit theorem for the sample autocorrelations of a Lévy driven continuous time moving average process. *Journal of Statistical Planning and Inference, 143*(8), 1295–1306.
66. Coursol, J., & Dacunha-Castelle, D. (1979). Sur la formule de Chernoff pour deux processus gaussiens stationnaires. *Comptes Rendus Hebdomadaires des Séances de l'Académie des Sciences, Série, A-B 288*, A769–A770.
67. Coursol, J., & Dacunha-Castelle, D. (1982). Remarques sur l'approximation de la vraisemblance d'un processus Gaussien stationary. *Theory of Probability and Its Applications, 27*(1), 162–167.
68. Cramér, H. (1946). *Mathematical methods of statistics*. Princeton University Press.
69. Cramér, H., & Leadbetter, M. R. (1967). *Stationary and related stochastic processes: Sample function properties and their applications*. Wiley.

70. Dahlhaus, R. (1983). Spectral analysis with tapered data, *Journal of Time Series Analysis, 4*, 163–174.
71. Dahlhaus, R. (1984). Parameter estimation of stationary processes with spectra containing strong peaks. In J. Franke, W. Hardle, & D. Martin (Eds.), *Robust and nonlinear time series analysis. Lecture notes in statistics* (Vol. 26, pp. 50–67). Springer.
72. Dahlhaus, R. (1985). A functional limit theorem for tapered empirical spectral functions. *Stochastic Processes and their Applications, 19*, 135–149.
73. Dahlhaus, R. (1988). Small sample effects in time series analysis: A new asymptotic theory and a new estimate. *Annals of Statistics, 16*, 808–841.
74. Dahlhaus, R. (1989). Efficient parameter estimation for self-similar processes. *Annals of Statistics, 17*, 1749–1766.
75. Dahlhaus, R. (1990). Approximations for the inverse of Toeplitz matrices with applications to stationary processes. *Linear Algebra and Its Applications, 127*, 27–40.
76. Dahlhaus, R. (1990). Nonparametric high resolution spectral estimation. *Probability Theory and Related Fields, 85*, 147–180.
77. Dahlhaus, R., & Janas, D. (1996). A frequency domain bootstrap for ratio statistics in time series analysis. *Annals of Statistics, 24*(5), 1934–1963.
78. Dahlhaus, R., & Künsch, H. (1987). Edge effects and efficient parameter estimation for stationary random fields. *Biometrika, 74*(4), 877–882.
79. Dahlhaus, R., & Wefelmeyer, W. (1996). Asymptotically optimal estimation in misspecified time series models. *Annals of Statistics, 24*, 952–974.
80. Davies, B. R. (1973). Asymptotic inference in stationary Gaussian time-series. *Advances in Applied Probability, 5*, 469–497.
81. Dembo, A., & Zeitouni, O. (1993). *Large deviations technique and applications*. Jones and Bartlett.
82. Djellout, H., Guillin, A., & Wu, L. (2006). Moderate deviations of empirical periodogram and non-linear functionals of moving average processes. *Annales Henri Poincaré, PR 42*, 393–416.
83. Dobrushin, R. L. (1979). Gaussian and their subordinated self-similar random generalized fields. *Annals of Probability, 7*(1), 1–28.
84. Dobrushin, R. L., & Major, P. (1979). Non-central limit theorems for non-linear functions of Gaussian fields. *Zeitschrift für Wahrscheinlichkeitstheorie und verwandte Gebiete, 50*, 27–52.
85. Doob, J. L. (1953). *Stochastic processes*. Wiley.
86. Doukhan, P., León, J., & Soulier, P. (1996). Central and non central limit theorems for strongly dependent stationary Gaussian field. *Rebrape, 10*, 205–223.
87. Dwight, H. B. (1961). *Tables of integrals and other mathematical data.* The Macmillan Company.
88. Dym, H., & McKean, H. P. (1976). *Gaussian processes, function theory, and the inverse spectral problem.* Academic Press.
89. Dzhaparidze, K. (1970). On the estimation of spectral parameters of a Gaussian stationary process with rational spectral density. *Theory of Probability and Its Applications, 15*(3), 531–538.
90. Dzhaparidze, K. (1974). On simplified estimators of unknown parameters with good asymptotic properties. *Theory of Probability and Its Applications, 19*, 347–358.
91. Dzhaparidze, K. (1981). *Parameter estimation and testing hypothesis in spectral analysis of stationary time series*. Tbilisi University Press.
92. Dzhaparidze, K. (1983). On iterative procedures of asymptotic inference. *Statistica Neerlandica, 37*, 181–189.
93. Dzhaparidze, K. (1986). *Parameter estimation and hypothesizes testing in spectral analysis of stationary time series*. Springer.
94. Dzaparidze, K. O., & Yaglom, A. M. (1983). Spectrum parameter estimation in time series analysis. In P. R. Krishnaiah (Ed.), *Developments in statistics* (Vol. 4, pp. 1–181). Academic Press.

95. Eichler, M. (1995). Empirical spectral processes and their applications to stationary point processes. *Annals of Applied Probability, 5*(4), 1161–1176.
96. Ellis, R. S. (1985). *Entropy, large deviations, and statistical mechanics*. Springer.
97. Farré, M., Jolis, M., & Utzet, F. (2010). Multiple stratonovich integral and hu-meyer formula for Lévy processes. *Annals of Probability, 38*(6), 2136–2169.
98. Flandrin, P. (1989). On the spectrum of fractional Brownian motions. *IEEE Transactions on Information Theory, 35*, 197–199.
99. Fox, R., & Taqqu, M. S. (1985). Non-central limit theorems for quadratic forms in random variables having long-range dependence. *Annals of Probability, 13*(2), 428–446.
100. Fox, R., & Taqqu, M. S. (1986). Large-sample properties of parameter estimation for strongly dependent stationary Gaussian time series. *Annals of Statistics, 14*, 517–532.
101. Fox, R., & Taqqu, M. S. (1987). Central limit theorem for quadratic forms in random variables having long-range dependence. *Probability Theory and Related Fields, 74*, 213–240.
102. Gamboa, F., Rouault, A., & Zani, M. (1999). A functional large deviations principle for quadratic forms of Gaussian stationary processes. *Statistics and Probability Letters, 43*, 299–308.
103. Gao, J. (2004). Modeling long-range dependent Gaussian processes with application in continuous time financial models. *Journal of Applied Probability, 41*, 467–482.
104. Gao, J. (2007). *Nonlinear time series: Semiparametric and nonparametric methods*. Chapman & Hall/CRC.
105. Gao, J., Anh, V., & Heyde, C. (2002). Parameter estimation of stochastic processes with long-range dependence and intermittency. *Stochastic Processes and their Applications, 99*, 295–321.
106. Gao, J., Anh, V., Heyde, C., & Tieng, Q. (2001). Parameter estimation of stochastic processes with long-range dependence and intermittency. *Journal of Time Series Analysis, 22*, 517–535.
107. Garnett, J. (2007). *Bounded analytic functions*. Springer.
108. Gel'fand, I. M., & Vilenkin, N. Ya. (1964). *Generalized functions: Volume 4 of applications of harmonic analysis: Equipped Hilbert spaces*. Academic Press.
109. Gel'fand, I. M., & Yaglom, A. M. (1959). Calculation of the amount of information about a random function contained in another such function. *American Mathematical Society Translations, 12*, 199–246.
110. Gihman, I. I., & Skorokhod, A. V. (1980). *The theory of stochastic processes I*. Springer.
111. Ginovyan, M. S. (1983). $\sqrt{n}$-approximation of the likelihood function. *Journal of Soviet Mathematics, 21*(1), 20–31.
112. Ginovyan, M. S. (1984). Asymptotic behavior of the logarithm of the likelihood function when the spectral density has polynomial zeros. *Journal of Soviet Mathematics, 25*(3), 1113–1125.
113. Ginovyan, M. S. (1984). Asymptotic behavior of the Toeplitz determinant. *Journal of Soviet Mathematics, 24*(5), 494–500.
114. Ginovyan, M. S. (1984). On the asymptotical estimation of the maximum likelihood of parameters of the spectrum of a stationary Gaussian time series. In P. Revesz (Ed.), *Limit theorems in probability and statistics* (Vol. 1). *Colloquia Mathematica Societatis János Bolyai, 36*, 457–497. North-Holland.
115. Ginovyan, M. S. (1986). Local asymptotic normality of a family of Gaussian distributions. *Journal of Soviet Mathematics, 33*(1), 706–715.
116. Ginovyan, M. S. (1986). On the asymptotic estimator of the maximum likelihood of parameters of the spectral density having zeros. *Acta Scientiarum Mathematicarum (Szeged), 50*, 169–182.
117. Ginovyan, M. S. (1988). Asymptotically efficient nonparametric estimation of functionals on spectral density with zeros. *Theory of Probability and Its Applications, 33*, 315–322.
118. Ginovyan, M. S. (1988). On estimate of the value of the linear functional in a spectral density of stationary Gaussian process. *Theory of Probability and Its Applications, 33*, 777–781.
119. Ginovyan, M. S. (1993). A note on central limit theorem for Toeplitz type quadratic forms in stationary Gaussian variables. *Journal of Contemporary Mathematical Analysis, 28*, 78–81.

120. Ginovyan, M. S. (1994). On Toeplitz type quadratic functionals in Gaussian stationary process. *Probability Theory and Related Fields, 100*, 395–406.
121. Ginovyan, M. S. (1995). Asymptotic properties of spectrum estimate of stationary Gaussian process. *Journal of Contemporary Mathematical Analysis, 30*, 1–17.
122. Ginovyan, M. S. (1996). Asymptotic upper bounds for the risk of estimators of linear functionals of a spectral density function. *Journal of Contemporary Mathematical Analysis, 31*(5), 1–9.
123. Ginovian, M. S. (1999). Asymptotic behavior of the prediction error for stationary random sequences. *Journal of Contemporary Mathematical Analysis, 4*(1), 14–33.
124. Ginovyan, M. S. (1999). Locally asymptotically normal families of Gaussian distributions. *Journal of Contemporary Mathematical Analysis, 34*(5), 18–29.
125. Ginovyan, M. S. (1999). Nonparametric estimation of the spectrum of homogeneous Gaussian fields. *Journal of Contemporary Mathematical Analysis, 34*(2), 1–15.
126. Ginovyan, M. S. (2000). Asymptotically exact bounds for minimax risk of estimators of linear functionals. *Journal of Contemporary Mathematical Analysis, 35*(3), 10–20.
127. Ginovyan, M. S. (2003). Asymptotically efficient nonparametric estimation of nonlinear spectral functionals. *Acta Applicandae Mathematicae, 78*, 145–154.
128. Ginovyan, M. S. (2003). Goodness-of-fit tests for continuous time stationary process. *Journal of Contemporary Mathematical Analysis, 38*(2), 1–13.
129. Ginovyan, M. S. (2011). Efficient estimation of spectral functionals for continuous time stationary models. *Acta Applicandae Mathematicae, 115*, 233–254.
130. Ginovyan, M. S. (2011). Efficient estimation of spectral functionals for Gaussian stationary models. *Communications on Stochastic Analysis, 5*, 211–232.
131. Ginovyan, M. S. (2018). Chi-square type goodness-of-fit tests for stationary Gaussian process. *Journal of Contemporary Mathematical Analysis, 53*(3), 121–128.
132. Ginovyan, M. S. (2020). Parameter estimation for Lévy-driven continuous time linear models with tapered data. *Acta Applicandae Mathematicae, 169*(1), 79–97.
133. Ginovyan, M. S. (2021). Goodness-of-fit tests for stationary Gaussian processes with tapered data. *Acta Applicandae Mathematicae, 171*(1), 1–12.
134. Ginovyan, M. S. (2024). Simplified Whittle estimators for spectral parameters of stationary linear models with tapered data. *Journal of Contemporary Mathematical Analysis, 59*(1), 29–37.
135. Ginovyan, M. S., & Sahakyan, A. A. (2005). On the central limit theorem for toeplitz quadratic forms of stationary sequences. *Theory of Probability and Its Applications, 49*, 612–628.
136. Ginovyan, M. S., & Sahakyan, A. A. (2007). Limit theorems for toeplitz quadratic functionals of continuous time stationary process. *Probability Theory and Related Fields, 138*, 551–579.
137. Ginovyan, M. S., & Sahakyan, A. A. (2012). Trace approximations of products of truncated Toeplitz operators. *Theory of Probability and Its Applications, 56*(1), 57–71.
138. Ginovyan, M. S., & Sahakyan, A. A. (2013). On the trace approximations of products of Toeplitz matrices. *Statistics and Probability Letters, 83*(3), 753–760.
139. Ginovyan, M. S., & Sahakyan, A. A. (2016). Robust estimation for continuous-time linear models with memory. *Theory of Probability and Mathematical Statistics, 95*, 75–91.
140. Ginovyan, M. S., & Sahakyan, A. A. (2019). Estimation of spectral functionals for Lévy-driven continuous time linear models with tapered data. *Electronic Journal of Statistics, 13*, 255–283.
141. Ginovyan, M. S., & Sahakyan, A. A. (2019). Limit theorems for tapered Toeplitz quadratic functionals of continuous time Gaussian stationary processes. *Journal of Contemporary Mathematical Analysis, 54*(4), 222–239.
142. Ginovyan, M. S., & Sahakyan, A. A. (2021). Statistical inference for stationary models with tapered data. *Journal of Contemporary Mathematical Analysis, 56*(6), 347–367.
143. Ginovyan, M. S., & Sahakyan, A. A. (2021). Statistical inference for stationary models with tapered data. *Statistics Surveys, 15*, 154–194.

144. Ginovyan, M. S., & Sahakyan, A. A. (2025). On moderate deviations for quadratic functionals in continuous time Gaussian stationary processes. *Journal of Contemporary Mathematical Analysis, 60*(1), 10–24.
145. Ginovyan, M. S., Sahakyan, A. A., & Taqqu, M. S. (2014). The trace problem for Toeplitz matrices and operators and its impact in probability. *Probability Surveys, 11*, 393–440.
146. Ginovyan, M. S., & Taqqu, M. S. (2022). Limit theorems for Toeplitz-type quadratic functionals of stationary processes and applications. *Probability Surveys, 19*, 1–64.
147. Giraitis, L., Koul, H., & Surgailis, D. (2012). *Large sample inference for long memory processes*. Imperial College Press.
148. Giraitis, L., & Surgailis, D. (1990). A central limit theorem for quadratic forms in strongly dependent linear variables and its application to asymptotical normality of Whittle's estimate. *Probability Theory and Related Fields, 86*, 87–104.
149. Giraitis, L., & Taqqu, M. S. (1997). Limit theorems for bivariate Appell polynomials: Part I. Central limit theorems. *Probability Theory and Related Fields, 107*, 359–381.
150. Giraitis, L., & Taqqu, M. S. (1998). Central limit theorems for quadratic forms with time-domain conditions. *Annals of Probability, 26*(1), 377–398.
151. Giraitis, L., & Taqqu, M. S. (1999). Convergence of normalized quadratic forms. *Journal of Statistical Planning and Inference, 80*(1), 15–35.
152. Giraitis, L., & Taqqu, M. S. (1999). Whittle estimator for finite-variance non-Gaussian long-memory time series. *Annals of Statistics, 27*(1), 178–203.
153. Giraitis, L., & Taqqu, M. S. (2001). Functional non-central and central limit theorems for bivariate Appell polynomials. *Journal of Theoretical Probability, 14*(2), 393–426.
154. Giraitis, L., Taqqu, M. S., & Terrin, N. (1998). Limit theorems for bivariate Appell polynomials. II. Non-central limit theorems. *Probability Theory and Related Fields, 10*, 333–367.
155. Gohberg, I. C., & Krein, M. G. (1969). *Introduction to the theory of linear nonselfadjoint operators in Hilbert space*. American Mathematical Society.
156. Goldstein, L., & Khas'minskii, R. Z. (1995). On efficient estimation of smooth functionals. *Theory of Probability and Its Applications, 40*, 151–156.
157. Granger, C. W. J., & Joyeux, K. (1980). An introduction to long-memory time series and fractional differencing. *Journal of Time Series Analysis, 1*, 15–29.
158. Graybill, F. A. (1983). *Matrices with Applications in Statistics* (2nd ed.). Wadsworth.
159. Grenander, U. (1952). On Toeplitz forms and stationary processes. *Arkiv for Matematik, 1*(37), 555–571.
160. Grenander, U., & Rosenblatt, M. (1954). An extension of a theorem of G. Szegő and its application to the study of stochastic processes. *Transactions of the American Mathematical Society, 76*, 112–126.
161. Grenander, U., & Rosenblatt, M. (1957). *Statistical Analysis of Stationary Time Series*. Wiley.
162. Grenander, U., & Szegő, G. (1958). *Toeplitz forms and their applications*. University of California Press.
163. Guégan, D. (2005). How can we define the concept of long memory? An econometric survey. *Econometric Reviews, 24*, 113–149.
164. Guyon, X. (1982). Parameter estimation for a stationary process on a d-dimensional lattice. *Biometrika, 69*, 95–105.
165. Guyon, X. (1995). *Random fields on a network: Modelling, statistics and applications*. Springer.
166. Hájek, J. (1970). A characterization of limiting distributions of regular estimates. *Zeitschrift für Wahrscheinlichkeitstheorie und verwandte Gebiete, 14*, 323–330.
167. Hájek, J. (1972). Local asymptotic minimax and admissibility in estimation. In L. M. LeCam, J. Neyman, & E. M. Scott (Eds.), *The Proceedings of the Sixth Berkeley Symposium on Mathematical Statistics and Probability* (Vol. 1, pp. 175–194). University of California Press.
168. Hannan, E. J. (1960). *Time series analysis*. Wiley.
169. Hannan, E. J. (1970). *Multiple time series*. Wiley.

170. Hannan, E. J. (1973). The asymptotic theory of linear time series models. *Journal of Applied Probability, 10*, 130–145.
171. Hasminskii, R. Z., & Ibragimov, I. A. (1986). Asymptotically efficient nonparametric estimation of functionals of a spectral density function. *Probability Theory and Related Fields, 73*, 447–461.
172. Henry, M., & Zaffaroni, P. (2003). The long range dependence paradigm for macroeconomics and finance. In P. Doukhan, G. Oppenheim, M. S. Taqqu (Eds.), *Long range dependence: Theory and applications* (pp. 417–438). Birkhüser.
173. Heyde, C., & Dai, W. (1996). On the robustness to small trends of estimation based on the smoothed periodogram. *Journal of Time Series Analysis, 17*(2), 141–150.
174. Heyde, C. C., & Gay, R. (1993). Smoothed periodogram asymptotics and estimation for processes and fields with possible long-range dependence. *Stochastic Processes and their Applications, 45*, 169–182.
175. Hida T., & Hidsuda, M. (1993). *Gaussian processes*. Translations of mathematical monographs (Vol. 120). AMS.
176. Hong, D., Wang, J., & Gardner, R. (2005). *Real analysis with an introduction to wavelets and applications*. Elsevier Academic Press.
177. Hosking, J. R. M. (2002). Fractional differencing. *Biometrika, 68*, 165–176.
178. Hosoya, Y., & Taniguchi, M. (1982). A central limit theorem for stationary processes and the parameter estimation of linear processes. *Annals of Statistics, 10*, 132–153. Correction: **21** (1993) 1115–1117.
179. Hunt, R. A., Muckenhoupt, B., & Wheeden, R. L. (1973). Weighted norm inequalities for the conjugate function and Hilbert transform. *Transactions of the American Mathematical Society, 176*, 227–251.
180. Ibragimov, I. A. (1963). On estimation of the spectral function of a stationary Gaussian process. *Theory of Probability and Its Applications, 8*, 366–401.
181. Ibragimov, I. A. (1964). On asymptotic behavior of the prediction error. *Theory of Probability and Its Applications, 9*(4), 695–703.
182. Ibragimov, I. A. (1965). A class of estimates for the spectral function of a stationary sequence. *Theory of Probability and Its Applications, 10*, 123–126.
183. Ibragimov, I. A. (1967). On maximum likelihood estimation of parameters of the spectral of stationary time series. *Theory of Probability and Its Applications, 12*, 115–119.
184. Ibragimov, I. A. (1968). On a theorem by Szegő. *Matematicheskie Zametki, 3*, 693–702.
185. Ibragimov, I. A., & Has'minskii, R. Z. (1981). *Statistical estimation: Asymptotic theory*. Springer.
186. Ibragimov, I. A., & Khas'minskii, R. Z. (1986). Estimating the value of a linear functional of a distribution density. *Zapiski Nauchnykh Seminarov Leningradskogo Otdeleniya Matematicheskogo Instituta, 153*, 45–59.
187. Ibragimov, I. A., & Khas'minskii, R. Z. (1991). Asymptotically normal families of distributions and efficient estimation. *Annals of Statistics, 19*, 1681–1724.
188. Ibragimov, I. A., & Linnik, Yu. V. (1971). *Independent and stationary sequences of random variables*. Wolters-Noordhoff Publishing.
189. Ibragimov, I. A., & Rozanov. Yu. A. (1978). *Gaussian random processes*. Springer.
190. Ihara, S. (2000). Large deviation theorems for Gaussian processes and their applications in information theory. *Acta Applicandae Mathematicae, 63*, 165–174.
191. Inoue, A. (2002). Asymptotic behavior for partial autocorrelation functions of fractional ARIMA processes. *Annals of Applied Probability, 12*(4), 1471–1491.
192. Inoue, A., & Kasahara, Y. (1999). On the asymptotic behavior of the prediction error of a stationary process. In N. Kono, N. R. Shieh (Eds.), *Trends in probability and related analysis* (pp. 207–218). World Scientific.
193. Itô, K. (1951). Multiple wiener integral. *Journal of the Mathematical Society of Japan, 3*(1), 157–169.

194. Ivanov, A. V., & Prikhod'ko, V. V. (2016). On the whittle estimator of the parameters of spectral density of random noise in the nonlinear regression model. *Ukrainian Mathematical Journal, 67*(8), 1183–1203.
195. Janas, D., & von Sachs, R. (1994). Consistency for non-linear functions of the periodogram of tapered date. *Journal of Time Series Analysis, 16*(6), 585–606.
196. Janson, S. (1997). *Gaussian Hilbert spaces* (Vol. 129). Cambridge tracts in mathematics. Cambridge University Press.
197. Jeffrey A., & Zwillinger D. (2007). *Table of integrals, series, and products* (7th ed.). Elsevier.
198. Kac, M. (1954). Toeplitz matrices, translation kernels and a related problem in probability theory. *Duke Mathematical Journal, 21*, 501–509.
199. Kakizawa, Y. (2007). Moderate deviations for quadratic forms in Gaussian stationary processes. *Journal of Multivariate Analysis, 98*, 992–1017.
200. Kolmogorov, A. N. (1941). Interpolation and extrapolation of stationary random sequences. *Izvestiya Akademii Nauk SSSR. Seriya Matematicheskaya, 5*, 3–14.
201. Kolmogorov, A. N. (1941). Stationary sequences in a Hilbert space. *Bulletin, Moscow State University, 2*(6), 1–40.
202. Koul, H. L., & Surgailis, D. (2000). Asymptotic normality of the Whittle estimator in linear regression model with long memory errors. *Statistical Inference for Stochastic Processes, 3*(1), 129–147.
203. Kutoyants, Yu. A. (1998). *Statistical inference for spatial poisson processes*. Springer.
204. Kutoyants, Yu. A. (2004). *Statistical inference for ergodic diffusion processes*. Springer.
205. Lavancier, F., & Philippe, A. (2011). Some convergence results on quadratic forms for random fields and application to empirical covariances. *Probability Theory and Related Fields, 149*, 493–514.
206. Le Cam, L. (1956). On the asymptotic theory of estimation and testing hypotheses. In *Proceedings of the third Berkeley symposium on mathematical statistics and probability* (Vol. 1, pp. 129–156).
207. Le Cam, L. (1960). Locally asymptotically normal families of distributions. *University of California Publications in Statistics, 3*, 27–98.
208. Le Cam, L. (1986). *Asymptotic methods in statistical decision theory*. Springer.
209. Leonenko, N. N., & Sakhno, L. M. (2006). On the Whittle estimators for some classes of continuous-parameter random processes and fields. *Statistics and Probability Letters, 76*, 781–795.
210. Levit, B. Ya. (1974). On optimality of some statistical estimates. In *Proceedings of the second Prague symposium on asymptotic statistics* (pp. 215–238). Charles University.
211. Levit, B. Ya. (1975). On the efficiency of a certain class of nonparametric estimators. *Theory of Probability and its Applications, 20*(4), 723–740.
212. Levit, B. Ya. (1978). Infinite-dimensional informational inequalities. *Theory of Probability and Its Applications, 23*(2), 371–377.
213. Levit, B. Ya., & Samarov, A. M. (1978). A remark on estimation of spectral function. *Problems of Information Transmission, 14*(2), 61–66.
214. Lieberman, O., & Phillips, P. C. B. (2004). Error bounds and asymptotic expansions for Toeplitz product functionals of unbounded spectra. *Journal of Time Series Analysis, 25*, 733–753.
215. Lieberman, O., Rousseau, J., & Zucker, D. M. (2001). Valid edgeworth expansion for the sample autocorrelation function under long dependence. *Econometric Theory, 17*, 257–275.
216. Lieberman, O., Rousseau, J., & Zucker, D. M. (2003). Valid asymptotic expansions for the maximum likelihood estimator of the parameter of a stationary, Gaussian, strongly dependent process. *Annals of Statistics, 31*(2), 586–612.
217. Lindgren, G. (2013). *Stationary stochastic processes.* Chapman & Hall/CRC texts in statistical science series. CRC Press.
218. Liu, W., & Wu, W. B. (2010). Asymptotics of spectral density estimates. *Econometric Theory, 26*, 1218–1245.
219. Loéve, M. (1978). *Probability theory II*. Springer.

220. Ludeña, C. (1999). Efficient estimation of functionals of the spectral density of stationary Gaussian Fields. *ESAIM: Probability and Statistics, 3*, 23–47.
221. Ludeña, C., & Lavielle, M. (1999). The whittle estimator for strongly dependent stationary gaussian fields. *Scandinavian Journal of Statistics, 26*, 433–450.
222. Major, P. (2014). *Multiple Wiener-Itô integrals*. Lecture notes in mathematics. Springer.
223. Mann, H., & Wald, A. (1943). On the statistical treatment of linear stochastic difference equations. *Econometrics, 11*, 173–220.
224. Millar, P. W. (1985). Non-parametric applications of an infinite dimensional convolution theorem. *Zeitschrift für Wahrscheinlichkeitstheorie und verwandte Gebiete, 68*, 545–556.
225. Muckenhoupt, B. (1972). Weighted norm inequalities for the Hardy maximal function. *Transactions of the American Mathematical Society, 165*, 207–226.
226. Natanson, I. P. (1955). *Theory of functions of a real variable* (Vol. I). Ungar.
227. Nikol'skii, S. M. (1975). *Approximation of functions of several variables and embedding theorems*. Springer.
228. Nikol'skii, N. K. (1986). *Treatise on the shift operator: Spectral function theory*. Springer.
229. Nourdin, I., & Peccati, G. (2009). Stein's method and exact Berry-Esséen asymptotics for functionals of Gaussian fields. *Annals of Probability, 37*, 2231–2261.
230. Nualart, D. (2006). *The Malliavin calculus and related topics* (2nd ed.). Springer.
231. Nualart, D., & Schoutens, W. (2000). Chaotic and predictable representations for Lévy processes. *Stochastic Processes and their Applications, 90*(1), 109–122.
232. Osidze, A. G. (1974). On a goodness of fit test in the case of dependence of spectral density of Gaussian processes on unknown parameters. *Reports of AN Georgian SSR, 74*(2), 273–276.
233. Osidze, A. G. (1975). On a statistic for testing the composite hypothesis regarding the form of a spectral density of a stationary Gaussian random process. *Reports of AN Georgian SSR, 77*(2), 313–315.
234. Palma, W. (2007). *Long-memory time series*. Wiley.
235. Parter, S. V. (1986). On the distribution of the singular values of Toeplitz matrices. *Linear Algebra and Its Applications, 80*, 115–130.
236. Parzen, E. (1957). On consistent estimates of the spectrum of a stationary time series. *The Annals of Mathematical Statistics, 28*, 329–348.
237. Parzen, E. (1958). On asymptotically efficient consistent estimates of the spectral density function of a stationary time series. *Journal of the Royal Statistical Society, C20*, 303–322.
238. Parzen, E. (1961). An approach to time series analysis. *The Annals of Mathematical Statistics, 32*(4), 951–989.
239. Parzen, E. (1967). *Time series analysis papers*. Holden-Day.
240. Peccati, G., & Taqqu, M. S. (2011). *Wiener chaos: Moments, cumulants and diagrams: as survey with computer implementation*. Springer.
241. Peccati, G., & Tudor, C. (2005). Gaussian limits for vector-valued multiple stochastic integrals. *Séminaire de Probabilités, XXXVIII*, (Vol.1857, pp. 219–245). Springer.
242. Petrov, V. V. (1975). *Sums of independent random variables*. De Gruyter.
243. Pfanzagl, J. (1982). *Contributions to a general asymptotic statistical theory*. Springer.
244. Pham-Dinh, T. (1977). Estimation of parameters of a continuous time Gaussian stationary process with rational spectral density. *Biometrika*, **64**, 385–399.
245. Pipiras, V., & Taqqu, M. S. (2010). Regularization and integral representations of Hermite processes. *Statistics and Probability Letters, 80*, 2014–2023.
246. Pipiras, V., & Taqqu, M. S. (2017). *Long-Range dependence and self-similarity*. Cambridge University Press.
247. Pisarenko, V. F. (1965). On the computation of the relation of likelihood for Gaussian processes with rational spectrum. *Theory of Probability and Its Applications, 10*, 299–303.
248. Pourahmadi, M. (1988). Remarks on extreme eigenvalues of Toeplitz matrices. *The International Journal of Mathematics and Mathematical Sciences, 11*(1), 23–26.
249. Priestley, M. (1981). *Spectral analysis and time series*. Academic Press.
250. Prohorov, Yu. V. (1956). Convergence of random processes and limit theorems in probability theory. *Theory of Probability and Its Applications, 1*(2), 157–214.

251. Protter, P. E. (2004). *Stochastic integration and differential equations* (2nd ed.). Springer.
252. Rao, C. R. (1965). *Linear statistical inference and its applications*. Wiley.
253. Resnick, S.(1999). *A probability path*. Springer.
254. Riesz, F., & Sz.-Nagy, B. (1955). *Functional analysis*. Frederick Ungar Publishing Co.
255. Robinson, P. M. (1995). Gaussian semiparametric estimation of long-range dependence. *Annals of Statistics, 23*, 1630–1661.
256. Robinson, P. M. (2003). Long memory time series. In P. M. Robinson (Ed.), *Time series with long memory* (pp. 4–32). Oxford University Press.
257. Rosenblatt, M. (1957). Some purely deterministic processes. *Journal of Mathematics and Mechanics, 6*(6), 801–810. (Reprinted In R. A. Davis, K.-S. Lii, D. N. Politis (Eds.), *Selected works of Murray Rosenblatt* (pp. 124–133). Springer, 2011).
258. Rosenblatt, M. (1961). Independence and dependence. *Proceedings of the Fourth Berkeley Symposium on Mathematical Statistics and Probability, 2*, 431–443.
259. Rosenblatt, M. (1962). Asymptotic behavior of eigenvalues of Toeplitz forms. *Journal of Mathematics and Mechanics, 11*, 941–950.
260. Rosenblatt, M. (1979). Some limit theorems for partial sums of quadratic forms in stationary Gaussian variables. *Zeitschrift für Wahrscheinlichkeitstheorie und Verwandte Gebiete, 49*, 125–132.
261. Rosenblatt M. (1985). *Stationary sequences and random fields*. Birkhäuser.
262. Rousseau, J., Chopin, N., & Liseo, B. (2012). Bayesian nonparametric estimation of the spectral density of a long and intermediate memory Gaussian processes. *Annals of Statistics, 40*, 964–995.
263. Rozanov, Yu. A. (1967). *Stationary Random Processes*. Holden-Day.
264. Rudin, G. (1987). *Real and complex analysis* (3rd ed.). McGraw-Hill.
265. Saitoh, S. (1997). *Integral transforms, reproducing Kernels and their applications*. Addison Wesley Longman.
266. Sakhno, L. M. (1977). Minimum contrast method for parameter estimation in th espectra domain. In V. Korolyuk et al.(Eds.) *Modern stochastics and applications*. Optimization and Its Applications (vol. 90, pp. 319–336).
267. Samarov, A. M. (1977). Lower bound on the risk for spectral density estimates. *Problemy Peredachi Informatsii, 13*(1), 67–72.
268. Samorodnitsky, G. (2016) *Stochastic processes and long range dependence*. Springer International Publishing.
269. Samorodnisky, G., & Taqqu, M. S. (1994). *Stable non-gaussian random processes*. Chapman & Hall/CRC.
270. Sato, T., Kakizawa, Y., & Tahiguchi, M. (1998). Large deviations results for statistics of short– and long–memory Gaussian processes. *The Australian & New Zealand Journal of Statistics, 40*, 17–29.
271. Saulis, L., & Statulevicius, V. A. (1991). *Limit theorems for large deviations*. Springer.
272. Seneta, E. (1976). *Regularly varying functions*. Springer.
273. Seneta, E. (2019). Slowly varying functions in the Zygmund sense and generalized regular variation. *Journal of Mathematical Analysis and Applications, 475*, 1647–1657.
274. Serra, S. (1998). On the extreme eigenvalues of hermitian (block) toeplitz matrices. *Linear Algebra and Its Applications, 270*, 109–129.
275. Serra Capizzano, S. (2002). Test functions, growth conditions and Toeplitz matrices. *Rendiconti del Circolo Matemàtico di Palermo. Serie II, Supplement, 68*, 791–795.
276. Sevastyanov, B. A. (1961). A class of limit distributions for quadratic forms of normal stochastic variables. *Theory of Probability Application, 6*, 337–340.
277. Shiryaev, A. N. (1984). *Probability*. Springer.
278. Simon, B. (2011). *Szegő's theorem and its descendants: Spectral theory for L^2 perturbations of orthogonal polynomials*. Princeton University Press.
279. Simon, B. (2015). *Real analysis. A comprehensive course in analysis, part 1*. AMS.
280. Simon, B. (2015). *Operator theory. A comprehensive course in analysis, part 4*. AMS.

281. Solev, V. N. (1982). Gaussian f-regular processes and asymptotic behavior of likelihood function. *Zapiski Nauchnykh Seminarov Leningradskogo Otdeleniya Matematicheskogo Instituta, 119*, 203–217.
282. Solev, V. N. (1983). Approximation of the likelihood function. *Theory of probability and its applications, 28*(1), 201–203.
283. Solev, V. N., & Gerville-Reache, L. (2006). Large toeplitz operators and quadratic forms generated by a stationary Gaussian sequence. *Journal of Mathematical Sciences, 139*(3), 6625–6630.
284. Solev, V. N., & Zerbet, A. (2003). Conditions of local asymptotic normality for Gaussian stationary processes. *Journal of Mathematical Sciences, 118*(6), 5635–5649.
285. Solo, V. (1992). Intrinsic random functions and the paradox of 1/f noise. *SIAM Journal on Applied Mathematics, 52*, 270–291.
286. Spangenberg, F. (2015). A central limit theorem for the sample autocovariance of a continuous time moving average process with long memory. Preprint. arXiv:1502.04851.
287. Stein, Ch. (1972). Efficient nonparametric testing and estimation. *Proceedings of the Third Berkeley Symposium on Mathematical Statistics and Probability, 1*, 187–194.
288. Stein, E. (1993). *Harmonic analysis. Real-variable methods, orthogonality, and oscillatory integrals.* Princeton University Press.
289. Surgailis, D. (1982). Zones of attraction of self-similar multiple integrals. *Lithuanian Mathematical Journal, 22*(3), 327–340.
290. Szegő, G. (1920). Beiträge zur Theorie der Toeplitzschen Formen, I. *Mathematische Zeitschrift, 6*(3–4), 167–202.
291. Takabatake, T. (2024). Corrigendum: Error bounds and asymptotic expansions for Toeplitz product functionals of unbounded spectra. *Journal of Time Series Analysis, 45*, 158–160.
292. Taniguchi, M. (1983). On the second order asymptotic efficiency of estimators of Gaussian ARMA processes. *Annals of Statistics, 11*, 157–169.
293. Taniguchi, M. (1986). Berry-Esséen theorems for quadratic forms of Gaussian stationary processes. *Probability Theory and Related Fields, 72*, 185–194.
294. Taniguchi, M. (1987). Minimum contrast estimation for spectral densities of stationary processes. *The Journal of the Royal Statistical Society, Series B (Statistical Methodology), 49*, 315–325.
295. Taniguchi, M., & Kakizawa, Y. (2000). *Asymptotic theory of statistical inference for time series.* Academic Press.
296. Taqqu, M. S. (1975). Weak convergence to fractional Brownian motion and the Rosenblatt process. *Zeitschrift für Wahrscheinlichkeitstheorie und Verwandte Gebiete, 31*, 287–302.
297. Taqqu, M. S. (1979). Convergence of integrated processes of arbitrary Hermite rank. *Probability Theory and Related Fields, 50*(1), 53–83.
298. Taqqu, M. S. (1987). Toeplitz matrices and estimation of time series with long-range dependence. In *Proceedings of the first world congress of the bernoulli society* (Vol. 1, pp. 75–83). VNU Science Press.
299. Taqqu, M. S. (2003). Fractional Brownian motion and long-range dependence. In P. Doukhan, G. Oppenheim, M. S. Taqqu (Eds.), *Long range dependence: Theory and applications* (pp. 5–38) Birkhüser.
300. Taqqu, M. S. (2011). The Rosenblatt process. In R. A. Davis et al. (Eds.), *Selected works of Murray Rosenblatt, selected works in probability and statistics* (pp. 29–45). Springer Science & Business Media.
301. Terrin, N., & Taqqu, M. S. (1990). A noncentral limit theorem for quadratic forms of Gaussian stationary sequences. *Journal of Theoretical Probability, 3*, 449–475.
302. Terrin, N., & Taqqu, M. S. (1991). Convergence to a Gaussian limit as the normalization exponent tends to 1/2. *Statistics and Probability Letters, 11*, 419–427.
303. Terrin, N., & Taqqu, M. S. (1991). Power counting theorem in Euclidean space. In R. Durrett, & H. Kesten (Eds.), *Random walks, Brownian motion, and interacting particle systems. Progress in probability* (Vol. 28, pp. 425–440). Birkhäuser.

304. Tsai, H., & Chan, K. S. (2005). Quasi-maximum likelihood estimation for a class of continuous-time long memory processes. *Journal of Time Series Analysis, 26*(5), 691–713.
305. Tukey, J. W. (1967). An introduction to the calculations of numerical spectrum analysis. In B. Harris (Ed.), *Advanced seminar on spectral analysis of time series* (pp. 25–46). Wiley.
306. Tyrtyshnikov, E. E. (1996). A unifying approach to some old and new theorems on distribution and clustering. *Linear Algebra and Its Applications, 232*, 1–43.
307. Veillette, M., & Taqqu, M. S. (2013). Properties and numerical evaluation of the Rosenblatt distribution. *Bernoulli, 19*, 982–1005.
308. Velasco, C., & Robinson, P. M. (2001). Edgeworth expansions for spectral density estimates and studentized sample mean. *Econometric Theory, 17*, 497–539.
309. Vitale, R. A. (1973). An asymptotically efficient estimate in time series analysis. *Quarterly of Applied Mathematics, 30*, 421–440.
310. von Sachs, R. (1994). Estimating non-linear functions of spectral density, using a data-taper. *Annals of the Institute of Statistical Mathematics, 46*(3), 453–474.
311. Vong, S-W., & Jin, X-Q. (2007). Unitarily invariant norms of Toeplitz matrices with Fisher-Hartwig singularities. *Annals of the Institute of Statistical Mathematics, 29*, 850–854.
312. Walker, A. M. (1964). Asymptotic properties of least-squares estimates of parameters of the spectrum of a stationary non-deterministic time-series. *Journal of the Australian Mathematical Society, 4*, 363–384.
313. Whittle, P. (1951). *Hypothesis testing in time series*. Hafner.
314. Whittle, P. (1953). Estimation and information in stationary time series. *Arkiv for Matematik, 2*, 423–434.
315. Widom, H. (1976). Asymptotic behavior of block Toeplitz matrices and determinants II. *Advances in Mathematics, 21*, 1–29.
316. Wu, W. B., & Shao, X. (2007). A limit theorem for quadratic forms and its applications. *Econometric Theory, 23*, 930–951.
317. Yaglom, A. M. (1957). Some classes of random fields in n-dimensional space, related to stationary random processes. *Bernoulli- Bayes-Laplace Anniversary Volume. Theory of Probability and Its Applications, 2*, 273–320.
318. Yaglom, A. M. (1986). *The correlation theory of stationary and related random processes* (Vol. 1). Springer.
319. Zamarashkin, N. L., & Tyrtyshnikov, E. E. (1997). Distribution of the eigenvalues and singular numbers of Toeplitz matrices under weakened requirements on the generating function. *Matematicheskiy Sbornik, 188*, 1191–1201.
320. Zhu, K. (2007). *Operator theory in function spaces*. American Mathematical Society.
321. Zygmund, A. (1959). *Trigonometric series* (Vol. 1, 2nd ed.). Cambridge University Press.

Index

M. S. Ginovyan, *Random Toeplitz Functionals and Their Applications*,
Frontiers in Probability and the Statistical Sciences,
https://doi.org/10.1007/978-3-032-00961-6

GPSR Compliance
The European Union's (EU) General Product Safety Regulation (GPSR) is a set of rules that requires consumer products to be safe and our obligations to ensure this.

If you have any concerns about our products, you can contact us on

ProductSafety@springernature.com

In case Publisher is established outside the EU, the EU authorized representative is:

Springer Nature Customer Service Center GmbH
Europaplatz 3
69115 Heidelberg, Germany

www.ingramcontent.com/pod-product-compliance
Lightning Source LLC
Chambersburg PA
CBHW051427031125
34911CB00001B/47
* 9 7 8 3 0 3 2 0 0 9 6 0 9 *